REVIEWS in MINERALOGY　　　　　　*Volume 27*

MINERALS AND REACTIONS AT THE ATOMIC SCALE: TRANSMISSION ELECTRON MICROSCOPY

PETER R. BUSECK, Editor

The authors:

Fred Allen
Engelhard Corporation
Iselin, New Jersey 08830

Alain Baronnet
CRMC^2 Campus Luminy, Case 913
13288 Marseille Cedex 09, France

Peter R. Buseck
Depts. of Geology and Chemistry
Arizona State University
Tempe, Arizona 85287

Harry W. Green, II
Dept. of Geology
University of California at Davis
Davis, California 95616

J.-P. Morniroli
Laboratoire de Métallurgie Physique
Université de Lille
59655 Villeneuve d'Ascq Cedex,
France

Gordon L. Nord, Jr.
U.S. Geological Survey
Mail Stop 969
Reston, Virginia 22092

Donald R. Peacor
Dept. of Geological Sciences
University of Michigan
Ann Arbor, Michigan 48109

Richard J. Reeder
Dept. of Earth and Space Sciences
State University of New York
Stony Brook, New York 11794

Peter Self
C.S.I.R.O.
Private Bag No. 2, Glen Osmond
South Australia 5064, Australia

John W. Steeds
University of Bristol
H.H. Wills Physics Laboratory
Tyndall Avenue
Bristol BS8 1TL, England

David R. Veblen
Dept. of Earth and Planetary Sciences
The Johns Hopkins University
Baltimore, Maryland 21218

Series Editor: **Paul H. Ribbe**
Department of Geological Sciences
Virginia Polytechnic Institute & State University
Blacksburg, Virginia 24061

MINERALOGICAL SOCIETY (
WASHINGTON, D

D1241916

REVIEWS IN MINERALOGY

(Formerly: SHORT COURSE NOTES)

ISSN 0275-0279

Volume 27: *Minerals and Reactions at the Atomic Scale: Transmission Electron Microscopy*

ISBN 0-939950-32-4

ADDITIONAL COPIES of this volume as well as those listed below may be obtained at moderate cost from the MINERALOGICAL SOCIETY OF AMERICA, 1130 Seventeenth Street, N.W., Suite 330, Washington, D.C. 20036 U.S.A.

MINERALS AND REACTIONS AT THE ATOMIC SCALE: TRANSMISSION ELECTRON MICROSCOPY

FOREWORD

The Mineralogical Society of America has been sponsoring short courses in conjunction with their annual meetings with the Geological Society of America since 1974, and this volume represents the proceedings of the eighteenth in the sequence. This TEM course was convened October 23-25, 1992, at Hueston Woods State Park, College Corner, Ohio, by Peter Buseck, who is the scientific editor of this volume. As series editor of *Reviews in Mineralogy* I thank Peter and the ten additional authors for their heroic efforts to maintain the high quality of publication we have come to expect in the *Reviews* volumes.

I am particularly grateful to Thomas DeBusk for long hours of work in assembling camera-ready copy for Volume 27. Margie (Strickler) Sentelle, who has contributed her secretarial skills to at least fifteen *RiM* volumes, is still doing a superb job, and Sharon Long rendered valuable editorial assistance.

Paul H. Ribbe
Series Editor
Blacksburg, VA
October 1, 1992

PREFACE

The transmission electron microscope (TEM) has become an integral part of much current mineralogical research. It combines many of the functions of standard mineralogical instruments such as (a) the petrographic microscope for optical and textural studies, (b) x-ray diffractometer for structural data, (c) electron microprobe analyzer for determining chemical composition, as well as (d) several other spectroscopic instruments and techniques for more specialized measurements. This blend of capabilities in a single instrument, permitting multiple types of measurements on the same mineral grain, allows unique insights to be obtained. Moreover, each of these observations can be made on a finer scale than is possible by any other technique. The utility of the TEM as a mineralogical, petrographic, and geochemical tool is obvious.

While extremely powerful, it is not surprising that electron microscopy is also complex. A familiarity with the instruments mentioned in the previous paragraph is desirable for utilizing the TEM and interpreting the data arising from it, but there are also features that are unique to electron microscopy that this short course and volume attempt to impart. Clearly, a short course would be impractical if our goal was to become thorough experts, knowledgeable in all aspects of theory and practice of electron microscopy. Fortunately, however, such expertise is not required. Modern TEMs are quite "user friendly," with many functions handled by tame, dedicated computers. Thus, an operator with modest knowledge and skills can obtain useful information using modern TEMs. At the same time, the range and quality of information obtained increases in direct proportion to the experience of the TEM operator.

The purposes of this short course are to provide a background to the TEM as a mineralogical tool, to give an introduction to the principles underlying its operation, and to explore mineralogical applications and ways in which electron microscopy can augment our knowledge of mineral structures, chemistry, and origin. Much time will be devoted to mineralogical applications. If successful, the short course will provide sufficient information to allow mineralogists and petrologists to have an informed understanding of the data produced by transmission electron microscopy and to have enough knowledge and experience to undertake initial studies on their own.

The opening chapters cover the principles of electron microscopy (Chapters 1, 2, and 3) and chemical analysis using the TEM (Chapters 4 and 5); the next seven chapters consider mineralogical, petrological, and geochemical applications and their implications, for both low- and high-temperature geological environments.

Chapter 1 introduces the TEM and provides a background for its use, including a survey of both theory and practice. Electron diffraction and its uses for understanding symmetry and structures are given in Chapter 2, while a more detailed understanding of the physics of the TEM and its imaging abilities are covered in Chapter 3. Taken together, these chapters summarize much of the theory of imaging and diffraction by electron microscopy. Chapters 4 and 5 consider chemical analysis on a micro-scale using emitted x-rays and transmitted electrons, respectively.

The remaining chapters detail specific geological applications, including the new types of mineralogical information that have been obtained through electron microscopy. Chapter 6 details the mechanisms of certain metamorphic reactions and mineralogical complexities, while the question of the rich variety of ways that layered minerals can be organized are considered in Chapter 7. The high resolutions afforded by the TEM have caused new questions to be raised, such as those relating to the definition of mineral species and phases, and these questions are explored in Chapter 8. Reactions and processes occurring at low temperatures in sedimentary rocks are considered in two chapters: Chapter 9 on diagenetic processes and their effects on clays and shales, and Chapter 10 on carbonate microstructures. The effects of pressure and tectonic processes are discussed in Chapter 11, and the effects of transformations at elevated temperatures are explored in Chapter 12. These last two chapters emphasize what is commonly called "amplitude contrast" or "conventional" microscopy; some of the earlier ones utilize "phase contrast" and "high-resolution" microscopy. Most of the mineralogical studies actually utilize several complementary techniques; it is important to realize that the TEM is only one of many powerful instruments available today for the study of mineralogical problems.

Many people gave generously of their time to help prepare this volume and short course. Both the JEOL and Philips Corporations kindly provided material support so that additional people could participate and so that there would be the possibility for "hands-on" experience on a modern electron microscope. The authors of the chapters in this volume all worked extremely diligently to produce their manuscripts on time (or, in a few cases, close to on time). In addition to the authors, the following people kindly reviewed one or more chapters or provided other types of assistance: D. Barber, D. Blake, J.M. Cowley, P. Crozier, D. Eisenhour, S. Guggenheim, P. Heaney, M.A. O'Keefe, J. Post, R. Reynolds, J.R. Smyth, J. Spence, and D. Williams. Finally, I thank Paul Ribbe, the Series Editor, and my assistants, J. Macias and D. Goldstein, for their concern and careful attention. The efforts of these people together has produced a book of which we can all feel proud.

Peter R. Buseck
Tempe, Arizona
July 14, 1992

TABLE OF CONTENTS

IMAGING and DIFFRACTION

Chapter 1   Peter R. Buseck

PRINCIPLES OF TRANSMISSION ELECTRON MICROSCOPY

Chapter 2   John W. Steeds & J.-P. Morniroli

ELECTRON DIFFRACTION—SAED & CBED

Chapter 3   **Peter Self**

HIGH-RESOLUTION IMAGE SIMULATION AND ANALYSIS

ANALYTICAL THEORY and APPLICATIONS

Chapter 4   **Donald R. Peacor**

ANALYTICAL ELECTRON MICROSCOPY: X-RAY ANALYSIS

Chapter 5   **Peter R. Buseck & Peter Self**

ELECTRON ENERGY-LOSS SPECTROSCOPY (EELS) AND
ELECTRON CHANNELLING (ALCHEMI)

MINERALOGY and CRYSTALLOGRAPHY

Chapter 6   David R. Veblen

ELECTRON MICROSCOPY APPLIED TO NONSTOICHIOMETRY, POLYSOMATISM, AND REPLACEMENT REACTIONS IN MINERALS

Chapter 7   **Alain Baronnet**

POLYTYPISM AND STACKING DISORDER

Chapter 8   **Fred Allen**

MINERAL DEFINITION BY HRTEM: PROBLEMS AND OPPORTUNITIES

PETROLOGY — LOW-TEMPERATURE REACTIONS

Chapter 9   **Donald R. Peacor**

DIAGENESIS AND LOW-GRADE METAMORPHISM
OF SHALES AND SLATES

Chapter 10   Richard J. Reeder

CARBONATES: GROWTH AND ALTERATION MICROSTRUCTURES

PETROLOGY — HIGH-TEMPERATURE and DEFORMATION-INDUCED REACTIONS

Chapter 11   Harry W. Green, II

ANALYSIS OF DEFORMATION IN GEOLOGICAL MATERIALS

Chapter 12   Gordon L. Nord, Jr.

IMAGING TRANSFORMATION-INDUCED MICROSTRUCTURES

Chapter 1. PRINCIPLES OF TRANSMISSION ELECTRON MICROSCOPY

Peter R. Buseck   Departments of Geology and Chemistry, Arizona State University, Tempe, Arizona 85287, U.S.A.

1.1   INTRODUCTION

The mineralogical use of transmission electron microscopes (TEMs) is so widespread today that it is difficult to find a current issue of a mineralogical journal that does not contain one or more papers that use electron microscopy as an essential component of the research. This situation is in stark contrast with that of two decades ago, when TEMs were rarely, if ever, used for mineralogical studies, and high-resolution studies essentially did not exist.

Although the first electron microscope was built in 1932 (Ruska, 1988), the impact of electron microscopy on mineralogy was negligible for several decades thereafter. Why? TEMs were widespread in the 1950s and 1960s, and access would certainly have been possible had there been the desire on the part of mineralogists. However, instrumental resolutions were inadequate to reveal important crystallographic details on the scale needed for many structural studies, and so TEMs were largely ignored for crystallographic research. This situation changed in the 1970s, when high-resolution imaging to reveal structural details at the unit-cell level produced major insights, resulting in the widespread use of electron microscopy in the study of minerals today. One of the goals of this short course and introductory chapter is to explore how and why this marked change occurred.

Resolutions of modern TEMs are sufficient that structural details at the unit-cell level of most minerals can be imaged directly, and the range of other electron microscopy applications is broad. They include studies that explore crystallography, defects in crystals, reaction remnants, reaction mechanisms, structural complexities like inter-calations, modulations, twinning, and other features that occur at or near the unit-cell scale. Another highly important set of studies uses the TEM as a powerful high-resolution petrographic microscope, so that petrography can be done on a far finer scale than was previously possible. In this mode, the TEM is used to study mineral relations, textures, intergrowths, replacements, and a myriad of other applications for which optical microscopes are used. Moreover, not only is it possible to obtain images and diffraction information, but compositional measurements on a scale finer than are possible with electron microprobe analyzers are also obtained routinely.

Isolated studies of minerals that illustrated the utility of high-resolution imaging appeared in the early 1970s (Yada, 1971; Yoshida and Suito, 1972; Iijima et al., 1973). A comprehensive demonstration of high-resolution images of silicates, and a comparison between experimentally obtained mineral images and their idealized structures as determined by x-ray diffraction, appeared shortly thereafter (Buseck and Iijima, 1974). Buseck and Iijima (1974) and Pierce and Buseck (1974) also illustrated the use of high-resolution imaging to show defects in minerals. A steadily increasing stream of mineral-ogical studies that utilize TEMs appeared thereafter.

TEM images are formed in two stages. The first consists of scattering of an incident electron beam by a specimen ('A' in Fig. 1-1); this scattered radiation passes through an objective lens, which focuses it to form the primary image that is subsequently magnified

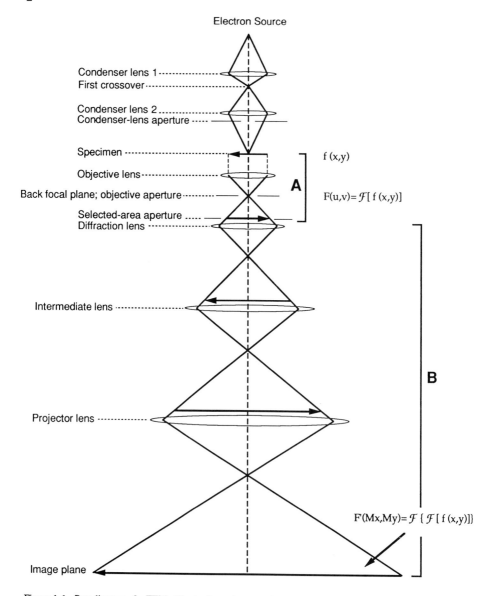

Figure 1-1. Ray diagram of a TEM. The horizontal arrows show the positions of the specimen and its images as they are progressively enlarged at various levels of the microscope. "A" indicates stage 1, and "B" indicates stage 2; see text for details. The formulae shown on the right side are discussed in §1.3.3. The bottom formula applies to the image plane.

by additional lenses to form a highly magnified final image ('B' in Fig. 1-1). In the process of forming the primary image, the objective lens produces a diffraction pattern at its back focal plane. This diffraction pattern is the Fourier transform of the scattered electron wave; in turn, the primary image is the Fourier transform of the diffraction pattern. This two-step process forms the basis of image formation during high-resolution transmission electron microscopy (HRTEM). The high-resolution image is, in effect, an interference pattern of the beams formed at the back focal plane of the objective lens.

Chemical analysis is related to image formation. A portion of the energy of the incident electron beam is typically absorbed by the specimen through the process of inelastic scattering, with the subsequent emission of x-rays exhibiting a range of energies (Fig. 1-2). These energies depend, among other things, on the composition of the sample, and so the x-rays can be used for chemical measurements. Such measurements form the basis for analytical electron microscopy (AEM), which is an increasingly important application of the TEM.

The power of the TEM is evident, but so are its limitations. (1) TEMs are expensive, with costs in the range of electron microprobes. The hundreds of thousands of dollars needed to buy a single TEM suffice to purchase a large number of petrographic microscopes. Nonetheless, TEMs are widely available. (2) TEMs are delicate and require care and experience for their use. However, they are becoming steadily easier to operate. Most modern TEMs are now under computer control, and for routine purposes the instruments are far more "user-friendly" than just a few years ago. Finally, (3) their size and complexity have made TEMs somewhat formidable for the new user or the non-user. One of the purposes of this short course is to minimize or even eliminate this concern.

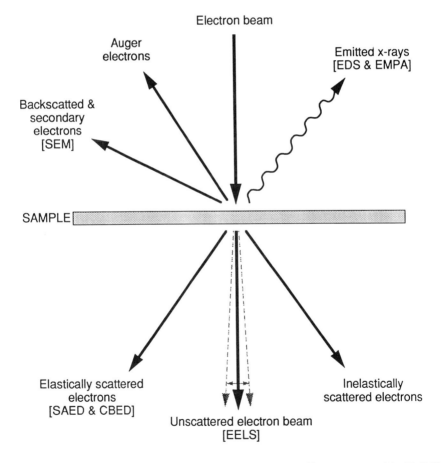

Figure 1-2. Signals produced from a sample irradiated by an electron beam. The acronyms, explained in Table 1 and the text, are the measurement techniques or instruments used to detect the respective signals. The electrons giving rise to the EELS signal actually deviate at a small angle from the direct, unscattered beam, but the angle is too small to show accurately on the scale of the diagram.

Table 1. Commonly used terms and acronyms in electron microscopy.

Term	Brief explanation	Section where cited and explained
AEM	Analytical Electron Microscopy	1.3.4
amplitude (A)	"peak" or highest point of a wave	1.2.3
amplitude contrast	contrast produced by excluding all but the direct electron beam	1.2.4; 1.4.4
analytical electron microscopy	chemical analysis using EDS or EELS	1.3.4
astigmatism	optical aberration resulting from asymmetry of the focussing field	1.5.2
Auger electrons	low-energy electrons used for surface studies	1.3.4
bright-field (BF) images	images produced by including the direct electron beam	1.2.4
CBED	Convergent-Beam Electron Diffraction	1.3.2
CCD	Charge-Coupled Device	1.5.3
chromatic aberration	optical distortion produced by spread in energy (or wavelength) of incident electrons	1.5.2
CL	Cathodoluminescence	
condenser aperture	circular aperture that blocks widely divergent electron beams; placed just below condenser lens	1.2.2
condenser lens	"demagnifying" lens above the specimen	1.2.2
conventional electron microscopy	microscopy using amplitude contrast; generally is "low magnification"	1.4.4
damping envelope/function	a function that corrects the transfer function for the effects of chromatic aberration, beam divergence, and coherence	
dark-field (DF) images	images produced by excluding the direct electron beam	1.3.5
dynamical diffraction/scattering	multiple diffraction/scattering	1.2.4
EDS	Energy Dispersive Spectrometry/spectroscopy	1.3.2
EELS	Electron Energy-loss Spectrometry/spectroscopy	
electron energy-loss spectroscopy	AEM measurements of the loss in energy produced when electrons are scattered inelastically	1.3.4
FEG	Field-Emission Gun	1.3.4
FFT	Fast Fourier Transform	1.2.2
Fourier images	similar images that repeat at several values of focus	1.5.3
Gaussian focus	exact optical focus if there were no aberrations	1.3.3
hkl	Miller indices of planes/diffractions	1.4.1; 1.4.2
holey-carbon films	carbon films containing holes, used on TEM grids to support samples	1.3.5
HRTEM	High-Resolution Transmission Electron Microscopy	1.6.2

Table 1, continued

Term	Brief explanation	Section where cited and explained
image plane	optical plane where images form	1.2.2
in phase/out of phase	two or more waves having the same phase/… having opposite phase	1.2.3
kinematical diffraction/scattering	single scattering events	1.3.2
multiple diffraction/scattering	repeated scattering events within one crystal	1.3.2
objective lens	lens producing maximum magnification	1.2.2
objective aperture	circular aperture that blocks widely divergent electron beams; placed just below objective lens	1.2.2
object plane	position of the specimen in the TEM	1.2.2
phase	advance or retardation of one wave relative to a reference wave	1.2.2
phase contrast	optical contrast produced by interference in thin samples	1.3.3
phase objects	thin specimens that yield phase contrast	1.4.1
PIMS	Precision Ion Milling System	1.4.1
projected charge-density (PCD)	images and theory for dynamical (multiple) scattering in thin samples	1.6.2
projected potential images	images from very thin specimens corresponding to weak phase-object theory	1.4.3
projector/intermediate lenses	lenses located between objective lens and image plane	1.4.2
REM	Reflection Electron Microscopy	1.2.2
SAED	Selected-Area Electron Diffraction	1.3.2
spherical aberration	optical distortion produced by stronger refraction of incident electrons at the outer edges of a lens than near its axis	1.5.2
secondary electrons	low-energy electrons produced near surfaces	1.3.4
Scherzer focus	optical focus value with optimal phasing for low-order diffracted beams	1.3.5
spatial frequencies	reciprocal spacings	1.3.5
STEM	Scanning Transmission Electron Microscopy	1.2.2
TF	Transfer Function	1.3.5
through-focus series (cf. § 1.3.6)	images produced at various focus settings	1.3.3
top-/side-entry stage	types of specimen stages	1.5.1
weak phase-object theory	theory for kinematical (single-event) scattering for very thin samples	1.4.2
WPOA	Weak Phase-Object Approximation	1.4.2

As with any new subject, there is considerable unfamiliar jargon. While daunting and sometimes unnecessarily complex and confusing, the terms are useful for understanding both the practitioners and the relevant literature. Table 1-1 summarizes terms that may be unfamiliar (they are italicized where first used in the text) and lists the section where they are explained or used. It also summarizes some common acronyms. The table will be most useful to people who have done limited electron microscopy and have already encountered and perhaps been confused by some of these terms. A more complete list is provided on pages *xviii* and *xix* of Buseck et al. (1988a).

Many of the principles of analytical electron microscopy will already be familiar to mineralogists through their use of electron microprobes. Image formation, and especially high-resolution images, will be less familiar and so will receive more attention in this chapter.

1.2   THE TEM: A COMPARISON TO OPTICAL MICROSCOPY

1.2.1   General

Light waves change direction when travelling between materials having different refractive indices; in general, refractive indices increase with specimen density. Such changes of direction or "bending" (the familiar "where is the fish?" effect observed in hobbyist's aquaria) is a result of a change in wavelength and velocity (§1.2.3) in the light waves. Similar bending occurs with the electron beam, which is more affected by heavy atoms than light ones. In regions of high atomic potential, occurring where there are heavy atoms, the refractive index is high and the wavelengths of the beam electrons are reduced, resulting in a phase advance relative to electrons passing through regions of lower atomic number or density. A challenge of electron microscopy is to utilize the changes in phase produced by interactions with the sample to produce contrast variations that correspond to the structure of the sample.

Table 1-2. Comparison between electron and petrographic microscopes

	Petrographic	Electron
Physical characteristics		
Size	Desktop	Room-sized
Cost	n\$100 to n\$1,000	n\$100,000 up
	($n = factor$)	
Environment	Air	Vacuum
Radiation		
Sources	Visible light	Electrons
Wavelengths	Polychromatic ("white light")	Almost monochromatic
Polarized option	Yes	Not normally
Lenses		
Construction	Glass	Electromagnetic
Continuously adjustable	No	Yes
Functions		
Imaging	Yes	Yes
Diffraction	No	Yes
Spectrometry	No	Yes
Sample position	Below objective lens	Within objective lens

ELECTRON   OPTICAL

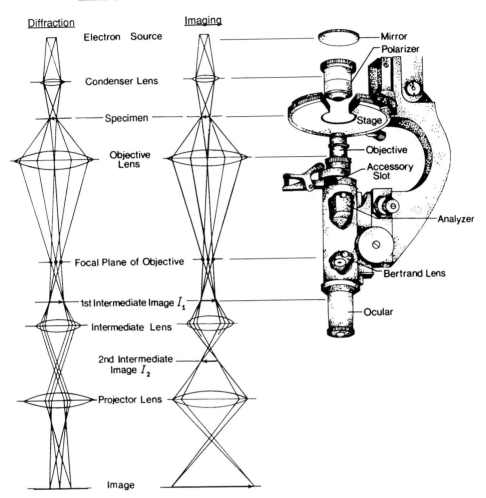

Figure 1-3. Comparison between a petrographic microscope ("optical") and a TEM ("electron"). The ray paths for the TEM are shown in the diffraction mode (left) and imaging mode (center). Note that the ray paths are identical until the intermediate lens, where the field strengths are changed, depending on the desired operation mode.

1.2.2   The TEM and the petrographic microscope

All mineralogists are familiar with the petrographic microscope. Therefore I will introduce the TEM by making a comparison between the two instruments. Selected details are given in Table 2, and major points of comparison are shown in Figure 1-3. Much of this discussion follows Buseck and Iijima (1974).

In their standard configurations, the radiation source is below the specimen in the petrographic microscope, whereas it is above the specimen in the TEM. The petrographic microscope has therefore been inverted in Figure 1-3 to illustrate corresponding elements. In the TEM there is an incident electron beam that passes through a condenser lens. It next is scattered by the specimen, which is positioned within the core of the objective lens.

After passing through the objective lens, the electrons form separate (diffracted) beams at its back focal plane. In order to form a highly magnified image, these beams are then recombined by additional lenses, so that rays from a given point in the specimen (the *object plane*) converge to corresponding points to form an image (at the *image plane*). This sequence of events is analogous to what occurs in the petrographic microscope. There are, however, important distinctions, only some of which are considered here (Table 1-2).

Intensity variations in specimens viewed with the petrographic microscope are produced by differences in light absorption by different minerals or regions within a given mineral. In the TEM, contrast is produced by either of two mechanisms. In the first method, called *amplitude* (or *aperture*) *contrast* imaging, scattering of electrons takes the place of absorption in producing contrast variations; the scattered electrons are removed from contributing to the image by an aperture placed below the objective lens (the *objective aperture*). The result can be viewed as an intensity deficit in the image (Spence, 1981). In the second, high-resolution method, called *phase contrast* imaging, more of the scattered electrons are allowed to pass through the objective aperture, and a defocus of the objective lens results in phase differences that give rise to contrast.

At the back focal plane of the objective lens (Figs. 1-1, 1-3) the beams form an interference pattern, analogous to the familiar interference pattern that is obtained during petrographic microscopy with the aid of the Bertrand lens. However, in optical microscopy we normally use "white" light (polychromatic radiation) and have a convergent incoming beam, which produces the well-known isogyres in the interference pattern of a mineral. In marked contrast, in the TEM the incoming electron beam is normally as close to parallel and monochromatic as possible (important exceptions include convergent-beam electron diffraction and microanalysis, described in Chapters 2 and 4, respectively). A consequence of the parallel, monochromatic radiation is that in the TEM the interference pattern at the back focal plane of the objective lens consists of discrete spots for a periodic object like a mineral; it is called a diffraction pattern.

The image plane in the petrographic microscope is observed with the set of lenses that, in combination, form the ocular; the top lens in this series is the eye lens. In the TEM the image plane is generally seen on a fluorescent viewing screen, observed through a port in the microscope column; the image is produced after the beam electrons have passed through one or more intermediate lenses and a projector lens. While the optical details differ, the effect of these lenses could be considered analogous to the ocular in the petrographic microscope.

An important feature of the TEM is that it uses electromagnetic lenses. These consist of tightly wound wire coils encased in soft-iron sheaths. When an electric current passes through the wires, the lenses produce highly concentrated magnetic fields; such lenses have the useful characteristic that their focal lengths and thus magnifications can be varied smoothly and almost continuously by altering the electric currents. Increasing the objective lens current reduces the focal length of the lens and reduces the spherical aberration constant (see §1.5.2), thereby improving resolution. Magnifications can be adjusted over a wide range by changing the lens strengths using sensitive rheostats. Similarly, optical aberrations, although more severe than in simple glass lenses, can be minimized by iterative adjustments of the lens currents. The glass lenses in petrographic microscopes, on the other hand, have fixed focal lengths as well as aberrations that cannot be adjusted once a lens is constructed. Interchanges of multiple objective lenses and oculars are required for petrography because each lens has a unique magnification, whereas in the TEM changes in magnification are produced electronically.

Microscopists have names for different lenses in the TEM. Those between the electron source and the specimen are called *"condenser"* lenses and are designed to focus ("demagnify") the electron source so that when the electron beam reaches the sample it is small and intense. One or more condenser apertures limit the beam incident on the specimen, and an objective aperture is typically inserted in the back focal plane of the objective lens to exclude beams that have been scattered at large angles. The specimen is positioned within the center of the objective lens, which uses the electron beams (more properly the electron wavefield) exiting the specimen as the object being magnified. This lens is the single most critical part of the microscope: it provides the initial magnification, focuses the image, and produces the diffraction pattern.

A typical TEM has several additional lenses between the specimen and image plane (Fig. 1-1). The first, called the diffraction lens, can focus on the back focal plane of the objective lens and so can be used to produce an "image" of the specimen diffraction pattern. Alternatively, it can be used to focus on the image plane of the objective lens. Each lens magnifies the image produced by the preceding (relative to the specimen) lens. The lenses following the diffraction lens are called *"projector"* or *"intermediate"* lenses (the projector lens is the final one). The current passed through each lens controls its respective focal length and thus magnification. The specimen is successively magnified by each lens until a highly magnified image results at the final image plane, which is where the viewing screen. camera, or other detector is placed.

A related instrument called a scanning transmission electron microscope (STEM) is also widely used, although it will not receive primary attention here. A STEM uses a scanned electron beam and has as its major application any of a wide variety of analytical purposes. With a STEM, signals are collected and recorded serially, point-by-point, across a sample. The information is generally recorded digitally, so that the number of signal counts per small region of sample is determined. Some ambiguity exists around the STEM acronym because it has been applied to two different types of electron microscopes. One is a "standard" TEM to which deflector (scanning) coils have been added to scan the beam across the specimen, as in an SEM, but no fundamental optical changes have been made. The other type of instrument, also called a "dedicated STEM," has no post-specimen lenses such as projector lenses (cf. Fig. 1-1). In such a STEM, in order to obtain an adequate number of counts per sample area in a reasonable amount of time, it is necessary to generate a large electron current within a small probe. A field-emission gun (FEG) is the electron source of choice for this purpose; probes as small as a few tenths of a nanometer can then be attained while still maintaining useable current. See Cowley (1988) for more details.

Finally, most adjustments made during the operation of a modern TEM are made electronically and, in many cases, under computer control. The result is a highly stable instrument that is capable of use for high-resolution as well as analytical microscopy.

1.2.3   Waves and phases – a brief review

TEMs and petrographic microscopes have many similarities in their optical principles. However, rather than resembling polarized-light microscopy, image formation in high-resolution electron microscopy is more similar to phase-contrast microscopy, a technique that is widely used by biologists. In order to pursue the analogy to phase-contrast microscopy, it is useful first to explore some fundamentals of waves and their interactions.

It is convenient to represent waves by sine curves that show the wavelength, amplitude, and phase (wave retardation or advance) of one wave relative to another. With

10

visible light, we observe changes in wavelength as different colors, and changes in amplitude as different intensities (brightnesses). However, we are unable to observe changes in phase although, as we shall see, such changes are critical in microscopy. The interference of two or more waves can cause increases or decreases in amplitude, or even cancellation. If their wavelengths differ, changes in phase also result.

Figures 1-4a to 1-4g illustrate features of waves and their interactions. Figure 1-4a shows a simple wave and reference axes **x** and **y**. The wavelength, λ, and amplitude, A,

Figure 1-4. Schematic diagrams showing waves and the effects of their interference with one another.

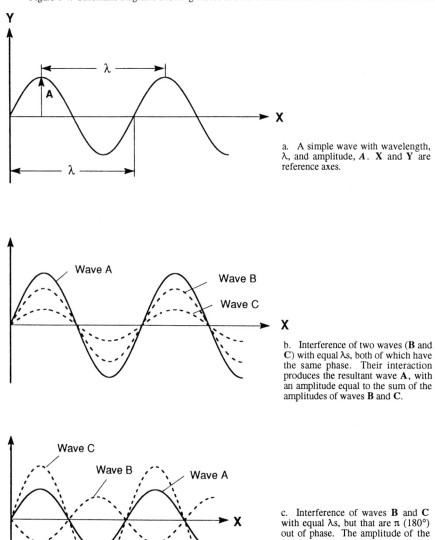

a. A simple wave with wavelength, λ, and amplitude, A. **X** and **Y** are reference axes.

b. Interference of two waves (**B** and **C**) with equal λs, both of which have the same phase. Their interaction produces the resultant wave **A**, with an amplitude equal to the sum of the amplitudes of waves **B** and **C**.

c. Interference of waves **B** and **C** with equal λs, but that are π (180°) out of phase. The amplitude of the resultant wave, **A**, equals the difference of the amplitudes of waves **B** and **C** because the phase difference of π introduces a relative shift of λ/2 between waves **B** and **C**.

1-4, cont. Schematic diagrams showing waves and the effects of their interference with one another.

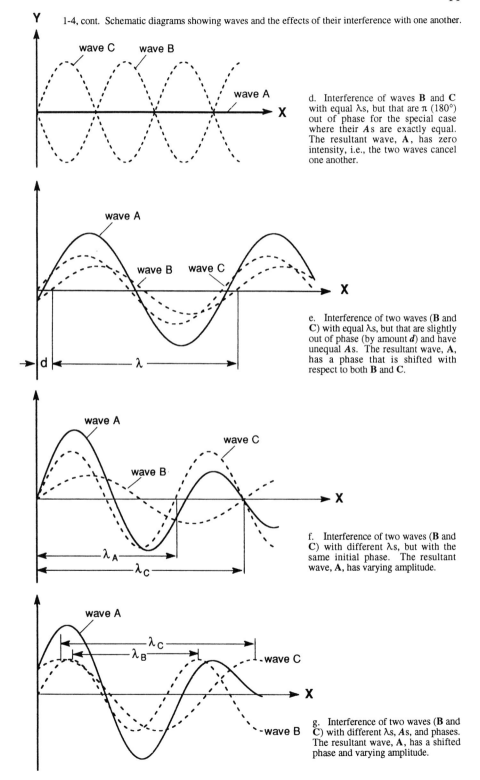

d. Interference of waves **B** and **C** with equal λs, but that are π (180°) out of phase for the special case where their *A*s are exactly equal. The resultant wave, **A**, has zero intensity, i.e., the two waves cancel one another.

e. Interference of two waves (**B** and **C**) with equal λs, but that are slightly out of phase (by amount *d*) and have unequal *A*s. The resultant wave, **A**, has a phase that is shifted with respect to both **B** and **C**.

f. Interference of two waves (**B** and **C**) with different λs, but with the same initial phase. The resultant wave, **A**, has varying amplitude.

g. Interference of two waves (**B** and **C**) with different λs, *A*s, and phases. The resultant wave, **A**, has a shifted phase and varying amplitude.

are indicated. Two or more waves are *in phase* if they have the same λs and are exactly synchronous. The waves will then interfere with one another to form a new wave having an *A* that is the sum of the individual waves (Fig. 1-4b). A phase difference can be considered to be a retardation or acceleration of a wave. When waves having different phases interact, changes in amplitude occur as a result of reinforcement or cancellation. If the waves are exactly *out of phase*, meaning one is accelerated or retarded by exactly π radians (180°) relative to the other, then they will interfere destructively (Fig. 1-4c); in the special case where their *A*s are exactly equal in magnitude and opposite in sign, the two waves cancel (Fig. 1-4d).

A more general case than the ones given so far is when two or more waves are out of phase by an amount other than π. In the TEM such differences in phase can range from 0 to 2π radians, whereas differences in λ will be small—generally of the order of a few parts per million—because the spread in energy of the electron beam at any point in time is small. When minor differences in phase occur, then the sum of the waves has an *A* and phase that differ from the original waves (Fig. 1-4e). Such changes occur when an electron beam passes through a specimen. Even more complex situations arise when two (or more) waves differ in both phase and λ. For example, Figure 1-4f could apply to a crystal where one wave goes through a vacuum and the other through atoms (e.g., Fig. 1-8), whereas Figure 1-4g could be used to explain periodic images in dynamical theory (cf. §1.4.3). As we shall see, for crystals thin enough for high-resolution transmission electron microscopy (HRTEM), changes in phase dominate changes in *A*, and such interactions have profound effects for producing good images.

1.2.4   Origin of contrast

Two methods are widely used for obtaining contrast with a TEM. One, used for "conventional" electron microscopy, produces "*amplitude contrast*" by excluding a significant fraction of electrons from reaching the image plane. The other method, used for high-resolution imaging, produces "*phase contrast*" by adjustment of the phases of the electron beams through judicious focus of the objective lens.

Amplitude contrast results from a filtering process whereby some of the electrons from the incoming electron beam are unable to reach the image plane. These electrons that are extracted from the exiting beam decrease the image intensity, and therefore the amplitude is reduced. Which electrons reach the image plane and which do not varies greatly with the experimental conditions, and the resulting images differ accordingly (Chapters 11 and 12). For example, in normal (low-resolution) *bright-field* electron microscopy, only the direct electron beam is allowed to reach the image plane; all scattered electrons are excluded by the objective-lens aperture. In normal *dark-field* electron microscopy, the direct electron beam is excluded from the image plane, and only one or more scattered beams are allowed to pass through the objective-lens aperture. The selection of scattered beams is accomplished either by electronically tilting the direct electron beam so it no longer travels down the optic axis of the TEM, or by moving the aperture so that the direct beam is blocked.

For high-resolution imaging it is desirable to obtain as much structural information as possible about the specimen, and this means allowing the maximum number of diffracted beams to pass through the objective-lens aperture to form an image. However, there is an unavoidable problem. Although information is carried by all the scattered electron beams, those scattered at high angles tend to be grossly misphased (see §1.3.5), so that their contributions to the image can be highly disruptive; such beams greatly complicate or even preclude intuitive interpretations of the image. Moreover, beams scattered at the highest angles (greater than 10^{-2} to 10^{-1} rad) fall outside of the physical limits of the optical system of the TEM and cannot be included. The mathematical analogue is that of reconstructing an object from a truncated Fourier series, where increasing severity of the truncations allows increasingly poor reconstructions (cf. §1.3.3). Additional information about the origin of contrast, and especially the difference between "phase contrast" and "amplitude contrast," is given in §1.4 and in Table 1-3.

Table 1-3.   Comparison of instrumental conditions for phase- and amplitude-contrast microscopy

| | Contrast type | |
	Phase	Amplitude
Type of microscopy	High-resolution	Conventional
Sample thickness	As thin as possible; a few nm	"Thick"; could be 10s of nm
Beam coherence	High	Not as critical
Objective aperture	Large	Small
Condenser (illuminating) aperture	Small	Large
Focus	Critical; under-focus by 40 to 90 nm	Not as crucial
$I_{direct\ beam}/I_{scattered\ beams}$	High	Low to moderate

1.2.5   The phase-contrast microscope

High-resolution transmission electron microscopy is in many ways analogous to the phase-contrast optical microscopy that is widely used by biologists (Rochow and Rochow, 1978; Culling, 1974). Thin biological specimens (like the thin crystals used for HRTEM imaging) have minimal differences in refractive index, so that light is not significantly refracted in passing through the sample. Details are then hard to see by conventional microscopy because of the lack of intensity variations (contrast). Although the amplitude of the transmitted wave has hardly changed, there may be significant changes in phase. With a TEM, contrast can be obtained by defocusing the electron beam, but in optical microscopy one cannot go out of focus because there would be a loss in resolution, and so other procedures are required. Special techniques were devised to take advantage of these changes in phase. Phase-contrast microscopy, for which F. Zernike won the 1953 Nobel Prize in Physics, utilizes this characteristic whereby even slight differences in optical density can cause significant alterations in the appearance of the image. Such contrast enhancement was accomplished by changing the phase of the direct beam (with a 1/4 λ phase plate that either retards or advances the direct beam) and then allowing interference to occur between it and the transmitted beam.

Techniques analogous to phase-contrast microscopy, essentially electron interferometry, are used for HRTEM imaging. For the HRTEM case, the samples behave as if their refractive indices vary proportionally to the atomic potential (or electron density) from point to point in the sample (Spence, 1981). Strong, highly coherent monochromatic radiation is required. Providing such electron beams was a major advance in TEM design and led to high-resolution imaging.

Another way of viewing this situation is to note that with the TEM we observe contrast differences, and these arise from intensity variations across a specimen. The intensity of any radiation (electrons, x-rays, visible light) is proportional to its amplitude squared. However, if the amplitude of the incoming radiation is unaffected during its passage through the specimen, as occurs for extremely thin TEM samples, then there can be no amplitude contrast and the specimen will appear devoid of details. On the other hand, the phases of the transmitted radiation will have been changed within the specimen, and it is these phase changes that appear as differences in amplitude (Fig. 1-4e) when diffracted beams interfere to produce images from very thin crystals, as occurs in HRTEM.

1.2.6   Resolution

High spatial resolution is one of the major attractions of the TEM. Two widely used methods for specifying resolution are relevant for our purposes, and it is important to distinguish between them. Point-to-point resolution (also called "Scherzer" or "structure-image" resolution) is the most useful; it is a measure of the ability of the TEM to faithfully reproduce the structure of the sample as projected in the direction of the incident electron beam.

Another measure, line (or fringe) resolution, considers the smallest distinguishable separations between regularly spaced fringes. Such fringes are generated from the interference between two diffraction spots and have no simple correlation with the crystal structure other than reflecting a periodicity perpendicular to one crystallographic direction. For certain focus conditions the fringes will be shifted by half a period, and there are even certain specimen thicknesses and focus conditions where half-period fringes will appear (O'Keefe, 1979; Spence, 1981), i.e., having half the periodicity of the structure. While line spacings are useful for measuring instrument stability, they provide no significant information about the ability of a TEM to image structures. Line resolutions are generally smaller than point-to-point resolutions. O'Keefe (1992) provides a good review of the different types of resolution and their significance.

The general expression for resolution, r, for bright-field point-to-point imaging is

$$r = 0.67 \ \lambda^{3/4} \ C_s^{1/4} \ ,$$

where λ is the electron wavelength and C_s is the coefficient of spherical aberration of the objective lens (Cowley, 1988). It is evident that improvements in resolution can be made by decreasing the wavelength and spherical aberration. An increase in accelerating potential decreases the wavelength but causes other problems such as power-supply and instrumental instabilities as well as greatly increased costs and TEM size (a 1991 rule of thumb: each increase of 100 kV in operating voltage costs roughly US$100,000). Achieving decreases in C_s requires the solution of extremely difficult pole-piece design and engineering problems, including the need for very small gaps into which to place the specimen (§1.5.1). A result is that there is little or no space for tilting the specimen to obtain proper orientations. Another view of resolution and its limitations can be understood by considering the transfer function, which is discussed in §1.3.5 and illustrated in Figure 1-7 (below).

It is likely that within the next several years we shall see significantly improved resolutions (the current goal is 0.1 nm) through the use of a new generation of electron microscopes and either of two techniques that are currently the subject of vigorous research. One is based on electron holography, the Nobel-winning method originally proposed by Gabor (1949), using fixed-beam TEMs and field-emission guns (Lichte, 1991; Tonamura, 1992). The other method utilizes focal-series, image reconstruction methods (van Dyck and Op de Beeck, 1990). These instruments and techniques will be of potential use for dense minerals and for imaging along low-order zone axes.

1.3   ELECTRON INTERACTIONS WITH MATTER:
A COMPARISON TO X-RAYS

1.3.1   General

As with the petrographic microscope, x-ray diffraction (XRD) is familiar to mineralogists, and therefore I will use it as a means of introducing electron diffraction in the TEM. Although the principles of diffraction are identical, there are prominent differences arising from the contrasting character of x-radiation versus electron radiation. The most prominent difference for our purposes is that electrons can be used to produce images whereas x-rays, except in very special cases, cannot. Also, electrons can be focussed into narrow, high-intensity beams that permit measurements confined to small specimen volumes, i.e., at high spatial resolutions. In addition, the ease of obtaining diffraction patterns by the two methods differ greatly.

Electron beams can be focused into far smaller regions than is normally possible with x-rays. Electron beams as small as 0.3 nm can be used, whereas x-ray beams are commonly 0.25 to 0.5 mm across. This difference gives rise to both advantages and disadvantages relative to x-ray instruments. TEMs routinely provide far higher spatial resolutions—in several respects—than do x-rays. In addition to the precise spatial positioning, the electron beam can produce diffraction from significantly smaller volumes than is possible with conventional XRD measurements. Moreover, TEMs can achieve high spatial resolution within crystals, which is why, for example, atom clusters and variations within unit cells can be detected, The disadvantage is that the averaging over substantial volumes of crystal ($>10^{12}$ nm^3) that typically occurs with x-rays is impractical with electrons, which typically average over volumes between 10^{-1} and 10^8 nm^3, a rather substantial difference! In electron microscopy there is inevitably a question about how representative a given measurement or observation is. It has been mentioned anecdotally among electron microscopists that all specimens examined to date could probably fit into a modest-sized pill box or perhaps even a thimble. So far, however, that statement has not been quantified.

X-rays interact with and are diffracted by orbital electrons, whereas electrons interact with and are diffracted by the electrostatic potential arising from both orbital electrons and the atomic nuclei. The result of this strong interaction is that electron scattering is far more efficient than for x-rays (by a factor of 10^2 or 10^3; Cowley, oral communication, 10/91), and so diffraction patterns are obtained much more rapidly with electron beams. In the TEM, diffraction patterns are produced almost instantaneously, simply requiring a flick of a switch to change the current and thus strength of the lenses between the objective focal plane and the image plane. By contrast, x-ray exposures typically require many hours or days in order to obtain good diffraction patterns.

Electrons have far shorter wavelengths than x-rays, and that difference greatly affects the diffraction conditions. For example, 200-keV electrons, which are commonly used in high-resolution electron microscopy, have $\lambda = 2.5 \times 10^{-3}$ nm. For comparison, Cu K_α radiation has $\lambda = 1.54 \times 10^{-1}$ nm, almost two orders of magnitude greater. It follows from the Bragg equation ($n\lambda = 2d \sin \theta$) that a decrease in λ results in a corresponding decrease in diffraction angles, and so Bragg angles for diffraction are far smaller for electrons than for x-rays. Moreover, increases in electron energies (decreases in λ) decrease the Bragg angles even more. For example, the scattering angles for 1-MeV electrons is roughly a quarter those of 100-keV electrons. As shown in the following section, such differences in scattering angles have major effects on the ways diffraction data are collected.

All materials diffract radiation; a unique feature of periodic objects such as crystals is that their diffracted beams give rise to sets of sharp spots, the relative positions of which have a direct geometric relation to the atomic spacings in the crystal. The spacings of these spots is unaffected by the aberrations of the TEM. However, as discussed in §1.4.1, the phases of the diffracted beams producing these spots are determined by the atomic arrangement within the sample as well as by lens aberrations within the TEM.

1.3.2   Diffraction patterns

A consequence of the difference in wavelength between x-rays and electrons is that the sphere of reflection (the Ewald sphere, relating real and reciprocal space) is far larger for electrons than for x-rays, and so diffraction patterns can be obtained much more simply with electrons. [For details regarding the sphere of reflection see standard references on x-ray diffraction such as Cullity (1956) or Stout and Jensen (1968).] With x-rays, the most convenient method for obtaining a photograph of a plane through reciprocal space is to use a precession camera, which requires extended exposures and complicated camera motions designed to maintain the film tangent to the x-ray sphere of reflection. With the TEM, on the other hand, the radius of the sphere is so large and the crystal so thin (producing extended diffraction spikes instead of spots at the reciprocal lattice sites), that the image plane is essentially tangent over a large region of reciprocal space, and many diffraction spots can normally be recorded without any detector motion (Fig. 1-5). That, in combination with the strong diffraction effects that occur with electrons, means that planes through reciprocal space can be viewed almost instantaneously when operating in electron-diffraction mode. In fact, because of the essentially immediate production of diffraction spots it is common practice to orient a crystal in the TEM by watching its diffraction pattern as the crystal is tilted.

Another consequence of the strong interaction of electron beams with atoms is that, in distinction to x-ray diffraction, diffracted electrons have a greater probability of themselves being scattered as they pass through the sample. Such multiple scattering is characteristic of transmission electron microscopy and is important for all but the thinnest of specimens. The multiple scattering produces major effects and so must be considered when interpreting electron-diffraction patterns and HRTEM images. Single-scattering diffraction is called *kinematical* and multiple-scattering diffraction is called *dynamical*. Kinematical theory is typically used for x-ray diffraction; it is also a useful first approximation for understanding events in the TEM, particularly when very thin crystals are being studied (cf. §1.4.2). Dynamical theory (cf. §1.4.3), while more complex, is required for understanding electron diffraction in all but the thinnest crystals (Cowley, 1981, 1988). O'Keefe and Buseck (1979) evaluated the effects of kinematical and dynamical scattering on the formation of images of minerals, and much of the following discussion parallels their treatment.

Dynamical diffraction gives rise to diffracted intensities that are difficult to interpret quantitatively. Even the slightest differences in path lengths, such as are produced by local variations in crystal thickness or crystal tilt (bending), can have significant effects on the intensities. Moreover, some energy can be lost through inelastic scattering independently of these diffraction events, further complicating the resultant intensities.

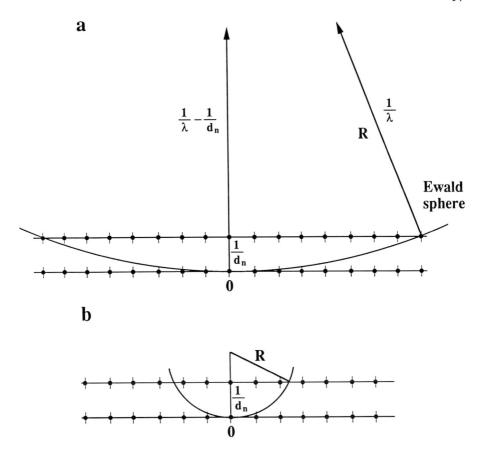

Figure 1-5. Schematic diagram of the Ewald sphere of radius R = 1/λ for (a) an electron beam and (b) an x-ray beam. The lines with spots on them represent planes in reciprocal space. Because of the far shorter wavelength of the electrons used in microscopy, the Ewald sphere in (a) intersects far more reciprocal-lattice points for a given orientation than is the case for the x-ray beam (b). The effect is accentuated by the extremely thin crystals, which produce extended reciprocal-lattice spikes [the thin streaks of intensity that lie perpendicular to the lattice planes and whose lengths are inversely proportional to the specimen thickness]. λ = wavelength of the incident radiation, d = separation of the planes of the reciprocal lattice.

Dynamical diffraction also complicates the calculation and interpretation of electron-diffraction patterns and images. The calculation techniques are described in Chapter 3. Here it suffices to note that x-ray intensities can be calculated more easily than electron intensities, which is a major reason why crystal-structure analyses are best done through x-ray measurements when a sufficiently large crystal is available. (In some cases structure analyses are done using convergent-beam electron diffraction because it is sensitive to a lack of inversion symmetry, whereas, to a first order, x-ray diffraction is not; see Chapter 2 for details.) Another consequence of the dynamical effects that are so prominent when using electron beams is that weak reflections such as those "forbidden" by symmetry commonly appear, and so great care must be used in interpreting electron-diffraction patterns.

In order to quantify the interactions of the electron beam with the sample, a number of variables have to be considered. These depend on the sample itself (such as absorption, fluorescence, and sample thickness) and a variety of lens and beam complexities in the TEM (such as specimen drift, lens astigmatism, chromatic aberrations, and beam instability). These effects are considered in §1.5.2.

Diffraction patterns can be produced using parallel electron beams (Fig. 1-3). Such patterns are called selected-area electron-diffraction (SAED) patterns; the areas of specimen to be studied are chosen with a "selected area" aperture inserted in the image plane of the objective lens (Fig. 1-1) or, alternatively, in the image plane of the diffraction lens (either position works). Generally the area of specimen used for an SAED pattern, perhaps a half micrometer or so in diameter, is significantly larger than the portion used for an HRTEM image. For many purposes it is helpful to use a focused electron beam, generated by converging the beam by suitable adjustments to the condenser-lens aperture (Fig. 1-1). These patterns are called convergent-beam electron-diffraction (CBED) patterns. Both SAED and CBED patterns are discussed in greater detail in Chapter 2.

1.3.3   Quantification of electron microscopy – the computed analogue

Neglecting complexities in order to illustrate basic principles, the diffraction pattern from a thin specimen can be considered as the Fourier transform of the specimen (more accurately, as the Fourier transform of the projected potential distribution of the atoms in the specimen). The relations are illustrated schematically in Figure 1-1. We shall assume the specimen is planar, with its features described as $f(x,y)$. Here x,y are real-space position parameters, and $f(x,y)$ represents the distribution of electron potential in the sample. Then $F(u,v)$, suitably adjusted for the variables considered in the previous section, describes the diffraction pattern; $F(u,v) = \mathcal{F}[f(x,y)]$, where $\mathcal{F}$ represents a Fourier transform. The variables u,v are the reciprocal-space analogues of x,y. With the TEM, the electron wave in the image plane can then be described by applying another Fourier transform, $F(x,y)$, so that $F(Mx,My) = \mathcal{F}\{\mathcal{F}[f(x,y)]\}$ (the inverse Fourier transform is used for calculations; for the lens it is a proper Fourier transform). The magnification produced by the projector lenses is represented by the term M (Cowley, 1981). The amplitudes of the Bragg reflections form the Fourier coefficients in a series whose sum gives the image amplitude.

While a comparable process is also conceptually feasible for x-rays, the second Fourier transform would need to be carried out on the diffracted-beam intensities. However, since the phases of the x-ray beams are lost, it is not normally possible to obtain the inverse Fourier transform, nor to obtain images using x-rays. [There are exceptions, such as shown by McNulty et al. (1992), who use synchrotron radiation to produce intense x-rays and then holography for imaging.] Methods for evaluating the various Fourier transforms to produce computed analogues of images are given in Chapter 3.

Computed images are critical for achieving proper interpretations of experimental results (O'Keefe et al., 1978). The necessary input variables, in addition to the fixed instrumental parameters for a given TEM, include the focus and specimen thickness, and the atom positions, site occupancies, and structure factors for the mineral of interest. Repeated iterations are typical, since commonly one or more of these variables are not known with high precision and must be estimated until good matches are obtained between experiment and calculation. By using a set of images obtained at a range of focus conditions (a "*through-focus series*"; cf. §1.3.6) it is generally possible to confirm that the approximations are reliable. An interesting effect is that for some structures, images of similar appearance repeat regularly at different thicknesses as a result of particular dynamical diffraction conditions. Also, repeating *Fourier images* (Cowley and Moodie, 1960) occur for fixed crystal thicknesses but for different values of focus. Fourier images are especially prominent in crystals having small unit cells. Although the periodic structure can be accurately represented in successive Fourier images, defects are not imaged sharply (Spence, 1981).

Figure 1-6 illustrates the effects of a lens and Fourier transform (optically, the diffraction pattern) as well as showing the effects of truncating a Fourier series. The Fourier transform of the dog (Fig. 1-6b), corresponding to the pattern that appears in the back focal plane of the objective lens, contains all the essential features of the image, except that the phases are lost if and when the intensities are recorded. The inverse Fourier transform of the diffraction pattern (Fig. 1-6c) produces the original image. It is possible to explore the information content of the Fourier transform by using various parts of the diffraction pattern. If the outermost portions of the diffraction pattern are excluded by a central aperture (= low-pass filter; Figs. 1-6d, 1-6e), the resulting image is degraded. When the direct beam is excluded by a central beam stop (= high-pass filter), as in a dark-field image (cf. §1.2.4), the contrast is reversed (Fig. 1-6f, 1-6g); in the TEM, the resolution is commonly lower in dark-field images, although there are exceptions (e.g., Pierce and Buseck, 1974). The combination of Figures [1-6d + 1-6f] gives 1-6a, the original image. Another type of dark-field image with the high frequencies excluded, produced by applying a donut-shaped (annular) aperture (= band-pass filter), is shown in Figure 1-6h. It contains the mid-range information, equivalent to the signal from Figures 1-6b minus those from Figures [1-6d + 1-6g]. Mineralogical examples of the exploration of diffraction patterns and their corresponding images are given by Buseck et al. (1988b).

1.3.4 Elastic versus inelastic scattering – analytical electron microscopy

Bragg diffraction occurs through elastic interactions, meaning that the emerging diffracted beam has the same λ as the incoming beam; there is no loss in energy, only a change in direction. There are, however, many other types of interactions during which the electrons from the incoming beam lose energy. Such interactions are grouped under the heading of inelastic scattering. They are important for mineralogy because they permit chemical analysis on an extremely fine scale. In general, measurements of inelastic scattering are made by spectroscopic techniques, where energies are determined (rather than the atom positions that are determined by HRTEM imaging).

Measurements analogous to those made with the electron microprobe analyzer are routinely made with many TEMs. As in the microprobe, the incoming electron beam excites x-rays that are characteristic of the elements in the specimen, and these can be monitored by using a detector placed above or at the same level as the specimen. Such *analytical electron microscopy* is discussed in Chapter 4. When the electrons in the incoming beam interact inelastically with the specimen atoms there is a transfer of energy, and the beam electrons lose energy. This decrease in energy can be monitored by a detector that is placed below the sample, so that it records electrons after they have traversed the sample. Such *electron energy-loss spectroscopy* (EELS) is useful for measuring elements that are too light for normal x-ray spectroscopy. EELS is discussed in Chapter 5.

The incoming beam can also excite *secondary electrons*, such as are used in the conventional scanning electron microscope (SEM). Similarly, low-energy *Auger electrons* are excited and can be measured to detect surface compositions. While these sorts of signals are not normally recorded within the TEM, there are instruments that are used for

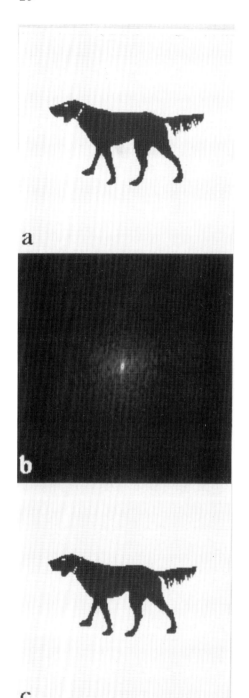

Figure 1-6. Effects of Fourier trans-forms on a familiar image. All images are, for illustrative purposes, of a point setter dog. The same Fourier transform (diffraction pattern) is used in all examples, but different apertures (filters) have been applied. These filtered diffraction patterns are shown in the right-hand sides of the figures, with the relevant apertures indicated, while their inverse Fourier transforms are shown on the left-hand sides. In the examples used here, the transforms and their inverses have been computed; in the TEM they are generated optically by the lens systems. See §1.3.3 for details.

a. The original image (a dog).

b. The Fourier transform of the dog. The transform does not resemble the well-spaced, regularly repeating spots typical of diffraction patterns of crystals because the dog is not a periodic object. The central beam has greatest intensity, and therefore the scattered beams appear suppressed (which is a result of the limited sensitivity—technically, the dynamic range—of the recording medium); if the direct beam is blocked (as by a beam stop; see Figs. 1-6f, g, and h), then the scattered beams are more visible in the display of the transform.

c. The inverse Fourier transform yields the original image.

Figure 1-6 is continued on the next two pages.

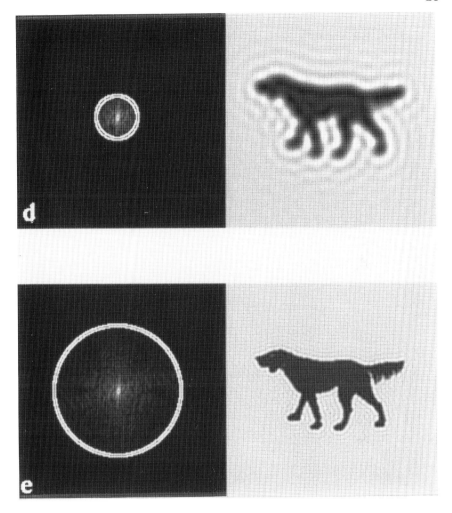

Figure 1-6. d. Addition of a field-limiting aperture to the pattern in Figure 1-6b excludes the outer diffracted beams from reaching the image plane. In a TEM, the exclusion occurs because (1) lenses are not infinitely large and so cannot include all the diffracted beams and (2) the outermost beams are generally misphased and so are intentionally excluded (by the objective aperture) to minimize their distorting effects. The resulting image has lower resolution and appears fuzzy because much information has been excluded. The ripple fringes surrounding the dog arise from the edge effects of the aperture used to produce this image.

e. When more of the diffraction pattern is included, the image is less degraded, although details are still missing. As larger portions of the pattern are utilized, the image becomes progressively more distinct (cf. 1-6c).

Figures 1-6 f,g,h are on the following page.

f. Addition of an opaque central beam stop excludes the direct beam and adjacent scattered beams from reaching the image plane. The result is a dark-field image, one that represents the high-frequency signal. The details at the edge of the dog are produced by the high-frequency signal and are greater than at the edge of Figure 1-6d. Exclusion of the bright direct beam makes it possible to see details in the diffraction pattern that are otherwise suppressed (cf. 1-6b). The situation is approximately opposite of that shown in Figure 1-6d.

g. Addition of a larger central beam stop than in Figure 1-6f results in a far less contrasty image, reflecting the important information that exists near the direct beam and that has now been excluded. The situation is approximately the opposite of that shown in Figure 1-6e.

h. Use of a donut-shaped (annular) aperture produces a band-pass filter that excludes both the direct beam and the outer diffracted beams from reaching the image plane, resulting in a composite type of dark-field image that differs from those illustrated in Figures 1-6f and 1-6g.

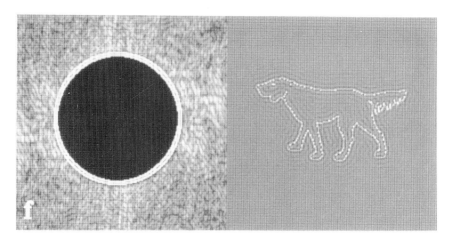

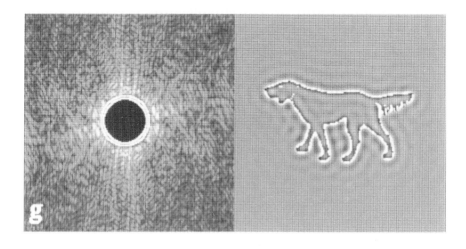

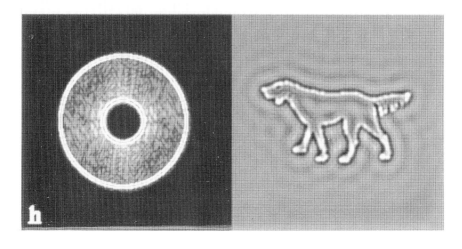

this purpose (Hembree et al., 1989, 1991). Spence (1988) provides a thorough review of the many effects of inelastic scattering as well as methods for their study.

1.3.5   Transfer function

Reference to a transfer function (TF) is commonly used to describe optimal resolutions and to understand factors underlying TEM adjustment for producing high-resolution images. The TF describes the imperfections in the lens system that result in modifications of the amplitudes and phases of the electron beam (wave) as it passes through the lenses to the image plane. The effect of these changes is that they prevent proper interference of the waves and result in distortions of the image. The TF is specific for particular TEMs used at specified values of defocus. Changing the defocus values changes the TF, expanding or contracting the region of optimal focus, as described below. Additional details are provided in §3.4.1

As an example, the TF for a good high-resolution TEM (e.g., JEOL JEM model 4000EX) at a defocus value of -46.8 nm (the Scherzer focus; see below) is given in Figure 1-7a. For very weakly scattering objects (§1.4.1), the TF can be expressed as $\sin \chi$, where χ is the phase difference of a diffracted beam relative to a beam along the optic axis of the TEM (the direct beam in bright-field microscopy). The vertical axis in the figure represents $\sin \chi$; the horizontal axis represents reciprocal distance, but is scaled in real-space units for convenience. Those diffracted beams having negative values for $\sin \chi$ are properly phased to maximize transfer of the corresponding reciprocal spacings (also called *spatial frequencies*) into the image and thereby minimize distortion. As detailed by O'Keefe and Buseck (1979), for values of $\sin \chi = -1$ the beams contribute to produce "normal" contrast, where atoms (regions of high potential) are dark and areas where atoms are absent (regions of low potential) are white. Where $\sin \chi = 1$, the contrast is reversed. For most practical purposes, beams falling at reciprocal spacings corresponding to values between approximately $\sin \chi = -1$ and -0.7 have sufficient amplitude to contribute significantly to the image and thus to permit intuitive interpretations of images; those between 0.7 and 1 have reverse contrast. The contributions to the contrast of beams falling between -0.7 and 0.7 are correspondingly weaker. Between -0.3 and 0.3 contributions are so weak as to be lost in the image "noise", and conditions under which significant reflections fall at such values of $\sin \chi$ should be avoided when possible.

The focus value with optimal phasing for low-order (small values of indices *hkl*) diffracted beams is called the *Scherzer focus*, and it is specific to particular TEMs. It is the focus value for which $\sin \chi$ is close to -1 for the broadest range of spatial frequencies (Scherzer, 1949). Maximum resolution is defined as occurring at the spatial frequency where the $\sin \chi$ curve = -0.7, at the right edge of the broad trough of the TF (e.g., ~0.18 nm in Fig. 1-7a). However, there are narrow troughs where higher spatial frequencies fall between $\sin \chi = -0.7$ and -1 (e.g., ~0.14 nm in Fig. 1-7a). If there is a reciprocal spacing for a given mineral that fortuitously falls within the region of one or more of these troughs, then a corresponding correct high-resolution component is possible. Thus, it is occasionally possible to obtain images showing "anomalously high" resolu-tions, but it is important to realize that these are unusual and so are special cases.

As discussed in §1.5.2, a variety of instrumental factors (lens aberrations, vibration, specimen height within the objective lens) degrade the idealized optimal performance of the TEM. The complex effects of these variables can be seen graphically by a damping envelope, shown in Figures 1-7b and 1-7c. The effect of these parameters becomes increasingly critical as resolutions increase, as can be seen by the suppression of the peaks and troughs of the TF by the damping envelope (Fig. 1-7d). In general, the TF goes to zero after only a small number of oscillations. In especially unfavorable cases, high resolutions are precluded by a damping envelope that goes to zero at a lower reciprocal spacing than that at which sin χ falls to -0.7 at the right edge of the broad trough.

1.3.6 Through-focus series of images and focusing

Obtaining a through-focus series of exposures is the best way of ensuring that each diffracted beam will have proper phasing for maximum transfer in at least one image. Such a series is acquired by recording images at a variety of focus settings around that for Scherzer focus. Details are summarized by O'Keefe and Buseck (1979). The TF trough has its greatest width at Scherzer focus (see previous section), and so this is the focus value that is generally the most useful for producing an intuitive structural interpretation of the image. It generally occurs at a known focus distance from the position of minimum contrast, which is readily recognizable. For this focus condition, atoms produce black spots and voids white ones; for a reverse-contrast focus, atoms are white and voids are black.

It is useful to compute a through-focus series (see §1.3.3) and compare it to the experimentally obtained images. If there are good matches at all conditions of defocus, then the operator can have confidence that a good structure model has been chosen, and image interpretation is likely to be correct. Additional details are provided in Chapter 3.

Figure 1-7 (on subsequent pages). The transfer function, TF, associated damping functions, and their effects on image interpretation. In all instances the horizontal scale is linear in reciprocal distance (in units of nm^{-1}), and is marked in nm. [Curves produced using the SHRLI routine (O'Keefe and Buseck, 1979); plots courtesy of M.A. O'Keefe.]

a. Plots of sin χ as a function of reciprocal distance, u. Two values of objective-lens defocus are shown: -46.8 nm, the "optimum" or "Scherzer" value, and -93.6 nm, the value where a second broad trough (passband) occurs. sin χ is the sine of the phase change, χ, produced by the objective lens on a diffracted electron beam of a given reciprocal spacing (and thus the weighting of the lens on the contribution of the beam to the image). Beams having spacings that fall within the passbands (the broad, shallow troughs marked by shading), give useful contrast (see text). Notice the different limits on the allowed spacings in the passbands under the two defocus conditions shown. At optimum defocus (left), a maximum resolution of ~0.17 nm is defined for crystal spacings producing beams that fall at the right edge of the broad passband.

b. Envelope (damping) functions that result from angular spreading of the electron beam (through either convergence or divergence). Such beam spreading blurs out some spatial frequencies and dampens their transfer into the image; areas above the upper curves and below the lower ones are suppressed. Three values of angular spread are plotted at the two defocus values of the transfer functions above. The lowest value (1 mrad) produces the least damping, and corresponds to the optimal operating conditions for the JEOL JEM 4000EX TEM that we use. The shape of this damping function varies with defocus.

c. Envelope (damping) functions that result from a spread of focus in the image that dampens higher spatial frequencies. Chromatic aberration in the objective lens converts any energy spread in the electron beam, or ripple in the high-voltage and objective-lens power supplies, to a spread of focus in the image. The three plotted values of spread of focus of 5.7, 9.3, and 13 nm correspond to electron-beam energy spreads of 1, 2, and 3 eV, with one part per million ripple in the microscope power supplies. The JEOL JEM 4000EX TEM is normally operated with between 1 and 2 eV energy spread. Obviously, best resolution is obtained with lowest energy spread. The shape of this damping function is independent of defocus.

d. When the appropriate envelope functions (Fig. 1-7b with 1 mrad and Fig. 1-7c with 2 eV) are applied to the transfer functions (Fig. 1-7a), a corrected region of proper beam phasing results. In this case, TF is the transfer function with the results of the blurring effects of incident beam convergence and chromatic aberration included. Compared with the undamped TF plots in Figure 1-7a, the passband in each damped TF is now slightly attenuated at its upper, or high-frequency side (on the right side of the troughs), whereas the narrow higher frequency peaks are severely diminished.

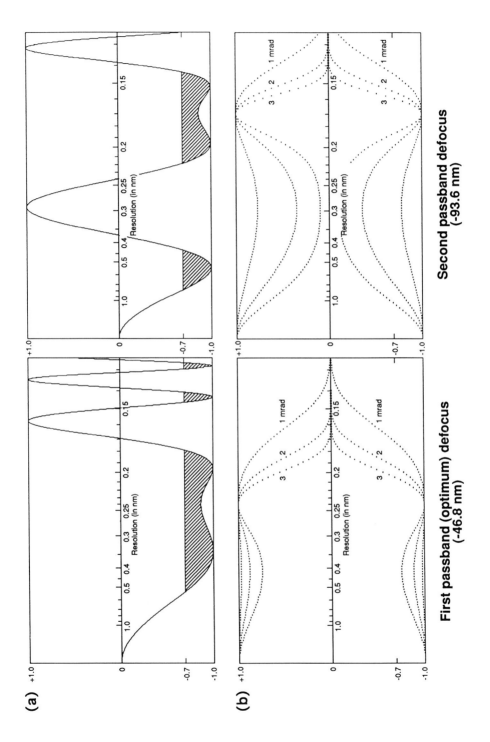

(a)

(b)

First passband (optimum) defocus
(-46.8 nm)

Second passband defocus
(-93.6 nm)

26

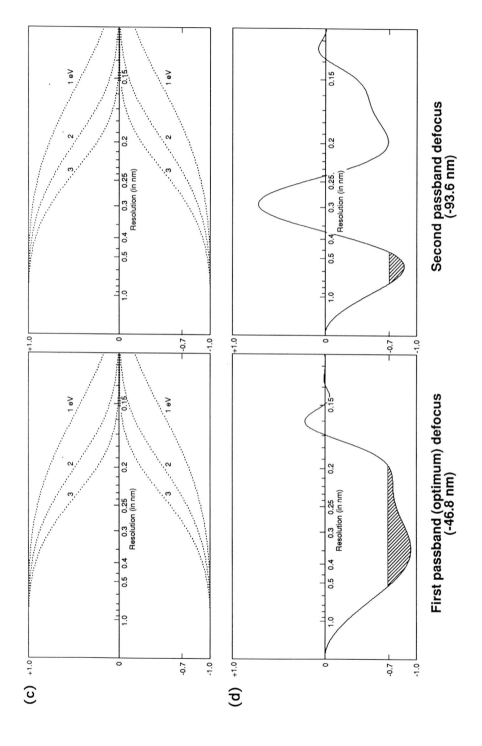

(c)

(d)

First passband (optimum) defocus
(-46.8 nm)

Second passband defocus
(-93.6 nm)

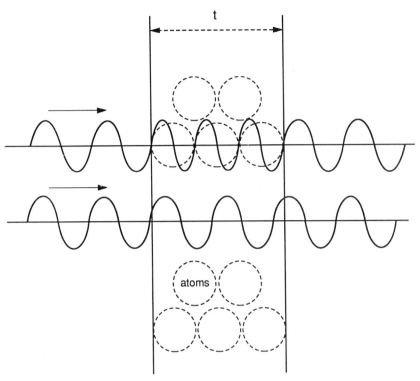

Figure 1-8. The effect of atomic potential on electron waves passing through a crystal of thickness t. The lower wave passes between atoms (regions of low potential) and so traverses the crystal with only minimal change in λ, whereas the upper wave passes through a region containing atoms, where it undergoes a large change in λ as a result of the change in refractive index. It has its exit-face phase changed as a result (after Spence, 1981).

1.4   MICROSCOPY OF THIN CRYSTALS:
PHASE *vs.* AMPLITUDE CONTRAST

1.4.1   General

We have seen that radiation experiences changes in phase, associated with changes in velocity and direction, when traveling between media having different indices of refraction (§1.2.1 and §1.2.4). Such phase changes occur when an electron beam passes from the vacuum of the TEM into a sample. Figure 1-8 shows an incoming electron wave impinging on a thin slab of crystal. The lower wave travels between atoms and so suffers no change in wavelength. The upper wave encounters atoms (regions of high potential) and experiences a change in wavelength within the crystal that produces a phase advance relative to the unperturbed wave as measured at the crystal exit face.

As mentioned in §1.2.5, thin crystals are required for HRTEM imaging. If we consider samples thin enough to minimize multiple scattering events so that only single scattering is significant, and if we neglect the effects of absorption produced by inelastic scattering, then only the phase of the incident electron beam is changed during passage through the specimen. Such an idealized material is called a pure or weak *phase object*, and it gives rise to the phase-object approximation described below.

A pure phase object at critical focus in an idealized microscope produces zero image contrast, which is obviously not useful for producing images. However, such lack of contrast only occurs if there are no limiting apertures that exclude some of the transmitted beams and if there are no instrumental aberrations that distort the phases of the waves. In a real TEM, these ideal conditions do not hold; moreover, a slight change in focus from the critical in-focus value, called Gaussian focus, also enhances contrast (§1.3.5). Nonetheless, conditions close to those produced for crystals thin enough to produce minimal dynamical diffraction produce what is called *phase contrast*.

The thin crystals used for HRTEM images give rise to different approximations used to describe the effects on the electron beam as it passes through and interacts with the sample. The projected potential and projected charge-density approximations consider the effects of kinematical and dynamical scattering (§1.3.2), respectively; they are briefly described below. If thicker crystals are used, in which case a substantial fraction of the incoming electron beam is scattered by the crystal, then contrast can also be developed by excluding these scattered beams from reaching the image plane, resulting in *amplitude contrast* (cf. §1.4.4).

1.4.2   The projected potential approximation – kinematical scattering

In §1.4.1 and §1.2.5 we considered phase objects and phase-contrast microscopy, respectively. In a weak phase object, for which only a single scattering event occurs as the electron beam traverses the specimen, the diffracted beams experience a phase change of $\pi/2$ relative to the unscattered direct (central) beam. However, in order to form an image that has a simple correspondence to the structure of the specimen, the beams must be π out of phase with the direct beam. Moreover, if it were not for the effects of lens aberrations, the sample would be invisible at Gaussian focus (the position of exact optical focus if the TEM were free of aberrations; operationally, the Gaussian focus is deduced from a knowledge of the aberrations and the observed position of minimum contrast).

In microscopy using visible light, a phase change of $\pi/2$ can be produced by a 1/4 λ waveplate such as is used in phase-contrast microscopy. In the TEM such a phase change can be approximately produced by a slight defocusing of the objective lens. In this pure kinematical case, which is highly idealized, all diffracted beams are in phase with one another and $\pi/2$ out of phase with the incoming beam (except for whatever misphasing is introduced by lens aberrations and is reflected by the sin χ function; §1.3.5). For this situation the direct beam has experienced little scattering and so has far greater intensity than the diffracted beams. Images produced under such conditions have been called *projected potential images*, with dark regions representing areas of structure containing relatively many atoms, relatively heavy atoms, or both, and light regions representing correspondingly fewer or lighter atoms. The basic theory for such images was formulated by Scherzer (1949); it has also been called the *weak phase-object approximation* (or WPOA). In practice, the WPOA is only valid in the thinnest of regions, such as immediately next to the thin wedge edge of a crystal. It does, however, provide a useful conceptual foundation for intuitive interpretations of image contrast in which atoms and regions of high potential (and density) appear dark (O'Keefe and Buseck, 1979).

1.4.3   The projected charge-density approximation – dynamical scattering

Most thin crystals are thick enough so that some multiple scattering of transmitted electrons occurs. Such dynamical effects result in diffracted beams whose phases differ from one another. In general, higher order beams are progressively more advanced in phase. Such crystals produce *projected charge-density* (PCD) images; the conditions for producing such images were defined by Cowley and Moodie (1957). Focus of the objective lens is adjusted so that, ideally, the highest order beams contributing to the image (i.e., allowed to pass through the objective lens aperture) are π out of phase with the direct beam. Interpretation of dark and light regions in such images is analogous to that for the projected potential approximation. The PCD approximation, since it includes the effects of dynamical scattering, generally provides more realistic results than the projected potential approximation. Most actual HRTEM images have mixed character between the projected potential and PCD approximations.

Computer-simulated images are important for all image interpretations (O'Keefe et al., 1978; §1.3.3 and Chapter 3). Beyond certain thicknesses, which vary according to differences in mineral compositions and structures, and for certain TEM operating conditions, the phase changes among the various beams are sufficiently complex that neither the projected potential nor the PCD approximations hold well. For such conditions, computer-simulated images become especially crucial for interpreting images (O'Keefe and Buseck, 1979; Stadelmann, 1987; Self and O'Keefe, 1988).

1.4.4 Amplitude contrast

"Amplitude contrast" is produced for all specimens when an aperture is used to exclude some diffracted beams. The decrease in beam intensity, i.e., the amount of energy by which the direct TEM beam is decreased by electron scattering, is proportional to the amount of scattered energy. For certain orientations, such as when a low-order zone axis is parallel to the optical axis of the TEM or for strongly scattering specimens (i.e., thick ones, or specimens with a high proportion of heavy atoms), a substantial fraction of the electrons is diffracted away from the direct beam. Insertion of an objective aperture excludes some of this scattered radiation, giving rise to strong amplitude contrast. Such contrast increases with decreasing aperture size and increasing sample thickness. Commonly produced by allowing only the direct beam to reach the image plane, such a configuration is used for most low-resolution or *"conventional"* transmission electron microscopy (Table 1-3; also Chapters 11 and 12).

1.5   INSTRUMENTAL DETAILS

1.5.1   Sample stages: top entry *vs.* side entry

The position of the specimen within the TEM greatly affects performance. The specimen is placed between the upper and lower halves (pole pieces) of the objective lens, so that it is effectively immersed within the lens. For maximum resolutions the pole-piece gap should be as narrow as possible; values of only a few millimeters are used in modern high-resolution TEMs such as the JEOL 4000*EX*. Since TEM grids are thin, as are the samples used for microscopy, this space restriction might seem minor. However, it actually places severe and conflicting constraints on lens design and TEM performance. There are two major complications.

For mineralogy, or any crystallographic work for that matter, we typically wish to view along specific crystallographic directions. Tilting of the specimen and its supporting grid is required to achieve such orientations. Clearly, a tilted grid requires a greater pole-piece separation than one that is perpendicular to the TEM axis. The tilting mechanism also requires space, and so an increased pole-piece separation is needed. The other problem arises if we desire to monitor the emitted radiation. Such spectroscopy is integral to many analytical purposes, but it requires a path by which the radiation can exit from between the pole pieces. Moreover, for x-ray detection, high take-off angles are desirable, and these are physically incompatible with small pole-piece gaps.

Clearly, the configuration of the lens pole pieces is critical, and it gives rise to two types of widely used sample stages: *top-entry* and *side-entry*. For minimum pole-piece separation, and thus maximum spatial resolution, the sample is inserted into the lens from the top. The sample-insertion mechanism translates the sample laterally from an exterior vacuum chamber to the central axis of the TEM and then lowers the sample and tilting stage into the objective lens. The sample is said to be immersed in the lens; such positioning facilitates a reduction of the lens focal length and thus lens aberrations, thereby making high resolutions possible.

When greater separation of pole pieces is required, such as for x-ray spectroscopy, a side-entry stage can be used. The sample-insertion mechanism then translates the sample directly into the pole-piece gap. Both types of stages require delicate mechanisms to produce such positioning. In general, however, side-entry stages are slightly simpler to build and to use; they are also less costly and more versatile since they can be used both for imaging and for monitoring emitted x-rays. An advantageous feature of some side-entry stages is that they can be made eucentric. In a eucentric stage the crystal remains in view throughout the tilting operation, thereby minimizing lateral motion during tilting. With top-entry stages, unless a crystal is located directly on the TEM optical axis—and such fortuitous positioning almost never happens—each time the sample stage is tilted the crystal suffers significant lateral displacement. For this case, the operator must continually reposition the crystal to keep it in the field of view.

1.5.2 Aberrations

Optical aberrations and instrumental instabilities such as power supply and specimen drift are the main limitations for problems in high-resolution work. Axial aberrations are the most important ones for TEMs, and the most important of these are spherical and chromatic (depth-of-focus) aberrations and astigmatism. They will be discussed in sequence.

The objective lens is commonly the strongest of the several lenses of the TEM. This means it has the greatest deflection angle (Fig. 1-1). A result is that the paths of many diffracted electrons extend further laterally from the optical axis of the TEM than occurs with the other lenses. As with glass lenses, distortions increase with distance from the optical axis, and thus aberrations are most severe for the objective lens (Spence, 1981). As shown graphically by O'Keefe and Buseck (1979, p. 33), the spherical aberration determines the number of diffracted beams that can be passed through the objective lens without production of unacceptable misphasing and thus image distortions. C_S, the spherical aberration constant, is commonly used as an indicator of the severity of the spherical aberration. Current high-resolution TEMs have C_S values of 1.6 ± ~1.2, with the lower values producing the best resolutions. The C_S can be decreased by closing the gap between pole pieces, narrowing the bore of the lens, or increasing the strength of the magnetic field. However, as discussed in §1.5.1, decreasing the size of the opening in the lens restricts sample handling, and increasing the field strength causes saturation of the magnetic material, effectively increasing the dimensions of the pole piece. Thus, difficult compromises are required in order to decrease the C_S. For most purposes aberrations can be neglected for the other TEM lenses.

Spherical aberration is commonly the dominating factor in limiting maximum resolution, but for thicker samples chromatic aberration is of greatest concern. Chromatic aberration arises from instabilities in the electron source and from interaction between the electron beam and the specimen (Spence, 1981). Both effects result in changes in the wavelength (or "color") of the electron beam, which explains why it is called chromatic aberration. A range in beam wavelengths produces a corresponding range of focal lengths together with inevitable blurring of the image. The same effect is produced by instabilities of the objective-lens current. Much effort is devoted to minimizing high-voltage and lens-current instabilities in the electron source of the TEM and the energy spread of the electrons as they leave the filament. However, even with the most stable sources and best designed filaments, there will be energy losses resulting from inelastic interactions as the beam passes through the specimen (Chapter 5). Thus, chromatic effects are unavoidable. Clearly, they increase with specimen thickness. Their effects also become more severe where the electron beam is diffracted at the largest Bragg angles, i.e., at the highest spatial frequencies.

The effects of the spherical and chromatic aberrations are influenced by the excitation of the objective lens and the height of the sample within the pole piece. There is thus a trade-off that must be made to balance these two aberrations, depending on whether thin or thick samples are to be studied, i.e., for high-resolution or "conventional" microscopy (Spence, 1981).

Astigmatism is produced by image distortions resulting when the magnetic field of a lens is not perfectly symmetrical. Here, again, the objective lens is of greatest significance. The slightest deviation in the symmetry of the casing and winding of the lens results in an asymmetric magnetic field, and such deviations are inevitable. They result in focal lengths that differ in different planes coincident with the TEM optical axis. To minimize the effects of astigmatism, the objective lens is fitted with a "stigmator," which consists of a set of auxiliary lenses that can be adjusted individually by electronically introducing weak compensating fields to produce improved cylindrical symmetry for the objective lens. Considerable operator experience and skill is required for proper correction of astigmatism, although automatic stigmation by computer is becoming possible (Fan and Krivanek, 1990); either way proper stigmation is critical for good high-resolution imaging.

1.5.3 Image recording

The primary method for recording images with the TEM is by photography; data for analytical microscopy is recorded by spectrometry, described in Chapters 4 and 5. Although glass plates are used in some TEMs for high fidelity, photographic film is the method of choice for most purposes; however, care has to be taken to minimize loss of spatial fidelity through uneven expansion or contraction of the film during processing. Also, films need to be put into evacuated vacuum chambers for several hours prior to placing them into the TEM or adsorbed vapors will produce contamination in the vacuum of the microscope. Film has a large format (a few cm across) with approximately 3000 pixels ("picture elements") per linear dimension. However, its dynamic range, meaning the number of intensity or "grey" levels it can record, is commonly limited to less than 1000, and it cannot be used to record data in real time.

In place of film, video methods are excellent for showing dynamic effects that occur within the TEM. Twinning, oxidation and reduction reactions, inversions, and other phase transitions are examples. An advantage of video recording is that a digital signal is produced, and this can be handled with a computer for image processing. However, the spatial distortions inherent in current video cameras restrict their use to qualitative or semi-quantitative measurements of distances, thereby limiting their use for quantitative crystallographic measurements.

A recent development, and one that may well replace film methods in the next few years, is the use of charge-coupled devices (CCDs) for recording images (e.g., Koster and de Ruijter, 1991). CCDs were developed and are widely used for capturing the weak signals encountered in astronomy. A CCD is a semiconducting chip consisting of a two-dimensional array of pixels, each of which can independently accumulate a charge as absorbed photons dislodge electrons. Each pixel acts much like a small phototube, except that a CCD can integrate the charge in its pixels over long periods of time and then move the charges out in a regimented way, digitally preserving the spatial pattern of light intensity falling on the chip. Light emitted from a phosphorous viewing screen can be focused via a lens onto the pixel array. More commonly, the CCD is coupled via fiber optics to a yttrium aluminum garnet (YAG) scintillator, which converts each incoming electron directly into several thousand usable photons.

Since CCDs are extremely sensitive, they can be used for low light intensities, which is obviously desirable in electron microscopy for materials that are sensitive to radiation and subject to radiation damage, such as many hydrous minerals. Their dynamic range ($>10^4$) is significantly greater than film. Thus, a CCD can be used to record far more subtle intensity differences than can film; such information can be highly useful, for example, for the quantitative interpretations of electron-diffraction patterns. A result is that the output from a CCD can be directly processed with a computer, with readout times of the order of seconds. Finally, the information is recorded directly in digital form, pixel by pixel, rather than in analog form such as occurs with film; and it can be recorded in real time so that reactions in progress can be recorded, provided they are not extremely rapid.

Digital data collection such as occurs with CCD recorders is useful for both instrument operation and data interpretation. The rapid recording of quantitative information from diffraction patterns allows on-line, automated TEM adjustment, leading to improved alignment procedures. The same information can permit indexing of diffraction patterns while the operator is at the TEM controls; such immediate orientation information can greatly reduce the time spent looking at crystals in undesired orientations or even the wrong crystals within multimineralic assemblages. Images recorded digitally can by analyzed on-line by computerized fast Fourier transform (FFT) analysis, thereby expediting image interpretation and thus greatly increasing the efficiency of the TEM operator.

1.6   SAMPLES

1.6.1   Samples for electron microscopy

Except for surface measurements, and these are not considered here, all electron microscopy samples must be thin. The details depend on whether the primary purpose is spectroscopy for analytical electron microscopy (AEM) or imaging. The number of interactions between beam electrons and the sample should be maximized for spectroscopy so that a large signal is produced. For example, x-ray production has a low efficiency, and therefore an appreciable sample volume is required to produce statistically meaningful results. For most materials, a thickness as great as 100 nm is still electron-transparent and suitable for AEM measurements.

The situation differs for imaging, in which case a small thickness is desirable. The strong interaction of electrons with matter means that especially for high-resolution imaging there is a premium in using thin specimens to minimize multiple scattering events. The mean free path in solids for elastic scattering ranges from a few nm for heavy elements to a few tens of nm for light elements. Thicknesses no greater than the mean free path are best for images in which we wish to see maximum structural details, and so thicknesses of a few nm and in most cases less than ~10 nm are commonly sought. Preparing such samples can be challenging.

1.6.2   Sample preparation

Sample preparation is a critical part of much mineralogical electron microscopy, and it is commonly a formidable problem. Some samples contain minerals with greatly different hardnesses and resistances to thinning. Others orient themselves perversely, so that the desired viewing direction almost never coincides with the way the mineral tends to lie; common examples are grains of layered minerals (micas, clays, chlorites), where we wish to to see how they are stacked (Chapters 7 and 9) and so want to view parallel to the layers.

The problem of sample preparation is not too severe for most analytical electron microscopy since there is considerable tolerance in crystal thickness; normally, sample orientation is not critical [an exception is ALCHEMI (Chapter 5), for which a specific crystallographic direction must lie parallel to the TEM optic axis].

The situation differs for high-resolution imaging, where several criteria must be fulfilled. Critical aspects are proper crystallographic orientation and appropriate thinness. Also, in order to avoid the blurring effects of a substrate, signals are generally recorded where part of a crystal projects over a hole in the support film (typically made of a thin layer of amorphous carbon containing holes—"holey-carbon films"). Although not directly related to specimen preparation, the crystal must be stable in the electron beam or, as is common, the operator must work sufficiently rapidly or at low enough beam intensities to obtain and record signals before the sample degrades (CCDs, §1.5.3, can help limit this problem).

Perhaps the simplest way of preparing samples for the TEM is by crushing them into small grains. This is the method we used for many years. The operator searches for wedge-shaped edges of grains, hoping that they lie close to a desired orientation. The crystal is then tilted so that a critical zone axis lies parallel to the optic axis of the TEM. For high-resolution imaging, the crystal must also lie over a hole in the support grid. Clearly, several fortuitous circumstances have to occur simultaneously, and much time is typically used searching for the proper grain for good HRTEM imaging. Use of crushed grains is inefficient, although in some instances it is the best method available.

Milling is the most common method currently used for preparing mineral samples. In an ion mill the sample is bombarded with a beam of ions. Uncharged atoms can also be used, in which case the unit is call an atom mill; for our purposes the distinction is unimportant, and for simplicity I will refer to both units as ion mills.

A standard petrographic thin section can be used for the initial sample, thereby allowing the operator to search for and select mineral grains having the desired crystallographic orientations. Thin sections also permit the observation of mineral fabrics and textures prior to selecting regions for electron microscopy. The only requirement is that the thin section be prepared with an adhesive that can be removed reasonably easily. Thus, epoxy is undesirable. Bonding agents such as Lakeside cement, Canada balsam, or

Crystalbond (AREMCO Products, Inc., P.O. Box 429, Ossining, NY 10562) work well because they can be softened by gentle heating, thereby permitting the region of interest to be removed from the thin section. Precision cutting devices (such as provided by M.U. Medenbach, Bruno-Heide-Straße 8, D-5810 Witten 3 (Herbede), Germany) can be used to cut out specific regions of interest prior to extraction. The chip of interest is then placed onto a small metal plate that is inserted into the ion mill.

The mill is evacuated, and then the sample is bombarded for 8 to 24 hours (at ~10^{-5} torr) by a stream of ions or atoms and thereby gently eroded. Typically the sample is placed at an angle between 10 and 15° to the ion or atom beam and then rotated around the beam axis so that thinning occurs relatively uniformly. Initial thinning is done with an energetic ion (or atom) beam (e.g., 5 keV), and then surface damage, if present, can be removed by a brief period of milling (10-30 min) with a lower energy beam (e.g., 1 to 2 keV). An optical microscope is available to observe the specimen within the milling chamber.

A precision ion milling system (PIMS) can be used for thinning of precisely located areas. In this instance, areas as small as 2 μm on a side can be selected and preferentially thinned. Observation of the area to be thinned can be done via a small SEM that is incorporated within the PIMS unit (available from Gatan Inc., Pleasanton, CA). The PIMS unit differs from the mills described above in that the specimen cannot be rotated, but it can be placed at a lower angle to the ion beam, resulting in the potential of more gentle thinning. The PIMS holder with sample grid intact is also directly insertable into some TEMs, so that transfer of the sample grid to a separate TEM holder can be avoided. However, the PIMS holder is not as versatile for tilting as a standard TEM stage.

Ultramicrotomy is used for samples that can be prepared by cutting them into extremely thin sections, such as is typically done with tissues in biological samples. Suitable minerals either have to be relatively soft, like many layer silicates (e.g., Iijima and Buseck, 1978), or very fine grained (e.g., Bradley and Brownlee, 1986). In those special instances, the minerals can be encased within a suitable epoxy and then sliced, much like a salami is sliced in a delicatessen. The result is a series of extremely thin wafers that can be floated onto sample grids and examined directly with the TEM.

1.7   FINAL THOUGHTS

1.7.1   Sign conventions

The symbols and sign conventions follow those on p. xvii in *High-Resolution Transmission Electron Microscopy and Associated Techniques* (Buseck, 1988a).

1.7.2   General references

Several good books provide thorough reviews of the principles and techniques of electron microscopy. Agar et al. (1974) is a good introduction to the TEM and its uses. Williams (1984) provides a careful introduction to the various methods of analytical electron microscopy. Spence (1981) gives a general overview of high-resolution transmission electron microscopy, and Cowley (1981) provides a comprehensive treatment of diffraction physics, while Krivanek (1988) and Cowley (1991) give good introductions to the principles and practice of electron microscopy. Buseck et al. (1988a) provides both principles and applications of HRTEM and associated spectroscopic techniques, including mineralogical applications. Hirsch et al. (1965) is the standard reference for discussions of defect analysis and "conventional" electron microscopy.

ACKNOWLEDGMENTS

I thank my ASU colleagues for help in many ways with this chapter. Reviews at various stages were provided by John Cowley, Don Eisenhour, Michael A. O'Keefe, and Peter Self. Debra Bolin, Sue Selkirk, Weida Jian, and Michael O'Keefe kindly helped with preparing figures. The work was supported in part by NSF grant EAR 8708529.

REFERENCES

Agar, A.W., Alderson, R.H., and Chescoe, E. (1974) Principles and Practice of Electron Microscope Operation. Elsevier/North-Holland, Amsterdam, The Netherlands, 345 pp.

Bradley, J.P. and Brownlee, D.E. (1986) Cometary particles: thin-sectioning and electron beam analysis. Science 231, 1542-1544.

Buseck, P.R. and Iijima, S. (1974) High resolution electron microscopy of silicates. Am. Mineral. 59, 1-21.

Buseck, P.R., Cowley, J.M., and Eyring, L. eds., (1988a) High-resolution Transmission Electron Microscopy and Associated Techniques, Oxford University Press, New York, 1-128.

Buseck, P.R., Rimsky, A, and Epelboin, Y. (1988b) Signal processing of high-resolution transmission electron microscope images using Fourier transforms, Acta Cryst. A44, 975-986.

Cowley, J.M. (1991) Principles and practice of electron microscopy. In Physical Methods of Chemistry, 2nd ed., Vol. 4. Microscopy, B.W. Rossiter and J.F. Hamilton, eds. John Wiley & Sons, New York, 1-50.

Cowley, J.M. (1988) Imaging, imaging theory, elastic scattering of electrons by crystals, and elastic-scattering theory. In High-resolution Transmission Electron Microscopy and Associated Techniques, P.R. Buseck, J.M. Cowley and L. Eyring, eds., Oxford University Press, New York, 1-128.

Cowley, J.M. (1981) Diffraction Physics, 2nd ed., North-Holland, Amsterdam, 410 pp.

Cowley, J.M. and Moodie, A.F. (1957) Fourier images I: The point source. Proc. Phys. Soc., London, Sect. B, 70, 486-496.

Cowley, J.M. and Moodie, A.F. (1960) Fourier images IV: The phase grating. Proc. Phys. Soc., London, Sect. B, 76, 378-384.

Culling, C.F.A. (1974) Modern Microscopy—Elementary Theory and Practice. Butterworths, London, England, 148 pp.

Cullity, B.D. (1956) Elements of X-ray Diffraction. Addison-Wesley, Reading, Massachusetts, 549 pp.

Fan, G.Y. and Krivanek, O.L. (1990) Computer-controlled HREM alignment using automated diffractogram analysis. Proc., 12th Int'l. Cong. on Electr. Microsc., Seattle, Peachey, L.D. and Williams, D.B., eds., San Francisco Press, San Francisco 1, 532-533.

Gabor, D. (1949) Microscopy by reconstructed wave-fronts. Proc. Roy. Soc. London A197, 454-487.

Hembree, G.G., Crozier, P.A., Drucker, J.S., Krishnamurthy, M., Venables, J.A. and Cowley, J.M. (1989) Biassed secondary electron imaging in a UHV-STEM. Ultramicrosc. 31, 111-115.

Hembree, G.G., Drucker, J.S., Luo, F.C.H., Krishnamurthy, M., and Venables, J.A. (1991) Auger electron spectroscopy and microscopy with probe-size limited resolution. Appl. Phys. Lett. 58, 1890-1892.

Hirsch, P., Howie, A., Nicholson, R.B., Pashley, D.W., and Whelan, M.J. (1977) Electron Microscopy of Thin Crystals, Robert E. Kreiger Publishing Co., Malabar, Florida, 563 pp.

Iijima, S. and Buseck, P.R., (1978) Experimental study of disordered mica structures by high-resolution electron microscopy. Acta Crystallog. 34, 709-719.

Iijima, S., Cowley, J.M., and Donnay, G. (1973) High resolution microscopy of tourmaline. Tschermaks mineral. petrograph. Mitt. 20, 216-224.

Koster, A.J. and de Ruijter, W. J. (1991) Practical autoalignment of transmission electron microscopes. Ultramicrosc. 40, 89-107.

36

Krivanek O.L. (1988) Practical high-resolution electron microscopy. In High-resolution Transmission Electron Microscopy and Associated Techniques, P.R. Buseck, J.M. Cowley and L. Eyring, eds., Oxford University Press, New York, 519-567.

Lichte, H. (1991) Electron image plane off-axis holography of atomic structures. Adv. Opt. Electr. Microsc. 12, 25-91.

McNulty, I., Kirz, J., Jacobsen, C., Anderson, E.H., Howells, M.R., and Kern, D.P. (1992) High-resolution imaging by Fourier transform x-ray holography. Science 256, 1009-1012.

O'Keefe, M.A. (1992) "Resolution" in high-resolution electron microscopy, Ultramicrosc., in press.

O'Keefe, M.A. (1979) Resolution-damping functions in non-linear linages. 37th Ann. Proc. Electr. Microsc. Soc. Am. 556-557.

O'Keefe, M.A. and Buseck, P.R. (1979) Computation of high-resolution TEM images of minerals. Trans. Am. Crystallogr. Assoc. 15, 27-46.

O'Keefe M.A., Buseck, P.R., and Iijima, S. (1978) Computed crystal structure images for HREM. Nature 274, 322-324.

Pierce, L. and Buseck, P.R. (1974) Electron imaging of pyrrhotite superstructures. Science 186, 1209-1212.

Rochow, T.G and Rochow E.G. (1978) An Introduction to Microscopy by Means of Light, Electrons, X-rays, or Ultrasound. Plenum Press, New York, 367 pp.

Ruska, E.E. (1988) The development of the electron microscope and of electron microscopy (Noble Lecture). Bull. Electr. Microsc. Soc. Am. 18, 53-61.

Scherzer, O. (1949) The theoretical resolution limit of the electron microscope. J. Appl. Phys. 20, 20-29.

Self, P.G. and O'Keefe, M.A. (1988) Calculation of diffraction patterns and images for fast electrons. In High-resolution Transmission Electron Microscopy and Associated Techniques, P.R. Buseck, J.M. Cowley and L. Eyring, eds., Oxford University Press, New York, 244-307.

Spence, J.C.H. (1981) Experimental High-resolution Electron Microscopy. Oxford University Press, New York, 370 pp.

Spence, J.C.H. (1988) Inelastic electron scattering: Part I, Inelastic electron scattering. In High-resolution Transmission Electron Microscopy and Associated Techniques, P.R. Buseck, J.M. Cowley and L. Eyring, eds., Oxford University Press, New York, 129-159.

Stadelmann, P. (1987) EMS - A software package for electron diffraction analysis and HREM image simulation in material science. Ultramicrosc. 21, 131-146.

Stout, G.H. and Jensen, L.H. (1968) X-ray Structure Determination – A Practical Guide. Macmillan Co., New York, 467 pp.

Tonomura, A. (1992) Electron-holographic interference microscopy. Adv. Phys. 41, 59-103.

van Dyck, D. and Op de Beeck, M. (1990) New direct methods for phase and structure revival in HREM. In Proc. 12th Int'l Cong. Electr. Microsc., Seattle, WA, Peachey, L.D. and Williams, D.B., eds., San Francisco Press, San Francisco 1, 26-26.

Williams, D.B. (1984) Practical Analytical Electron Microscopy in Materials Science. Philips Electronic Instruments, Mahwah, NJ, 153 pp.

Yada, K. (1971) Study of the microstructures of chrysotile asbestos by high resolution microscopy. Acta Crystallogr. 5, 119-124.

Yoshida, T. and Suito, E. (1972) Interstratified layer structure of organo-montmorillonites as revealed by electron microscopy. J. Appl. Crystallogr. 5, 119-124.

Chapter 2.

SELECTED AREA ELECTRON DIFFRACTION (SAED) and CONVERGENT BEAM ELECTRON DIFFRACTION (CBED)

John W. Steeds University of Bristol, H.H. Wills Physics Laboratory, Royal Fort, Tyndall Avenue, Bristol, BS8 1TL, England

J.-P. Morniroli Laboratoire de Métallurgie Physique, Université de Lille, 59655 Villeneuve d'Ascq Cedex, France

2.1   INTRODUCTION

2.1.1   General introduction

Electron diffraction is a rich source of extremely accurate data that can be used to deduce reliable crystallographic information. Of course, the quality of the data depends on the quality of the specimen under study, and this can be significantly affected by specimen preparation techniques and beam-induced degradation (see §2.1.3). At the lowest level of data collection it should be possible to deduce the crystal system, its Bravais lattice, and the unit cell dimensions. In the case of high-quality data one can uniquely identify the space group of a crystal structure, from a single zone axis pattern in a favorable case. It is also possible to have accurate information about a particular polytype or about non-stoichiometry of the specimen.

This chapter concentrates on two common techniques of electron diffraction, selected-area electron diffraction (SAED) and convergent-beam electron diffraction (CBED), both of which are easily performed on a modern analytical transmission electron microscope (TEM). The main skill required for effective exploitation of electron-diffraction techniques is the ability to orient, with control, the region of interest that is under study. These techniques, together with the related techniques of microdiffraction and large-angle convergent-beam electron diffraction (LACBED), are discussed in §2.1.4.

Electron diffraction has not been widely used in scanning electron microscopy (SEM), but this situation is now changing rapidly as electron back scattered diffraction (EBSD) apparatus is being installed on these instruments. It seems likely that this technique will increase greatly in popularity in the future, and as it is of considerable interest in mineralogical studies, an introductory section on this topic is included in §2.1.5.

There are two common reasons for the relatively slow uptake of interest in electron diffraction as a technique for solving practical problems of phase identification. It is evident from the recent growth in the use of CBED (as in, for example, the identification of crystal structures of high-temperature superconductors) that neither of these represent significant barriers in reality. One of these reasons is that diffraction patterns, occurring in the back focal plane of the objective lens, exist in reciprocal space. Thus, diffracting planes that are close together give widely separated diffraction spots, while widely separated diffraction planes give diffraction spots that are close together. The second reason is that electrons, being charged, interact strongly with matter, and thus, in general, multiple scattering (dynamical diffraction) occurs rather than single scattering (kinematical diffraction). This state of affairs carries with it the important consequence that the diffracted intensities are no longer simply related to the Fourier transforms (structure factors) of the charge distribution of the crystal under investigation. While it might seem that this is a disadvantage we shall emphasize that, on the contrary, all sorts of benefits arise from this situation. By learning the rules of dynamical diffraction it is possible to obtain much new information not available under conditions of kinematical diffraction. The details of the diffraction theory on which this approach is based are beyond the scope of this chapter, but references are given for those wishing to understand the background we assume.

2.1.2 A brief history of electron diffraction

Selected-area electron diffraction has been widely used since the early 1960s when transmission electron microscopy became a common tool for studying inorganic materials. Until the mid-1970s it was generally accepted (with a few notable exceptions) as the only practical electron microscopic tool for crystallographic investigation of materials (Hirsch et al., 1965). The development of convergent-beam electron diffraction waited for the availability of small-probe-forming TEMs (Steeds, 1979) following the remarkable achievements of Crewe's STEM in the early 1970s. The generation of analytical TEMs that followed was generally suitable for CBED experiments, although some models suffered severely from specimen contamination and poor electron optics for electron diffraction experiments. In fact, excellent CBED patterns had been published by Kossel and Möllenstedt in 1938 (see Möllenstedt, 1989) but the technique was limited to sufficiently flat, parallel-sided specimens for the probe size then available. Widespread use of CBED has really only occurred during the past few years and, for example, since the time of the discovery of high-temperature superconductors the technique has been used routinely in many laboratories for crystal structure identification.

Methods for performing convergent-beam electron diffraction were proposed towards the end of the 1970s and have become increasingly used throughout the 1980s, although still in relatively few laboratories. No doubt this technique will be used more extensively in the future as its strengths are more widely appreciated.

Back-scattered electron diffraction is a development of the work of Coates. Electron-channelling patterns were discovered by Coates in 1967 and developed as a rocking-beam method in Oxford (Joy, 1974). Back-scattered diffraction was first proposed by Venables and Harland (1973), but its widespread implementation in SEMs awaited the development by Dingley (1984) of a simple system for its effective exploitation. Its impact on SEM in many laboratories has yet to come but surely cannot be delayed much longer.

2.1.3 Specimen considerations

There are a number of aspects of specimen preparation and preservation that deserve particular attention if good results are to be obtained by convergent-beam electron diffraction. It is a very sensitive technique and calls for the exercise of appropriate care.

2.1.3.1 Specimen preparation

Clean, contamination-free surfaces are required for good results. Fracture chips formed by crushing the sample are, in this respect, ideal. Samples prepared by ion milling invariably have a surface layer of amorphous material that can be of variable thickness. Such layers inevitably degrade the quality of information available and should be as thin as possible. Quite apart from the diffuse scattering caused by the amorphous layers, they frequently create stresses that cause bending of thin regions and strains in the thicker regions. These strains can distort the pattern symmetry and complicate lattice parameter determination. One technique that has proved useful for precipitate identification when a suitable etching procedure exists, is extraction of replicas on to carbon films.

2.1.3.2   *Time-dependent changes*

Time-dependent changes may occur as a result of specimen contamination during exposure to the beam or degradation of the sample. Most such contamination results from carbon formation by the decomposition of hydrocarbons by the electron beam. The hydrocarbons may be present in the vacuum system or introduced by the specimen itself. In order to minimize this effect it is important to pay attention to the maintenance of a high-quality vacuum system on the microscope with a low hydrocarbon partial pressure and to go to considerable care in the preparation of hydrocarbon-free surfaces. Surface diffusion is generally involved in the deposition process, that therefore becomes very much more serious as the probe size is reduced (it has been suggested that the contamination rate is proportional to the square of the reciprocal of the probe diameter).

Sample degradation may be the result of sample heating, ionization damage, or atomic displacement. In each case the effect is reduced by using the lowest possible beam exposure of the sample. Increased irradiated area (as in LACBED; see §2.1.4) or decreased convergence angle (in CBED) are two techniques that help to reduce this problem. The beam-heating effect is a complicated issue dependent on the sample thickness, thermal conductivity, and its thermal contact with the supporting metal grid or specimen holder. In extreme cases of small particles, not well thermally connected to a supporting grid, temperature rises of several hundred degrees have been recorded by CBED, but under more normal conditions the increase in temperature should not exceed about 50°C. The cross section for ionization damage is lower at higher microscope operating voltages while atomic displacement does not occur below a certain well-defined threshold. There may therefore be a conflict between the wish to operate at as high a voltage as possible to reduce ionization damage while remaining below the displacement threshold. The compromise arrived at for optimum specimen lifetime is likely to depend on both the specimen purity and its temperature. The presence of impurities in a sample can greatly enhance the damage rate while both cooling (to prevent displaced atoms from diffusing away) and heating (to increase the repair rate) can be advantageous in different circumstances.

2.1.3.3   *Use of hot and cold stages*

Double-tilting heating and cooling stages can be beneficial to particular diffraction studies. Both heating and cooling are known to reduce the contamination rate: heating causes the removal of hydrocarbons adsorbed on the surface, and cooling reduces the rate of surface diffusion. Cooling also reduces the thermal diffuse scattering and improves the large-angle diffraction (reduction of the Debye-Waller factor). Cooling reduces the thermal damage of the specimen, and either heating or cooling may reduce the effect of radiation damage. The thermal drift of hot or cold stages can be a problem and should be investigated prior to purchase.

2.1.4   Different forms of transmission electron diffraction and related terminology

Although many terms are used to describe electron diffraction in transmission microscopy, they essentially refer to one of two basic approaches. In one, generally referred to as selected-area diffraction, the sample is illuminated with approximately parallel

40

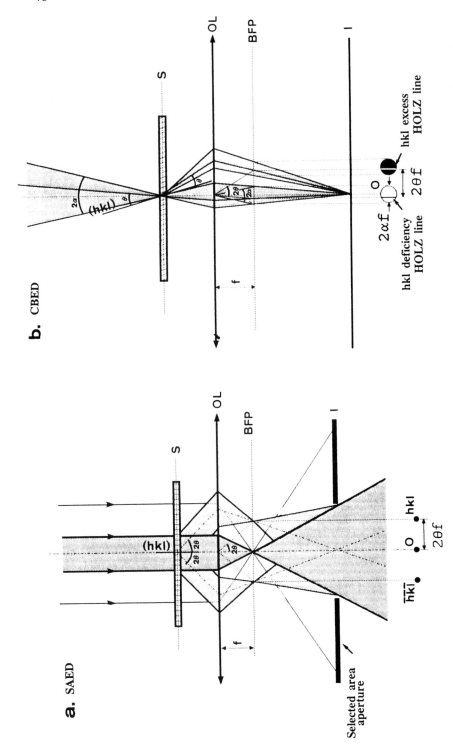

Figure 2-1. Schematic ray diagrams for selected-area electron diffraction and convergent-beam electron diffraction (S = specimen, OL = objective lens, BFP = back focal plane of objective lens, I = intermediate or selected-area aperture). Shaded regions represent electron trajectories; heavy shading is used for the direct paths, lighter shading for diffracted paths.

illumination, and the area from which the diffraction occurs is selected by insertion of a small aperture in the lens system of the microscope in a plane that is conjugate with the specimen plane. This approach produces spot diffraction patterns (SAED) from areas whose smallest dimensions are limited by the lens aberrations of the microscope to something approaching 1μm. The second approach, called convergent-beam electron diffraction (CBED), selects the area of interest by focussing the illumination to a small spot on the chosen location. This technique involves a convergent cone of electrons covering a range of angles of incidence and generates disc diffraction patterns. The focussed spot may be even smaller than 1 nm in diameter with field-emission electron sources, and the angular range in the incident cone can be quite large. Ray diagrams illustrating these two different approaches are given in Figure 2-1. As each point within a disc corresponds to a fixed angle of incidence within the cone, CBED patterns are maps of intensity as a function of angle of incidence for a fixed specimen thickness. SAED patterns, lacking the spatial resolution of the CBED technique, inevitably involve thickness averaging over the selected area and angular averaging as a result of buckling of the thin foil: as a result they generally contain much more diffuse scattering and less information than micro-diffraction patterns.

Microdiffraction is a name that has been introduced to describe convergent-beam electron diffraction when the individual discs lack internal detail (Fig. 2-2). This lack of

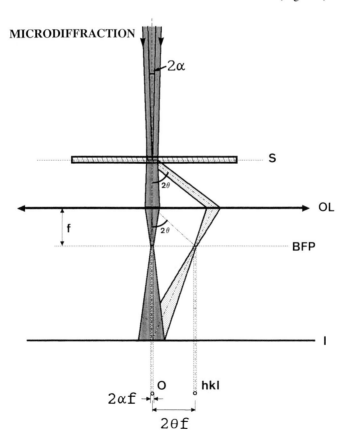

Figure 2-2. Schematic ray diagram for microdiffraction (symbols defined in caption for Fig. 2-1).

detail is generally the result of diffraction from small particles or from highly disordered or strained crystals. This situation is common for small precipitates and inclusions in minerals. Such particles are often in a state of considerable strain, made inhomogeneous by partial stress relief as a result of the creation of the free surfaces during specimen preparation. As the discs are featureless, the convergence angle is generally made small by use of a very small second condensor aperture, and the patterns at first glance resemble SAED patterns. However, unlike the latter, they have been formed with a very small focussed probe, thereby minimizing the chance of double diffraction with matrix reflections and maximizing the precipitate diffraction intensity relative to the matrix diffraction. The thickness and angular averaging is also generally negligible as in conventional CBED. There is a distinct enhancement of the quality of the results obtained as compared with SAED patterns. The latter not only normally contain rather strong matrix diffraction but also average over the orientations present in the partially strain-relieved particle in such a way as to wipe out HOLZ (higher-order Laue zone—see §2.1.6) diffraction entirely and to destroy any useful symmetry information in the diffracted intensities. Another problem with selected area diffraction from small particles is that the intensity in the diffraction plane is so low that one has to 'shoot blind' and study the results of long exposures later. This makes accurate alignment on a zone axis extremely difficult or even impossible. Nano-diffraction is a term sometimes used to describe diffraction of this sort performed with a nanometer-sized probe, but then the possibility exists that the diffraction is coherent rather than incoherent. This article is concerned with incoherent diffraction. When convergent beam diffraction discs overlap it is generally found that intensities rather than amplitudes are added. Coherent nano-diffraction is not discussed here and is indeed in its infancy for practical applications.

There is a relatively small, but nevertheless important, distinction between CBED and Kikuchi patterns. Kikuchi patterns are the result of diffuse (largely thermal-diffuse or point-defect-diffuse) large-angle scattering of the high-energy electrons passing through the specimen. Therefore, the Kikuchi patterns are not limited to the interiors of convergent beam discs and give pattern details between the discs. Some of the features evident in CBED discs are continuous with Kikuchi-like features outside the discs. In CBED we concentrate on the information within the discs and assume that it is largely the result of elastic scattering by the specimen. The limitations of this assumption are evident when energy-filtered CBED patterns are generated, when the thermal diffuse scattering is reduced by specimen cooling, or when CBED is performed on thick specimens and the diffuse background washes out the detail in the individual discs. However, if we restrict our attention to specimen thicknesses where disc outlines remain clearly visible, then it is still common to find relatively strong Kikuchi patterns between the discs. The distinction between the detail within the discs and the Kikuchi patterns is that the electrons within the discs are present as a result of elastic scattering and may be associated with a well-defined thickness (the thickness at the place of incidence) while the Kikuchi patterns are essentially thickness averaged and are the result of a host of phonon scattering events that occur at all depths within the specimen. The information content of Kikuchi patterns is therefore reduced relative to CBED.

Large-angle convergent-beam electron diffraction (LACBED) is a technique for overcoming the angular limitation imposed by the need to prevent adjacent discs from overlapping. In a double-rocking method technique proposed by Eades (1980), LACBED remains a true reciprocal-space method. In an alternative but simpler implementation proposed by Tanaka and co-workers (1980), the probe is no longer focussed on the specimen but above or below it. Unwanted diffracted beams are eliminated by a small area-selecting aperture in a plane conjugate with the focal plane of the probe (no longer the image plane of the specimen—see Fig. 2-3). The resulting diffraction patterns are obtained from a considerably larger area of the specimen and are superimposed on a shadow image of the specimen. Because of the coexistence of real and reciprocal space in such patterns (a technique once called real-space crystallography) they are well-adapted to studying extended defects and interfaces. Because of the large illuminated area the technique is restricted to larger crystals, say ~1 μm in diameter, but note that, as a result of the defocus, the dose is lower, comparable with that in SAED, and the technique is therefore particularly effective for beam-sensitive specimens.

2.1.5   Electron back-scattered diffraction (EBSD)

The back scattering coefficient for normal incidence is low, approximately the ratio of the work function to the accelerating voltage, i.e., in the range 10^{-3}-10^{-4}. However, it is greatly enhanced as the angle to the surface is reduced, and this effect is exploited by EBSD. The sample is normally tilted to an angle of 45° to the incident beam, large enough to give greatly enhanced back-scattering but not so great as to cause significant fore-shortening of the image along the line of greatest slope. A static (not a rocked) focused beam is incident on a chosen region of the specimen, and the back-scattered electrons are detected by a phosphor dispersed on a glass plate. This plate is situated close enough to the specimen so as to subtend a very large angle (often greater than 90°) at the place of incidence (Fig. 2-4). An intensified silicon vidicon tube scans the back surface of the glass plate and produces, in real-time (i.e., short compared with the eyes' response time), the EBSD pattern on a video monitor. For better signal/noise ratios, frame integration may be performed, and software exists as an aid to pattern indexing. Relative orientations of adjoining grains in polycrystalline samples may be rapidly evaluated, and a new and more accurate form of texture map can be created (Dingley and Randle, 1992).

The electrons forming the EBSD patterns are generated near to the specimen surface. As a result, the beam spreading in the specimen is quite small, and spatial resolutions of the order of 0.2 μm can be achieved. Like Kikuchi patterns, the EBSD patterns are thickness averaged and are not therefore as detailed as CBED patterns, but the information is of sufficiently high quality that useful phase identification can often be performed (Baba-Kishi and Dingley , 1989). As it is a surface-sensitive technique, the state of surface preparation is critical, but cleaved mineral samples are ideal. Beam damage and contamination can be important just as in CBED. The quality of the EBSD patterns is a measure of the density of extended defects in the chosen region of the specimen (Wilkinson and Dingley , 1991).

2.1.6   Geometrical considerations

High-energy electrons have short associated wavelengths and hence small associated Bragg angles. According to the well-known Bragg's Law

$$2d \sin \theta = \lambda \tag{2.1}$$

For $\lambda \sim 10^{-3}$ Å and $d \sim 2$ Å,

$$\theta \cong \sin \theta = \lambda /2d = 2.5 \cdot 10^{-4} \text{ rad} \tag{2.2}$$

For such small Bragg angles, it is much more convenient to discuss electron diffraction in terms of the Ewald sphere construction in the reciprocal lattice (Figs. 2-1to 2-5) than to use Bragg's Law itself. The reciprocal lattice is constructed from the real lattice by a well-known procedure. Suppose the primitive translation vectors of the real lattice (not necessarily orthogonal) are a_1, a_2, and a_3. Then the reciprocal lattice with primitive

44

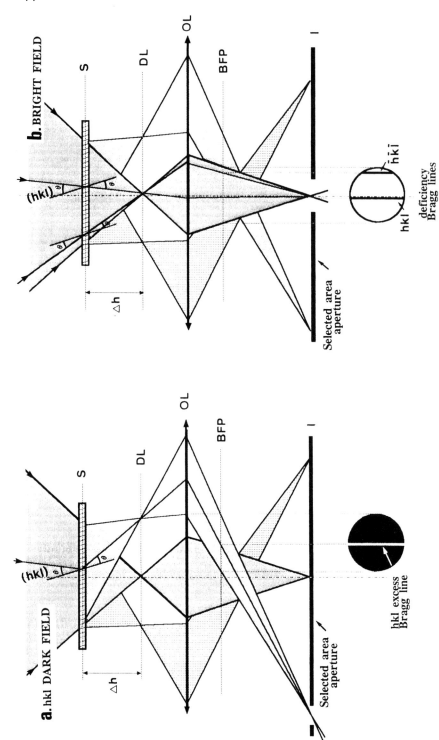

Figure 2-3. Schematic ray diagrams for large-angle convergent-beam electron diffraction (DL = disc of least confusion; other symbols as defined in the caption to Fig. 2-1).

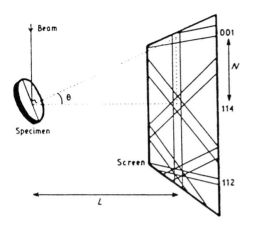

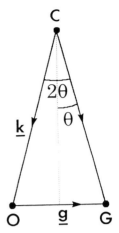

Figure 2-4 (left). Schematic diagram of the arrangement for generating electron back scattered diffraction patterns in an SEM using a fixed beam of small diameter.

Figure 2-5 (right). Reciprocal space construction to demonstrate the equivalence of the Ewald sphere construction and Bragg's Law.

translations $\mathbf{b}_1$, $\mathbf{b}_2$, and $\mathbf{b}_3$ is constructed from the real lattice translations according to the formulae

$$\mathbf{b}_1 = \frac{\mathbf{a}_2 \times \mathbf{a}_3}{V}, \quad \mathbf{b}_2 = \frac{\mathbf{a}_3 \times \mathbf{a}_1}{V}, \quad \mathbf{b}_3 = \frac{\mathbf{a}_1 \times \mathbf{a}_2}{V} \tag{2.3}$$

where V is the volume of the primitive unit cell of the lattice given by $\mathbf{a}_1 \cdot (\mathbf{a}_2 \times \mathbf{a}_3)$. Note that the dimensions of the primitive translations of the reciprocal lattice are reciprocal lengths: if the real lattice translations are in nm the reciprocal lattice translations are in nm^{-1}. Having created, in this way, the reciprocal lattice, the Ewald sphere construction specifies that we choose any point in this lattice as the origin (O). Construct a vector of length k $(= 1/\lambda)$ in the direction of incidence $\mathbf{k}$ (which may, for example, be a particular direction within the cone of incidence of a convergent beam incident on the specimen or the direction of propagation of the plane waves in the case of SAED) that ends on the origin. The other end of this vector is taken as the center (C) on that a sphere is constructed. Any reciprocal lattice point, G, intersecting this sphere corresponds to a Bragg reflection of the incident reflections. To see this (Fig. 2-5) let the point G be displaced by the vector $\mathbf{g}$ from the origin. Then by the properties of the sphere,

$(OC)^2 = (CG)^2$ or $k^2 = (\mathbf{k} + \mathbf{g})^2$

Then $\quad 0 = 2 \mathbf{k} \cdot \mathbf{g} + g^2$

But $\quad \mathbf{k} \cdot \mathbf{g} = -kg \sin \theta$ (see Fig. 2-5; the angle between $\mathbf{k}$ and $\mathbf{g}$ is obtuse).

So finally $\quad 2k \sin \theta = g$.

As $k = 1/\lambda$, we find $g = 1/d$ for consistency with Bragg's Law. It follows that $\mathbf{g}$ is the normal for the Bragg reflecting planes with spacing d equal to the reciprocal of the distance of the lattice point G from O. The angle of incidence on, and reflection from, these planes is equal to θ, as expected.

Because of the large radius of the Ewald sphere relative to the spacing of the individual reciprocal lattice points a large number of reflections is normally excited. It is precisely for this reason that the Ewald sphere construction is so effective in solving electron diffraction patterns as compared with direct application of Bragg's Law itself. Let us consider the case where the beam is incident along a zone axis direction, that is along the line of intersection of a number of crystal planes. Note in passing that although this situation is the same direction as a 'pole' for a cubic crystal, this is not generally the case, and the term pole, although often used, should be avoided. Each plane with a common line of intersection along the chosen direction of incidence will have a normal **g**, and the individual normals can be considered to give a two-dimensional mesh of points that is, in fact, a plane of the reciprocal lattice. The large Ewald sphere has this plane as its tangent at the origin O, and a large number of reflections close to this sphere near O will be excited as Bragg reflections. The details of whether or not a nearby point generates a reflection that is appreciably excited is beyond the scope of this article but is related to the structure factor of the reflection and the specimen thickness. In general terms, the larger the structure factor and the thinner the specimen the more likely it is that the reflection will be excited. As a rough order of magnitude, if the minimum reciprocal lattice repeat in the plane is g_{min} the reflection is likely to be excited if the sphere approaches as close as $0.1\ g_{min}$ to a reciprocal lattice point.

This plane of excited reflections is known as the *zero-layer plane*, and it is this that is conventionally studied in SAED patterns. In the case of CBED, a series of orientations within the incident cone have to be considered. In the zero-layer plane, the effect of this range of angles of incidence on whether or not a reflection is excited is scarcely significant, so that a series of discs is formed with the geometrical arrangement of the reciprocal lattice plane. The discs are each of the same diameter, governed by the angular range within the cone of incidence, itself dictated by the diameter of the second condensor aperture. On studying a CBED pattern it may not be immediately evident how to relate it to Bragg's Law. If we bear in mind that any point within the disc corresponding to the direct beam corresponds to a given angle of incidence within the focused cone of illumination, then we realize that this point is coupled to points in exactly equivalent positions in each of the diffracted discs. If the illumination were, on the other hand, coherent then there is the possibility of interference between neighboring points. However, this situation does not exist in the case of heated tungsten filaments or LaB_6 sources. As the distance between the Ewald sphere and the zero-layer plane increases, the zero-layer reflections eventually switch off and there is then an annular region without reflections. At greater scattering angles the Ewald sphere crosses the next plane of the reciprocal lattice, exciting a circular ring of reflections in what is known as the first-order Laue zone (FOLZ). Successive planes are intersected in a series of concentric circles of increasing diameter known, in turn, as the second-order Laue zone (SOLZ), third-order Laue zone, and so on (Fig. 2-6). Collective-ly, all the diffraction out of the zero-order plane is known as higher order Laue zone (HOLZ) diffraction, and the zero-layer diffraction is often referred to as the zero-order Laue zone (ZOLZ). While it is not common to observe HOLZ diffraction in SAED patterns, perhaps because of the sensitivity of this large angle diffraction to specimen curvature and the strong diffuse scattering that is often present, it is a common feature of CBED patterns and is widely used. The large angle required to reach the FOLZ makes HOLZ diffraction effects relatively weak compared with ZOLZ diffraction and, if they are ignored, to a first approximation, then this is equivalent to the projection approximation discussed in §1.4.2. In any case, HOLZ diffraction effects do not normally contribute to high-resolution imaging because of the cut-off in the contrast transfer function (see §1.3.5) at these angles (see Fig.1-7). They are, nevertheless, very important for electron-diffraction experiments and often clearly visible, as will be discussed. The strength of three-dimensional (HOLZ) diffraction with respect to two-dimensional (projection approximation) diffraction can be enhanced by specimen cooling (producing a reduction of lattice vibrations and hence the Debye-Waller factor) and by reduction of the microscope operating voltage. HOLZ diffrac-tion is weakened by point disorder in the specimen and affected by extended defects.

By measurement of the HOLZ circle radii it is possible to deduce the spacing of the reciprocal lattice planes along the zone axis direction. Let us call this spacing H and the HOLZ circle radius G, then by Pythagoras (Fig. 2-7) we have

$$k^2 = G^2 + (k-H)^2$$
$$0 = G^2 - 2kH + H^2$$

Since $k \gg H$ we can write, to a good approximation,

$$G = (2kH)^{1/2} \qquad (2.4)$$

Since successive reciprocal planes are equally spaced H apart we see that the ratios of their diameters are $1:\sqrt{2}:\sqrt{3}$ to achieve an accurate determination of H from a measurement of G, it is important to correct for any lens distortions in the diffraction pattern by a standard procedure (Steeds, 1983). To obtain a well defined HOLZ circle, it is necessary to use a large convergence angle in forming the CBED pattern and to view the result at low camera length. The Ewald sphere intersects the higher order Laue zones in well-defined circles, but these may be ill-defined if the convergence angle is too small. By use of a large convergence angle, the circle in that the Ewald sphere intersects the higher order plane will be revealed as a series of small segments in each of the neighboring orders of diffraction (Fig. 2-7). In fact, additional fine structure exists within these HOLZ reflections, that is useful for structure determination, as discussed briefly at the end of this chapter.

The use of HOLZ diffraction in accurate determinations of the lattice constants is discussed in §2.4. In SAED it is normal to talk about the camera length (L) as defined in Figure 2-8. Here x is the distance on a negative of a particular diffraction spot from the direct beam. We see that

$$x = 2L\theta \qquad (2.5)$$

but since from Equation (2.2), $2\theta = \lambda/d$,

we have $\qquad\qquad\qquad d = L\lambda/x. \qquad (2.6)$

The quantity $L\lambda$ is approximately constant (known as the camera constant) and may be determined for a particular instrument and diffraction-lens setting. Once determined, the relevant spacing of diffraction planes, d, may be arrived at from a measurement of x. Unfortunately, such measurements are not generally very accurate because they depend critically on a number of a factors, particularly the 'specimen height' (i.e., the exact position in the objective pole piece gap, that varies from place to place, and with eucentric height control. They are also strongly dependent on the diffraction-lens setting (usually variable). It is not uncommon for errors of 5% to occur in measurements made in this way. Accuracies of 1% can be achieved if considerable care is exercised, and accuracies of 0.1% have been achieved where there is an internal standard at the same location as the particle of interest. The internal standard may be a thin evaporated reference layer on the specimen or, better, the matrix adjoining a precipitate if its lattice constant is known from x-ray measurements.

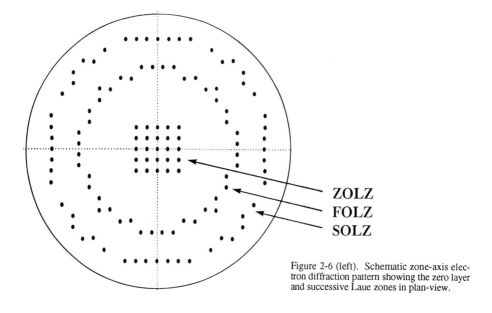

ZOLZ
FOLZ
SOLZ

Figure 2-6 (left). Schematic zone-axis electron diffraction pattern showing the zero layer and successive Laue zones in plan-view.

Figure 2-7 (right). Cross-section through a zone-axis diffraction pattern such as that illustrated in Figure 2-6, showing the intersection of the Ewald sphere through the reciprocal lattice.

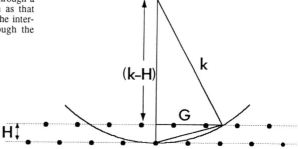

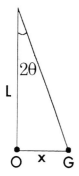

Figure 2-8. Diagram illustrating the relationship between the distance x of a diffracted beam (G) from the direct beam (O) and the camera length L.

2.1.7 Bravais-lattice determination

Any well-formed three-dimensional diffraction pattern can, in principle, be used to deduce the Bravais lattice in reciprocal space, and from that the real-space Bravais lattice follows. In practice, the task can offer quite a geometrical challenge, but it may be greatly simplified by examining the crystal at one of its principal zone axes (axes of high symmetry, where they exist) so long as at least the FOLZ is clearly visible. For this purpose, a large convergence angle is a distinct disadvantage, and it is preferable to use a small condenser aperture. Let us suppose then that we have a CBED or microdiffraction pattern generated at a principal zone axis of the crystal under examination (see, for example, Fig. 2-6). The pattern will consist of a central patch of zero-layer reflections surrounded by an annular ring of FOLZ reflections. A lattice net is then created that intersects at all the nodes of the zero-layer pattern and another formed from the FOLZ reflections. The relative displacement of the two nets determines the reciprocal lattice centering. Zero displacement corresponds to a primitive lattice; A, B, or C centering or body centering are all clearly distinguished. The only exception is the distinction between f.c.c. and b.c.c. lattices at ⟨001⟩ zone axes. In these cases we define the minimum spacing of the zero-layer net as g_{min}. Simple geometry reveals that if $2H = g_{min}$ the reciprocal lattice is I centered, but if $\sqrt{2}H = g_{min}$ the reciprocal lattice is F centered. A similar exercise can be used to decide whether or not a trigonal crystal has a primitive rhombohedral or a primitive hexagonal unit cell.

2.1.8 Kinematical and dynamical diffraction theory

Kinematical diffraction is a particular case of weak scattering that meets the following criteria:

(i) The intensities of each diffracted beam must be much less than the incident intensity.

(ii) Back scattering into the original beam or multiple scattering between diffracted beams must be negligible (Fig. 2-9a).

(iii) The diffracted intensity of each of the diffracted beams is proportional to the square of the specimen thickness and the square of the magnitude of the associated structure factor $|F_g|^2$ (as in the customary x-ray case).

Note that kinematical diffraction is distinct from but somewhat related to the weak phase object approximation (§1.4.2).

In the case of dynamical diffraction strong scattering occurs from the direct beam into the diffracted beams, and thickness dependent ('dynamical') multiple diffraction occurs between the individual diffracted beams themselves and between them and the direct beam (Fig. 2-9b). This form of (elastic) diffraction is controlled by characteristic lengths, called extinction lengths (ξ_g). According to the simplest form of dynamical diffraction theory, the so-called two beam theory, we can write the diffracted intensity

$$I_g \propto \sin^2 \frac{\pi t}{\xi_g} \, , \quad \text{where} \quad \xi_g \equiv \frac{2\pi}{F_g} \qquad (2.7a,b)$$

Note that for crystal thicknesses that are much less than ξ_g we can write

$$\sin^2 \frac{\pi t}{\xi_g} \cong \frac{\pi^2 t^2}{x_g^2} = \frac{t^2 F_g^2}{4} ,$$

i.e., we recover the kinematical result.

50

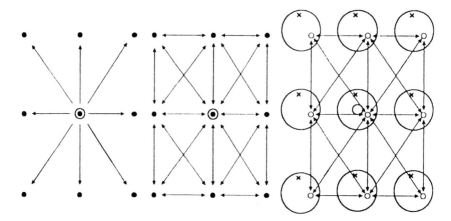

Figure 2-9. Schematic diagrams illustrating electron diffraction in different situations: (a) for kinematical diffraction; (b) for dynamical diffraction; (c) for dynamical diffraction in CBED. In the latter case all equivalent points (o or x) are related by dynamical diffraction, but different orientations are unrelated (o is not related to x).

As typical extinction lengths in electron diffraction are a few tens of nanometers in magnitude, the kinematical condition requires specimen thicknesses considerably less than 10 nm. Such thin specimens are rarely encountered, and so dynamical diffraction conditions may generally be assumed in electron-diffraction experiments. Note that this is not the disadvantage that is often claimed. It is precisely because of the dynamical nature of electron diffraction that the presence or absence of an inversion center in a crystal structure can readily be identified. As a result, it is normally possible to deduce the crystal point group without difficulty rather than the Laue group as in the x-ray case, where Friedel's rule applies. Again, while it is true that multiple diffraction effects cause so-called forbidden reflections to occur, and this would seem to over-rule the use of electron diffraction for space group determinations, this is in fact far from being the case. Properly aligned CBED patterns show characteristic dynamic absences (i.e., Gjønnes-Moodie lines), that will be discussed in the next section and can be used to rapidly and unambiguously identify the presence of glide planes or screw axes in crystal structures.

More detailed discussions of dynamical diffraction effects in CBED can be found elsewhere, and the interested reader is referred to the articles of Bird (1989), Humphreys (1979) or Steeds (1984) on this subject. By careful control of the specimen orientation, special diffraction conditions can be set up where long extinction lengths are operative so that kinematical diffraction effects can be exploited. It is of interest to note that HOLZ diffraction effects can often be treated by a pseudo-kinematic approximation (Bird, 1989) and hence used in *ab initio* crystal structure determination (see §2.4.4).

2.2 SYMMETRY DETERMINATION

2.2.1 General principles of symmetry determination

It is evident that CBED patterns often contain a large amount of symmetry information. More restricted symmetry information can also be observed in micro-diffraction patterns.

The question is how to extract this information in order to deduce the symmetry of the crystal responsible for creating the diffraction pattern. We wish to exploit the advantages of dynamical diffraction theory and hence avoid weak scattering approxi-mations. Fortunately, powerful symmetry arguments can be used that are valid for rather general conditions of dynamical diffraction from thin foils (Buxton et al., 1976). The chief assumption, that should be borne in mind, is that the sample takes the form of a parallel-side plane slab of material perpendicular to the incident electron beam. Fortunately this condition need not be applied rigorously and is remarkably tolerant of abuse. Neither highly tilted flat samples nor reasonably severe wedge angles lead to any significant reduction of symmetry. The one important situation to avoid, if the methods outlined in this section are to be used, is that of CBED patterns generated with a relatively large probe (tens of nm) from highly buckled and moderately inclined (>30°) samples. It is vitally important that in the case of, for example, a precipitate, the electron beam should only traverse the particle of interest in propagating through the specimen and not some of the adjoining matrix. Finally, it should be emphasized that the diffraction symmetry that is recorded is that of the whole irradiated region. If, for example, the specimen contains a stacking fault through that the electron beam propagates it is likely to have a marked effect on the resulting symmetry and to take away symmetry that is present in an unfaulted crystal. As a general rule it is important to examine carefully regions used for CBED experiments to ascertain that they are defect-free. In the case of extended planar faults, these should be oriented parallel to the electron beam, when possible, in order to minimize their contribution on the diffraction patterns.

2.2.2   The diffraction groups

As a result of detailed theoretical investigation using two independent methods, it has been established that under the conditions outlined above there exist thirty-one different diffraction groups describing all possible CBED pattern symmetries (Table 2-1). It has also been shown how to relate these thirty-one diffraction groups to the thirty-two crystal point groups (Table 2-2), and a table has been constructed to give the expected diffraction symmetry at any particular zone axis of each of the 32 crystal point groups (Table 2-3).

Tables 2-2 and 2-3 are essentially self-explanatory once the symmetry properties of the thirty-one diffraction groups has been explained. Reference to Table 2-1 reveals the properties of each of the diffraction groups under a series of column headings. Each diffraction group (column 1) has a unique set of properties as described by the columns. We briefly discuss each in turn. The bright-field symmetry (column 2) is that of the central disc or direct beam of the CBED pattern. In order to examine this particular property, it is necessary to describe the point symmetry of the intensity distribution by one of the ten two-dimensional point groups. Note that it is essential to have accurate axial alignment of the pattern. This may be achieved by careful specimen orientation in a goniometer stage, followed by final alignment using a small displacement of the second condenser aperture or by use of the dark-field deflectors. For effective use of this column of the table it is assumed that three-dimensional diffraction effects occur. These are usually distinctive, taking the form of fine dark lines, that are known as deficiency (or defect) HOLZ lines, that cross the pattern. Where these lines cannot be observed, they may often be made visible by use of a lower microscope operating voltage, by cooling the specimen in a liquid-nitrogen stage, or by increasing the convergence angle (perhaps using the LACBED technique if necessary—e.g., see Figure 2-11 (below). Additional symmetry can be present in the bright-field disc as compared with the rest of the pattern as a result of the unique properties of the direct beam. The symmetry of the pattern as a whole (column 3) will either be the same as that of the bright-field symmetry or it will be of lower symmetry. In the case of

Table 2-1. Pattern Symmetries.

Where a dash appears in column 7, the special symmetries can be deduced from columns 5 and 6. [Used by permission of the editor of *Philosophical Transactions of the Royal Society*, from Buxton et al. (1976), Table 2, p. 186.]

diffraction group	bright field	whole pattern	dark field general	special	$\pm G$ general	special	projection diffraction group
1	1	1	1	none	1	none	
1_R	2	1	2	none	1	none	1_R
2	2	2	1	none	2	none	
2_R	1	1	1	none	2_R	none	21_R
21_R	2	2	2	none	21_R	none	
m_R	m	1	1	m	1	m_R	
m	m	m	1	m	1	m	$m1_R$
$m1_R$	2mm	m	2	2mm	1	$m1_R$	
$2m_Rm_R$	2mm	2	1	m	2	—	
2mm	2mm	2mm	1	m	2	—	
2_Rmm_R	m	m	1	m	2_R	—	$2mm1_R$
$2mm1_R$	2mm	2mm	2	2mm	21_R	—	
4	4	4	1	none	2	none	
4_R	4	2	1	none	2	none	41_R
41_R	4	4	2	none	21_R	none	
$4m_Rm_R$	4mm	4	1	m	2	—	
4mm	4mm	4mm	1	m	2	—	
4_Rmm_R	4mm	2mm	1	m	2	—	$4mm1_R$
$4mm1_R$	4mm	4mm	2	2mm	21_R	—	
3	3	3	1	none	1	none	
31_R	6	3	2	none	1	none	31_R
$3m_R$	3m	3	1	m	1	m_R	
3m	3m	3m	1	m	1	m	$3m1_R$
$3m1_R$	6mm	3m	2	2mm	1	$m1_R$	
6	6	6	1	none	2	none	
6_R	3	3	1	none	2_R	none	61_R
61_R	6	6	2	none	21_R	none	
$6m_Rm_R$	6mm	6	1	m	2	—	
6mm	6mm	6mm	1	m	2	—	
6_Rmm_R	3m	3m	1	m	2_R	—	$6mm1_R$
$6mm1_R$	6mm	6mm	2	2mm	21_R	—	

whole pattern symmetry, the presence of HOLZ reflections can be observed directly, and it is generally the case that three-dimensional diffraction information is available if a small enough camera length is employed. The absence of a HOLZ ring in the pattern implies either such a large diameter HOLZ ring that the Debye-Waller factor has extinguished three-dimensional diffraction or, more usually, the absence of three-dimensional order in the crystal under examination. In the former case a lower operating voltage or cold stage should be used; in the latter, another crystal region should be found (highly faulted regions do not have a three-dimensional crystal structure).

The first pair of columns to the left of the table labelled Dark Field (columns 4 and 5) refer to the internal symmetry of particular diffracted beams. In order to examine this symmetry it is necessary to perform a small specimen reorientation such that the Bragg point lies at the center of that particular diffraction disc. As the Bragg point is the perpendicular bisector of the line joining the center of the reflected disc to the center of the direct beam, of an axis-centered pattern, the reorientation is small for diffracted beams that lie close to the direct beam. For close diffracted beams, aperture displacement or dark-field deflection can be used. The diffraction pattern may contain mirror lines. When chosen

Table 2-2. Relation between the diffraction groups and the crystal point groups.

[Used by permission of the editor of *Philosophical Transactions of the Royal Society*, from Buxton et al. (1976), Table 3, p. 188.]

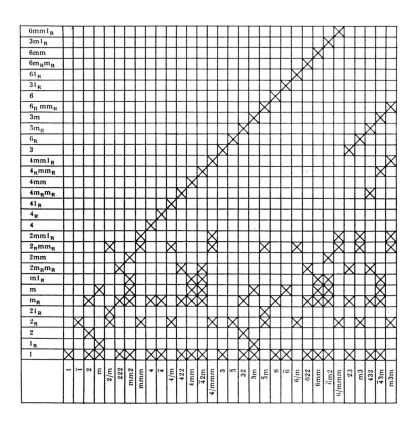

Bragg reflection lies on the mirror line this is referred to as a "special dark field" (column 5) otherwise it is referred to as "general" (column 4)The entry "none" in column 5 means that there are no special reflections in this case.

The pair of columns labelled ±G (columns 6 and 7) call for a slightly more elaborate operation comparing the symmetry observed in the +G reflection centered on its Bragg point with that observed in the -G reflection centered on its Bragg point. The terms "special" (column 7) and "general" (column 6) have the same meaning as described above. A new and probably unfamiliar symbol, R, is used at this point (and also in column 1) to describe a rotation of 180° about the Bragg point of the particular reflection G. The symbol 2 means that the ±G patterns are related by a two-fold (180° rotation) about the pattern center. The symbol 2_R means that the two-fold rotation about the pattern center is coupled with a two-fold rotation about the Bragg point as part of one operation. The result is that the patterns in +G and -G appear to be related by a sideways displacement through the origin. The symbol 21_R means that the patterns are related by a separate two-fold rotation about the pattern center and a two-fold rotation about the Bragg point. The — (dash) in the "Special" column means the absence of symmetry of this type.

Table 2-3. Zone Axis Symmetries.

[Used by permission of the editor of *Philosophical Transactions of the Royal Society*, from Buxton et al. (1976), Table 4, p. 189.]

point group	$\langle111\rangle$	$\langle100\rangle$	$\langle110\rangle$	$\langle uvo\rangle$	$\langle uuw\rangle$	$[uvw]$
m3m	$6_R mm_R$	$4mm1_R$	$2mm1_R$	$2_R mm_R$	$2_R mm_R$	2_R
$\bar{4}3m$	$3m$	$4_R mm_R$	$m1_R$	m_R	m	1
432	$3m_R$	$4m_R m_R$	$2m_R m_R$	m_R	m_R	1

point group	$\langle111\rangle$	$\langle100\rangle$	$\langle uvo\rangle$	$[uvw]$
m3	6_R	$2mm1_R$	$2_R mm_R$	2_R
23	3	$2m_R m_R$	m_R	1

point group	$[0001]$	$\langle11\bar{2}0\rangle$	$\langle1\bar{1}00\rangle$	$[uv.o]$	$[uu.w]$	$[u\bar{u}.w]$	$[uv.w]$
6/mmm	$6mm1_R$	$2mm1_R$	$2mm1_R$	$2_R mm_R$	$2_R mm_R$	$2_R mm_R$	2_R
$\bar{6}m2$	$3m1_R$	$m1_R$	$2mm$	m	m_R	m	1
6mm	$6mm$	$m1_R$	$m1_R$	m_R	m	m	1
622	$6m_R m_R$	$2m_R m_R$	$2m_R m_R$	m_R	m_R	m_R	1

point group	$[0001]$	$[uv.o]$	$[uv.w]$
6/m	61_R	$2_R mm_R$	2_R
$\bar{6}$	31_R	m	1
6	6	m_R	1

point group	$[0001]$	$\langle11\bar{2}0\rangle$	$[u\bar{u}.w]$	$[uv.w]$
$\bar{3}m$	$6_R mm_R$	21_R	$2_R mm_R$	2_R
3m	$3m$	1_R	m	1
32	$3m_R$	2	m_R	1

point group	$[0001]$	$[uv.w]$
$\bar{3}$	6_R	2_R
3	3	1

point group	$[001]$	$\langle100\rangle$	$\langle110\rangle$	$[uow]$	$[uvo]$	$[uuw]$	$[uvw]$
4/mmm	$4mm1_R$	$2mm1_R$	$2mm1_R$	$2_R mm_R$	$2_R mm_R$	$2_R mm_R$	2_R
$\bar{4}2m$	$4_R mm_R$	$2m_R m_R$	$m1_R$	m_R	m_R	m	1
4mm	$4mm$	$m1_R$	$m1_R$	m	m_R	m	1
422	$4m_R m_R$	$2m_R m_R$	$2m_R m_R$	m_R	m_R	m_R	1

point group	$[001]$	$[uvo]$	$[uvw]$
4/m	41_R	$2_R mm_R$	2_R
$\bar{4}$	4_R	m_R	1
4	4	m_R	1

point group	$[001]$	$\langle100\rangle$	$[uow]$	$[uvo]$	$[uvw]$
mmm	$2mm1_R$	$2mm1_R$	$2_R mm_R$	$2_R mm_R$	2_R
mm2	$2mm$	$m1_R$	m	m_R	1
222	$2m_R m_R$	$2m_R m_R$	m_R	m_R	1

point group	$[010]$	$[uow]$	$[uvw]$
2/m	21_R	$2_R mm_R$	2_R
m	1_R	m	1
2	2	m_R	1

point group	$[uvw]$
$\bar{1}$	2_R
1	1

The column at the far right of Table 2-1 (column 8) indicates the symmetry that may be expected in the absence of three-dimensional diffraction information (projection). For the coupled diffraction groups, showing a particular symbol in this column, the diffraction symmetry is given by that symbol. This symbol is always the same as the lowest row of those coupled together. Rows in the group above the lowest row take on the higher symmetry properties described by the lowest row.

Some concluding remarks may be helpful. Perfect symmetry probably does not exist. It is always a matter of judgment what symmetry a particular pattern has. Many factors reduce the symmetry below what it would be in the ideal case. If the symmetry increases towards a particular point symmetry as the perfection of the crystal structure improves then it is likely that the highest symmetry consistent with this trend is the required result. As against this, any single feature of a pattern that consistently breaks a higher symmetry, irrespective of area or crystal quality, must be regarded as symmetry breaking. For crystallographic purposes, near-symmetry is meaningless.

2.2.3   Examples of point-group determination

We first illustrate some of the ways in which symmetry properties may be detected in CBED. Three-dimensional bright-field symmetry can be clearly evident in the central disc as elaborate patterns of deficiency HOLZ lines crossing a zero-layer pattern with much broader pattern details. Figures 2-10a,b,c illustrate, as examples, the $3m$ symmetry of the [0001] axis of α-Al_2O_3 (corundum), the 3 symmetry of the ‹111› zone axis of FeS_2 (pyrite), and the $4mm$ bright-field symmetry of spinel at the ‹100› axis. In cases where the closeness of the reflection prevents a significant field of view without overlap of the diffraction orders, LACBED patterns can be used to reveal the internal symmetry of any chosen order of reflection. Figures 2-11a,b,c illustrate the $6mm$ bright-field symmetry of [0001] beryl, the $2mm$ symmetry of the [010] axis of mullite (a = 0.758, b = 0.768, c = 0.289 nm; space group $Pbam$), and the 2 symmetry of the [010] axis of diopside (a = 0.972, b = 0.891, c = 0.525 nm; $\beta = 105°$; space group $C2/c$).

Whole-pattern symmetry may be determined in one of three different ways. It may be revealed simply by the intensity differences between individual featureless diffraction discs, as in the case of apatite (6 symmetry) in Figure 2-12. It may also be judged on the basis of detail within the individual ZOLZ reflections, as in the $4mm$ case of ‹100› spinel in Figure 2-10c above. Finally it may be determined by careful examination of the individual reflections in a particular HOLZ ring. It is harder to illustrate such examples than it is to study the negatives, but Figure 2-13 illustrates a case for ‹110› FeS_2 where the ZOLZ pattern would indicate, at first sight, $2mm$ symmetry, but the FOLZ ring clearly has a horizontal mirror but no vertical mirror.

For several further examples obtained by CBED of minerals, the interested reader is referred to the review article by Champness (1987). For an early example of this form of analysis, see the work of Steeds and Eades (1973) on muscovite mica.

2.2.4   Gjønnes-Moodie lines and space-group determination

When a crystal structure contains screw axes or glide planes (of which there are many different types) so-called 'forbidden reflections' occur, according to kinematical diffraction theory, whenever the structure factor vanishes. By identification of these forbidden reflections in x-ray crystallography, it is possible to deduce the crystal space group. For electron diffraction, with multiple scattering, 'forbidden reflections' often appear quite strongly. Nevertheless, the existence of screw axes or glide planes can

56

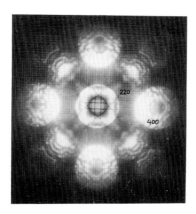

Figure 2-10. CBED patterns. (a) [0001] α–Al$_2$O$_3$ [used by permission of Adam Hilger from Mansfield and Steeds in *Convergent Beam Electron Diffraction of Alloy Phases* (1984) Fig. 19.2, p. 83]. (b) ⟨111⟩ FeS$_2$ [used by permission of the editor of *Microscopy, Microanalysis and Microstructures*, from Steeds and Vincent in *Journal de Microscopie et de Spectroscopie Electroniques* (1983) Fig. 1, p. 420]. (c) ⟨100⟩ spinel (courtesy of Dr. T.A. Bielicki).

usually be deduced quite straightforwardly by use of carefully aligned zone-axis CBED. The reason for this fortunate situation is the cancellation of the scattered amplitudes in pairs along principal lines in the pattern, leading to the occurrence of dynamic absences (sometimes called dark bars or black crosses), as was appreciated many years ago by Gjønnes and Moodie (1965). Consider a perfectly aligned zone-axis pattern with a perpendicular 2_1 screw axis (or its equivalent in the screw axes 4_1, 4_3, 6_1, 6_3, and 6_5) or a glide plane parallel to the direction of incidence. The line of the screw axis or the line of intersection of the glide plane corresponds to one of the principal lines in the pattern (often called the A line). The second principal line (the B line) is perpendicular to the A line and intersects the A line for a particular 'forbidden' reflection at the Bragg point for that reflection. Under two-dimensional diffraction conditions, every forbidden reflection will have lines of absent intensity along both the A and B lines. For an axially aligned pattern the Bragg points are not visible, and the 'forbidden' reflections appear as a line in which every odd member in the zero-layer pattern has a line of absence (dark bar) along the A

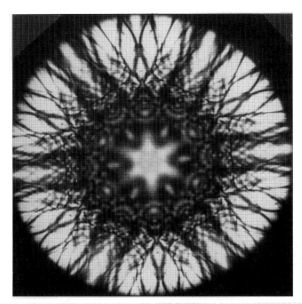

Figure 2-11. LACBED patterns (a) [0001] beryl (courtesy of Dr. J.C. Walmsley). (b) [010] mullite (courtesy of W.J. Vine). (c) [010] diopside — at top of next page.

Figure 2-12. CBED of [0001] apatite [used by permission of the editor of *Physics and Chemistry of Minerals*, from Lang and Walmsley (1983) Fig. 2, p. 7].

Figure 2-13. ‹110› CBED of FeS$_2$ at 120kV. Details of the first Laue zone intensity distribution are shown in enlargements to left and right of the figure. The pattern has a horizontal mirror line passing through its center. [Used by permission of the editor of *Microscopy, Microanalysis and Microstructures,* from Steeds and Vincent in *Journal de Microscopie et de Spectroscopie Electroniques* (1983) Fig. 2, p. 421].

direction (see Fig. 2-14c). Every route (arrowed) on one side of the A line contributing a diffracted amplitude ϕ_g to the 'forbidden' reflection g is matched by an equivalent route on the opposite side of the A line that contributes an amplitude $-\phi_g$, so that perfect cancellation occurs. This effect is independent of specimen thickness or the microscope operating voltage. It also occurs in all odd reflections along the zero-layer systematic row and cannot therefore be confused with accidental absences arising from other causes. An exactly equivalent argument explains the origin of the B line (Fig. 2-14d), but in this case the so-called principle of reciprocity (between source and observer) has to be used to explain the absence. In the case of ZOLZ diffraction, both A and B lines will be present for either the glide plane (horizontal intersection of figure) or screw axis (horizontal). In cases of domination by three-dimensional diffraction, such as when double diffraction from the FOLZ occurs to a zero-layer kinematically-forbidden reflection, then only the A line is observed in the case of the glide plane and only the B line in the case of the screw axis. A final confirmation can be achieved by orienting each 'forbidden' reflection in turn to its Bragg point (Fig. 2-14b), when the black cross illustrated in Figure 2-14a becomes visible in the case of ZOLZ diffraction. Alternatively LACBED patterns in each forbidden reflection clearly show the features characteristic of this behavior (Steeds et al., 1982).

Distinguishing glide planes from screw axes is generally simple, although the details vary depending on the quality of the information available. If only zero-layer detail is present (no HOLZ diffraction), it is necessary to perform tilting experiments along the axis parallel or perpendicular to the A line. Rotation about the A line as axis will preserve the 2_1 axis (or its symmetry equivalent) perpendicular to whichever zone axis arises, and so similar dynamic absences will occur. In the case of the glide plane, rotation about the axis perpendicular to the A line will preserve the glide plane parallel to the direction of incidence, and so the existence of dynamic absences would then be preserved. For a screw axis together with a glide plane intersecting along the screw axis direction, dynamic absences would be retained for both rotations.

60

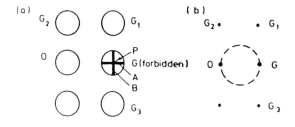

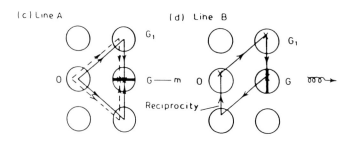

Figure 2-14. Schematic diagram of dynamic absences: (a) black cross with A and B arms intersecting at the Bragg point P, (b) the Laue circle with center P, (c) mirror-related pairs of routes giving cancelling contributions along A, (d) reciprocity relates routes giving cancellation along B when a 2_1 axis is present along A. [Used by permission of the Scottish Universities Summer School in Physics, from Steeds (1984) Fig. 8, p. 68].

When the CBED contains HOLZ information it is important to investigate the presence of whole-pattern mirror lines in relation to the A line in the pattern. Where bright- field HOLZ lines are visible the situation is straightforward. A bright-field mirror-line perpendicular to the A line indicates a screw axis, while a bright field mirror line parallel to the A line indicates a glide plane. The various possibilities were summarized by Steeds and Vincent (1983) and are illustrated in Figures 2-15a,b.

2.2.5   Examples of space-group determination

Before giving the details of some particular cases we first illustrate the characteristic and familiar form of Gjønnes-Moodie dynamic absences in CBED. Figures 2-15a,b,c give three examples. In Figure 2-16a the ⟨110⟩ zone axis pattern of cristobalite reveals a line of alternate reflections, in a vertical row through the center of the pattern, that have the characteristic line of absence running through them. These reflections, $\pm001, \pm003, \pm005$ are kinematically and dynamically absent in consequence of the 4_1 or 4_3 screw axis along c. The second example shows two lines of absences at right angles in Figure 2-16b. In this case, for the [001] zone axis of mullite, the responsible symmetry elements are the b and a glides in the space group $Pbam$. The third example, for the ⟨101⟩ axis of SnO_2 (cassiterite), deficiency HOLZ lines are clearly present in the bright-field disc (see Fig. 2-16c): the vertical mirror unambiguously identifies a screw axis as the reason for the absences.

The four examples of space group determination that follow all differ and have been chosen to illustrate a variety of different approaches that may be used.

61

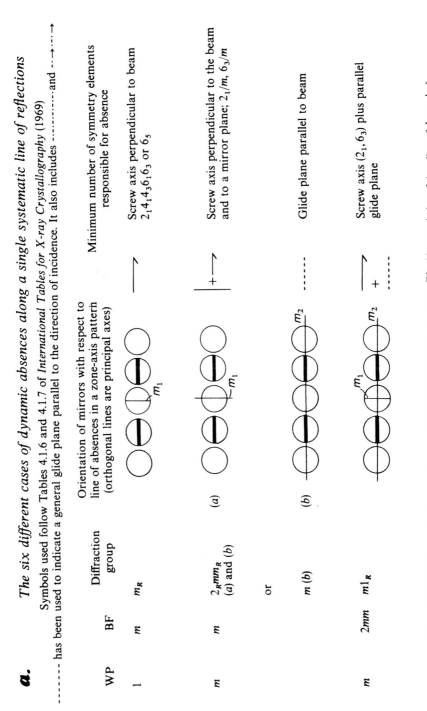

Figure 2-15. (a) Systematic dynamic absences (b) orthogonal dynamic absences. [Used by permission of the editor of *Journal of Applied Crystallography*, from Steeds and Vincent (1983) Tables 5 and 6, pp. 322 and 323]. WP = whole pattern symmetry; BF = bright field symmetry; DG = diffraction group.

62

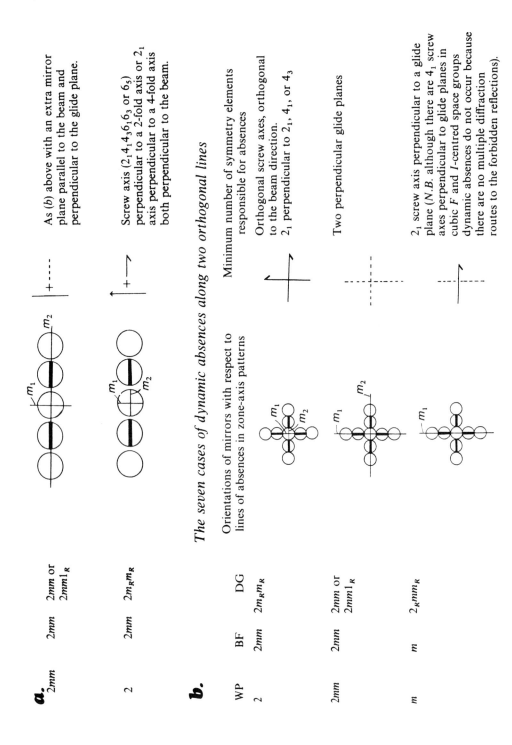

a.

WP	BF	DG		
$2mm$	$2mm$	$2mm$ or $2mm1_R$		As (b) above with an extra mirror plane parallel to the beam and perpendicular to the glide plane.
2	$2mm$	$2m_Rm_R$		Screw axis ($2_1, 4_1, 4_3, 6_1, 6_3$ or 6_5) perpendicular to a 2-fold axis or 2_1 axis perpendicular to a 4-fold axis both perpendicular to the beam.

The seven cases of dynamic absences along two orthogonal lines

b.

WP	BF	DG		Minimum number of symmetry elements responsible for absences
				Orientations of mirrors with respect to lines of absences in zone-axis patterns
2	$2mm$	$2m_Rm_R$		Orthogonal screw axes, orthogonal to the beam direction. 2_1 perpendicular to 2_1, 4_1, or 4_3
$2mm$	$2mm$	$2mm$ or $2mm1_R$		Two perpendicular glide planes
m	m	2_Rmm_R		2_1 screw axis perpendicular to a glide plane (*N.B.* although there are 4_1 screw axes perpendicular to glide planes in cubic F and I-centred space groups dynamic absences do not occur because there are no multiple diffraction routes to the forbidden reflections).

b.

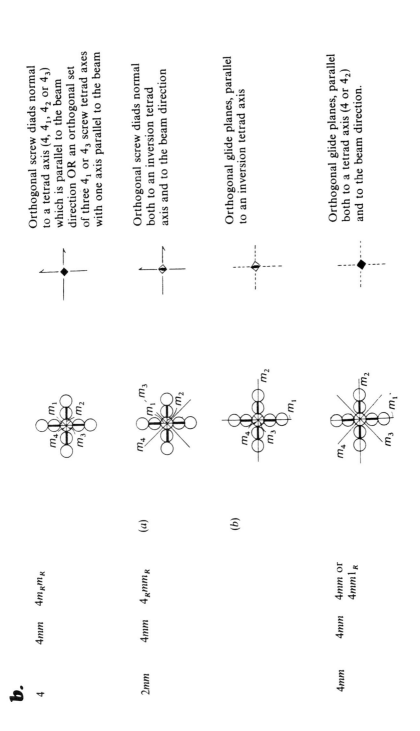

Orthogonal screw diads normal to a tetrad axis $(4, 4_1, 4_2$ or $4_3)$ which is parallel to the beam direction **OR** an orthogonal set of three 4_1 or 4_3 screw tetrad axes with one axis parallel to the beam

4    $4mm$ $4m_Rm_R$

m_1 m_2 m_3 m_4

Orthogonal screw diads normal both to an inversion tetrad axis and to the beam direction

(a)

$2mm$ $4mm$ 4_Rmm_R

m_1, m_3 m_2 m_4

Orthogonal glide planes, parallel to an inversion tetrad axis

(b)

m_1 m_2 m_3 m_4

Orthogonal glide planes, parallel both to a tetrad axis $(4$ or $4_2)$ and to the beam direction.

$4mm$ $4mm$ $4mm$ or $4mm1_R$

m_1' m_2 m_3 m_4

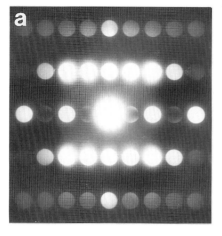
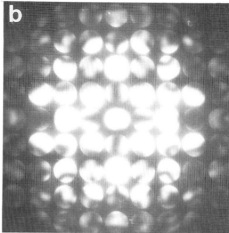
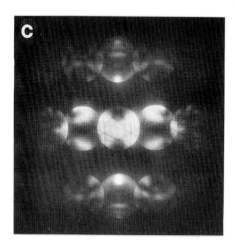

Figure 2-16. Examples of dynamic absences: (a) ‹110› zone axis of cristobalite (b) [001] zone axis of mullite (c) ‹101› zone axis of SnO_2 [courtesy of Dr M.D. Shannon]. Figures (a) and (b) are used by permission of the Institute of Physics, from Vine, Wong-Leung, Steeds and Vincent (1991) Figs. 1 and 3, pp. 268 and 269, Conf. Series No. 119.

The first example is that of FeS_2, (pyrite) studied at the ‹110› axis (Fig. 2-13). By measurement of the ratio of the spacings of the reflections in the two orthogonal directions in the zero layer, a factor of $\sqrt{2}$ is obtained. This, together with the spacing in the perpendicular direction deduced from measurement of the HOLZ ring diameter and substituted into Equation 2.4, indicates the ‹110› zone axis of a cubic crystal. The existence of 3-fold zone axes, such as that illustrated in Figure 2-10b, after ±30° rotation about one of two orthogonal axes in the ‹110› pattern, is firm confirmation of this tentative conclusion. The whole-pattern symmetry of Figure 2-13 is m, as is evident from the HOLZ detail shown to left and right of the central CBED pattern. Examination of the central disc, overexposed in the print, also reveals m symmetry (the apparent lack of a horizontal mirror in the ZOLZ pattern is an artefact of dodging during printing). Examination of Table 2-1, in association with Table 2-3, reveals that the only cubic point group to have ‹110› whole-pattern and bright-field m symmetry is $m3$. The position of the FOLZ mesh when projected onto the ZOLZ mesh indicates a primitive lattice. The first step in space-group determination is to note that the central reflection in the segment of the HOLZ ring enlarged to the right of Figure 2-13 has a dynamical absence running through its center in a radial line coinciding

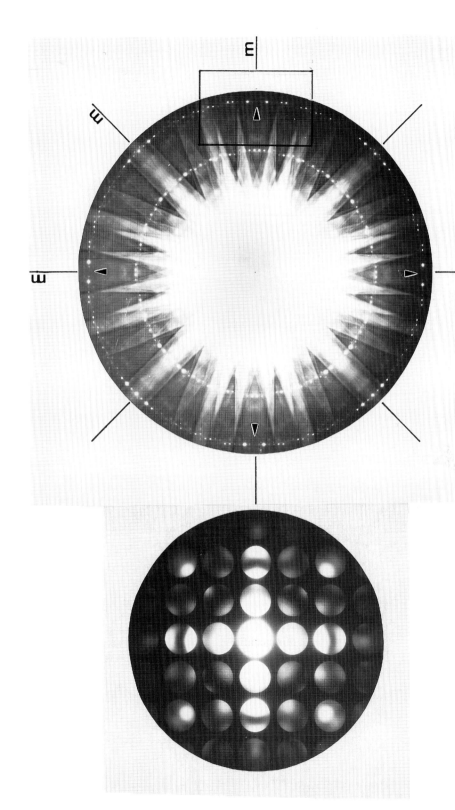

Figure 2-17. CBED of garnet at <001>: (a) detail of ZOLZ field pattern (crystal increases in thickness from top to bottom of the pattern) (b) large angle microdiffraction pattern (c) LACBED bright-field pattern (d) detail of (b) indicated by rectangular box; indexing of the SOLZ reflections (specimen kindly provided by Prof. J.C. Donkhan).

66

$\overline{43}$ 23 2 $\overline{40}$ 28 2 $\overline{36}$ 32 2 $\overline{32}$ 36 2 $\overline{28}$ 40 2 $\overline{23}$ 43 2

m

$\overline{34}$ 34 2

hhl forbidden reflection
with 2h+l≠4n
d glide plane //(110)

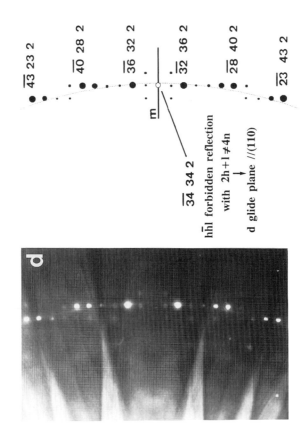

with the mirror line through the center of the pattern. This reflection may be indexed as 11,10,0 (see §2.4.2) and is in marked distinction to the diametrically opposite reflection, 10,11,0 to the left of the pattern. Inspection of the zero-layer pattern indicates dynamic extinctions in reflections 0,0,2n+1 and (2n+1), -(2n+1), 0 (n integer). These extinctions unambiguously identify the space group as Pa3.

The second example is that of garnet, illustrated by the CBED results in Figure 2-17. Figures 2-17a,b illustrate the 4mm whole-pattern symmetry. LACBED of the direct beam in Figure 2-17c indicates 4mm bright-field symmetry. Study of Table 2-1 leads us to identify either diffraction group 4mm or 4mm1$_R$. The projection of the FOLZ mesh onto the zero layer indicates a body centered lattice, and measurement of the FOLZ and SOLZ ring diameters indicates a cubic lattice. From Table 2-3 we find that there is no cubic crystal that gives ‹001› diffraction with 4mm symmetry and that the only point group giving 4mm1$_R$ is the point group in m3m (at ‹100›). We conclude thus far Im3m. The extra reflection in the FOLZ and SOLZ relative to the ZOLZ indicates that {100} is a glide plane, i.e., Ia3m. The final step, identifying the d glide in the structure, comes from a careful study of the FOLZ/SOLZ reflections. What this study reveals is the absence of just those reflections, on the two mirror lines in the pattern, with indices that correspond to a diamond glide in the structure. Figures 2-17d,e show a detail of the FOLZ and SOLZ reflections and a diagram giving the indices of the reflections. Absent reflections are observed and indexed as shown. We conclude a space group of Ia3d.

The third example is that of spinel. The ‹100› CBED results already illustrated (Fig. 2-10c) give bright-field and whole-pattern symmetries of 4mm. The projection of the FOLZ net onto the zero-layer net indicates a face-centered cell. The diameter of the FOLZ ring indicates the ‹100› layer spacing of a cubic crystal. As in the case of garnet, this, together with the 4mm symmetry, indicates a point group m3m, but in this case an F-centered cell. The final step of space-group identification is of some subtlety and is worth noting. Weak 020 and 002 reflections are present in the ‹100› diffraction patterns. Study of the CBED pattern reveals that the diffraction in the 020 and 002 discs takes the form of fine bright lines, like HOLZ excess lines (Fig. 2-18). They can, in fact, be traced to double diffraction out to the FOLZ and back again to the zero-layer, as indicated in Figure 2-19. There is no zero-layer diffraction route to 002 or 020. According to the treatment of Gjønnes and Moodie (1965), the diamond glide planes that lead to the 002 and 020 absences should also lead to radial lines of dynamic absence. As there is no detail there in Figure 2-18, it is not possible to confirm or deny this conclusion. However, since FOLZ diffraction is sensitive to the operating voltage, the voltage may be varied so as to cause the excess lines to move together to meet on the mirror line. If the diamond glide is present, Gjønnes-Moodie lines of absence will remain at all voltages as the lines come together at the mirrors, and this is confirmed by detailed experiments such as that illustrated in Figure 2-20. We conclude that the space group is Fd3m.

The final example is that of diopside, previously identified as having two-fold bright-field symmetry (Fig. 2-11c), and from the microdiffraction pattern of Figure 2-21 we also deduce two-fold whole-pattern symmetry. This symmetry, together with the information from the ZOLZ and FOLZ meshes, indicates a monoclinic crystal with point group 2 or 2/m. The analysis then proceeds according to the systematic approach indicated in Figure 2-22. Create the FOLZ net from the microdiffraction pattern and superimpose it on the ZOLZ (Fig. 2-22a). From the nodes of this mesh and the nodes of the ZOLZ mesh (Fig. 2-22b) we identify the smallest primitive translations that could generate the combined pattern (Fig. 2-22c). Finally, compare the result with the ten possibilities that exist for the monoclinic system (Fig. 2-23). The solution is complicated in this particular example

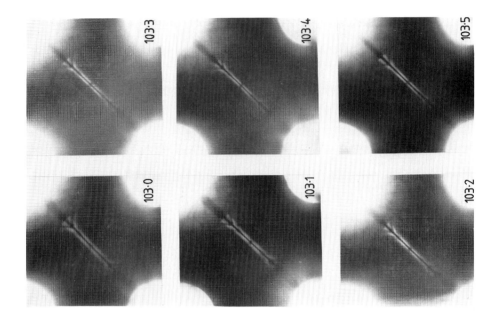

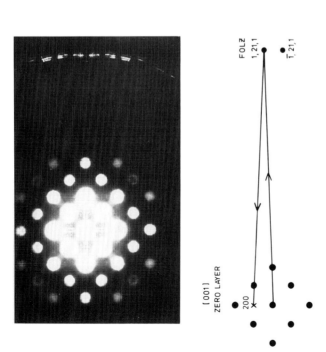

Figure 2-18 (upper left). ⟨001⟩ spinel CBED showing a portion of the FOLZ ring [used by permission of the editor of the *Proceedings of the Annual Meeting of the Electron Microscopy Society of America*, from Steeds and Evans (1980) Fig. 4, p. 190].

Figure 2-19 (lower left). Diagram indicating a double-diffraction route from the direct beam to the spinel FOLZ reflection 1, 21, 1 and back to the zero-layer 200 reflection in Figure 18.

Figure 2-20 (right). Detail of the spinel 200 ZOLZ reflection for small changes in the microscope operating voltage as indicated [courtesy of Dr. T.A. Bielicki]. The minimum separation (lines come together) occurs at 103.25 kV.

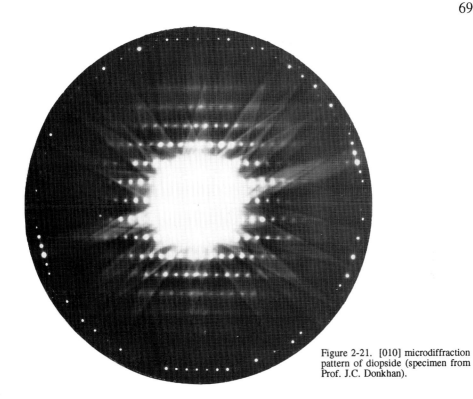

Figure 2-21. [010] microdiffraction pattern of diopside (specimen from Prof. J.C. Donkhan).

because of a chance relationship between **a** and **c** (c = 2a cos β). As a result, no unique choice of primitive translations (Fig. 2-22c) is possible. For one choice (Fig. 2-22d), we obtain C1c1, and for the other I1a1. We conclude, therefore, that the space group is either Cc (or C2/c) or Ia (or I2/a). To distinguish between point group 2 and 2/m an experiment has to be performed at an axis perpendicular to the unique axis of the cell.

2.2.6   Special cases: polarity, chirality and modulated structures

Electron diffraction is probably the most effective and sensitive technique for symmetry determination that exists in many practical cases. It can easily tackle problems of polarity or chirality. Having said this, it is one thing to use CBED to identify an inversion boundary and to note the discontinuous change of the CBED on crossing a boundary, and quite another to relate the observation to a particular atomic arrangement in the crystal. The case of GaAs must surely be the most widely studied, and a host of different prescriptions for deciding how to distinguish the gallium from the arsenic site have been given, some with one set of advantages, another with others (Spellward and James, 1991). Experience indicates that every case has to be the subject of detailed investigation, and it is unlikely that simple rules can be given. Some prefer a kinematical approach without recourse to elaborate calculations. Others, with well tried computer programs, see no advantage in the extra experimental trouble required to identify a reliable set of reflections and orientations for the application of kinematical theory. Chirality is a less well-explored problem, but once again it has been demonstrated on a number of occasions (Vincent et al., 1986) that it is relatively trivial to identify the boundary between enantiomorphic pairs. However, the problem of deciding which pattern corresponds to the right-handed screw is much more difficult to solve, and simple solutions are still under investigation.

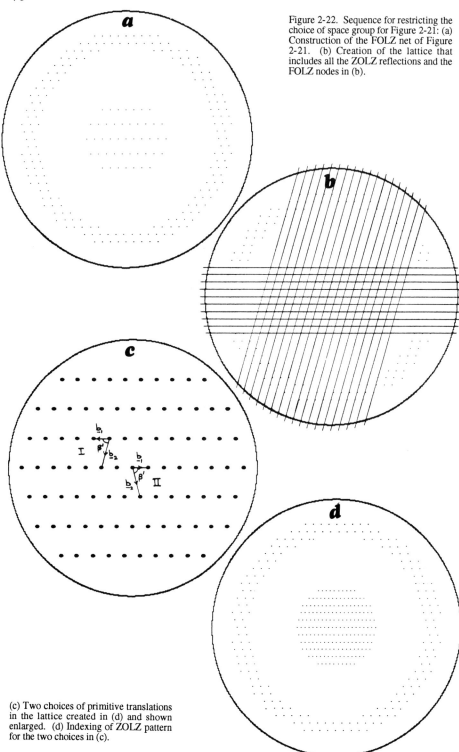

Figure 2-22. Sequence for restricting the choice of space group for Figure 2-21: (a) Construction of the FOLZ net of Figure 2-21. (b) Creation of the lattice that includes all the ZOLZ reflections and the FOLZ nodes in (b).

(c) Two choices of primitive translations in the lattice created in (d) and shown enlarged. (d) Indexing of ZOLZ pattern for the two choices in (c).

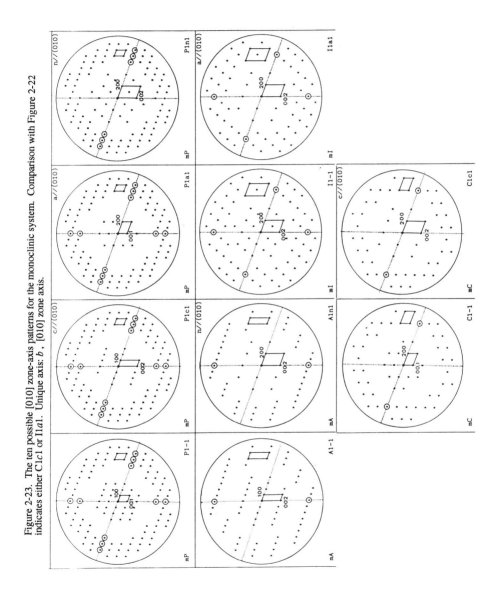

Figure 2-23. The ten possible [010] zone-axis patterns for the monoclinic system. Comparison with Figure 2-22 indicates either C1c1 or I1a1. Unique axis: b, [010] zone axis.

72

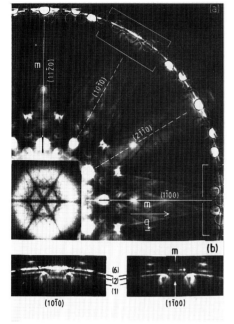

Figure 2-24. Part of the [0001] CBED from NiAs (niccolite) at a temperature of 100K and a microscope operating voltage of 90kV. Broken hexagonal symmetry is evident in the bright-field disc, shown enlarged in the lower left corner, and in the boxed regions of the FOLZ, shown enlarged below the pattern. [Used by permission of the editor of *Philosophical Magazine*, from Vincent and Withers (1987) Fig. 2, p. 60].

Long-period modulated structures might at first sight seem to present a difficult challenge to CBED, and one that is better taken up by SAED, or preferably microdiffraction. It is certainly the case that all the examples of modulated minerals of which we are aware have been tackled using SAED until now. As against this there are a few notable exceptions, not minerals, in the literature where LACBED has been spectacularly successful in solving long-period modulated structures. In one of these, NiGeP, a modulated NiAs structure, was shown to have a displacive modulation, and the precise nature of the modulation and even the magnitude of the displacements was deduced from details of the LACBED observations (Vincent and Pretty, 1986). This problem was studied with the long-period modulation perpendicular to the direction of incidence. A remarkable and little-known conclusion that resulted from extending this study to niccolite minerals was that all of these supposed classical NiAs structures are modulated (Vincent and Withers, 1987). The true symmetry of these minerals, as a result of these modulations, is in fact orthorhombic and not hexagonal. The evidence for this is illustrated in Figure 2-24. An analysis of this modulation was possible in terms of a specific soft mode of the Ni subcell. Even more surprising, perhaps, is the recent success of LACBED rocking curves generated from specimens with long-period modulations along the direction of incidence (Cherns, 1991). These results on synthetic materials created from semiconductors, encourage the speculation that similar methods could be used in the future to study long-period modulated minerals.

2.3   IMPERFECT CRYSTALS

2.3.1   Point disorder

The presence of large densities of point defects, non-stoichiometry, and related effects in minerals produces a good deal of additional diffuse scattering in electron-diffraction patterns and a corresponding marked reduction in the strength of large-angle elastic diffraction (particularly HOLZ diffraction). The best way to study such samples is generally by SAED where the diffuseness of what would otherwise be sharp diffraction maxima is immediately apparent. Quite often the point defects order, and then the diffuse scattering will be structured, sometimes in quite elaborate and detailed patterns (Hoier, 1973; Welberry and Withers, 1987, 1991), from which useful information can be obtained. It is worth noting that the most detailed patterns are normally obtained slightly away from exact zone-axis directions, where the intense multiple scattering, that would otherwise smear out the diffuse intensity, is somewhat reduced. Any broadening of the diffraction spots, into even the small discs of microdiffraction patterns, can reduce the ease with that these effects can be detected.

A little explored area, that perhaps merits further investigation in the future, is the quantitative measure of the degree of point disorder from a study of HOLZ diffraction in CBED. The ability of HOLZ diffraction to separate sub-lattices in a crystal structure and to study their effects independently, a topic to which we return in §4.3, makes this a potentially valuable area of study. An early result, obtained by comparing the isostructural compounds SnO_2 and $VSbO_4$, showed relatively small changes in the FOLZ intensities of these two structures studied at their ‹111› axes, but significant differences in the SOLZ intensities (Shannon and Eades, 1978). As the reflections in the FOLZ ring, at this zone axis, depend only on the oxygen structure factors, the implication is of a well-ordered cation sub-lattice. However, the SOLZ diffraction, being dominated by the metallic atoms, indicates marked disorder in the anion sub-lattice in the case of the V and Sb alloy as compared with the Sn sub-lattice. Other preliminary results of this sort are known to the authors but remain for more systematic investigation.

2.3.2 Planar disorder and polytypism

As planar disorder is common in minerals, it is a particularly important topic to discuss. Where the disorder is of planes that are parallel to the incident electron beam, SAED or high-resolution electron microscopy give a good means of studying the nature and degree of disorder [Treacy et al. (1991) provide an interesting recent example of analysis of SAED patterns]. Unfortunately, the disorder frequently occurs in planes perpendicular to the incident beam and parallel to the specimen surface; micas and clays are just two well-known examples of this type. It has been said that electron diffraction is incapable of investigating such situations on account of the projection of the structure that occurs to a good first approximation. Fortunately, the advent of CBED studies has completely revised this situation, and the state of order along the beam direction is now well studied by HOLZ diffraction. 2-, 3-, 4-, 6-layer and even greater periods have been studied routinely in layer structures of transition metal chalcogenides and other related examples (Steeds, 1979, Steeds and Vincent, 1983). Single faults cause severe changes of CBED symmetry (Jesson and Steeds, 1989) and splitting of single HOLZ lines into two components. Analysis of these effects can lead to fault displacement vector determination (Tanaka et al., 1988, Jesson and Steeds, 1990). Repeated faulting parallel to the surface produces several characteristic effects, including reduced HOLZ diffraction intensity, enhanced diffuse scattering, multiple (rather than doubled) HOLZ lines, and extended zones of HOLZ and ZOLZ diffraction on account of the streaking of reciprocal-lattice points along the beam direction. In the extreme case of disorder parallel to the surface, HOLZ rings are entirely absent, and ZOLZ reflections extend out to large angles until they are lost against the strong diffuse background (Steeds, 1979). On tilting such samples well away from the horizontal, 'rubber' HOLZ rings are observed that deform continuously with change of orientation and from that no meaningful HOLZ lattice can be created.

2.3.3   Dislocation Burgers vector determination

Well-established methods exist for Burgers vector determination in conventional diffraction contrast electron microscopy (Hirsch et al., 1965), and these have been applied to minerals where the origin and motion of dislocations in the specimens is of interest. However, there are a number of difficulties with this so-called $\mathbf{g} \cdot \mathbf{b} = 0$ method that are now apparent. These include some or all of the following:

(a) the difficulty of establishing good two-beam conditions in complex crystal structures (to determine $\mathbf{g}$).

(b) failure of isotropic elasticity theory.

(c) precipitation on the dislocations giving rise to supplementary strain fields (the $\mathbf{g} \cdot \mathbf{b} = 0$ approach assumes elastic homogeneity).

(d) dissociation into closely spaced partial dislocations.

(e) radiation damage of the material during study, leading either to additional strain fields when the core of the dislocation is the first region to change (see (c) above) or worse, complete loss of diffraction from the area of interest before sufficient electron micrographs can be recorded.

These problems can generally be overcome by a relatively new, but now well established, alternative method using LACBED. As mentioned in §2.1.4, this technique provides simultaneous reciprocal-space, and through a shadow image, real-space information. For this purpose we require relatively weak, kinematical reflections rather than the strong reflections used in the $\mathbf{g} \cdot \mathbf{b} = 0$ diffraction-contrast technique. Any orientation away from strongly dynamical reflections may be used and indexed with the aid of one of the many versions of the simple PC-compatible computer programs in existence. Where a Bragg extinction line $\mathbf{g}$ intersects the dislocation line, a splitting occurs with $n = \mathbf{g} \cdot \mathbf{b}$ subsidiary maxima of intensity in direct-beam (bright-field) LACBED patterns or minima in dark-field patterns (Cherns and Preston, 1989; Tanaka et al., 1988). In principle, only two intersections with linearly independent reflections $\mathbf{g}$ are required to determine the direction of $\mathbf{b}$, and three to determine the magnitude. The sense in which the splitting occurs fixes the sign of $\mathbf{b}$. Note that all reflections $\mathbf{g}$ provide information in this case, and it is not necessary to discover that subset of reflections for which the Bragg line is unsplit ($\mathbf{g} \cdot \mathbf{b} = 0$). The high density of high-order reflections in reciprocal space makes the experiment quick to perform, since only small adjustments of orientation are required to cause three distinct intersections between the dislocation and the Bragg lines. The basis of the method relies on the long-range topological discontinuity of the Burgers circuit around a dislocation line and is therefore independent of problems such as precipitation, anisotropic elasticity and dissociation (Bird and Preston, 1988). The chief limitations of the method are that the specimen should be in the thickness range 60 to 120 nm, it should be relatively flat, and with well-separated dislocations, more than 100 nm apart. Such limitations are not normally significant in the case of minerals. Where a fairly dense dislocation arrangement occurs, it is often found easier to work in dark field rather than bright field because the shadow images of the dislocations in the array are more readily discerned in this case. The electron dose given to the specimen while carrying out this operation is minimal, as will be illustrated by the following example (work of P. Cordier, Lille).

Rather few determinations of dislocation Burgers vectors have been performed in quartz. $\mathbf{a}$, $\mathbf{c}$, and $\mathbf{a} + \mathbf{c}$ dislocations are believed to exist, depending on the mode of deformation. However, the rate of radiation damage in quartz at room temperature is so rapid that it is hard to obtain the minimum of three diffraction contrast images in different

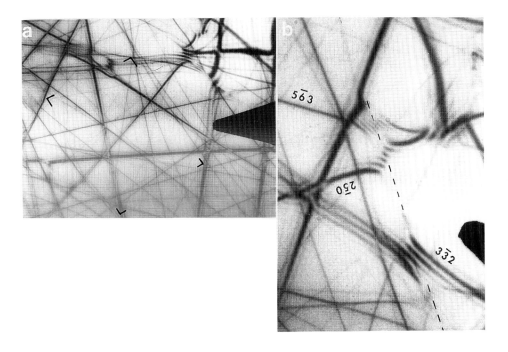

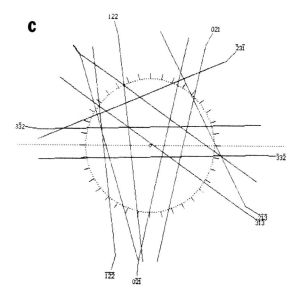

Figure 2-25. Burgers vector determination in quartz by LACBED: (a) shows a large angular view: the pointer (inserted at righthand side) indicates the position of the dislocation, (b) detail in the region indicated by the rectangle whose corners are superimposed on (a). The dislocation has been moved to intersect these Bragg contours, and its presence is indicated by the dashed line drawn on the figure. (c) Diagram indicating some of the reflections present in (a). (b) is taken close to the [516] zone axis. [Material for this illutration was kindly provided by Dr. P. Cordier].

76

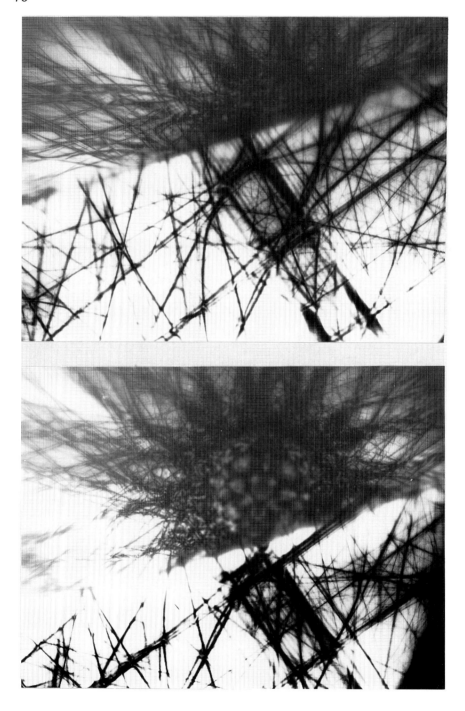

Figure 2-26. Interface boundary between garnet at ‹100› (above) and orthopyroxene below. Two slightly different orientations of the specimen are shown (specimen kindly provided by Prof. J.C. Donkhan).

reflections before destruction of the crystal occurs. In addition, radiation damage commences at the dislocation core, leading to ambiguity of the results (Cherns et al., 1980). By contrast, the task is readily performed by the LACBED technique (see Fig. 2-25). Ten individual LACBED images can be formed before any sign of radiation damage is detected. The preparation of these regions by ion milling produces some distortion of the LACBED pattern near the edge of the specimen, but not sufficient to complicate the analysis. A few of the results are given in the figure, and by analyzing these results it was concluded that the dislocation was of a type.

2.3.4   Interfacial studies

There are many reasons for wishing to study the interfaces between two dissimilar minerals such as epitaxy, relative displacement between unit cells, strain, and so on. LACBED may be applied to such studies effectively. It is normally advantageous to have the interfaces of interest aligned parallel or perpendicular to the incident electron beam by appropriate goniometry prior to commencement of the LACBED experiments.

Epitaxial relationships may be determined by the continuity of certain Bragg lines across interfaces that are parallel to the beam, and strain may be detected by the distortion of Bragg lines in the vicinity of such an interface. The relative displacement of unit cells on either side of the interface may be studied when the interface is aligned perpendicular to the beam (Al-Khafaji et al., 1991). It is possible to investigate the interfacial strain in this case (Cherns et al., 1991) by tilting through about 45°. Such studies are at a relatively early stage of development but could be explored in mineralogical investigations in the future. An example of their application is to the boundary between garnet and orthopyroxene (Fig. 2-26). The garnet was in the form of long whiskers about 3 μm in cross-section. The figure shows two LACBED patterns at the fourfold axis of the garnet and an undetermined axis of the orthopyroxene. A small additional rotation was performed between the two exposures. It is evident that such results contain a wealth of crystallographic information.

2.4   HIGHER ORDER LAUE ZONE LINES

2.4.1   Excess and deficiency lines

There has already been a good deal of reference to HOLZ diffraction in this chapter, and it has been used extensively in exercises of crystal symmetry determination. We now consider HOLZ diffraction in more detail. We take as our typical situation for more detailed analysis the case of a ZOLZ pattern composed of fairly well-spaced reflections, i.e., ~5 nm^{-1} apart rather than 1 nm^{-1}, and form a CBED pattern using a relatively large convergence angle so that the individual ZOLZ discs practically touch but do not overlap. We also envisage a FOLZ ring of reflection in a zone about 5 nm^{-1} away from the zero layer. For a specimen thickness giving well-defined HOLZ lines without excessive diffuse scattering we will probably be working in the range 80 to 150 nm. Under these circumstances the direct beam will be crossed by a mesh of well-defined straight or nearly straight dark HOLZ lines, while the FOLZ will be composed of a series of bright segments approximately tangential to the Laue circle in that the Ewald sphere intersects the FOLZ plane. Already some details have been deliberately ignored in the interests of simplicity. Some HOLZ lines may have hybridized near the point of intersection of a pair of lines. Under these circumstances one line of the hyperbola thus formed will be enhanced and the other diminished, relative to the line strengths away from the intersection. In addition, it is common to see complicated meshes of additional weaker line segments, often no longer

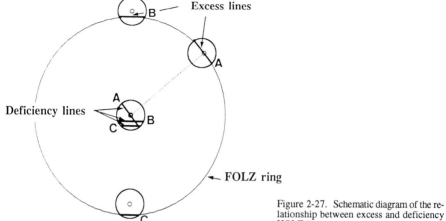

Excess lines

Deficiency lines

FOLZ ring

Figure 2-27. Schematic diagram of the relationship between excess and deficiency HOLZ lines.

straight, outside the innermost straight line of the FOLZ ring. This detail is referred to as HOLZ fine structure in §2.4.3. We ignore these finer points for the present.

The dark lines within the direct beam are called deficiency (or defect) HOLZ lines; the bright lines in the FOLZ ring (or in additional HOLZs, if present) are known as excess lines. The terms deficiency or excess refer to reduction or enhancement of intensity. The intensity initially present in the direct beam at the entrance surface of the specimen is Bragg reflected away to each of the reflections excited in the FOLZ ring for orientations corresponding to the position of the dark line. The length, orientation, and position in the direct-beam CBED disc of each deficiency line is exactly matched by the corresponding excess line in each FOLZ reflection (Fig. 2-27).

Usually, not much attention is given to the HOLZ lines that may sometimes be observed in the ZOLZ reflections clustered round the direct beam. One exception is where it is required to distinguish the bright-field and whole-pattern symmetries. The HOLZ lines in these reflections are generally weaker than those in the direct beam and are also deficiency lines resulting from diffraction out to the HOLZs. One important exception is the case, as in ⟨001⟩ diffraction from spinel (see §2.2.4), where the ZOLZ reflection is kinematically forbidden and only comes about by double diffraction, first out to the FOLZ and then back to the forbidden reflection in the ZOLZ. In this case there would not otherwise be intensity present, and so the ZOLZ lines are necessarily of excess character.

2.4.2   HOLZ line indexing and their pseudo-kinematic character

Although we shall revise what is written in this section in the final sections of this chapter in order to have an accurate understanding of HOLZ diffraction, we shall for the moment regard HOLZ lines as being the result of kinematical diffraction. At first sight this is an eminently reasonable approximation. When we calculate the two-beam extinction length from the FOLZ-reflection structure factors according to Equation 2.7b, the result turns out much greater than the specimen thicknesses for which we normally perform our experiments, indicating that something is wrong. Nevertheless, the approximation allows us to arrive at a simple and useful explanation for the observed HOLZ patterns and is widely used in their simulations. We shall concentrate for the present on the deficiency line pattern in the direct beam.

From the Ewald sphere construction we know that, in order to satisfy the Bragg condition for a particular FOLZ reflection $\mathbf{g}$, the wave in the direction of incidence $\underline{k}$ must satisfy the condition (Fig. 2-28a), $\mathbf{k'} = \mathbf{k} + \mathbf{g}$.

It is well-known that, for a given electron-beam energy, the locus of points C for all $\mathbf{k}$ that satisfy this condition is a circle described in the plane that is the perpendicular bisector of OG (called the Brillouin zone boundary)—see Figure 2-28b. An alternative and equivalent way of arriving at this conclusion is to draw spheres (called dispersion spheres) on O and G as origins with radius |k| (note the distinction from the Ewald sphere, that is drawn with C as origin). All possible directions of propagation are vectors CO originating on the surface of the dispersion sphere centered in O. That subset that diffracts into the reflection $\mathbf{g}$ lies in the rather large small-circle that is created by the intersection of the two spheres (Fig. 2-28c). This circle is the same circle as in Figure 2-28b.

Consider now another reflection $\mathbf{g'}$ corresponding to another HOLZ (not necessarily FOLZ) reflection G'. The intersection of the spheres on O and G' will generate another rather large small-circle, as indicated in Figure 2-28d. The radii of the two small-circles on the surface of the sphere on O, created by the spheres on G and G', will not generally be quite the same. Next, let us add all the other HOLZ reflections that are excited, to generate a host of small-circles on the central dispersion sphere. All the angles of incidence in a particular convergent cone of a beam focussed on the specimen can also be represented on the surface of the central dispersion sphere, this time as a very small circle, with center the zone axis direction. This small circle is so small that it can safely be treated as planar. All those large small circles, created by the HOLZ dispersion sphere intersections that lie within the near planar CBED small circle, are the HOLZ lines. Computer programs based on the geometry described may be written and operated rapidly on a simple personal computer. The use of such computer programs is one common method for indexing HOLZ line patterns. However, it should be noted that the microscope operating voltage (that is required in order to determine |k|) must be regarded as a slightly variable parameter (±2% or so). The reason for this perhaps surprising statement is related to the kinematical approximation made at the outset of this section, that is not strictly accurate. A more accurate approximation is discussed briefly in the next section.

There exists a completely different and in some ways much easier way to index HOLZ deficiency lines based on the situation described by Figure 2-28. The centers of the individual ZOLZ reflections may be regarded as the nodes of a two-dimension net that may be continued out to the FOLZ ring. Any chosen FOLZ reflection may be reached by an integral number of steps $n_1\mathbf{g}_1 + n_2\mathbf{g}_2$, where n_1 and n_2 are integers and $\mathbf{g}_1$ and $\mathbf{g}_2$ are the primitive translations of the ZOLZ net. Once the closest node in the ZOLZ net to the FOLZ reflection is reached, the final step is to add an out-of-plane vector $\mathbf{g}_3$ to reach the chosen reflection (if it is not known already, $\mathbf{g}_3$ can be deduced from measurements made on the pattern, making use of Equation (2.4). The indices of the FOLZ reflection are therefore $n_1\mathbf{g}_1 + n_2\mathbf{g}_2 + \mathbf{g}_3$ (Fig. 2-29).

2.4.3 Dynamical diffraction theory, HOLZ lines revisited and sub-lattice selection

The assumption made in the last section, of the geometrical construction based on HOLZ dispersion spheres of radius K, is valid to a very good degree of approximation.

80

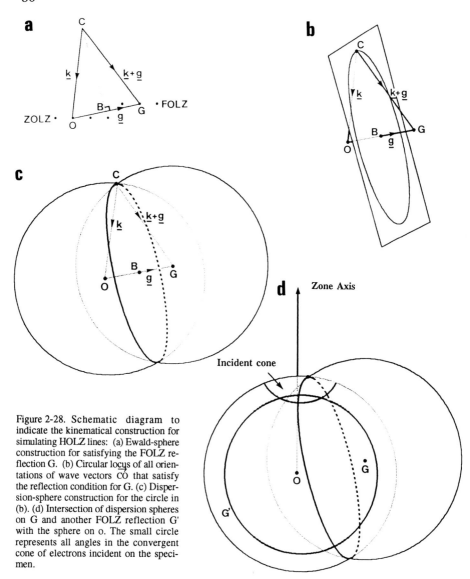

Figure 2-28. Schematic diagram to indicate the kinematical construction for simulating HOLZ lines: (a) Ewald-sphere construction for satisfying the FOLZ reflection G. (b) Circular locus of all orientations of wave vectors CO that satisfy the reflection condition for G. (c) Dispersion-sphere construction for the circle in (b). (d) Intersection of dispersion spheres on G and another FOLZ reflection G' with the sphere on o. The small circle represents all angles in the convergent cone of electrons incident on the specimen.

The shortcoming of this simple approach is that of a dispersion sphere on the origin. This sphere may not only be used to describe the direction of propagation of the incoming electrons; it also has another significance. Since the electron energy, E, is given by the well-known formula,

$$E = \frac{\hbar^2 k^2}{2m},$$

the radius of this sphere is also related to the incident electron energy. Within the crystal a more complicated situation exists. The high-energy electrons are subject to elastic interactions with the atomic potentials, that can affect the kinetic energy given above by a potential energy contribution. Physically speaking, ZOLZ diffraction gives rise to complicated beating effects between incident and diffracted waves as they propagate through the

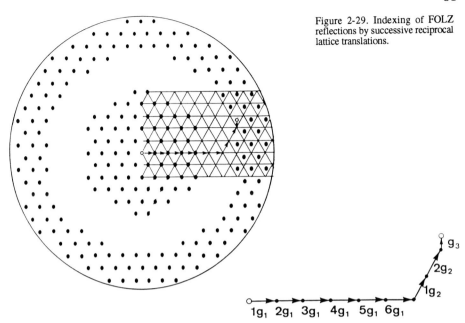

Figure 2-29. Indexing of FOLZ reflections by successive reciprocal lattice translations.

crystal. The electrons are no longer smoothly distributed within the illuminated region but acquire modulations on the atomic scale as a result of nodes and antinodes formed by the beating effects. These modulations have the periodicity of the projected atomic potential and are known as Bloch waves. One solution, corresponding to antinodes in the positions of the projected atomic positions, has particularly low potential energy (and therefore high kinetic energy) as a result: this state is conventionally labelled (1) and is called Bloch wave (1). Another, having nodes on the atomic positions is kept well away from the low potential regions and therefore has lower kinetic energy: this is called Bloch wave (2). In general there may be several important Bloch waves but only rather few for zone-axis incidence.

To express it another way, what we have left out of the story in §2.4.2 is the dispersion spheres centered on the ZOLZ reflections neighboring the direct beam. It is the inclusions of these ZOLZ reflections that gives rise to the beating effects and the creation of Bloch waves described in the previous paragraph. As the situation becomes geometrically complicated and difficult to draw if we include a two-dimension arrangement of reflections in the ZOLZ, let us simply consider a line of reflections that includes the direct beam. The result of a detailed dynamical calculation shows that the individual HOLZ spheres centered on each of the reflections in the line is converted into a series of separated constant-energy surfaces called branches of a dispersion surface. The situation is illustrated schematically in Figure 2-30. The relationship between the two constructions is evident, but gaps are introduced in Figure 2-30b relative to the construction in Figure 2-30a. In the rectangular box, the intersection between the circles (spheres in three dimensions) on O and G is split into hyperbolic sections that are asymptotic to the spheres in Figure 2-30a. This splitting is the result that may be familiar from the two-beam theory mentioned in §2.1.7. The magnitude of k in Figure 2-30a becomes one of several values $k^{(1)}$ $k^{(2)}$, etc. in Figure 2-30b, where $k^{(1)} > k > k^{(2)}$. Branch (1) of this dispersion surface is related to Bloch waves (1), and they have low potential energy, as mentioned above, and high kinetic energy because $k^{(1)}$ is large. Likewise branch (2), is related to Bloch waves (2), with high

82

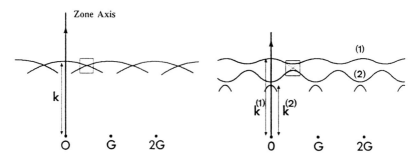

Figure 2-30. Diagram to indicate the creation of a dispersion surface (b) for systematic diffraction (a single reciprocal lattice now), from dispersion spheres (a) centered on each of the reciprocal-lattice points in the row. The incident electrons have wave number k but acquire wave vectors $k^{(1)}$ and $k^{(2)}$ as a result of diffraction.

potential energy and low kinetic energy ($k^{(2)} < k$). There exists a forbidden range of values between branches (1) and (2) of the dispersion surface.

When a dynamical diffraction calculation is performed for the whole ZOLZ net of excited reflections, branch (1) becomes flat, almost planar, but branch (2) and higher integer branches have oscillations of increasing amplitude. We are left, therefore, with a revision of the picture described in the last section. Instead of a dispersion sphere on the origin we should, in reality, consider the intersection of HOLZ spheres with branch (1) and a few other branches of the dispersion surface. Branch (1), that is almost planer, lies just outside the rather flat dispersion sphere on the origin. HOLZ sphere intersection will be displaced slightly on account of this difference. It is as though the sphere on the origin had a slightly larger radius, that is, the electrons have a slightly higher energy. It is for this reason that the operating voltage must be treated as a variable in the kinematic program. Sometimes branch (1) intersections are strongly absorbed, since the electrons are highly concentrated on the projected atom position. Then the strongest lines to be observed are branch (2) lines, and these lie just inside the dispersion sphere on the origin. For these, an artificially reduced operating voltage would have to be used for good fit using a kinematical simulation program. Intersections with higher integer branches give weaker lines, and these are the origin of the fine structure observed in excess HOLZ lines outside the HOLZ ring, as discussed in §2.4.1.

The different distribution of the electrons associated with each of the lines in the HOLZ fine structure means that it may be possible for different Bloch waves to be associated with different sub-lattices of a complex crystal structure. This is the basis of sub-lattice selection, that can be turned into a powerful technique for crystal structure determination (as introduced in the final section of this chapter), in ALCHEMI investi-gations (see §5.2), and in investigations of point disorder (see §2.3.1).

2.4.4 *Ab initio* crystal-structure determination

This topic is introduced more to give an indication of the current state of development of this aspect of electron diffraction than because it is likely to be of direct interest to many readers. The theory of high-energy electron diffraction, although more complex than that of x-ray diffraction, is now really rather well understood, and accurate computations based on reliable computer codes are now undertaken routinely in a large number of laboratories throughout the world. However, with a non-linear theory, it is one thing to compute the anticipated results from a particular model, but quite another to be able to deduce an atomic arrangement directly from experimental results. Nevertheless, significant developments are under way on this inverse problem. In particular, Vincent and his co-workers have been making use of the fine structure information in HOLZ reflections to arrive at quite remarkably accurate solutions for unknown structures.

The basic principle of the approach is that of the pseudo-kinematical HOLZ approximation with sub-lattice selection via the separate Bloch states (Bird, 1989). By obtaining CBED patterns at a range of operating voltages in the range 100-300 keV, some hundreds of HOLZ reflections can be obtained in one or more zones (Vincent and Exelby, 1991). Each reflection may have two or more fine-structure lines. The next step is to generate so-called conditional Patterson maps from each line of the fine structure. This method parallels the classical x-ray technique, except that the data set is limited to a HOLZ (hence the term conditional). From a knowledge of the chemistry of the phase under investigation (obtained by EDX or EELS, see Chapters 4 and 5), and from the properties of the Bloch states it is possible to deduce inter-atomic separations from the conditional Patterson maps. Because the reflections used in this operation represent large-angle scattering, accurate interatomic separations can be deduced. For example, we know that the structure factor F_g for a HOLZ reflection g takes the form

$$F_g = \sum_{\substack{j \\ \text{unit} \\ \text{cell}}} f_j \exp[2\pi i g \cdot (r_j + \Delta r_j)] \, ,$$

where f_j is the j atom scattering form factor. Since g is large, any uncertainty, Δr_j in the atomic position r_j has a large effect on the magnitude of F_g. Atomic positions have been fixed by this technique to an accuracy of about one part in 1,000 (Vincent et al., 1984).

REFERENCES

Al-Khafaji, M., Cherns D., Rossouw C.J. and Woolf D.A. (1992) Determination of rigid shift across the Al (001) GaAs (001) interface. Phil. Mag., in the press.

Baba Kishi, K.Z. and Dingley, D.J. (1989) Backscatter Kikuchi diffraction in the SEM for identification of crystallographic point groups. Scanning 11, 305-312.

Bird, D.M. (1989) Theory of zone axis electron diffraction. J. Electron Microsc. Tech. 13, 77-97.

Bird, D.M. and Preston, A.R. (1988) Observation of Berry's geometrical phase in electron diffraction from a screw dislocation. Phys. Rev. Lett. 61, 2863-2866.

Buxton, B.F., Eades, J.A., Steeds, J.W. and Rackham, G.M. (1976) The symmetry of electron diffraction zone axis patterns. Phil. Trans. Royal Soc. London A281, 171-194.

Champness, P.E. (1987) Convergent beam electron diffraction. Mineral. Mag. 51, 33-48.

Cherns, D. (1991) Convergent beam diffraction and microscopy of single and multiple quantum-well structures. Microsc. Semicond. Mater. Conf., Inst. Phys. Conf. Ser. No. 117, 549-558.

Cherns, D., Jenkins, M.L. and White, S. (1980) The structure of dislocations in quartz under electron irradiation. Inst. Phys. Conf. Ser. 52, 121-124.

Cherns, D. and Preston, A.R. (1989) Convergent beam diffraction studies of interfaces, defects, and multi-layers. J. Electron Microsc. Tech. 13, 111-122.

Cherns, D., Touaitia, R., Preston, A.R., Rossouw, C.J., and Houghton, D.C. (1991) Convergent beam electron diffraction studies of strain in Si/SiGe superlattices. Phil. Mag. 64A, 597-612.

Coates, D.G. (1967) Kikuchi-like reflection patterns obtained with the scanning electron microscope. Phil. Mag. 16, 1179-1184.

Dingley, D.J. (1984) Diffraction from sub-micron areas using electron backscattering in a scanning electron microscope. Scanning Elect. Microsc. II, 569-575.

Dingley, D.J. and Randle, V. (1992) Microtexture determination by electron backscatter diffraction. J. Mat. Sci., in the press.

Eades, J.A .(1980) Zone-axis patterns formed by a new double-rocking technique. Ultramicroscopy 5, 71-74.

Gjønnes, J. and Moodie, A.F. (1965) Extinction conditions in the dynamical theory of electron diffraction. Acta Crystallogr. 19, 65-67.

Hirsch, P.B., Howie, A., Nicholson, R.B., Pashley, D.W. and Whelan, M.J. (1965) Electron Microscopy of Thin Crystals. Butterworth, London (revised ed., 1977, Krieger, New York).

Høier, R. (1973) Multiple scattering and dynamical effects in diffuse electron scattering. Acta Crystallogr. A29, 663-672.

Humphreys, C.J. (1979) The scattering of fast electrons by crystals. Rep. Prog. Phys. 42, 1825-1887.

Jesson, D.E. and Steeds, J.W. (1989) High energy electron diffraction from transverse stacking faults in the projection approximation. Ultramicroscopy 31, 399-430.

Jesson, D.E. and Steeds, J.W. (1990) Higher-order Laue zone diffraction from crystals containing transverse stacking faults. Phil. Mag. 61A, 385-415.

Joy, D.C. (1974) In Quantitative Scanning Electron Microscopy. D.B. Holt, M.D. Muir, P.R. Grant and I. M. Boswarva, eds. Academic Press, London, 131-182.

Möllenstedt, G. (1989) My early work on convergent beam electron diffraction. Phys. Stat. Sol. (a) 116, 13-22.

Shannon, M.D. and Eades, J.A.(1978) Information from rutile catalyst particles by convergent beam electron diffraction. Electron Diffraction 1927-1977, Inst. Phys. Conf. Ser. No. 41, 411-415.

Spellward, P. and James, R. (1991) The use of electron diffraction in the determination of the polarity of II-VI and III-V semiconductors. EMAG 91, Inst. Phys. Conf. Ser. No. 119, 375-378.

Steeds, J.W. (1979) Introduction to Analytical Electron Microscopy. J.J. Hren, J.I. Goldstein and D.C. Joy, eds. Plenum Press, New York and London, 387-422.

Steeds, J.W. (1983) Further developments in the analysis of CBED data. EMAG 83, Inst. Phys. Conf. Ser. No. 68, 31-36.

Steeds, J.W. (1984) In Scottish Universities Summer School in Physics, J.N. Chapman and A.J. Craven, eds. 49-93.

Steeds, J.W., Baker, J.R. and Vincent, R. (1982) Crystallographic data from zone axis patterns. Proc. Electron Microscopy 1932-1982, Deutsche Ges. Elektronenmikroskopie, Hamburg, 617-624.

Steeds, J.W. and Eades, J.A. (1973) Crystallography and defects in the epitaxial growth of titanium on mica. Surface Science 38, 187-196.

Steeds, J.W. and Vincent, R. (1983) Use of high-symmetry zone axes in electron diffraction in determining crystal point and space groups. J. Appl. Crystallogr. 16, 317-324.

Tanaka, M., Saito, R., Ueno, K. and Harada, Y. (1980) Large angle convergent beam electron diffraction. J. Electron Microsc. 29, 408ff.

Tanaka, M., Terauchi, M. and Kaneyama, T. (1988) Convergent Beam Electron Diffraction II. JEOL Ltd., Tokyo.

Treacy, M.M.J., Newsam, J.M. and Deem, M.W. (1991) A general recursion method for calculating diffracted intensities from crystals containing planar faults. Proc. Royal Soc. London A433, 499-520.

Venables, J.A. and Harland, C.J. (1973) Electron back-scattering patterns—a new technique for obtaining crystallographic information in the scanning electron microscope. Phil. Mag. 27, 1193-1200.

Vincent, R., Bird, D.M. and Steeds, J.W. (1984) Structure of AuGeAs determined by CBED. Phil. Mag. 50A, 765-786.

Vincent, R. and Exelby, D.R. (1991) Structure of metastable Al-Ge phases determined from HOLZ Patterson transforms. Phil. Mag. Lett. 63, 31-38.

Vincent, R., Krause, B. and Steeds, J.W. (1986) Determination of crystal chirality by electron diffraction. Prox. XIth Int'l Cong. Electron Micros., Kyoto, 695-696.

Vincent, R. and Pretty, S.F. (1986) Phase analysis in the Ni-Ge-P system by electron diffraction. Phil. Mag. 53A, 843-862.

Vincent, R. and Withers, R.L. (1987) Analysis of a displacive superlattice in nickel arsenide, Phil. Mag. Lett. 56, 57-62.

Welberry, T.R. and Withers, R.L. (1987) Optical transforms of disordered systems displaying intensity loci. J. Appl. Crystallogr. 20, 280-288.

Welberry, T.R. and Withers, R.L. (1991) The role of phase in diffuse diffraction patterns and its effect on real-space structure. J. Appl. Crystallogr. 24, 18-29.

Wilkinson, A.J. and Dingley, D.J. (1991) Quantitative deformation studies using electron backscatter patterns. Acta Metall. Mater. 39, 3047-3055.

Chapter 3. HIGH-RESOLUTION IMAGE SIMULATION AND ANALYSIS

Peter Self CSIRO, Private Bag No. 2, Glen Osmond,

South Australia 5064, Australia

3.1   INTRODUCTION

3.1.1   Why and when

The transmission electron microscope is a complex and subtle (but always beautiful) beast. Chapter 1 gives an outline of the processes involved in image formation in the TEM. That chapter demonstrates how the microscope optics and dynamical scattering in the specimen conspire to "distort" the image, so that features within the image are not necessarily directly interpretable as true structural features of the specimen. An example of this conspiracy is given in Figure 3-1 which shows, in a general way, the effects of objective lens focus and specimen thickness on high-resolution transmission electron microscopy (HRTEM) images of silicon. Many of the images shown in Figure 3-1 can be directly correlated with the structure, but without an *a priori* knowledge of the silicon structure the variability of the images makes it difficult (if not impossible) to deduce anything definite about atomic structure.

Fortunately for microscopists, the theory of dynamical electron diffraction and image formation is built on a sound foundation, and it is a fairly straight forward matter to simulate the images that can be expected when viewing a particular structure. However, the computation of HRTEM images can be time consuming, and so it is worthwhile knowing when it is necessary to calculate images. For specimens with periodic (i.e., crystalline) structure it is necessary to calculate images when seeking detail at resolutions smaller than the unit cell. For aperiodic objects it is necessary to carry out some computational analysis in every case. Even fairly large-scale aperiodic objects, such as line and plane defects examined at relatively low resolution, require image computation to quantify their nature. Indeed, much of the pioneering work on the computation of transmission electron microscopy (TEM) images was to quantify the nature of macroscopic defects (Head et al., 1973).

These sweeping generalizations on the necessity to calculate HRTEM images warrant some form of explanation. After all, it is every electron microscopist's ardent desire to obtain insights into atomic arrangements directly from a small number (preferably only one) of electron micrographs. The computation of images is considered to be a nuisance, and many scientific articles have devoted considerable space to arguments as to why image calculation was not necessary in a particular case. For periodic structures the need to compute can be explained, not by considering the image, but rather the diffraction pattern. From a diffraction pattern it is possible to measure unit cell geometry and scattered amplitude, but it is not possible to measure the phase of the scattered waves. (For TEM these waves are electrons). This inability to measure phase is known as the "phase problem" (Cowley, 1981, p. 126), and virtually the whole field of structure analysis by electron, x-ray, and neutron diffraction has been devoted to finding ways to solve this problem. As explained in Chapter 1, HRTEM images are related to the diffraction pattern by a Fourier transform (FT). One of the properties of a FT is that a phase shift in reciprocal (diffraction) space causes a position shift in real space. Hence, it is the relative phases of the diffracted waves that determine the positions of features in the image. Thus,

86

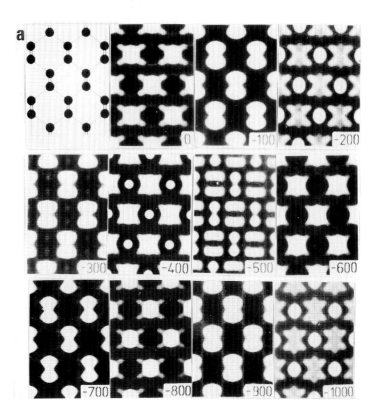

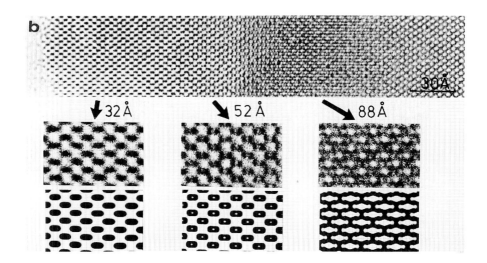

Figure 3-1. HRTEM images of [110] silicon showing (a) effect of defocus and (b) effect of crystal thickness. In a) computed images at various values of defocus (in Å) are shown for a crystal of constant thickness. (A projection of the atomic structure is shown in the top left corner.) The upper section of (b) shows a HRTEM image of a wedge-shaped crystal where thickness increases from left to right. In the lower section of (b), matches between experimental and computed images are shown for three thicknesses (in Å). (Courtesy of R.W. Glaisher.)

in the top and bottom row of Figure 3-1a, images of the same form can be seen but shifted relative to a fixed origin; this shift is a manifestation of a change of relative phase of the diffracted beams. The effect of dynamical diffraction, lens focus, and lens aberration is to change the phases of the scattered waves to such an extent that they no longer correspond to the phases of the FT of the structure (cf. §1.3.1 and §1.3.3). Hence, in a HRTEM image, features at the sub-unit cell level are shifted relative to each other so that distances between features do not necessarily represent true inter-atomic spacings. The same argument is applicable to aperiodic structures with the added complication that, because these structures do not form a spot diffraction pattern, not even geometry can be readily determined.

3.1.2 How

Having established that in order to get the maximum information from HRTEM images some form of computation is required, it is necessary to ask "what is the best way to go about this image computation process?" The simplest way is to buy one of the commercial packages for the computation of HRTEM images and diffraction patterns. Examples of these packages are SHIRLI (O'Keefe et al., 1978) and EMS (Stadelmann, 1987). These packages have several advantages including that they are well tested and therefore give reliable results. They also save many hours of software engineering for which there is little kudos as the engineering involves programming known formulae only. The disadvantages of the packages include price and that they can require specialized hardware. Perhaps their biggest disadvantage is that, because of their "black box" nature, they do not necessarily give the user insight into the processes involved in HRTEM. Even if the microscope user does not want to get into the detailed processes, the holistic microscopist will want to have some knowledge of these processes. It is the aim of this chapter to provide this knowledge or at least give reference to places where it can be found.

3.2 METHODOLOGY FOR IMAGE MATCHING

3.2.1 What is being modelled?

The old adage "rubbish in, rubbish out" is often applied to computational systems, and the calculation of HRTEM images and diffraction patterns is no exception. To calculate images and diffraction patterns it is necessary to know about both the microscope and the specimen. Figure 1-1 shows a detailed ray diagram for a transmission electron micro-scope. For the purposes of computation this is simplified to the form shown in Figure 3-2.

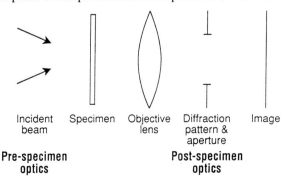

| Incident beam | Specimen | Objective lens | Diffraction pattern & aperture | Image |

Pre-specimen optics   **Post-specimen optics**

Figure 3-2. Simplified microscope used for HRTEM image and diffraction amplitude calculations. The microscope is considered in three parts: pre-specimen, specimen, and post-specimen.

Table 3-1. Parameters that need to be specified for the computation of HRTEM images and diffraction patterns.

Pre-specimen:

> Beam-electron energy (i.e., accelerating voltage)
> Spread in beam-electron energy
> Convergence (i.e., angular range of the incident electron beam)
> Coherence (i.e., phase relations, if any, between different parts of the incident beam)
> Angle of incident beam relative to optic axis of post-specimen (objective) lens

Specimen:

> Unit cell
> Space group
> Unit group (i.e., atom positions)
> Site occupancies
> Atomic scattering factors for relativistic electrons
> Debye-Waller factor (thermal vibration)
> Absorption
> Orientation relative to incident beam
> Thickness

Post-specimen:

> Defocus
> Coefficient of spherical aberration (C_S)
> Coefficient of chromatic aberration (C_C)
> Diameter of diffraction aperture
> Astigmatism.

This simplified form breaks the process into three parts: (1) pre-specimen (electron gun and condenser lenses), (2) specimen (dynamical diffraction), and (3) post-specimen (in particular the objective lens). The information that needs to be known about these three parts to simulate HRTEM images and diffraction patterns is shown in Table 3-1.

The astute reader will realise that if all the parameters in Table 3-1 are known, there is little point in carrying out any experiment. The purpose of experiments is to ascertain those parameters in Table 3-1 that are unknown. Normally, the unknown parameters will be those relating to the specimen such as atomic positions (usually a sub-set of the unit group) or site occupancies. In some cases (e.g., when calibrating a microscope), a "known" specimen will be used and parameters such as the coefficient of chromatic aberration (C_C) and the coefficient of spherical aberration (C_S) (see Chapter 1) will be determined. The aim of calculation is to facilitate the process of determining the unknown parameters.

The methodology for image matching is to ascertain values for as many parameters listed in Table 3-1 as possible and then hypothesize values for the remainder. Calculations are then carried out with the hypothesized values and an attempt is made to match a range of computed images to a set of experimental images (usually a through-focus series as detailed in §1.3.6). If there is a match, then the hypothesized values are assumed to be correct. If there is no match, the hypothesized values are modified (either with foresight or at random) and the process repeated until there is a match.

Sources for the values of the parameters listed in Table 3-1 are varied. Values for electron energy, energy spread, C_S, and C_C can be taken from manufacturers values or determined experimentally using techniques such as diffractogram analysis (Spence, 1981).

The values of these parameters are generally refined over a period of time after a series of image-matching experiments.

In general, HRTEM is not used to determine space group and unit cell parameters. These are best determined by selected-area electron diffraction and convergent-beam electron diffraction (see Chapter 2). Additional information from x-ray or neutron diffraction on unit cell parameters, space group, unit group (i.e. atom positions within the unit cell), and site occupancies is a great help to HRTEM experiments. Site occupancies can also be determined by ALCHEMI (see Chapter 5).

Beam convergence, crystallographic orientation, and diffraction aperture size can be determined experimentally at the time of recording the HRTEM images. The effects of astigmatism and misalignment of the electron beam relative to the optic axis should be eliminated by careful experimentation (Saxton et al., 1983b).

Values for atomic scattering factors (i.e., the electron scattering power of an atom as a function of scattering angle) are readily available in tabulated form (e.g., *International Tables for X-ray Crystallography,* Vol. IV, 1974) or, more conveniently for computation, parameterised as the sum of Gaussians (e.g., Doyle and Turner, 1968; Fox et al., 1989). Debye-Waller factors account for the attenuation of electron scattering strength caused by thermal vibration of the atoms in a solid. Debye-Waller factors are also available in tabulated form (e.g., *International Tables for X-ray Crystallography,* Vol. III, 1974). In general, HRTEM images and diffraction patterns are fairly insensitive to small changes in scattering factor and Debye-Waller factor values. For example, attempts to determine atomic ionicities by HRTEM imaging have, for the most part, been fairly inconclusive. Hence, when setting up a HRTEM calculation, although it is necessary to obtain good values for atomic scattering and Debye-Waller factors, it is not necessary to go to great lengths to obtain highly accurate values for these parameters (i.e., the tabulated values are usually sufficient).

For the purposes of calculation, absorption is usually treated as an extra, imaginary term in the structure factor. The traditional value used for the extra term has been a percentage (between 1% and 10%) of the structure factor. Although this practice was shown some time ago to be incorrect (Humphreys and Hirsch, 1968) it has persisted. Approximating absorption by a percentage of the structure factor tends to over-estimate absorption between the atomic sites and under-estimate absorption at the atomic sites. More realistic values for absorption are now readily available in the literature (e.g., Bird, 1990; Bird and King, 1990; Weickenmeier and Kohl, 1991).

3.2.2 No cheating!

There remain parameters in Table 3-1 that vary from specimen to specimen and must be determined for each set of micrographs. These parameters include specimen thickness, objective-lens defocus, and coherence. With these free parameters it is tempting to choose values that give the best match without worrying whether the values are realistic. For example, a through-focus series of images is generally taken at fixed defocus steps and any matching procedure must recognise this. Image matches can only be considered to be correct if the defocus step size in computed images matching an experimental series is constant. To assist this process, the step size of the microscope objective-lens focus control should be calibrated.

Another factor that must not be ignored when matching images is image contrast.

Experimental electron micrographs in a through-focus series are taken under the same illumination conditions and developed under the same chemical conditions. This means that film sensitivity will be unchanged for all micrographs in the through-focus series. On the other hand, the equivalent of film sensitivity (i.e., contrast) for computed images can be adjusted from image to image, and there is a temptation to falsely boost the contrast of a computed image when it has features (albeit weak) that are seen in an experimental image. A true match of computed and experimental images is only achieved if the parameters used to set image contrast in the computed images are kept constant.

Another poor practice in image matching is using unrealistic values for structural parameters. This includes the use of negative Debye-Waller factors, site occupancies greater than 1, or excessive absorption. Similarly, even though specimen thickness is one of the hardest parameters to estimate, effort should be made to obtain a realistic value for specimen thickness. Specimen thickness can be estimated by a number of techniques including the use of thickness fringes or CBED patterns (e.g., see Champness, 1987).

3.3 COMPUTATION OF DIFFRACTED AMPLITUDES

3.3.1 Preliminaries

There are several methods of calculating the scattered amplitudes for fast electrons (Goodman and Moodie, 1974). Only two are in common use: the multislice method and Bloch wave (eigenstate) method. The reasons why these two methods predominate are that the multislice method is computationally fast (especially where large numbers of beams are involved) while the Bloch wave method gives good insight into electron-diffraction processes.

Before summarising these methods three points need to be stressed: (1) all computational methods applied to HRTEM are for steady-state conditions, (2) the computations require adherence to a strict sign convention, and (3) the effect of back-scattering is neglected.

Sometimes the progressive building up of the scattered amplitudes as a function of specimen thickness is confused with the progress of an electron particle through the specimen. This is not correct as the formulations of high-energy electron diffraction describe solutions to Schrödinger's wave equation in the equation's time-independent form. The building up of scattered amplitudes with specimen thickness is only a mathematical tool akin to the numerical integration of a function by summing areas under a curve.

As HRTEM is a wave phenomenon, sign conventions for forward and backward travelling waves must be defined. The sign assigned to the waves in turn determines the sign used in calculating structure factors, the phase introduced by the objective lens and even the form of the Fourier transforms used to transform between real and reciprocal space and vice-versa. The correct usage of sign for high-energy electron wave functions is given by Saxton et al. (1983a) and Buseck et al. (1988, pp. xvii-xviii). Another consequence of the wave nature of electron diffraction is that it is common to find terms such as electron wavefunction, wavefield, and scattered amplitudes used interchangeably.

Back-scattering can be neglected because of the high energy of the electrons used in HRTEM; for low-energy electron diffraction (LEED), backscattering processes cannot be neglected and a different formulation of electron scattering is used for LEED than HRTEM. A consequence of neglecting back-scattering is that computational methods can build up the scattered amplitude of high-energy electrons by starting at the entrance surface (i.e., where the beam electrons go into the specimen) and working through to the exit surface (i.e., where the electrons come out of the specimen); the absence of back-scattering means that any scattering at a given depth in the specimen has no effect on what has occurred closer to the entrance surface.

3.3.2   What is it?

When a material is examined by HRTEM what is it that is seen in the image? Electrons carry an electric charge and they interact with the electrostatic potential of the sample. It is the electrostatic potential of the sample that determines the form of electron diffraction patterns and HRTEM images. The starting point of all calculations of scattered amplitudes for HRTEM is the evaluation of the electrostatic potential of the sample. This potential is the total of the electrostatic potential for all the nucleii (positively charged) and all the electrons (negatively charged) in the sample.

Even for the simplest of structures, the electrostatic potential has a fairly complex form and so, to model the potential, it is preferrable to describe it in the simplest way possible. For the calculation of electron diffraction amplitudes, the simplicity is achieved by describing the electrostatic potential as a Fourier series and thereby working in reciprocal space where, for crystals at least, it is only necessary to consider scattering to the reciprocal lattice points. The Fourier coefficients of the electrostatic potential are linearly related to the electron structure factors for the sample, F(hkl), by the formula,

$$\Phi(hkl) = h^2 F(hkl)/2\pi m_0 e V_c, \tag{3.1}$$

where h is Planck's constant, m_0 the rest mass of the electron, e the electron charge, and V_c the unit-cell volume. If V_c is evaluated in units of nm^3, then $\Phi(hkl)$ is given as $0.047878 F(hkl)/V_c$ volts. Values of $\Phi(000)$ range from around +5V for specimens containing only light elements to around +30V for specimens containing mostly heavy elements.

The structure factor represents the electron scattering strength of a unit cell and for the hkl reciprocal lattice point is defined as

$$F(hkl) = \sum_j {}^e f_j(s_{hkl}) \exp(-B_j s_{hkl}^2) \exp[+2\pi i(g_{hkl} \cdot r_j)] \tag{3.2}$$

The sum over j indicates a sum over all atoms in the unit cell, with the fractional coordinates of the jth atom described by the vector r_j. Other parameters in the formula for the structure factors are g_{hkl}, the reciprocal lattice vector with indices hkl, and B_j, the Debye-Waller factor for the jth atom. The term s_{hkl} defines the scattering angle from the undiffracted beam to the reciprocal lattice point g_{hkl} and is given by the formula

$$s_{hkl} = \sin\theta_{hkl}/\lambda = 1/2|g_{hkl}| \tag{3.3}$$

where θ_{hkl} is the Bragg angle of the hkl reflection and λ is the electron wavelength. Finally, ${}^e f_j(s_{hkl})$ is the electron scattering factor of the jth atom at the Bragg angle for the hkl reflection.

Values for the electron scattering factors and other terms required to evaluate the electrostatic potential can be obtained from the sources listed in §3.2.1. Although the formulas (3.1) and (3.2) are not overly complicated, their evaluation requires knowledge of the majority of the specimen parameters listed in Table 3-1. The evaluation of these formulas is the most exacting task in any calculation of electron-diffraction amplitudes.

3.3.3   Multislice

The multislice method of evaluating scattered amplitudes for HRTEM is an approximation to the exact formulation of electron diffraction given by the physical-optics approach of Cowley and Moodie (Cowley and Moodie, 1957; Buseck et al., 1988, Chapters 1-4). This formulation, and therefore the multislice method, is both elegant and simple. The physical-optics approach uses Huygen's principle, Fresnel diffraction and the weak-phase-object approximation (see Chapter 1, §1.4.2) to successively evaluate the electron wavefunction starting from the entrance surface of the sample through to the exit surface. This approach is essentially an integration of the effects of electron scattering over the thickness of the sample. The multislice method approximates the physical optics approach by using a sum over thickness steps of finite (but necessarily small) size instead of the more rigourous integration. A detailed explanation of the methodology of the numerical evaluation of scattered amplitudes by the multislice method is given in Buseck et al. (1988, Chapter 8).

The multislice method evaluates the wavefield at the exit surface of a crystal of thickness H (in the incident beam direction) by considering the crystal as a number (m) of slices such that $\quad H = \sum_{i=1}^{m} \Delta z_n$,

where Δz_n is the thickness of the nth slice. The change in the electron wavefunction in traversing the distance from one slice to the next is given by the Fresnel propagation function (Buseck et al., 1988, §2.1.1). The change in wavefunction because of the electrostatic potential in the slice is given by the transmission function (Buseck et al., 1988, §2.1.2). Following Huygen's principle, the wavefunction after the nth slice, $\psi_n(xy)$, is given by

$$\psi_n(xy) = [\ \psi_{n-1}(xy) * p_n(xy)]\ q_n(xy)\ , \qquad (3.4)$$

where $q_n(xy)$ is the transmission function of the nth slice, and $p_n(xy)$ is the propagation function for the distance between the n-1 and nth slices. The symbol * represents convolution (Buseck et al., 1988, §2.1.1), and (xy) represents the two-dimensional, real-space coordinates perpendicular to the beam direction. The wavefunction at thickness H is given by the successive evaluation of (3.4) for m slices. The calculation is initiated by setting

$$\psi_1(xy) = \psi_0(xy)\ q_1(xy)\ , \qquad (3.5)$$

where $\psi_0(xy)$ is the incident wavefunction as determined by the pre-specimen optics. The simplest form of $\psi_0(xy)$ is a plane wave.

As the thickness of each slice is small (usually between 0.1 and 0.4 nm) the transmission function is given by the weak-phase object approximation. That is,

$$q_n(xy) = \exp[-i\sigma\phi_p(xy)\Delta z_n]\ , \qquad (3.6)$$

where $\phi_p(xy)$ is the nth-slice projected potential per unit length for a projection direction parallel to the incident-beam direction. Equation (3.6) derives from the change of wavelength experienced by the electrons as they pass through the electrostatic potential of the material. These changes in wavelength are manifest as changes in the relative phase of the electrons (e.g., Fig. 1-8). The essence of Equation (3.6) is that the phase change experienced by an electron passing through a material is proportional to the electrostatic potential of the material. The constant of proportionality, the interaction constant (σ), is given by the formula

$$\sigma = [2\pi m_0 e\lambda(1 + eE/m_0 c^2)]/h^2 \tag{3.7}$$

where λ is the free-space (relativistically corrected) electron wavelength for electrons of energy E, and c is the speed of light. The constants m_0, e , and h are as for Equation (3.1). The interaction constant describes the strength of the interaction between the beam electrons and the electrostatic potential of the sample. It includes corrections for the changes in the strength of the scattering processes that arise because of relativistic effects.

The projected potential can be obtained by integrating the potential of the sample over the distance Δz_n in the direction parallel to the incident beam (i.e., perpendicular to xy). Considerable computational advantage can be gained if the slice thicknesses are constant and (for crystalline specimens) equal to the repeat distance of the specimen parallel to the incident-beam direction. In this case, the projected potential is

$$\phi_p(xy) = \sum_{hk} \Phi(hk0)\exp[-2\pi i(hx/a+ky/b)] \tag{3.8}$$

assuming that the reciprocal lattice vector, $\mathbf{c}^*$, is parallel to the incident-beam direction and the reciprocal lattice vectors $\mathbf{a}^*$ and $\mathbf{b}^*$ are perpendicular to the beam direction.

In the majority of cases Equation (3.8) is an adequate description of the projected potential. Equation (3.8) can be used even when the slice thickness is smaller than the repeat distance, $\mathbf{c}$, provided there is no significant scattering to the first or higher-order Laue zone (Chapter 2). When scattering to diffraction points outside the zeroth-order Laue zone must be included (e.g., when the repeat distance parallel to the incident beam is large), the projected potential per unit length must be derived from the formula

$$\phi_p(xy) = \sum_{hkl} \Phi(hkl)[\sin(\pi l\Delta z/c)/\pi l\Delta z/c]\exp(2\pi ilz_0/c)\exp[-2\pi i(hx/a + ky/b)] \tag{3.9}$$

for a slice of thickness Δz centered on z_0. The calculation of HRTEM diffraction amplitudes for systems including higher-order Laue zones is discussed in Buseck et al. (1988, §8.3.1).

Absorption can be included as an imaginary component of the projected potential. The values for this imaginary component can be calculated from equations similar to (3.8) and (3.9) using structure factors derived from the tabulated values of the "absorption scattering factors" (§3.2.1). It should be noted that, although absorption removes electrons from a calculation, these "absorbed" electrons are not removed from experimental images but rather contribute to a background intensity that lowers the contrast in experimental images.

The Fourier coefficients of the propagation function for a distance Δz_n are given by

$$P_n(hk) = \exp[-2\pi i\zeta(hk)\Delta z_n] , \tag{3.10}$$

where $\zeta(hk)$ is the excitation error of the hk-reciprocal-lattice point. The excitation error is a scalar measure of the distance from a reciprocal lattice point to the Ewald sphere in a direction perpendicular to the specimen surface (Fig. 3-3). The propagation function records the change in the relative phase of the electron waves passing through the specimen at different angles.

The excitation error is a parameter common to all formulations of electron diffraction. The standard approximation used to evaluate the excitation error is that the plane of specimen surface is perpendicular to the incident-beam direction. In this case the formula for the excitation error of the hk-reciprocal lattice point is

$$\zeta(hk) = \{\lambda^{-2} - [(h-u)^2 a^{*2} + (k-v)^2 b^{*2} + 2(h-u)(k-v)a^* b^* \cos\beta^*]\}^{1/2} -$$

$$\{\lambda^{-2} - [u^2 a^{*2} + v^2 b^{*2} + 2uva^* b^* \cos\beta^*]\}^{1/2} \qquad (3.11)$$

where a^* and b^* are the lengths of the reciprocal lattice basis vectors, β^* is the angle between these vectors, and (uv) are the coordinates of the center of the Laue circle (Fig. 3-4). The formula for the excitation error as given in Equation (3.11) only describes the phase change relative to the undiffracted beam (cf. $\zeta(00) = 0$). Thus, multislice calculations based on (3.11) will not give absolute phases, but this is of little consequence as it is only the effects of relative phase that are observed in HRTEM images.

The Equations (3.1) through to (3.11) provide the complete set of information to calculate the scattered electron amplitudes for any situation. Computationally it is a straight forward matter to evaluate these equations, dutifully summing the sums and taking a Fourier transform at the appropriate place. Although this will lead to a correct evaluation it would be a very slow one. Over the years several computational efficiencies have been implemented. The most outstanding of these efficiencies has been the evaluation of the multislice equation (i.e., Equation (3.4)) using the fast Fourier transform (FFT) (Ishizuka and Uyeda, 1977). This is achieved by rewriting Equation (3.4) in the form

$$\Psi_n(hk) = \mathcal{F}\{\mathcal{F}^{-1}[\Psi_{n-1}(hk)P_{n-1}(hk)]\mathcal{F}^{-1}[Q_n(hk)]\} \qquad (3.12)$$

where $\mathcal{F}$ and $\mathcal{F}^{-1}$ represent the forward and inverse Fourier transform respectively. In this form of the multislice equation the convolution is replaced by a forward and inverse Fourier transform and a multiplication. The FFT version of multislice is implemented by evaluating the Fourier transforms using the FFT algorithm such that

$$\Psi_n(hk) = \mathbf{F}\{\mathbf{F}^{-1}[\Psi_{n-1}(hk)P_{n-1}(hk)]\mathcal{F}^{-1}[Q_n(hk)]\} \qquad (3.13)$$

where the FFT is represented by $\mathbf{F}$. This formulation makes the multislice very fast and simple. One minor problem with the FFT approach to multislice is that the evaluation of convolutions by Fourier transformation can lead to aliasing and false periodicity effects. These effects are readily accounted for by the methods described by Buseck et al. (1988, §8.2.3).

The multislice method as described above may seem to be no more than a collection of obtuse mathematical formulas. It does, however, embody the essence of dynamical diffraction. The limit of the weak-phase object approximation for very small slice thicknesses is single (kinematical) scattering. For thicker slices the transmission function describes a multiple scattering situation that is the basis of dynamical diffraction. The propagation function introduces slight phase differences in the scattering from each slice thereby causing interference between the scattering from each slice. This interference leads to dynamical diffraction effects such as beam extinction and pendellosung (thickness

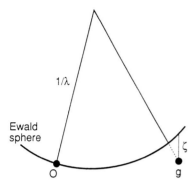

Figure 3-3. The excitation error of a reciprocal lattice point **g** for the case where the plane of the specimen surface is perpendicular to the incident beam direction. The origin of reciprocal space is marked 0. Note the radius of the Ewald sphere is considerably reduced relative to the true situation for high-energy electrons.

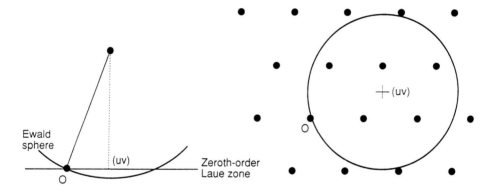

Figure 3-4. Plane (a) and perpendicular (b) view of the intersection of the zero-order Laue zone with the Ewald sphere. The circle of intersection of the Ewald sphere with the zero-order Laue zone defines the Laue circle. The center of the Laue circle is marked (uv) and the origin of reciprocal space is marked 0. In (b) the dots represent the reciprocal lattice points in the zero-order Laue zone.

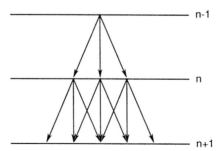

Figure 3-5. Ray diagram of the multiple scattering processes in dynamical electron diffraction. The figure shows the progressive "spread" in the electron beam as it progress through a crystal (in this case from top to bottom). For a multislice calculation the scattering would be progressively modelled at each layer n-1, n, n+1, etc.

fringes). The ray diagram (Fig. 3-5) demonstrates how the transmission and propagation functions interact to give rise to the many-beam nature of dynamical diffraction.

3.3.4   Bloch waves

The Bloch-wave method is based on direct solution of the time-independent Schrödinger equation using Bloch's theorem (Metherell, 1977, §3.2). This theorem states that any solution to the wave equation for a periodic potential must have the same periodicity as the potential. Furthermore, Bloch's theorem shows that the wave function of a periodic potential that is described by N Fourier coefficients (i.e., diffraction points) can be represented as the linear combination of N mutually orthogonal wave functions, or Bloch waves, which are the eigensolutions of the Schrödinger equation. Detailed explanation of the Bloch wave formulation is given by Hirsch et al. (1977) and Metherell (1975).

When written as a linear combination of Bloch waves, the real-space, electron wavefunction has the form

$$\psi(\mathbf{r}) = \sum_{j} \alpha_j \sum_{\mathbf{g}} C_\mathbf{g}^j \exp[2\pi i(\mathbf{k}^j+\mathbf{g})\cdot\mathbf{r}] \tag{3.14}$$

where $\mathbf{r}$ is the real-space position vector and α_j is the excitation coefficient of the jth Bloch wave. The sum over the vector $\mathbf{g}$ is a sum over all reciprocal lattice points that are to be included in the calculation. The terms $C_\mathbf{g}^j$ and $\mathbf{k}^j$ are the $\mathbf{g}$th Fourier coefficient and the wavevector ($|\mathbf{k}^j|=1/\lambda^j$) of the jth Bloch wave respectively.

The value of each of the Fourier coefficients of the Bloch waves can be found by substituting Equation (3.14) into the Schrödinger equation,

$$\partial^2\psi(\mathbf{r})/\partial^2\mathbf{r} + [\ K^2 + u(\mathbf{r})\ ]\ \psi(\mathbf{r}) = 0 \tag{3.15}$$

The influence of the electrostatic potential of the crystal is represented in this equation by the term $u(\mathbf{r})$ that has Fourier coefficients such that

$$U(hkl) = [\ F(hkl)/\pi V_c\ ]\ (m/m_o)\ , \tag{3.16}$$

where m is the relativistic mass of the beam electrons and the other terms are as defined in § 3.3.2. The term K in Equation (3.15) is the wavenumber for the beam electrons in the mean inner potential of the crystal (i.e., $K^2 = 1/\lambda^2 + U(000)$). The substitution of (3.14) into (3.15) gives a set of simultaneous equations that can be written as

$$\begin{bmatrix} K^2-(\mathbf{k}^j+\mathbf{h})^2 & U_{g-h} & \cdots & U_{l-h} & \cdots \\ U_{h-g} & K^2-(\mathbf{k}^j+\mathbf{g})^2 & \cdots & U_{l-g} & \cdots \\ \cdot & \cdot & & \cdot & \\ \cdot & \cdot & & \cdot & \\ \cdot & \cdot & & \cdot & \\ U_{h-l} & U_{g-l} & \cdots & K^2-(\mathbf{k}^j+\mathbf{l})^2 & \cdots \\ \cdot & \cdot & & \cdot & \end{bmatrix} \begin{bmatrix} C_h^j \\ C_g^j \\ \cdot \\ \cdot \\ \cdot \\ C_l^j \\ \cdot \end{bmatrix} = 0 \tag{3.17}$$

If there are N beams included in the calculation, there are N equations in the set (3.17) for the N Fourier coefficients of the Bloch waves. These equations will have non-trivial solutions only if the determinant of the N by N matrix is zero. The characteristic equation of the matrix is a vector polynomial equation of order N. The roots of this polynomial equation define N concentric surfaces collectively known as the dispersion surface (see §2.4.3). Each point on the dispersion surface defines one value of the vector k^j and a corresponding set of Fourier coefficients, C_g^j. The dispersion surface therefore defines an infinite number of Bloch-wave solutions; the particular set of solutions for a given situation are specified by the condition that the momentum of the incident electron parallel to the crystal surface must be conserved. This means that if all wavevectors, including that of the incident beams, are drawn with one end at the origin of reciprocal space (as in the Ewald sphere construction), then their other end must lie on a line that is normal to the entrance surface of the crystal (Fig. 3-6). Except in cases where the incident beam direction is nearly parallel to the entrance surface, this line will intersect the dispersion surface at N points (ignoring back reflection) and define N unique wavevectors that, in turn, define N unique Bloch wave solutions.

Although the above gives a true representation of the Bloch wave method, it would be an onerous task to solve the characteristic equation of the matrix in Equation (3.17) and then construct the appropriate surface normal to specify the Bloch wave solutions. It is much more practical to apply the boundary conditions at the time of solving (3.17). The most widely used (and simplest) boundary condition is that the surface normal is perpendicular to the reciprocal lattice plane of interest. In this case, the wavevectors of the Bloch waves can be split into two components; one lying in the plane of the zeroth-order Laue zone and the other perpendicular to this plane (Fig. 3-7). The component of the wavevector lying in the plane of the zeroth-order Laue zone will be the vector from the center of the Laue circle to the origin. This vector is shown as u_o in Figure 3-8.

The splitting of the wavevectors into two components allows the diagonal terms of Equation (3.17) to be rewritten in the form

$$K^2 - (k^j + g)^2 = K^2 - k_\perp^{j2} + (g + u_o)^2 \qquad (3.18)$$

where $k_\perp^j$ is the component of the wavevector perpendicular to the zeroth-order Laue zone as shown in Figure 3-7. The vectors g and u_o lie in the same plane and can be readily calculated from geometrical considerations of the reciprocal lattice (Fig. 3-8). It is useful to define the vector quantity $(g+u_o)$ as u_g. Equation (3.17) can then be rewritten as

$$\mathbf{M} \cdot \begin{bmatrix} C_h^j \\ C_g^j \\ \cdot \\ \cdot \\ \cdot \\ C_l^j \\ \cdot \end{bmatrix} = (k_\perp^{j2} - K^2) \begin{bmatrix} C_h^j \\ C_g^j \\ \cdot \\ \cdot \\ \cdot \\ C_l^j \\ \cdot \end{bmatrix} \qquad (3.19)$$

where

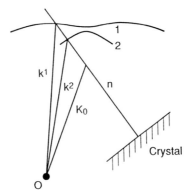

Figure 3-6. Construction defining the wavevector of Bloch waves. The in-vacuum incident wavevector is K_O and n is the normal to the crystal surface. The points of intersection of the surface normal and the dispersion surface define the wavevectors of the Bloch waves (k^1 and k^2). For this construction all wavevectors have the same component in the direction parallel to the crystal surface.

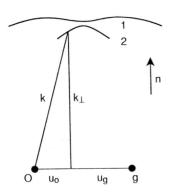

Figure 3-7. Separation of the wavevector into components perpendicular and parallel to the surface normal for the boundary condition that the normal to the crystal surface is perpendicular to the reciprocal lattice plane of interest.

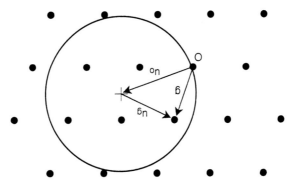

Figure 3-8. Vectors from the center of the Laue circle as used to construct the matrix **M** of Equation (3.19).

$$M = \begin{bmatrix} -u_h^2 & U_{g-h} & \cdots & U_{l-h} & \cdots \\ U_{h-g} & -u_g^2 & \cdots & U_{l-g} & \cdots \\ \cdot & \cdot & & \cdot & \\ \cdot & \cdot & & \cdot & \\ \cdot & \cdot & & \cdot & \\ \cdot & \cdot & & \cdot & \\ U_{h-l} & U_{g-l} & \cdots & -u_l^2 & \cdots \\ \cdot & \cdot & & \cdot & \end{bmatrix}$$

The advantages of choosing the surface normal perpendicular to the reciprocal lattice plane of interest are readily apparent from Equation (3.19). This equation contains only scalar terms of which all are fairly easy to calculate. Furthermore, Equation (3.19) defines an eigensystem with solutions such that $(k_\perp^{j2} - K^2)$ are the eigenvalues of the matrix M. The eigenvectors of M define the Bloch wave coefficients.

Equation (3.19) does not give any information about the excitation coefficients of the Bloch waves; that is α_j. These coefficients are obtained by making the wavefunction at the top surface of the crystal (i.e., at a crystal thickness of 0) equal to the incident-beam wavefunction. This equality generates a set of N simultaneous equations containing the Bloch wave coefficients, the Fourier coefficients of the incident-electron wavefunction, and the excitation coefficients of the Bloch waves. Because the eigenvectors are mutually orthogonal, the solution of the simultaneous equations is fairly straight forward (Self et al., 1983). In the case where a plane wave is incident on the crystal:

$$\alpha_j = C_0^j \cdot \qquad\qquad\qquad (3.20)$$

Absorption, as always, is included as an imaginary term in the electrostatic potential. The inclusion of this term considerably increases the mathematical difficulty of finding the eigensolutions to the matrix M in Equation (3.19). Without absorption M is a Hermitian because, since the electrostatic potential of the specimen must have only real values, $U_{-g} = U_g^*$. This property is no longer applicable if absorption is included, and so, to solve Equation (3.19) it is necessary to find the eigensolutions of a general matrix with complex terms. Algorithms for such solutions in general require considerable computation, and so, the standard approach for including absorption in the Bloch wave method is to treat it as a perturbation of the case without absorption. Using this approach $k_\perp^j$ should be replaced by $k_\perp^j + iq^j$ where

$$q^j = (\sum_g \sum_h U'_{g-h} C_h^j C_g^{j*})/2K . \qquad\qquad (3.21)$$

The term U'_{g-h} is related to the Fourier coefficient of the absorption potential in the same manner as U_{g-h} is related to the electrostatic potential of the crystal [see Equation (3.16)]. The Bloch wave coefficients remain unchanged from the case where absorption is not included. Although Equation (3.21) describes only an approximate solution to Equation (3.19), it is more than adequate for all realistic values of the absorption potential.

Equation (3.19) is in its simplest form when only two beams are included in the calculation. In this case the equation is reduced to

$$\begin{bmatrix} \gamma^j - u_o{}^2 & U_g \\ U_{-g} & \gamma^j - u_g{}^2 \end{bmatrix} \begin{bmatrix} C_o{}^j \\ C_1{}^j \end{bmatrix} = 0 \, , \tag{3.22}$$

where $\gamma^j = K^2 - k_\perp{}^{j2}$ and u_o and u_g are one dimensional vectors as shown in Figure 3-7. The solutions to Equation (3.22) are:

$$\gamma^1 = \{ u_o{}^2 + u_g{}^2 - [(u_o{}^2 - u_g{}^2)^2 + 4|U_g|^2]^{1/2} \}/2$$

$$\gamma^2 = \{ u_o{}^2 + u_g{}^2 + [(u_o{}^2 - u_g{}^2)^2 + 4|U_g|^2]^{1/2} \}/2$$

$$C_o{}^j = U_g/[U_g{}^2 + (\gamma^j - u_o{}^2)^2]^{1/2}$$

$$C_1{}^j = (\gamma^j - u_o{}^2)/[U_g{}^2 + (\gamma^j - u_o{}^2)^2]^{1/2}$$

$$\alpha^j = C_o{}^j \cdot \tag{3.23}$$

A special case of the two beam situation is when the Ewald sphere passes through both the origin of reciprocal space and the diffraction point **g**. That is, when the crystal is oriented at the Bragg angle for the reciprocal lattice point **g**. In this case $u_o = u_g = g/2$ and the solutions become:

$$\gamma^1 = u_o{}^2 - |U_g| \qquad\qquad \gamma^2 = u_o{}^2 + |U_g|$$

$$C_o{}^1 = 1/\sqrt{2} \qquad\qquad C_o{}^2 = 1/\sqrt{2}$$

$$C_g{}^1 = -1/\sqrt{2} \qquad\qquad C_g{}^2 = 1/\sqrt{2}$$

$$\alpha^1 = 1/\sqrt{2} \qquad\qquad \alpha^2 = 1/\sqrt{2} \tag{3.24}$$

Thus, for the two beam case, when the crystal is oriented to the exact Bragg angle the solutions for the Bloch wave coefficients and the Bloch wave excitation coefficients are completely independent of any crystallographic parameters; only the wave vectors contain crystallographic information. Using the approximation

$$\gamma^j = K^2 - k_\perp{}^{j2} = (K + k_\perp{}^j)(K - k_\perp{}^j) \approx 2K(K - k_\perp{}^j) \, , \tag{3.25}$$

the amplitude for each reflection as a function of crystal thickness (H) can be written as

$$\begin{aligned} \psi_0(H) &= \tfrac{1}{2} \exp\{ 2\pi i H[K - (u_o{}^2 - |U_g|)/2K]\} + \tfrac{1}{2} \exp\{ 2\pi i H[K - (u_o{}^2 + |U_g|)/2K]\} \\ &= \exp[2\pi i H(K - u_o{}^2/2K)] \cos(2\pi H|U_g|/2K) \\ \psi_g(H) &= \tfrac{1}{2} \exp\{ 2\pi i H[K - (u_o{}^2 - |U_g|)/2K]\} - \tfrac{1}{2} \exp\{ 2\pi i H[K - (u_o{}^2 + |U_g|)/2K]\} \\ &= \exp[2\pi i H(K - u_o{}^2/2K)] \sin(2\pi H|U_g|/2K) \, , \tag{3.26} \end{aligned}$$

and the intensities as

$$|\psi_0(H)|^2 = I_0(H) = \cos^2(\pi H|U_g|/K) \, ,$$

$$|\psi_g(H)|^2 = I_g(H) = \sin^2(\pi H|U_g|/K) \, . \tag{3.27}$$

The sinusoidal nature of the intensities is the mathematical description of thickness fringes. The period of the intensity variation is given by the extinction distance

$$\xi_g = K/|U_g| \tag{3.28}$$

The extinction distance is the crystal thickness for which the intensity of the diffracted beam goes from zero intensity (at the entrance surface of the crystal), through a maximum (at $\xi_g/2$) and back to zero (at ξ_g). Over the same crystal thickness, the intensity of the undiffracted beam goes from a maximum, through to zero and back to a maximum again. This behaviour, as shown in Figure 3-9a, is a direct result of the interference of two equally excited Bloch waves of slightly different period.

For diffraction conditions away from the exact Bragg angle, the form of the intensities are similar, but the period is shorter than ξ_g and the intensity of the undiffracted beam never goes to zero (Fig. 3-9b). This is because one Bloch wave is excited more than the other so that when the two Bloch waves interfere they can never completely cancel. The difference in relative excitation of the Bloch waves is also reflected in the absorption behaviour of crystals. The current density of Bloch wave 2 is peaked on the atom positions while the current density of Bloch wave 1 is peaked between the atom positions. Hence, in situations where Bloch wave 2 is strongly excited, the absorption of beam electrons will be high. This is observed experimentally as the channelling phenomenon that is the basis of analytical techniques such as ALCHEMI (see Chapter 5).

The solutions given above show the utility of the Bloch wave method. The method allows general trends in dynamical electron diffraction to be modelled by giving descriptions of wavefunctions for small numbers of beams. This ability to model wavefunctions that include only a small number of beams is because, for the Bloch wave method, scattering is only possible to those beams included in the calculation and so the diffraction processes remain internally consistent. Calculations that include only a small number of beams are simply not possible with the multislice method because this method allows scattering to beams not included in the calculation and so, if too few beams are included, the total intensity soon drains away and the wavefunction becomes inaccurate.

A utility of the Bloch wave method that is perhaps more important than internal consistency is its ability to give an analytical form of the wavefunction that spans all crystal thickness. This analytical form can be used to calculate other important diffraction parameters such as current density (i.e., $|\psi(\mathbf{r})|^2$) and the strength of characteristic x-ray emissions (Cherns et al., 1973; Chapter 5).

Before leaving the subject of Bloch waves, it is worthwhile asking the philosophical question "are Bloch waves real?". The above gives a mathematical description of the Bloch wave method in which the Bloch waves may seem no more than expedient tools for the solution of the Schrödinger equation. The weight of experimental evidence is that Bloch waves are real. Their presence can be observed experimentally in a variety of well known effects. For example, thickness fringes occur as a result of the interference of Bloch waves. Electron channelling and anomalous absorption can be attributed to the excitation of Bloch waves. The Bloch waves can also be observed directly from crystals where the entrance and exit surfaces are not parallel. For such crystals, the Bloch waves components of the wavefunction become separated at the exit surface of the crystal and the diffraction spots develop a fine structure that is indicative of the Bloch wave components. By examining the fine structure at a range of crystal tilts it is even possible to experimentally map the dispersion surface (Lehmpfuhl and Reissland, 1968).

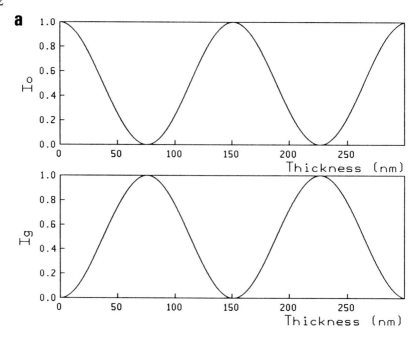

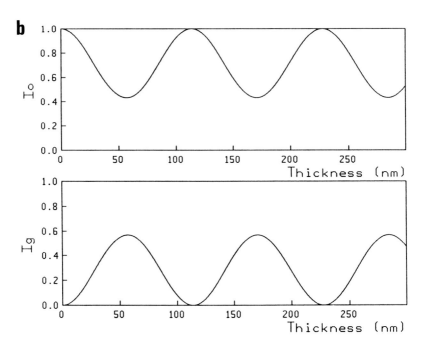

Figure 3-9. Two-beam solutions for the intensity of the (000) and (220) beams from cristobalite at an accelerating voltage of 100 kV. (a) Crystal oriented to the exact Bragg angle of the (220) beam. (b) Crystal oriented so that the center of the Laue circle is at $0.4\ g_{220}$ (i.e., $\theta = 0.8\ \theta_B$ or $\zeta = -5.8 \times 10^{-3}\text{Å}^{-1}$).

3.4   IMAGING

3.4.1   Theory

The multislice and Bloch wave formulations enable the diffracted amplitudes and phases to be calculated at the exit surface of the specimen. Other than squaring the amplitudes, the evaluation of the intensities in a diffraction pattern requires no further calculation. The evaluation of image intensities on the other hand requires that the phase changes introduced by the objective lens be added to the diffracted amplitudes. These lens-induced phase changes are described by the transfer function, TF (see also §1.3.5), of the lens that has the form

$$T(uv) = \exp[i\chi(uv)] , \tag{3.29}$$

where (uv) are the coordinates of the reciprocal lattice plane perpendicular to the optic axis of the lens. In a well aligned microscope the optic axis and the incident beam direction will be parallel.

The function $\chi(uv)$ describes the phase change experienced by electrons as they pass through the lens at different angles relative to the optic axis. This phase change may be written as

$$\chi(uv) = \pi \lambda \, \Delta f \, (u^2+v^2) \; + \; \pi \, C_s \, \lambda^3 \, (u^2+v^2)^2/2 , \tag{3.30}$$

where Δf is the lens defocus (positive for overfocus) and C_s is the spherical aberration constant of the lens. The wave function at the image plane is

$$\psi_{im}(Mx,My) = \mathcal{F}[\; \mathcal{F}(\psi_e(xy)) \; T(uv) \;] , \tag{3.31}$$

where M describes the magnification factor and $\psi_e(xy)$ is the exit-surface wavefunction of the specimen. As computed images are usually scaled to match experimentally recorded images, the magnification factor (often 10^6 or more) is ignored.

The form of Equation (3.31) does not make it conducive to intuitive interpretation. One case that can be readily interpreted is what on first sight may appear to be the optimum imaging conditions: a perfect lens ($C_s = 0$) at the in-focus condition ($\Delta f = 0$). In this case

$$T(uv) = 1, \; \psi_{im}(xy) = \psi_e(xy) \; \text{and} \; I_{im}(xy) = |\psi_e(xy)|^2 , \tag{3.32}$$

where $I_{im}(xy)$ is the intensity at the image plane of the microscope. For a thin crystal, $\psi_e(xy)$ is described by the weak-phase object approximation (WPOA) [see Equation (3.6)]. Hence

$$I_{im}(xy) = |\exp[-i\sigma\phi(xy)\Delta z]|^2 = 1 . \tag{3.33}$$

That is, for an in-focus perfect lens there is no contrast (i.e., no information) in the image.

The objective of HRTEM is to image the potential of the specimen, $\phi(xy)$. It is therefore reasonable to ask "under what imaging conditions will an experimental micrograph contain directly interpretable information about $\phi(xy)$"? If it is assumed that $\sigma\phi(xy)\Delta z$ is small, the wavefunction given by the WPOA can be expanded as

$$\psi_e(xy) = \exp[-i\sigma\phi(xy)\Delta z] = 1 - i\sigma\phi(xy)\Delta z , \tag{3.34}$$

and

$$\mathcal{F}[\psi_e(xy)] = \delta(uv) - i\sigma\Delta z\Phi(uv) , \tag{3.35}$$

where $\Phi(uv)$ is the Fourier transform of $\phi(xy)$, and $\delta(uv)$ is the Kronecker delta function. This expansion describes the kinematical approximation of electron diffraction in which the diffraction amplitudes are weak and $\pi/2$ out of phase relative to the undiffracted beam. For an in-focus, perfect lens the image intensity will be

$$I_{im}(xy) = 1 - [\sigma\phi(xy)\Delta z]^2 . \tag{3.36}$$

Given that quadratic terms are considered negligible, Equation (3.36) describes a no-contrast condition.

Contrast can be achieved if the TF imparts an additional $\pm\pi/2$ phase to the diffracted beams so that

$$\mathcal{F}(\psi_e(xy)) T(uv) = \delta(uv) \pm \sigma\Delta z\Phi(uv) , \tag{3.37}$$

and therefore

$$I_{im}(xy) = 1 \pm 2\sigma\Delta z\Phi(uv) . \tag{3.38}$$

Thus an ideal TF for HRTEM is

$$T(uv) = \delta(uv) \pm i(1 - \delta(uv)) . \tag{3.39}$$

The actual form of the TF as given in Equation (3.29) is

$$T(uv) = \cos[\chi(uv)] + i \sin[\chi(uv)] , \tag{3.40}$$

so that

$$\Psi_{im}(uv) = \delta(uv) + \sigma\Delta z\Phi(uv)\sin[\chi(uv)] - i\sigma\Delta z\Phi(uv)\cos[\chi(uv)] . \tag{3.41}$$

The wavefunction at the image plane will therefore approximate the wavefunction given in Equation (3.37) when $\sin[\chi(uv)] \approx \pm 1$ for as large a range of (uv) as possible. The defocus value at which $\sin[\chi(uv)] \approx \pm 1$ for the largest possible range is called Scherzer defocus (cf. §1.3.5). A plot of $\sin[\chi(uv)]$ at Scherzer defocus for a JEOL 4000EX microscope is shown in Figure 1-7. The range for which $\sin[\chi(uv)] \approx -1$ is from 0.5 to 0.17 nm. Sub-unit cell detail for image features with spacing outside this range will not give a true representation of the specimen potential.

For Scherzer defocus $\sin[\chi(uv)] \approx -1$, and so, the contrast in the image intensity will be proportional to $-\phi(xy)$. That is, places in the specimen with high potential (the atoms) will show up as dark regions on the image and places with low potential will show as light regions. Other defocus values can be found where $\sin[\chi(uv)] \approx \pm 1$ for an extended range of (uv). These defocus values are collectively known as the broad-band focussing conditions and occur when

$$\Delta f_n = \Delta f_s(1 + 3n/2)^{1/2} \tag{3.42}$$

where $n = 0,1,2,...$ and Δf_s is Scherzer defocus. That is

$$\Delta f_s = -(4C_s\lambda/3)^{1/2} \tag{3.43}$$

The value of $\sin[\chi(uv)]$ at the first broad band-value of defocus ($n = 1$) for the same microscope as Figure 1-7 is shown in Figure 3-10a.

The above analysis highlights a paradox in HRTEM imaging; spherical aberration is necessary for obtaining intuitively interpretable images of a structure (i.e., structure images). That is, an imperfect lens assists the interpretation of HRTEM images. This, of course, is not a strict truism. A perfect lens will give high quality, interpretable images. This is especially the case for crystalline structures because the defocus value can be tuned so that each diffracted beam experiences a $\pm\pi/2$ phase change. There are several defocus values where this phase change in the diffracted beams can be achieved and the resulting images are related to the Fourier images as described by Buseck et al. (1988, §1.4) and mentioned briefly in §1.3.3 of this volume.

The dependence of the image contrast on $\sin[\chi(uv)]$ as given in Equation (3.41) demonstrates why it is usual to plot only $\sin[\chi(uv)]$ rather than the full TF. For conditions where the WPOA does not hold (say for specimens thicker than 1 nm), Equation (3.41) will not apply and a full calculation of the image intensity must be used. For the thick specimen case, HRTEM image contrast will not be proportional to the specimen potential and so the image matching strategy outlined in §3.2.1 must be followed.

In all of the above analysis, it has been assumed that, in order to obtain images that show contrast that is proportional to the specimen potential, a $\pm\pi/2$ phase change must be imparted to the diffracted beams. There are, however, other methods of achieving images that have contrast directly interpretable in terms of the specimen potential. One such method is to exclude the undiffracted beam from the image plane. This gives an image intensity proportional to $[\phi(xy)]^2$ for conditions where $T(uv) \approx \pm1$. Such a condition occurs at the in-focus setting of the objective lens. Imaging with the undiffracted beam excluded gives slightly less resolution that at Scherzer defocus but potentially gives a more interpretable image because of a greater contribution to the image from beams scattered to low-angles. Unfortunately the experimental difficulties of only excluding the undiffracted beam make the technique impractical. However, comparable results can be achieved by using high-resolution dark-field imaging (Cowley, 1973a,b; Pierce and Buseck, 1974).

Equations (3.29) and (3.30) do not provide a complete representation of the lens transfer function. These equations neglect the effects of astigmatism, specimen drift, objective aperture, beam divergence, chromatic aberration, and coherence. The modelling of these effects is discussed in detail by Buseck et al. (1988, §8.4). Astigmatism is not hard to model, but the extra variables introduced by astigmatism make the image matching process much more difficult and it is best to experimentally reduce astigmatism to as low a level as possible. Similarly specimen drift or vibration can be modelled but the loss of resolution caused by these parameters is so great that all effort should be made to eliminate them whilst recording HRTEM images.

The objective-lens aperture (Fig. 1-1) is a useful tool for increasing image contrast and highlighting interplanar spacings of interest. Given that the Fourier coefficients of the exit-surface wavefunction are in 1 to 1 correspondence to the diffraction pattern formed in the back-focal plane of the objective lens, the objective aperture is modelled by removing the appropriate Fourier coefficients from the electron wavefunction at the exit surface of the specimen.

When chromatic aberration is discussed in connection with HRTEM, it encompasses two temporal effects. These effects are that both electron energy and lens defocus (i.e., lens current) are not constant over the time taken to record a HRTEM image. The variation in energy and lens current is roughly 1 in 10^5. To allow rigorously for variations in electron energy, exit-surface wavefunctions should be calculated for each energy value of the electron beam and then the images from these wavefunctions should be added together incoherently (i.e., as intensities) according to the distribution of energies in the electron beam. Obviously such a process is time consuming and it can be simplified by making the assumption that the exit-surface wavefunction is invariant over the range of energies in the electron beam. This assumption allows the effects of energy variation to be combined with those of lens-current instability and modelled as a distribution of defocus values with standard deviation

$$\mathbf{D} = C_c [(\Delta E/E)^2 + 4(\Delta J/J)^2]^{1/2} \tag{3.44}$$

In this equation ΔE is the standard deviation of the energy spread for an incident beam of energy E, and ΔJ is the standard deviation for variations in the objective lens current (J). The coefficient of chromatic aberration is C_c. HRTEM images that allow for chromatic aberrations can therefore be calculated as the average of images formed from the same exit-surface wavefunction but for a distribution of defocus values that has a standard deviation of $\mathbf{D}$. As the fluctuation in energy and lens current are random, it is usual to assume a Gaussian distribution in defocus values.

Beam divergence is another way of saying that the incident electron beam is not a plane wave; it is incident on the specimen over a range of angles. The angular range in the incident beam can be directly measured from the diffraction pattern which consists of disks rather than sharp spots. To correctly allow for beam divergence, computed HRTEM images should be the average of the images formed at each angle of incidence in the electron beam. As with chromatic aberration, the process can be simplified by making the assumption that the exit-surface wavefunction is invariant to tilt over the range of angles in the incident beam. Unlike the case for chromatic aberration, this assumption is not always applicable. This is especially the case when imaging crystal structures that contain glide planes or screw axes (Self et al., 1985).

An added complication with beam divergence is coherence. With chromatic aberration the intensities of many images are averaged to form the final image. For beam divergence, it is necessary to average intensities for a fully incoherent electron source (such as a thermionic gun), to average amplitudes for a fully coherent source and to average some mixture of the two for partially coherent sources (such as field emission guns). The averages for fully incoherent and fully coherent electron sources can be written as

$$I_{D\alpha}(uv) = \int\int E(f,\Delta f) \, S(\mathbf{s}) \, I_e(uv) \, d^2s \, df \tag{3.45}$$

and

$$I_{D\alpha}(uv) = \int E(f,\Delta f) \, | \int S(\mathbf{s}) \, \psi_e(uv) \, d^2s \, |^2 \, df \tag{3.46}$$

respectively. In these equations $E(f,\Delta f)$ is the distribution of defocus values (f) around an average value, Δf. The function $S(\mathbf{s})$ is the angular distribution of intensity of the incident beam. The assumption that the effects of beam divergence and chromatic aberration are separable has been made in defining these equations.

The integrals of (3.45) and (3.46) can be calculated analytically by expanding $\chi(uv)$ [see Equation (3.30)] as a Taylor series. The analytical expression for the averaged intensities shows that

$$I_{D\alpha}(uv) = \sum_{\mathbf{u'}} \psi_e(\mathbf{u+u'}) \, \psi_e(\mathbf{u'})^* \, \exp\{i[\chi(\mathbf{u+u'})-\chi(\mathbf{u'})]\} \, A(\mathbf{u+u'}) \, A(\mathbf{u'}) \, D(\mathbf{u+u',u'}),$$

$$\tag{3.47}$$

where **u** represents the reciprocal space vector (uv) and A(**u**) represents the objective aperture function. The function D(**u**+**u**',**u**') is known as the damping function and is a complex function that describes the effects of chromatic aberration, beam divergence, and coherence. The form of D(**u**+**u**',**u**') is different for coherent and incoherent illumination (O'Keefe and Saxton, 1983). A full description of the damping function is given by Buseck et al. (1988, §8.4.2).

The expression (3.47) can be simplified if the linear-imaging approximation is made. That is, if it is assumed that the majority of the intensity is carried by the undiffracted beam so that the only non-negligible terms in the summation of (3.47) are those for which either **u**+**u**'=0 or **u**'=0. In this case the Fourier coefficients of the final image intensity are given by

$$I_{D\alpha}(uv) = \psi_e(\mathbf{u}) \exp[i\chi(\mathbf{u})] A(\mathbf{u}') D(\mathbf{u}) \tag{3.48}$$

for both coherent and incoherent illumination. The damping function, D(**u**), may be written as the product of two functions (know as envelope functions) that represent the effect of chromatic aberration and beam divergence. Hence,

$$D(\mathbf{u}) = C(\mathbf{u}) E(\mathbf{u}) , \tag{3.49}$$

where C(**u**) represents the chromatic aberration envelope function and E(**u**) the divergence envelope function. Assuming Gaussian distributions for both chromatic aberrations and beam divergence, the envelope functions can be written as

$$C(\mathbf{u}) = \exp[-(\pi\lambda D\mathbf{u}^2)^2/2]$$
$$E(\mathbf{u}) = \exp[-\pi^2\alpha^2(D|\mathbf{u}|+C_s\lambda^2|\mathbf{u}|^3)^2] \tag{3.50}$$

for a divergence half-angle of α.

The transfer function modified for beam divergence and chromatic aberration under linear-imaging conditions is shown in Figure 3-10b. By comparing the undamped transfer function (Fig. 3-10a) with the damped transfer function (Fig. 3-10b), it can be seen that the effect of D(**u**) is to damp (hence damping function) the high-order Fourier coefficients (i.e., the higher spatial frequencies) of the image intensity and hence limit resolution. This damping is essentially brought about by the smearing of the image because of the variation in electron energy and the different angles of incidence. This smearing will be least pronounced for regions of **u** where $\chi(\mathbf{u})$ varies slowly. Such regions occur at Scherzer and other broad-band focussing conditions, which re-enforces the concept that these are the optimum focussing conditions.

3.4.2 Computational tricks (and traps)

The preceding discussion gives a summary of the techniques used for the simulation of diffracted amplitudes for HRTEM. Although efforts were made to make the discussion general and to apply to as wide a range of specimens as possible, many points are only applicable to crystalline specimens. This is particularly so for the Bloch wave formulation. What then, if the specimen of interest is not crystalline? For example, how can the diffracted amplitudes for amorphous materials and defect structures be evaluated? The simple answer is to make the specimen crystalline: not physically but mathematically by using the method of periodic continuation. This mathematical construct is achieved by

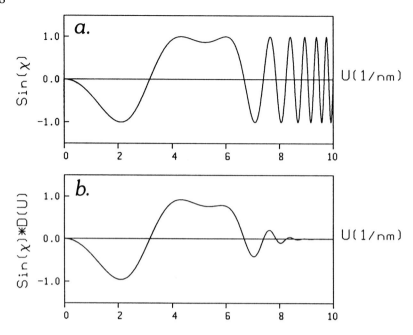

Figure 3-10. TF for 1st broad-band focus conditions for a JEOL 4000EX (cf. Fig. 1-7). (a) Imaginary part of TF. (b) Imaginary part of TF corrected for chromatic aberration (2eV) and beam divergence (0.5 mrad.).

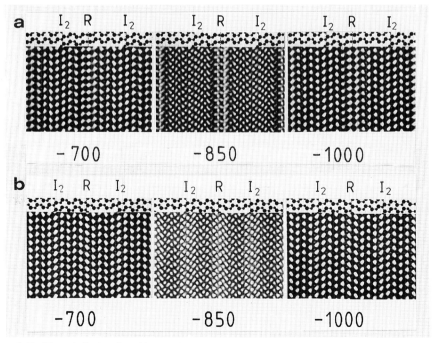

Figure 3-11. Through-focal series of images for an intrinsic (I_2) stacking fault in CdSe. In (a) atoms at the supercell edge (marked R) lie unrealistically close together. In (b) the supercell edge has been extended slightly so that the atoms at the edge are arranged much more realistically. The image contrast for the I_2-stacking fault is much easier to discern in (b). Defocus values are shown in Å. (Courtesy of R.W. Glaisher.

defining a supercell that is both representative of the structure being studied and as large as possible for the available computer hardware and software. The structure is made crystalline by laying the supercells together on a regular lattice. The wavefunction calculated for the extended cell samples reciprocal space at the Bragg angles for the periodically continued lattice (Wilson and Spargo, 1982; Buseck et al., 1988, §8.3.2)

An example of a periodic continuation calculation is shown in Figure 3-11 (above). To calculate the image of an intrinsic (I_2) stacking fault in CdSe, the fault has been embedded in a supercell with dimensions 10 times that of the crystallographic unit cell of CdSe. Some care must be taken in setting up the supercell so that edge effects between adjoining cells do not contribute excessive contrast to the image. In Figure 3-11a atoms from adjoining cells lie close together and cause an extended intensity modulation that interferes with the contrast from the I_2-stacking fault. In Figure 3-11b the situation is remedied by better matching of adjoining cells. In the latter case the matching is achieved by placing a defect exactly opposite to the I_2 defect at the edge of the supercell. (In the CdSe structure it turns out that the opposite defect to an I_2-stacking fault is itself an I_2-stacking fault.) Periodic continuation can also be applied to interfaces and the imaging of steps on crystal surfaces (Marks and Smith, 1983)

A technique that is related to periodic continuation is the column approximation. This approximation is generally applied to large defect structures that extend for many tens or even hundreds of nanometers. The technique consists of dividing the region of interest into columns parallel to the incident-beam direction and in which the structural variation from one column to the next is fairly small. Images are calculated for each column and the overall image formed as a mosaic of the images from the individual columns. The column approximation has been successful in calculating images for stacking faults, screw and edge dislocations and the strain fields associated with these faults (Hirsch et al., 1977, Chapter 11).

Throughout much of the discussion in the previous sections on the simulation of HRTEM images, it has been assumed that the incident radiation is in the form of a plane wave. Although plane-wave illumination is applicable in many situations, it is not applicable for convergent-beam electron diffraction (CBED) and scanning-transmission electron microscopy (STEM). For numerical calculation, CBED and STEM are treated in much the same way as the column approximation except that the mosaic is built up in reciprocal (i.e., angle) space. That is, CBED patterns are generated by computing diffraction amplitudes for plane-wave illumination over a range of beam incidence angles and then forming a mosaic from these calculated patterns. When the CBED disks overlap, the wavefunction at the overlapping points is either added as intensities or amplitudes depending on the coherence of the incident beam. STEM images are formed by taking the intensities of the CBED pattern that fall on the reciprocal space area defined by the STEM detector. If STEM images are the main interest of a CBED calculation it is usually quicker to use the reciprocity of HRTEM and STEM (Spence and Cowley, 1978) to calculate the images. A fuller discussion of the computation of CBED patterns and STEM images is given in Buseck et al. (1988, §8.3.3).

3.5   LOOKING FORWARD AND GOING BACKWARDS

Although the theory of HRTEM imaging is based on solid foundations, it is by no means a static subject. As display techniques improve and computer speeds increase, the method of using HRTEM imaging theory is changing from one where experimental micrographs are recorded and then matched laboriously over a period of months to one where computed images are displayed in real-time at the microscope console. Such systems are already installed in some laboratories (M.A. O'Keefe, personal comm.). The increases in computer speed has also meant that HRTEM imaging theory is no longer being used to obtain only specific information about specific samples but is also being used to give broad general predictions about a whole gambit of specimens (e.g., Spinnler et al., 1984; Ahn and Buseck, 1990; Guthrie and Veblen, 1990).

Studies such as these enable a database of HRTEM images to be built up so that the microscopist can obtain an intuitive feel for what is and is not interpretable detail in HRTEM images. Computational image studies have also been used to guide the microscopist in the way the microscope should be operated in order to obtain the best possible image detail (Smith et al., 1983). In turn this information is passed back to the manufacturers so as to improve the performance of high-resolution electron microscopes.

Not only have there been improvements in display technology, but also in the technology of capturing digital images (i.e., storing images to computer disc directly from the microscope). The ability to "grab" images from the microscope and process them in a computer in real-time has lead to systems for the automatic focussing, stigmation, and alignment of electron microscopes (Saxton et al., 1983b).

The advances in hardware beg the question, "Can the process be made to go backwards?" That is, can a series of images be captured in digital form from the micro-scope and then processed to give the atomic structure of the specimen? The answer would seem to be, "Yes, but not quite yet." Algorithms for taking a series of images and processing them to obtain the wavefunction at the exit surface of the crystal have existed for quite some time (Gerchberg and Saxton, 1971). However, knowing the exit-surface wavefunction does not uniquely specify the crystal structure. In fact, the only condition where the exit-surface wavefunction does give structural information directly is when the WPOA holds. Even for this condition, because of the low contrast in the image, fine detail within an image is hard to characterise.

The effects of fine details (i.e., small but significant variations) in the specimen potential are enhanced by the multiple scattering that occurs in thick crystals. Thus, if fine structure is to be characterized, it is necessary to work with the wavefunction from a thick crystal. Pragmatically, the crystal potential can be uniquely determined by measuring the exit-surface wavefunction for a large number of crystal thicknesses and using the measure-ments to specify the Bloch-wave Fourier coefficients and wavevectors. This approach requires accurate measurement of crystal thickness and tilt and cannot be considered a truly viable method.

A more realistic approach of determining the crystal potential, is to use the wavefunction from a very thin crystal (approximately 5nm or less thick) to obtain a general picture of the crystal structure and then use the wavefunction from a thick crystal to fill in the detail. The process would involve running a multislice calculation in reverse; starting from the exit-surface and aiming to get a plane wave at the entrance surface. In order to implement this process it is still necessary to know the majority of parameters in Table 3-1. At the present time only preliminary investigations have been made in this area of "direct methods" for HRTEM. However, if the progress in direct methods for x-ray diffraction is any guide, electron microscopist should be confident that direct methods for HRTEM will be achieved in the not too distant future.

ACKNOWLEDGMENTS

The author thanks Dr. R.W. Glaisher for access to his knowledge and his collection of HRTEM images. The author also thanks the Publications Unit of the CSIRO, Division of Soils, and in particular Mr. J. Coppi, for their assistance in preparation of the figures.

REFERENCES

Ahn, J.H. and Buseck, P.R. (1990) Layer-stacking sequences and structural disorder in mixed-layer illite/smectite: Image simulations and HRTEM imaging. Am. Mineral. 75, 267-275.

Bird, D.M. (1990) Absorption in high-energy electron diffraction from non-centrosymmetric crystals. Acta Crystallogr. A46, 208-214.

Bird, D.M. and King, Q.A. (1990) Absorptive form factors for high-energy electron diffraction. Acta Crystallogr. A46, 202-208.

Buseck, P.R., Cowley, J.M. and Eyring, L. (1988) High-resolution transmission electron microscopy and associated techniques. Oxford University Press, Oxford, 645 p.

Champness, P.E. (1987) Convergent beam electron diffraction. Mineral. Mag. 51, 33-48.

Cherns, D., Howie, A. and Jacobs, M.H. (1973) Characteristic x-ray production in thin crystals. Z. Naturforsch. 28a, 565-571.

Cowley, J.M (1973a) High-resolution dark-field electron microscopy. I. Useful approximations. Acta Crystallogr. A29, 529-536.

Cowley, J.M (1973b) High-resolution dark-field electron microscopy. II. Short-range order in crystals, Acta Crystallogr. A29, 537-540.

Cowley, J.M. (1981) Diffraction Physics, 2nd ed. North-Holland, Amsterdam, 430 p.

Cowley, J.M. and Moodie, A.F. (1957) The scattering of electrons by atoms and crystals. I. A new theoretical approach. Acta Crystallogr. 10, 609-619.

Doyle, P.A. and Turner, P.S. (1968) Relativistic Hartree-Fock x-ray and electron scattering factors, Acta Crystallogr. A24, 390-397.

Fox, A.G., O'Keefe, M.A. and Tabbernor, M.A. (1989) Relativistic Hartree-Fock x-ray and electron atomic scattering factors at high angles. Acta Crystallogr. A45, 786-793.

Gerchberg, R.W. and Saxton, W.O. (1971) Phase determination from image and diffraction plane pictures in the electron microscope. Optik 34, 275-284.

Goodman, P. and Moodie, A.F. (1974) Numerical evaluation of n-beam wave functions in electron scattering by the multi-slice method. Acta Crystallogr. A30, 280-290.

Guthrie, Jr., G.D. and Veblen, D.R. (1990) Interpreting one-dimensional high-resolution transmission electron micrographs of sheet silicates by computer simulation Am. Mineral. 75, 276-288.

Head, A.K., Humble, P., Clarebrough, L.M., Morton, A.J. and Forwood, C.T. (1973) Computed Electron Micrographs and Defect Identification. North-Holland, Amsterdam, 400 p.

Hirsch, P., Howie, A., Nicholson, R.B., Pashley, D.W. and Whelan, M.J. (1977) Electron Microscopy of Thin Crystals. Krieger, Huntington, New York, 563 p.

Humphreys, C.J. and Hirsch, P.B. (1968) Absorption parameters in electron diffraction. Phil. Mag. 18, 115-122.

International Tables for X-ray Crystallography (1974) Kynoch Press, Birmingham, 558 p.

Ishizuka, K. and Uyeda, N. (1977) A new theoretical and practical approach to the multislice method. Acta Crystallogr. A33, 740-749.

Lehmpfuhl, G. and Reissland, A. (1968) Photographical record of the dispersion surface in rotating crystal electron diffraction pattern. Z. Naturforsch. 23a, 544-519.

Marks, L.D. and Smith, D.J. (1983) Direct surface imaging in small metal particles. Nature 303, 316-317.

Metherell, A.J.F. (1975) Diffraction of electrons by perfect crystals. In Electron microscopy in materials science, Vol. 2, eds. U. Valdre and E. Ruedl, p. 397. NATO Publications, Luxembourg, 397-552.

O'Keefe, M.A., Buseck, P.R. and Iijima, S. (1978) Computed crystal structure images for high resolution electron microscopy. Nature 274, 322-324.

O'Keefe, M.A. and Saxton, W.O. (1983) The 'well-known' theory of electron image formation. In 41st Annual Proc. Electron Microscopy Society of America, ed. G.W. Bailey, San Fransisco Press, San Fransisco, CA, 288-289.

Pierce, L. and Buseck, P.R. (1974) Electron imaging of pyrrohtite superstructures. Science 186, 1209-1212.

Saxton, W.O., O'Keefe, M.A., Cockayne, D.J.H. and Wilkens, M. (1983a) Sign conventions in electron diffraction and imaging. Ultramicroscopy 12, 75-78.

Saxton, W.O., Smith, D.J. and Erasmus, S.J. (1983b) Procedures for focussing, stigmating and alignment in high resolution electron microscopy. J. Microscopy 130, 187-201.

Self, P.G., Glaisher, R.W. and Spargo, A.E.C. (1985) Interpreting high-resolution transmission electron micrographs. Ultramicroscopy 18, 49-62.

Self, P.G., O'Keefe, M.A., Buseck, P.R. and Spargo, A.E.C. (1983) Practical computation of amplitudes and phases in electron diffraction. Ultramicroscopy 11, 35-52.

Smith, D.J., Saxton, W.O., O'Keefe, M.A., Wood, G.A. and Stobbs, W.M. (1983) The importance of beam alignment and crystal tilt in high resolution electron microscopy. Ultramicroscopy 11, 263-282.

Spence, J.C.H. (1981) Experimental High-resolution Electron Microscopy. Clarendon Press, Oxford, 370 p.

Spence, J.C.H. and Cowley, J.M. (1978) Lattice imaging in STEM. Optik 50, 129-140.

Spinnler, G.E., Self, P.G., Iijima, S. and Buseck, P.R. (1984) Stacking disorder in clinochlore chlorite, Am. Mineral. 69, 252-263.

Stadelmann, P.A. (1987) EMS - A software package for electron diffraction analysis and HREM image simulation in materials science. Ultramicroscopy 21, 131-146.

Weickenmeier, A. and Kohl, H. (1991) Computation of absorptive form factors for high-energy electron diffraction. Acta Crystallogr. A47, 590-597.

Wilson, A.R. and Spargo, A.E.C. (1982) Calculation of the scattering from defects using periodic continuation methods. Phil. Mag. A46, 435-449.

Chapter 4.   ANALYTICAL ELECTRON MICROSCOPY: X-RAY ANALYSIS

Donald R. Peacor   Department of Geological Sciences,
University of Michigan, Ann Arbor, Michigan 48109, U.S.A.

4.1 INTRODUCTION

Inelastic scattering of the electron beam by a sample gives rise to many effects, one of which is emission of x-rays whose energies are characteristic of the elements in the specimen. Figure 4-1 illustrates the principal components of an analytical system consisting of electron beam, sample, detector, and associated electronics system. Beam-specimen interaction gives rise to characteristic x-ray spectra that are detected with an energy dispersive spectrometer (EDS) system consisting of a Li-drifted Si detector and associated electronics. The resulting signal is processed through a multichannel analyzer (MCA) and stored in digital form. On-line display on a cathode ray tube (CRT) gives rise to spectra as shown in Figure 4-2. Modern systems are computer controlled with software that permits on-line identification of elements and their approximate proportions. Alternatively, spectra can be stored, retrieved at a later time, and analyzed with software that determines peak areas (intensities), makes appropriate corrections, and calculates compositions. Although chemical information can be obtained in many different ways using TEM/STEM, relatively complete chemical analytical data can be obtained only with such a system, and the process is therefore commonly referred to by the term analytical electron microscopy (AEM).

The marriage of x-ray detectors with TEMs, combined with advances in electron optics that permitted formation of scanning images and the development of STEMs, was a natural outcome of developing technology (see §1.2.2 for a description of TEM/STEM and dedicated STEMs). Primarily because of space limitations, energy dispersive spectrometers (EDS) rather than wavelength dispersive spectrometers (WDS) have been

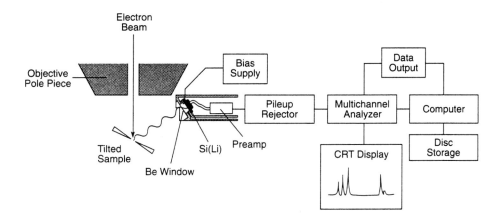

Figure 4-1. Principal components of an AEM system. Characteristic x-radiation and bremsstrahlung emitted by the sample is detected by an energy dispersive Si(Li) detector, collected in a multichannel analyzer, displayed on a cathode ray tube, and/or stored; computer software permits on-line or later analysis and display of spectra.

114

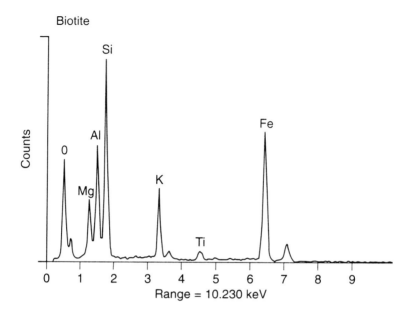

Figure 4-2. EDS spectrum of biotite. K_α peaks are labeled. Courtesy of W.-T. Jiang.

universally utilized. However, the first analytical instruments were little more than TEMs with spectrometers added without consideration of the special needs of such systems. They therefore gave rise to scattering of the beam with production of x-rays from areas of the specimen much larger than the effective beam diameter and from a great variety of scattering events in the column/specimen environment.

The early instruments ca. 1980 had two principal disadvantages: (1) the spectra were "dirty," with contributions from many sources other than the area of direct beam/specimen interaction, and (2) alignment of TEM/STEMs in TEM and STEM modes was separately carried out, so that shifts from one mode to the other were difficult and time-consuming, with resultant difficulty in recognition of areas in STEM mode that had been previously characterized in TEM mode. The recognition of the power inherent in combined TEM observations and accurate, high-resolution chemical analysis rapidly led to the development of TEM/STEM systems which were specifically engineered to optimize intensity, resolution, and sensitivity of x-ray spectra (i.e., a "clean" instrument) and, to provide easy transfer between TEM and STEM modes with direct correspondence between easily interpreted images. It is now possible to routinely obtain chemical analyses with resolutions approaching 20 nm whose quality approaches those obtained through electron microprobe analysis (EMPA).

The fundamental difference between electron microprobe analysis (EMPA) and TEM/STEM conditions is in the thickness of the sample. In EMPA the sample is thick enough to cause complete absorption of the incident beam; resultant spreading of the beam causes resolution to be approximately 2 μm. However, in TEM/STEM samples, little beam spreading occurs as the beam traverses the thin, electron-transparent sample. Advantages of EMPA include high precision and sensitivity and the ability to obtain absolute concentrations; however, EMPA data suffer from atomic number (Z), absorption

(A) and/or fluorescence (F) effects and have low point-to-point resolution. By contrast, STEM analyses have resolutions two orders of magnitude smaller than those of EMPA and usually require no ZAF corrections. However, they are of lower precision and sensitivity. The principal drawback however, is that absolute concentrations cannot be obtained; rather, only ratios of concentrations can be measured. Utilizing procedures introduced by Cliff and Lorimer (1975), mineral formulae are obtained by normalizing such ratios to a suitable absolute value. It is shown below that comparison of EMPA and TEM/STEM analyses of the same samples demonstrates that, under ideal circumstances, the quality of TEM/STEM analyses approach those of EMPA in most respects.

4.2 PRODUCTION OF X-RAYS

4.2.1 Inelastic scattering

Inelastic scattering, in which incident electrons lose energy, takes many forms. However, two processes dominate typical x-ray spectra as illustrated in Figure 4-2: (1) The continuous background, or "bremsstrahlung" radiation, and (2) The peaks super-imposed on the background, each of which corresponds to an energy, or equivalent wavelength, that is characteristic of the element that gave rise to the radiation. Such peaks are therefore referred to as "characteristic" peaks. Their intensity, or peak area, is proportional to the concentration of the corresponding element in the sample.

The continuous background is produced by deceleration of incident beam electrons through interaction with the charge field of the atom core and inner-shell electrons of specimen atoms. A given loss of velocity for a specific electron corresponds to an energy loss which is converted into an x-ray photon of wavelength λ:

$$E = h\upsilon = hc/\lambda .$$

Specific energy losses can vary from zero to E_0, where E_0 is the full energy of an electron ($E_0 = eV$, where V is the acceleration potential) that loses all energy in a single scattering event. The high energy cutoff of continuous background therefore varies with E_0. Bremsstrahlung intensity gradually increases with decreasing energy, at an increasing rate as $E = 0$ is approached.

Characteristic x-radiation is produced by ionization of specimen atoms by the incident beam as shown in Figure 4-3. If the energy, E_0, of the incident beam electron exceeds the binding, or ionization, energy of an inner-shell electron of a specimen atom, the inner-shell electron may absorb energy equal to or greater than the ionization energy and thus be ejected from the atom. The result is an ionized atom with outer electrons in an unstable state. Resultant relaxation occurs by transition of outer electrons from high potential energy states to the energy level of the ejected electron. The excess energy ($E_L - E_K$, as shown in Fig. 4-3) may be converted to an x-ray photon with wavelength

$$E_L - E_K = h\upsilon = hc/\lambda.$$

Because the energy difference is quantized and characteristically different for each element, the energy, or wavelength, of the x-ray photon is characteristic of the element. Although it is not relevant to AEM, it should also be noted that relaxation can occur through ejection of an Auger electron rather than through production of an x-ray photon.

Ionized atoms may undergo many possible transitions as illustrated in Figure 4-4 for Fe, and as determined by selection rules. They are identified with the Siegbahn nomenclature which is, unfortunately, not entirely logically based on quantum number

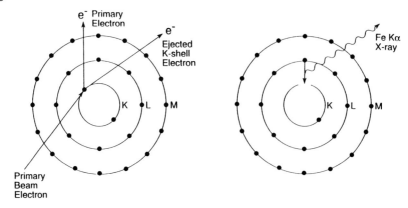

Figure 4-3. Production of characteristic x-radiation from an Fe atom. Transfer of energy of the primary beam electron (E$_o$) to a K-shell electron of Fe causes the K-shell electron to be ejected, leaving the Fe atom ionized. Replacement of the K-shell electron by an L-shell electron can give rise to a photon of x-radiation of energy E$_L$–E$_K$, labeled FeK_α (after Potts, 1987).

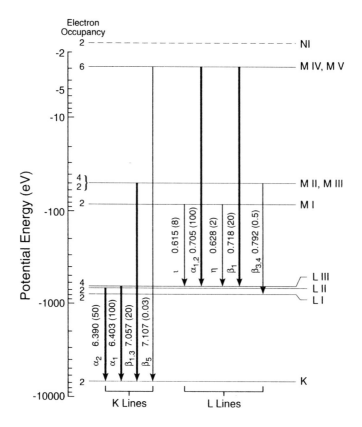

Figure 4-4. Characteristic emission lines of Fe. Relative intensities of lines are represented by the thickness of the arrows that represent electron transitions (after Potts, 1987).

considerations. As an example of the nomenclature, one Fe emission line is identified as FeK_{α_1}. The letter K identifies the K shell as having been ionized, and the symbol α corresponds to the energy level from which the replacing electron originated. The subscript 1 identifies a line that is one of two or more that are ordinarily observed as a single unresolved peak.

The energy of a given series of lines (e.g., K_α lines) increases as a linear function of Z^2 (Z = atomic number). Thus Z of the K_α lines in the EDS spectrum of Figure 4-2 increases from left to right. The energy of the lines of a given element decrease in the order $K > L > M$. Figure 4-4 illustrates the relative intensities of emission lines for Fe. The K_α line is always the strongest; they decrease in intensity in the orders: $I(K) > I(L)$...; $I(\alpha) > I(\beta)$...; $I(1) > I(2)$.... .

Few of the many emission lines are useful in AEM for two reasons: (1) most are too weak to be observed or to give rise to count rates high enough to yield precise measurements, and (2) there are limits to the practical detection range of EDS spectrometers (> 20 keV), simply because the Li-drifted Si is largely transparent to higher energy radiation, although Ge detectors are now available with improved detector efficiences in the range 20-40 keV. The K_α lines of low atomic number elements are used as they are the most intense and are within the detectable range, whereas L lines are utilized for elements of intermediate Z.

The continuous radiation is generally undesirable as it must be subtracted from the spectrum before measurement of the areas of characteristic peaks. High backgrounds therefore diminish the precision of characteristic peak area measurements. The background may be perturbed by a variety of factors (see below) that cause errors in the subtraction process. It is therefore essential that both the peak area (intensity) and peak to background ratio (P/B) be maximized.

4.3 ENERGY DISPERSIVE SPECTROMETERS (EDS)

4.3.1 Integrated EDS system

Space limitations usually preclude the use of wavelength dispersive spectrometers (WDS) in TEMs, but solid-state x-ray detectors are ideal because of their compact size. Such energy dispersive spectrometers are efficient because of relatively large collection angles. They simultaneously detect a wide range of energies, generally sufficient to include the characteristic lines of most elements, as illustrated in Figure 4-2. Their principal disadvantages include: (1) poor energy resolution over most of their useful range compared to WDS, resulting in overlap of peaks that are typically completely separated in WDS. As a result, peak stripping functions are especially important in EDS. (2) Absorption of low energy radiation by the window of the detector results in inefficient detection of elements with $Z < 11$ (Na). Detectors have been designed to circumvent that problem. (3) Artifacts in the spectrum, as discussed below.

The principal component of a solid state detector is a crystal of Si containing traces of Li (Li-drifted Si crystal) as diagrammed in Figure 4-5. X-radiation entering the detector causes electron-hole pairs to be generated whose number is proportional to the energy of the radiation. The resulting charge pulse is amplified and stored in a channel of a multi-channel analyzer that is calibrated for the given charge (energy of incident radiation). Intensity of incident radiation is determined by the number of pulses

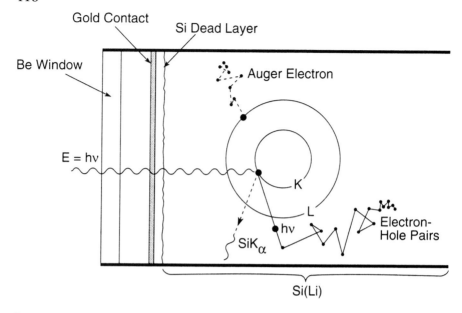

Figure 4-5. Representation of a part of a typical EDS detector, showing the Be window, gold contact layer, Si layer, and the Si(Li) crystal. The energy of the incident x-ray photon is dissipated in the crystal by a variety of inelastic interactions, the sum of which results in a pulse with current proportional to the energy of the incident photon (after Goldstein et al., 1981)

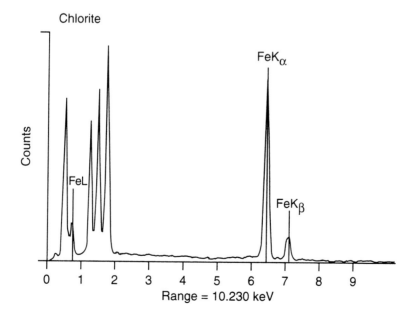

Figure 4-6. Typical EDS spectrum of chlorite, with characteristic Fe emission peaks labeled. Unlabeled peaks correspond, left to right, to O, Mg, Al, and Si. This spectrum was obtained using a Kevex *Quantum* detector. Courtesy of W.-T. Jiang.

accumulated in a given channel. When sufficient pulses have been obtained to define the spectrum, the contents of the MCA can be displayed on a CRT.

Modern spectrometer systems are computer-controlled with software that permits, among a wide range of options, immediate identification of all characteristic peaks, as shown in Figure 4-6, where the lines of Fe are identified. The contents of the MCA can be processed on-line or stored for later analysis. In either case, the manufacturer's software will have provision for processing the spectra by (1) subtracting the background curve that is primarily due to bremsstrahlung radiation, (2) "deconvolution" of overlapping peaks and determination of integrated intensities, (3) comparison of observed intensity ratios with those of standards in order to determine atomic ratios, and (4) calculation of normalized formulae from atomic ratios.

4.3.2 Types of detectors

The most common type of detector is constructed with a thin window of Be on the order of 10 μm thick in order to protect the crystal from contamination from the column environment. However, even such thin windows absorb incident x-radiation of low energy. The K lines of elements with $Z < 11$ are effectively entirely absorbed. Even NaK_α is significantly affected, with the result that peaks are severely diminished in intensity, with resulting imprecision in measurement of intensity; Na occurring in low concentrations may be entirely undetected. Although absorption of K_α lines decreases rapidly with increasing Z, the MgK_α, AlK_α, and SiK_α ($Z = 12$-14) are affected, and as they are characteristic emission lines of elements of critical importance in geological samples, the Be-windowed detector is not ideal.

The K_α lines of low atomic number elements are commonly measured with either a windowless or ultra-thin-window detector. Such detectors must be constructed with complex baffles to protect them from the atmosphere when not under vacuum. They are installed with low take-off angles, as is true for some other detectors, so that they can be placed near the specimen to give rise to as large a solid detection angle as possible, while avoiding the objective pole piece. They may be coupled with windowed detectors that are installed with high take-off angles, located above specially constructed objective pole pieces, and that are used for measurement of higher energy radiation.

Alternatively, relatively new windows made of a boron composite material have recently become available. Detectors with such windows can be installed at low angles to maximize solid detection angle and can be used for the full spectrum of energies. Nevertheless, even such detectors are relatively insensitive to radiation from low-Z elements, and results are only semiquantitative for O. Excellent results are obtained for Na, Mg, Al, and Si, however.

4.3.3 TEM/spectrometer systems

It is essential to obtain optimum count rates in order to cause measurements of integrated intensities to be precise, especially for K lines of low atomic number elements. For a given emission intensity, the count rate depends on two factors: (1) The active area of the detector, which is typically 30 mm^2, but may be as small as 10 mm^2 in early boron composite-window detectors. (2) Distance of the detector from the sample. The solid angle of detection is proportional to the inverse of the square of that distance. Efficient detection demands that it be optimized. That is accomplished by installation of the detector axis normal, or nearly normal, to the electron beam. The detector can thus avoid the objective pole piece.

The take-off angle, as measured from the specimen surface to the detector, is an important parameter. If the counter is nearly normal to the electron beam and the specimen surface is normal to the beam, then x-rays produced in the sample must travel along relatively long paths before exiting the sample, and absorption is severe, even though ultra-thin samples are used. The sample must therefore be tilted toward the detector where detectors are oriented normal to the column. That is a time-consuming operation that may require specimen reorientation. On the other hand, as detector angles increase, interference with the pole piece requires that distance to sample be increased with resulting loss of counting efficiency. Take-off angles of approximately 20° are reasonable compromises.

High-angle, windowed detectors can therefore be permanently emplaced and utilized with horizontal samples, but must be combined with, say, ultra-thin-window detectors at low angles for light element detection. On the other hand, the single detector at low take-off angles, especially the mechanically-simple boron-composite windowed detector, is a great convenience in measuring the full range of x-radiation, but with the inconvenience of tilting the sample prior to analysis. A compromise in which the specimen need not be tilted can be achieved with a 20° take-off angle.

4.4 TEM/EDS SYSTEM CONCERNS

There are a number of factors that can cause x-rays to be generated outside of the area of direct beam-specimen interaction, or cause an increase in intensity or irregularities in background, or a decrease in intensity of characteristic peaks, and occurrence of detector artifacts. Although such factors are usually of little consequence in qualitative analysis, it is critical that the operator be aware of them and take appropriate precautions, especially as they may give rise to peaks of elements that are not in the sample. Three sources of problems are briefly discussed in this section: (1) spurious sources of radiation, (2) false peaks arising through the detector system, and (3) electron channeling.

4.4.1 Spurious sources of radiation

Characteristic and bremsstrahlung radiation can be produced at many sites within a TEM other than in the area of direct specimen-beam interaction. Those sources can be induced either by pre- or post-specimen environments.

Electrons that are poorly collimated can excite any area of a specimen. In addition, x-rays, principally bremsstrahlung, are produced in the C2 aperture (see Fig. 1-1) and can, if the aperture is thin enough, be transmitted to the specimen and excite emission of characteristic radiation. Such sources of spurious radiation can be detected using a "hole count," obtained by centering the beam on a small hole in an otherwise continuous specimen. Ideally, no counts should be obtained. Modern analytical instruments now normally have special apertures designed to minimize bremsstrahlung and have been engineered so that stray electrons have essentially been eliminated. It is essential that the correct analytical aperture be used during analysis, however.

Stray radiation is also produced through beam-specimen interaction, resulting in production of x-rays from many sources. There are three principal kinds of sources:

(1) Back-scattered electrons cause x-rays to be produced from the specimen chamber. (2) Incident beam electrons are scattered or diffracted into the post-specimen environment, causing direct production of x-rays, and back-scattered electrons. X-rays so-produced may be directly detected or may in turn cause fluorescence of lower energy x-radiation in the specimen. (3) The characteristic and bremsstrahlung radiation from the specimen itself cause fluorescence of x-rays from other areas of the specimen, or if the specimen is mounted on a grid, from the grid. Cu peaks are inevitable for samples mounted on Cu grids, a practical demonstration of production of x-rays from extraneous sources. For this reason, low atomic number grids (e.g., Be) that are readily available through commercial sources should be used for crushed grain mounts; the characteristic radiation so-produced cannot cause fluorescence of characteristic radiation in the sample. As the specimen is tilted to higher angles, such interaction becomes more significant. Some radiation of these kinds is unavoidable, although it can be minimized by construction of the specimen environment with low atomic number materials.

Tests for spurious sources of radiation, as well as of the effective diameter of interaction of beam and specimen, can be carried out using a thin specimen with two minerals of markedly different composition, but having an interface parallel to the beam, and with the area occupied by one approaching that of the effective beam diameter. Interaction of the beam with the latter phase should not give rise to x-rays from the former.

4.4.2 Detector system-related spectrum features

The EDS spectrum that is collected in the MCA and ultimately displayed should ideally contain only the characteristic peaks superimposed on the bremsstrahlung with an intensity distribution that is in direct proportion to the actual x-ray intensity distribution. Several processes inherent in the detection system give rise to changes in that ideal configuration. As illustrated in Figure 4-7, The most significant of these include: (1) escape peaks, (2) sum peaks, (3) asymmetry in background on the low energy side of peaks, and (4) background distortion due to high energy x-rays and back-scattered electrons.

The detector pulse caused by a photon of x-radiation that enters the detector may be activated by ionization of a K-shell electron of a Si atom. The ionized Si atom may relax through production of a photon of SiK_α radiation of energy 1.74 keV as shown in Figure 4-5. That photon is ordinarily absorbed before exiting the detector, in which case that energy is retained and counted as part of the original excitation event. However, the SiK_α photon may be lost from the detector. In that case, the energy retained is E(exc)-$E(SiK_\alpha)$, and a peak—called the escape peak—occurs in the spectrum at 1.74 keV less than that of the incident exciting radiation. That is, if MnK_α (E = 5.90 keV) enters the detector, a small escape peak will occur at (5.90-1.74) keV, in addition to the large peak at 5.90 keV. Such peaks are small, being on the order of magnitude of 1% of the principal characteristic peak. They may be erroneously attributed to trace amounts of some other element, however. They are generally subtracted from the spectrum in proportion to the intensity of characteristic peaks during standard spectrum processing.

Sum peaks occur at energies that are the sum of two of the energies of incident radiation. They occur when two photons that enter the detector at the same time cannot be differentiated by the counting circuitry, and are treated as a single photon. That is, if two SiK_α photons (E = 1.74 keV) were simultaneously counted, a small sum peak with E

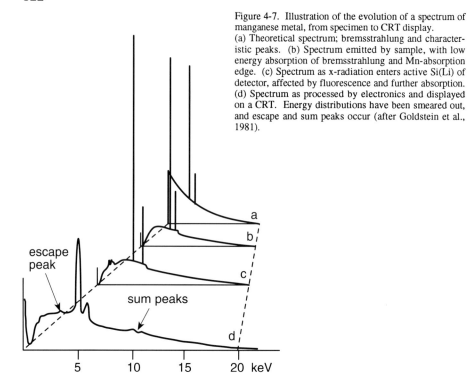

Figure 4-7. Illustration of the evolution of a spectrum of manganese metal, from specimen to CRT display.
(a) Theoretical spectrum; bremsstrahlung and characteristic peaks. (b) Spectrum emitted by sample, with low energy absorption of bremsstrahlung and Mn-absorption edge. (c) Spectrum as x-radiation enters active Si(Li) of detector, affected by fluorescence and further absorption. (d) Spectrum as processed by electronics and displayed on a CRT. Energy distributions have been smeared out, and escape and sum peaks occur (after Goldstein et al., 1981).

$= 2 \times 1.74$ keV would occur. Such peaks are generally not observed in AEM owing to the low count rates from thin specimens.

Peaks are slightly asymmetric with enhancement of the low energy limb and with increased background. This is caused by loss of some of the energy of some incident photons during the detection process. Such "shelving" and "incomplete charge collection" give rise to a slightly enhanced background on the low energy side of the peak, with the effect decreasing to lower energy values.

The accelerating voltage used in a TEM is much higher than the value of approximately 15 kV that is typically used in an SEM or EMPA. The bremsstrahlung therefore has a high energy component that gives rise to Compton scatter in the Si of the detector, resulting in enhancement of the background at high energies. In addition, relatively high energy back-scattered electrons may enter the detector and cause distortion in the background. Both of these effects alter the background relative to its theoretical form.

There are two other minor features that ordinarily go unnoticed. One is a very small Si peak caused by fluorescence of a neutral layer of pure Si at the leading edge of the Li-drifted Si crystal. It would give rise to an apparent Si content of 0.2 wt % in a Si-free sample. The other is two absorption edges in the bremsstrahlung spectrum. One is caused by a thin film of Au (used to apply a bias across the detector) on the Si crystal, and the other by Si itself. The latter is the result of the same process giving rise to the Si emission peak.

4.5 SPECTRUM ANALYSIS

4.5.1 Background subtraction

The final spectrum stored in the MCA consists of the characteristic peaks super-imposed on the continuous bremsstrahlung that has been modified as described above. In order to obtain the integrated intensities of peaks it is necessary to remove the background. That is accomplished in one of two ways: (1) filtering, using a "top-hat" function, or (2) direct calculation of background.

The top hat filter is a function in which the counts in a series of adjacent channels of the MCA are averaged. The averaged value is assigned to the central channel. The series is shifted one channel and the process repeated until the entire spectrum has been traversed. The top hat actually has three sections, a central series of channels and two adjacent series. The filtered value is the difference between the average central value and the average side value. The filtered spectrum is a smoothed function in which linear sloping backgrounds are entirely suppressed. The advantage of such a method is that it is independent of the detailed specific aberrations in the spectrum, giving rise to a spectrum for which the peaks can be directly interpreted. Interpretation is normally carried out by comparison of filtered spectra with a set of stored "library" peaks.

Alternatively, the background can be calculated and subtracted from the experimental spectrum. The background is generally calculated for a specific region of the spectrum of interest and then normalized to the specific spectrum. In computerized systems using such functions, the calculated background curve in the area of interest can be directly superimposed on the spectrum displayed on the CRT. If the operator observes that it is apparently not at the proper level, the operator may adjust the relative vertical position of the background curve until an appropriate fit is obtained.

It is essential that manufacturer's software systems be tested using well-defined standards, and not treated as "black-boxes." We have recently found, for example, that software-generated backgrounds may be normalized to values that are too high where several characteristic peaks are adjacent, as for Na, Mg, Al, and Si. For example, the small Mg concentrations in Na-containing micas may be inaccurate by factors of 2 or 3.

4.5.2 Determination of peak intensities

It is a trivial matter to determine the integrated intensity of a peak that is entirely separated from other peaks. However, when peaks overlap, one of several different procedures must be used to unravel the combinations. These have such names as deconvolution, curve fitting, filtered least-squares fitting, etc.

So-called deconvolution may be accomplished by modeling the peaks of individual elements in terms of amplitude, width, and position. Such individual peaks can then be combined and compared with the overlapping peaks of the observed spectrum. The relative contributions of the separate contributions can be varied until a best fit is obtained. Alternatively, "library" spectra of the separate elements, with properly weighted consideration of the multiple lines occurring for a given element (e.g., K_α and K_β), can be directly compared with the observed spectrum, optimizing the fit. The operator chooses the elements for which the search is to be made based on a combination of an observation of the spectrum that usually results in identification of all elements of interest, or on knowledge of which elements are to be expected in the mineral being analyzed. The result is a set of integrated intensities for each element that can be used in the determination of a normalized chemical formula. It is essential that the EDS spectrum energy scale be properly calibrated (an adjustment that can usually be made with peaks that occur at both the high- and low-energy region of the spectrum), so that the energy values for stored peaks may be correlated with the scale of observed spectra.

4.6 INSTRUMENTAL CONDITIONS

4.6.1 Instrument mode

It is, in general, possible to obtain EDS analyses in one of two ways: (1) In TEM mode by focusing the beam onto the area of interest. This technique gives the largest point-to-point resolution of several tens of nanometers. Beam damage can be minimized by defocusing the beam, but at the expense of a larger area of analysis. (2) In STEM mode in which the beam is focused to a fine probe whose size can be varied, but which can be on the order of 2.5 nm in diameter at the specimen. Analysis can be carried out with a stationary beam, or the beam can be rastered over the sample, with the length and width of the analyzed area controlled by the operator. One advantage of the method is that it permits the effects of the beam on any one point to be minimized, thus diminishing the effect of element diffusion (see below). In addition, element maps can be obtained (Fig. 4-8 below) for elements with relatively high concentrations, and compared with secondary, back-scattered, or transmitted electron images.

Operation in STEM mode was difficult in the earliest instruments, as they required separate alignments in TEM and STEM modes. Although features could be identified at a given site of a sample as viewed in TEM mode, transfer to STEM mode required lengthy alignment, commonly with loss of location in the specimen. The lens currents of modern computer-controlled STEMs can be stored, however, and transfer between TEM and STEM modes in well-aligned instruments requires only a few seconds.

4.7 METHODS OF DETERMINATION OF COMPOSITION

4.7.1 Cliff-Lorimer relation

The intensity of a given emission line emitted by a sample is proportional to the concentration, c_i, of a given element, with proportionality factors that relate to the physics of beam-sample interaction. In EMPA, the beam penetrates the sample to considerable depth, energy decreasing with penetration (Fig. 4-9). The intensity of x-ray production is therefore a function of atomic number (Z) effects (electron backscatter and retardation). The resulting emitted x-rays must pass through the sample and are subject to absorption (A) and fluorescence (F) effects. The observed intensity must therefore be corrected for these effects, each of which is a function, in part, of the composition of the sample.

However, in the thin electron-transparent films that are generally (but not always) used in AEM analysis, only a small proportion of the beam energy is lost through retardation in the sample and back scattering is inconsequential, so the beam may be considered to be of constant energy as it passes through the sample. Paths of emitted x-radiation from points of emission through the specimen to the surface are so short that absorption and fluorescence can usually be neglected.

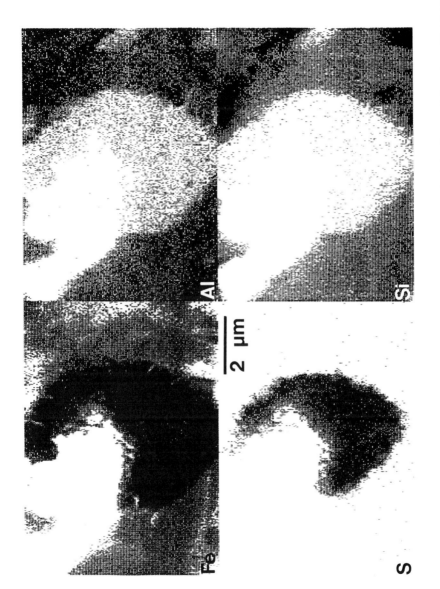

Figure 4-8. X-ray map of a pyrite framboid crystal in a matrix of illite, with the ion-milled thin edge bordering the upper left quadrant. The high Fe and S areas correspond to pyrite, but the upper left portion of the crystal has high Fe and no S contents, corresponding to magnetite that has replaced pyrite. The Si,Al-rich illite has a significant phengitic component as represented by the Fe map. Courtesy of W.-T. Jiang.

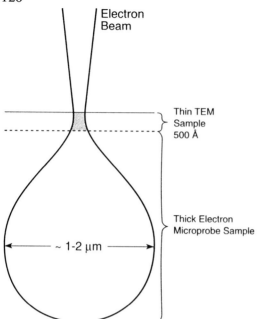

Electron
Beam

Thin TEM
Sample
500 Å

Thick Electron
Microprobe Sample

~ 1-2 μm

Figure 4-9. Diagram showing spreading of the electron beam within a thick electron microprobe or scanning electron microscope sample. Very little increase in effective beam width occurs for the first few tens of nanometers, corresponding to thin AEM samples. The analyzed area is thus only slightly larger than the beam diameter for AEM.

X-rays must penetrate the window of the detector, however, and are absorbed, especially in the case of K_α lines of low atomic number elements. In a standard Be-window detector, absorption occurs through the Be, a thin, dead layer of inactive Si, and a Au film. Note that if the same detector is used for standards and unknowns, the window absorption factor is negated.

Cliff and Lorimer (1975) applied these relations and formulated the equation:

$$c_a/c_b = k_{ab}(I_a/I_b)$$

where c_a and c_b are the atomic concentrations of elements a and b in a given compound, I_a and I_b are the intensities of emission lines of those elements, and k_{ab} is a proportionality factor that is a function of parameters concerned with beam/specimen interaction. If the compound contains only elements a and b, then absolute concentrations can be obtained by normalizing using the relation:

$$c_a + c_b = 1$$

If k_{ab} is experimentally determined for a standard compound of known composition, then the measured intensity ratio for a compound of unknown composition can be directly converted to a concentration ratio. The disadvantage of this technique is that absolute concentrations cannot be directly determined as they can by EMPA because the effective volume over which emission occurs cannot be determined. The principal advantage, however, is that ZAF corrections need not be applied as in EMPA, as the volume over which emission occurs is usually so small as to cause those factors to be inconsequential. Even where there is some ZAF effect, if the effects are for elements of similar atomic number, the effects largely are canceled through ratioing of the intensities.

4.7.2 k-values

It is traditional to determine the ratio of intensity of an x-ray line of a given element to that of Si because Si is the element most likely to occur in minerals, and its intensity

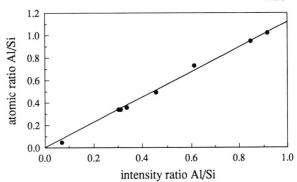

Figure 4-10. Plot of Al/Si atomic ratios versus measured Al/Si intensity ratios for several standards in the author's laboratory, with a least-squares-determined line fo best fit. The slope of the line is the k-value. Courtesy of W.-T. Jiang.

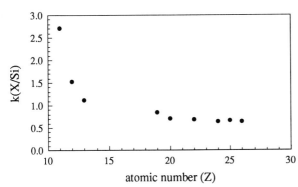

Figure 4-11. Plot of k-values ($k_{X,Si}$) used in the author's laboratory as a function of atomic number (Z) of element X. Courtesy of W.-T. Jiang.

can be accurately determined (as opposed to that of oxygen). It is possible to determine the intensity ratios for an element, say Al, in several standards having different ratios c_{Al}/c_{Si} (Fig. 4-10). The k-value, $k_{Al,Si}$ in this case, is determined by the slope of the line. The multiple determinations of k provide an internal check on the validity of the values (e.g., confirming that measurements were made in the true thin-film condition) as they should all plot on the same line.

The k values vary regularly as a function of atomic number. Figure 4-11 shows some values used in the author's laboratory. Where standards are not available for some element, approximate k-values can be obtained by interpolation between known values. If the concentration of the element in question is small, the resulting error in absolute concentration is negligible.

The k-values that are obtained by ratioing to Si can be used in the calculation of other k-values. The relation

$$k_{a,Si}/k_{b,Si} = k_{a,b}$$

can be used to calculate k-values for combinations of elements a and b that occur in the absence of Si.

4.7.3 Standardless analyses

The k-values are dependent on various instrumental factors, most notably beam potential, and thus vary with changes in electronics. The changes are small, however, and because they occur in an approximately proportional way for factors in both numerator and denominator of ratios, the k-factors are, for all intents and purposes, constants that can be stored and utilized long after they have been measured. This is valid only for values obtained under specific conditions for a specific instrument. The one caveat to this has to do with the counter characteristics. Should the counter window or crystal be replaced, large changes in k-values are possible and standard k-factors must be redetermined.

Because the k-factors are permanently stored, they can be retrieved at the time of analysis and utilized to obtain concentration ratios in unknowns. Although intensity data for unknowns are processed using standard k-factors, no standardization is usually necessary at the time of analysis. Indeed, some use the term "standardless" for such AEM analyses for that reason, although the term is also used where the k-factors are calculated from first principles, as reviewed by Goldstein et al. (1986). Because data for standards need not usually be obtained, AEM analysis has a distinct advantage as compared with EMPA, for which much time is required for determination of standard data during the session in which analyses are obtained. Indeed, a small number of quantitative analyses can generally be obtained very quickly by utilizing crushed grains on formvar films on grids, as compared with the time required for EMPA. Rapid preparation of samples for which no polishing is required, combined with "standardless" analysis, gives rise to very rapid analysis.

There is one important exception to the general rule that data for standards need not be measured in a given session. That occurs when beam damage causes loss of an element from the beam-specimen interaction volume. Alkali elements, particularly K and Na, are especially subject to this problem. Special experimental techniques that are required are described below.

4.7.4 Normalization procedures

The concentration ratios obtained using the Cliff-Lorimer method must be normalized to some quantity in the chemical formula. It is common practice to normalize EMPA data to mineral formulae by normalizing to the numbers of O atoms in a standard formula unit, or to a number of O atoms that would be present if all anions such as OH were present as O. Thus, mica formulae may be normalized to 11 (or 22) O atoms   (10 O + 2 OH pfu).

In the cases of minerals such as most olivines and orthopyroxenes for which there are few analytical problems, such a procedure may be appropriate. However, there may be uncertainties that cause such a procedure to give rise to misleading results: (1) Elements such as Fe may occur with more than one valence. Although valence can often be independently determined where analyzed volumes are large, the ultra-small volumes typical of AEM analyses negate even that procedure. (2) There is a tacit assumption that all elements have been accounted for in the analysis. Trace amounts of several elements may be unaccounted for, the levels of detectability being greater for AEM than EMPA. Light elements (e.g., N in NH_4) may remain undetected. (3) Diffusion may cause concentrations of alkali elements to be too low.

For those reasons, it is recommended that, whenever possible, normalization be carried out on the basis of some number of cations where such a space-group-determined number can reasonably be assumed on the basis of crystal chemical relations. For example, the sum of tetrahedrally-coordinated cations in alkali-containing tektosilicates is generally a constant, and normalization to such a value avoids the problem of alkali diffusion. In those structures that contain Fe, charge balance is subsequently obtained by adjusting the Fe^{3+}/Fe^{2+} ratio.

There is no convenient number of cations that is known on the basis of crystal structure refinements to be constant for some formulae. Amphiboles are particularly troublesome in that regard, for example, because of variation in A-site occupancy. Such problems are discussed for phyllosilicates in more detail in Chapter 9. In general, for those structures where there is some question about the constancy of some number of cations, normalization to cations introduces an error.

Thus, for structures such as amphibole where there may be both Fe^{3+} and Fe^{2+}, or Li that has not been detected, normalization to anions may be in error; however, because of ambiguity in A-site occupancy, normalization to cations is problematic for many, but not all, amphiboles. Each case must therefore be considered on its own merits, but it is essential that the choice of normalization be based on a full interpretation of the reasonable crystal chemical relations for the mineral in question while considering the analytical problems applicable to that mineral. Although I generally prefer cation normalization, there are many cases where anion normalization is preferable. In any event, normalization should not be automatically carried out with a single method, especially one that has been made available in "black box" computer software.

It is possible to compute weights percent from normalized formulae and to present an analysis in the form of oxides in the traditional fashion. Such a process is artificial however, and may be misleading for those who are not familiar with AEM techniques; it derives from analytical techniques for which individual weights percent represent directly observed data on an absolute scale, and for which standard errors can be easily estimated; by contrast, the errors associated with weights percent derived from AEM data are dependent on many assumptions in the normalization process in addition to the errors of analysis. Although it is seldom done, it is the concentration ratios that should normally be presented as the basic AEM data.

4.8 EXPERIMENTAL CONSTRAINTS

4.8.1 Specimen preparation

Samples of minerals are ordinarily prepared in one of three ways: (1) Fine particles spread on a formvar or C film over a metal grid. (2) Slicing with a diamond microtome, a process often used with clay mineral samples. (3) Ion milling, a process generally used when one wishes to retain original sample textures (see §1.6.2, this volume for detailed description).

The beam of high energy Ar ions not only removes atoms from the specimen surface, but damages the surface structure. Ar ions may be implanted in the surface, as frequently observed through a small Ar peak in EDS spectra of ion milled samples. More importantly, displaced surface ions may be reimplanted at small distances from original sites, resulting in apparent trace amounts of ions. The thinning rate varies with angle of the Ar beam to the surface, being greater at high angles. It is therefore advisable to thin rapidly at a high angle, but to change to a low milling angle or beam of lower energy at the end of thinning in order to minimize surface contamination. Because the sample is heated in the ion mill and therefore subject to damage, it may also be advisable to utilize sample cooling with liquid N_2.

4.8.2 Sample damage

The effects of direct atom displacement, ionization, and heating give rise to a variety

data for K, Al from muscovite, for Na from paragonite

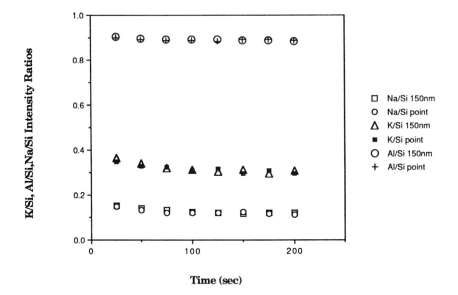

Figure 4-12. K/Si, Na/Si, and Al/Si intensity ratios for increasing time of analysis for constant beam and sample position. Data are plotted for both point (STEM mode) and 150 × 150 nm raster. The effect of diffusion (decreasing intensity) is large for K, intermediate for Na, and small for Al. Courtesy of G. Li.

of kinds of specimen damage that can cause serious AEM analysis errors. Figure 4-12 shows the variation in counts for K, Na, and Al relative to Si, obtained from ion-milled muscovite and paragonite with a stationary beam in STEM mode as a function of counting time. The change in count ratio is caused by diffusion, as can be detected by moving the beam to an area adjacent to beam-specimen interaction, where elevated counts are detected. The decrease in intensity is greatest for K, intermediate for Na, and smallest (but still measureable) for Al.

Figure 4-13 shows that the effect varies as a function of thickness. The data for K/Si and Al/Si counts were obtained with increasing distance from the edge of an ion-milled sample. Count rate increases because specimen thickness increases away from the edge. However, the K/Si intensity ratio continues to increase long after the Al/Si ratio has reached a constant value, implying that K diffusion occurs even for relatively thick samples.

Figure 4-13 also indicates that the intensity ratios vary as a function of area of analysis. At low count rates (thin edges), point analyses for both K (especially) and Al are subject to diffusion, whereas rasters 300 × 300 nm appear to be minimally affected by diffusion.

The effect increases with increasing count time, but sufficient counts must be obtained in order to optimize precision; thus, increasing accuracy (minimum diffusion) must be balanced against decreasing precision. In addition, the effects are minimized in relatively thick areas, but here too, accuracy is sacrificed due to absorption and fluorescence effects.

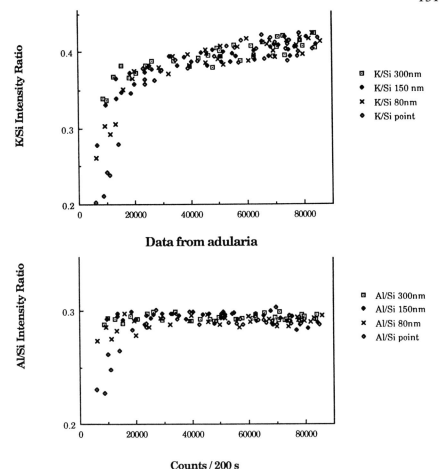

Data from adularia

Counts / 200 s

Figure 4-13. K/Si and Al/Si intensity ratios for increasing thickness of an ion-milled sample; increasing count rate is a measure of increasing thickness. The Al/Si intensity ratio approaches a constant, diffusion-free value, but that of K/Si is affected by K-diffusion at all thicknesses. Intensity ratios for different raster areas show that point analyses, especially, are subject to severe diffusion problems at very thin edges. Courtesy of G. Li.

These competing factors can be optimized by obtaining analyses in electron-transparent areas, but distant from ultra-thin edges in ion-milled samples, and by using a raster that is as large as possible with respect to the size of the grain to be analyzed and its relative homogeneity; the latter is determinable using a focused stationary beam in STEM mode, albeit in a qualitative way. If there is any doubt regarding the occurrence of diffusion, it should be validated by observing counts as a function of time; a decrease in number of counts per unit time may be ascribed to diffusion.

Mackinnon (1990) has shown how the rate of diffusion varies as a function of temperature. He showed that there was negligible diffusion of Na and Al in albite when a liquid N_2 stage was used, and he gave a detailed analysis of the effect. Diffusion can be approximately accounted for by obtaining data on a standard during the same work session and under the same conditions as for the unknown, the effects of diffusion being canceled in the ratioing process. Because diffusion is thickness-dependent, count rates for both standard and unknown should be identical (assuming identical compositions), as an indicator of similar thicknesses, assuming identical microscope parameters. Because the effect is structure-dependent (and even orientation-dependent in micas) the standard used should be of the same structure type as the unknown. Serious errors in alkali contents can still occur in analyses of clay minerals such as smectite or illite, as the alkali diffusion rate is found, qualitatively, to be dependent on defect state. Thus, even muscovite, which is relatively defect-free, is an imperfect standard for the relatively defect-rich illites occurring in low-grade pelites.

4.8.3 Electron channeling

ALCHEMI (atom location using channeling-enhanced microanalysis) is a technique whereby the electron beam is concentrated on different levels of planes of indices hkℓ when a crystal is in diffracting condition for hkℓ, depending upon incident beam orientation (see §5.3 of this volume for a detailed discussion). Thus, if such alternate planes have different electron densities (e.g., consist of alternating planes of two different kinds of cations), the proportion of x-rays generated by one set of atoms or the other will depend on small differences in orientation. The technique is a useful one for determining ordering in alternate sites of certain phases. However, it can introduce serious errors into ordinary EDS analysis. Analyses should therefore always be carried out when the sample is not in a strong diffracting orientation.

4.8.4 Limits of thin film criterion

Goldstein et al. (1986) have reviewed the theoretical form of absorption and fluorescence corrections. Although they can be calculated to a first approximation, in practice such corrections are inaccurate. They do give an estimate of the relative significance of the corrections however. Such calculations show that absorption is by far the more significant factor, varying with absorption coefficient (which generally increases with increasing atomic number) and specimen thickness. The effective specimen thickness is in turn a function of tilt angle and sample shape. Absorption is significant primarily when one sample element has an absorption edge with energy slightly less than the higher energy emission line of another element. Absorption will occur through ionization of the inner shell corresponding to the absorption edge, causing emission (fluorescence) of characteristic radiation. Such absorption and emission occur for elements with atomic numbers slightly less than that of the element that gave rise to the initial emission radiation. High values of fluorescence for one element (lower atomic number) are thus correlated with high values of absorption for the higher atomic number element. The fluorescence correction factor is also a function of thickness.

In order to calculate correction factors for fluorescence and absorption, sample thickness must be determined. Measurements of thickness are difficult, time-consuming, and subject to error. Moreover, they must be made for each analysis point. The fluorescence and absorption functions are therefore best used with approximations of thickness to determine if absorption (or fluorescence) is significant; that is, approximate correction factors can be calculated for a given estimated composition in order to determine if they are significant within the precision of counting statistics. If correction factors are within precision as calculated with standard counting statistics, correction is not necessary.

It is much easier to make a simple, practical test of thin-film limits using ion-milled standards having continuously increasing thickness away from an edge, of known homogeneous composition, and with composition (and therefore absorption coefficient) similar to that of unknowns. Spectra should be measured in a traverse extending normal to the thin edge. Count rates should increase in proportion to thickness and, more importantly, intensity ratios should remain constant. As the thin-film criterion is exceeded, the rate of increase in count rate will diminish and intensity ratios change. The latter effect is best tested with intensity ratios involving both low- and high-energy lines, for which absorption would be optimally different; intensity ratios are minimally affected for an intensity ratio involving two emission lines of similar energy (assuming that absorption edges do not intervene) as for FeK_α and MnK_α, for example—a distinct advantage of the ratio technique. The operator rapidly becomes familiar with the qualitative relation between sample transparency and the thin film criterion for typical rock-forming silicates. Such measurements show that the the thin-film criterion is fulfilled for rock-forming silicates to limits of thickness of several tens of nanometers.

4.8.5 Limits of spatial resolution

A definition of spatial resolution is ambiguous because the limits of the beam itself are ill-defined. A practical definition is one based on inclusion of 90% of the generation of x-rays. Modern STEMs are capable of production of probes with effective diameters on the order of 2.5 nm. Nevertheless, even though beam spreading is minimized by the use of thin films, some beam spreading is inevitable, resulting in some loss of resolution. Beam spreading is a function of many variables, but the most important is specimen thickness. Various determinations of the effect indicate that order of magnitude broadening of 10 nm can be expected for mineral sample thicknesses of 100 nm.

Even 90% of x-ray production for areas with diameters on the order of magnitude of 10 to 20 nm may not be acceptable. A test of resolution can be carried out by utilizing a specimen containing two adjacent minerals having a boundary parallel to the beam, each with at least one significant element not present in the other. Calcite (Ca) and quartz (Si) are two such minerals. Counting for Ca, for example, should be carried out as the boundary is approached from the quartz side. As the calcite begins to interact with the limits of the beam, small amounts of Ca will be detected, with the amount increasing to a constant value when the beam is entirely within calcite. Such experiments indicate that resolutions on the order of 20 nm are possible in thin areas. Indeed, field emission STEMs have probe diameters as small as 1 nm with resultant resolution of a few nanometers.

4.8.6 Limits of sensitivity

Two factors determine the minimum detectable concentration of an element: (1) The presence of peaks that overlap the peak of interest or other factors that locally affect the background in the region of the spectrum that is of interest. Such factors were discussed above. Discussion of minimum detectable concentrations generally are made with the assumption that there are no overlapping peaks and background is a smooth, easily interpretable function. (2) As the concentration of a given element decreases, a point is reached at which the peak can no longer be discriminated from the background. The minimum concentration that can be detected is therefore determined by a statistical analysis of the precision with which such a peak can be detected.

X-ray counts obey Gaussian statistics; the standard deviation, σ, of a given number of counts, N, is given by $\sigma = \sqrt{N}$. At the 3σ confidence level, it is therefore necessary that

134

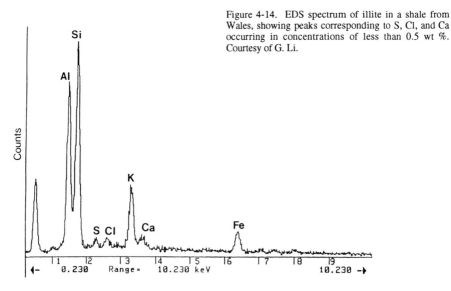

Figure 4-14. EDS spectrum of illite in a shale from Wales, showing peaks corresponding to S, Cl, and Ca occurring in concentrations of less than 0.5 wt %. Courtesy of G. Li.

the number of counts in a peak be greater than $3\sqrt{N}$. As the number of counts, N, increases, the relative error, $3\sqrt{N}/N$, decreases.

The minimum concentration is therefore determined by the following three parameters: (1) counting time, which should be maximized in order to minimize relative error within the bounds of convenience, sample damage, and sample position. (2) Peak counting rate, as determined by sample thickness and beam current. The need for high count rates must be balanced against absorption and specimen damage. (3) Peak-to-background ratio; subtraction of background from (peak + background) gives rise to an error in the peak intensity that is greater for higher backgrounds. For a given accelerating voltage, this is optimized at high take-off angles.

The actual limits of detection are best determined in a practical way by measuring intensity for a given element with decreasing concentration or with decreasing count times and peak count rates for low levels of concentration. Whereas EMPA data generally have minimum detection limits in the range of 0.01 to 0.1 wt %, AEM data are usually assumed to have limits on the order of a few tenths of one wt %. The latter value is valid for instruments designed several years ago, but modern instruments are capable of much better performance. Figure 4-14 exemplifies analysis for a sample with very small concentrations of several elements. The sample was muscovite from a Welsh slate, with small, but unusual S, Ca, and Cl peaks, and with minor amounts of Ti, and Cr. The S, Ca, and Cl are not generally expected in muscovite but were validated in several ways as occurring in this case. The formula obtained from a single analysis is

$$\left(K_{1.77}, Ca_{0.15}\right)_{1.92}\left(Al_{3.37}, Fe^{3+}_{0.55}, Cr_{0.05}, Ti_{0.03}\right)_4\left(Si_{6.01}, Al_{1.99}\right)_8 O_{20}\left(OH_{3.74}, Cl_{0.19}, S_{0.07}\right)_4.$$

The calculated wt % of S is 0.26%. Figure 4-14 shows that the minimum detectable limit is considerably below that value. In addition, the small but measurable Ti and Cr peaks correspond to 0.31 and 0.47 wt %, respectively.

Measurements such as the latter clearly show that AEM is routinely capable of detecting amounts on the order of 0.1 wt %, although for low atomic number elements

$(Z < 10)$ that limit is much larger. Nevertheless, the detection limits for most elements of geological significance now approach those typical of EMPA, with a limit near 0.1 wt %.

4.8.7 Accuracy and precision

The precision of a value of concentration of a given element relative to that of Si, c_a/c_{Si}, is determined simply by statistical relations based on the Cliff-Lorimer relation, where

$$c_a/c_{Si} = k_{a,Si} \times (I_a/I_{Si}) .$$

The standard error in c_a, $\sigma(c_a)$, is therefore the sum of the standard errors in $k_{a,Si}$, I_a, and I_b. The errors in k-values can be measured to relative precisions of better than $\pm 1\%$ if multiple standards are used, and especially if least-squares analysis is used to obtain k-values. The precision of the relative concentrations can then be readily calculated on the basis of counting statistics. Such standard errors are a normal component of the output of interpretation of spectra by commercial software. They in no way reflect accuracy, however.

Accuracy is affected by a number of factors, including beam-induced diffusion (especially of K; see above), absorption, and fluorescence. Because precision (total counts) increases with thickness, optimum conditions correspond to maximum thickness (assuming that the count-rate limit of the detector is not exceeded) as limited by the constraints of absorption and fluorescence. As with EMPA data, formulae should be normalized as soon as possible relative to the time of data measurement, as the basic requirement for analyses is that they produce formulae that are sensible relative to a known structure; possible errors should be detected at a time when they can be corrected. Such an initial test is available for EMPA data in that weights percent determined automatically just subsequent to analysis must sum to 100%, but AEM data can only be judged on the basis of the normalized formulae.

Tests of true accuracy of AEM results can best be obtained by data from standards. The relative accuracy of AEM data and the more familiar EMPA data can be determined from the same samples. Table 4-1 compares pyrophyllite and muscovite data as obtain-

Table 4-1. Structural formulae determined by EMPA and AEM of coexisting pyrophyllite, muscovite, and illite in Witwatersrand and Pennsylvania pelites.[1] (After Jiang et al., 1990)

	Witwatersrand, Republic of South Africa				Blackridge, Pennsylvania		
	Muscovite		Pyrophyllite		Muscovite	Pyrophyllite	NH$_4$-illite
	EMPA[2]	AEM[5]	EMPA[3]	AEM[5]	AEM[5]	AEM[5]	AEM[5]
Si	6.09(0.06)[4]	6.11(0.10)	7.96(0.04)	7.94(0.23)	6.09(0.16)	7.92(0.16)	6.37(0.27)
Al^{IV}	1.91(0.06)	1.89(0.10)	0.04(0.04)	0.06(0.23)	1.91(0.16)	0.08(0.16)	1.63(0.27)
Al^{VI}	3.93(0.08)	3.91(0.09)	3.99(0.02)	3.94(0.12)	3.54(0.13)	3.93(0.15)	3.78(0.31)
Ti	--	0.01(0.02)	--	--	0.06(0.01)	0.01(0.03)	0.01(0.02)
Fe	0.03(0.05)	0.03(0.02)	0.01(0.01)	0.01(0.01)	0.24(0.05)	0.04(0.03)	0.11(0.09)
Mg	0.03(0.02)	0.05(0.05)	0.01(0.01)	0.05(0.04)	0.16(0.06)	0.02(0.02)	0.10(0.11)
Ca	--	--	0.01(0.01)	--	0.02(0.03)	0.02(0.01)	0.02(0.03)
Na	0.21(0.03)	0.17(0.11)	0.01(0.01)	0.06(0.11)	0.04(0.01)	0.04(0.03)	0.03(0.04)
K	1.66(0.08)	1.72(0.19)	0.01(0.01)	0.04(0.07)	2.03(0.20)	0.05(0.05)	0.63(0.53)

[1]All formulae normalized to a total of 12 octahedral + tetrahedral cations.
[2]Average of 6 analyses.
[3]Average of 7 analyses
[4]Number in parentheses corresponds to two standard deviations.
[5]AEM analyses are from several rastered areas.

ed by EMPA and AEM for pyrophyllite and muscovite (see below for discussion). The AEM analyses have been converted to weights percent in order to facilitate comparison. The results show that the AEM-determined values for individual elements are identical to EMPA values, within error; i.e., the ranges of values for each of the methods are effectively the same. For those carefully obtained analyses, the accuracy of AEM results approaches that of the EMPA data. Although this example is for data for which special precautions were made, it nevertheless is not atypical of data that can be obtained with ordinary operating conditions.

4.9 APPLICATIONS

4.9.1 Qualitative exploratory analysis

The SEM is increasingly used in place of the optical microscope to characterize minerals in thin sections. Back-scattered electron images reflect differences in average atomic number of individual mineral grains, thus giving rise to contrast differences that permit easy definition of the individual grains. Once the contrast differences are defined, it is a simple matter to focus the beam on a given point or area having a specific contrast feature to obtain a qualitative EDS analysis. Experienced operators learn to recognize the spectra of specific minerals as surely as they can identify a mineral in hand specimen.

The AEM can be used in the same way for samples of rocks such as shales that consist of a variety of minerals with various textures, either in ion-milled samples or in fine-grained separates. As different areas are observed in TEM mode, contrast differences often lead directly to identification of given grains as being of interest. It is a simple matter to focus the beam on the area of interest and obtain an EDS analysis that generally is of sufficient quality to lead to immediate identification of the mineral in question. Figure 4-15 shows a low resolution TEM image of a shale for which contrast indicates the presence of an unusual mineral; the inset EDX spectrum immediately identifies it as florencite, a rare mineral, but one of great significance in this case because it is an important source of rare earth elements that are significant in Nd/Sm radioisotope systematics. Where there is ambiguity in identification based only on AEM data, as between iron oxides or polymorphs of TiO_2, additional data such as electron diffraction patterns can be immediately obtained. Alternatively, the operator may decide to switch to STEM mode if that option is available, in order to observe a scanning image, e.g., a transmitted electron image (TEI) or secondary electron image (SEI) image that displays contrast features that are different than those obtained in TEM mode. Areas with interesting contrast can immediately be qualitatively analyzed either with a stationary focused beam, or with a raster, with the resultant EDS spectrum leading directly to recognition of the mineral in question, as in an SEM. In either TEM or STEM mode, the sample can be rapidly surveyed until the operator is confident that all of the different minerals have been catalogued, and then the operator can proceed to focus on specific relations of interest.

4.9.2 Examples of analyses

4.9.2.1 Coexisting muscovite and pyrophyllite.

Jiang et al. (1990) studied coexisting pyrophyllite and muscovite in a low-grade pelite from Witwatersrand, South Africa. Both muscovite and pyrophyllite occurred as packets of layers in porphyroblasts that were large enough to analyze by EMPA, but fine-

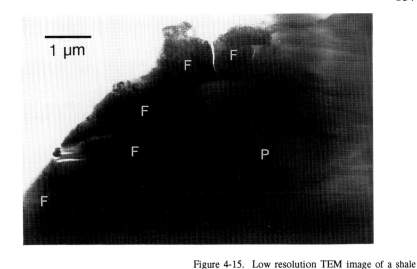

Figure 4-15. Low resolution TEM image of a shale from Wales, showing florencite, F, at an ion-milled thin edge in a matrix of phyllosilicates, P = chlorite plus illite. Courtesy of W.-T. Jiang.

grained matrix material consisted of finely intergrown pyrophyllite and muscovite for which EMPA analyses inevitably showed overlap. Qualitative AEM analyses indicated that muscovite and pyrophyllite in both matrix and porphyroblasts were essentially identical in composition. Table 4-1 shows typical compositions as determined by electron microprobe for porphyroblasts and by AEM for matrix phases. The AEM analyses were obtained using a Philips CM12 instrument operated in STEM mode, with the beam rastered over less than 100 nm on an edge. All analyses are reasonable with respect to crystal chemical relations for such minerals; e.g., interlayer cation totals show minor deficiencies. The EMPA and AEM analyses for a single phase are identical within standard errors of individual elements. Analyses of coexisting muscovite, pyrophyllite, and NH4-containing illite from a Pennsylvania slate are also given in Table 4-1. The Fe and Mg (phengitic) components that preferentially occur in the muscovite are clearly indicated; by comparison the small amounts of Fe and Mg in the Witwatersrand muscovite, although at the limits of detectability, are probably real. The interlayer cation contents of the NH_4-illite are low only because no analysis could be obtained for N. Nevertheless, its presence is clearly indicated by low interlayer cation and charge totals. The high ^{IV}Si value, as compared to muscovite, is typical of metastable illite that formed early during diagenesis, whereas the muscovite has the nearly ideal composition typical of muscovite that formed under conditions of equilibrium.

4.9.2.2   Five-amphibole assemblage in blueschist

Smelik and Veblen (1989) described an assemblage of five coexisting amphiboles that occur within a blueschist from Vermont. The specific materials that were analyzed occur in an area of only approximately 100 μm of a thin section, demonstrating the high resolving power of the method, in combination with high resolution imaging. The results provide rather dramatic definitions of amphibole solvi. Smelik and Veblen utilized a Philips EM420 TEM fitted with a high resolution pole piece to minimize probe diameter, obtaining analyses in TEM mode.

Table 4-2 lists average normalized chemical formulae, obtained by normalizing to 23 O atoms, and assuming all Fe is Fe^{2+}. The glaucophane is a relatively coarse primary mineral, with an inclusion of and rimmed by calcic amphiboles. Some of the originally heterogeneous glaucophane contained no lamellae, whereas some exsolved locally to cummingtonite in lamellae no wider than 50 nm, and some exsolved to winchite lamellae. The analyses, apparently without significant overlap, demonstrate the minimal point to point resolution. Formulae of the glaucophane associated with exsolved cummingtonite and winchite are very similar to that of the unexsolved glaucophane, demonstrating that exsolution was controlled by local variations in original glaucophane composition. The calcium-amphiboles consist of actinolite and hornblende, demonstrating a solvus between those phases. Even though the normalized analyses must be in error to some degree based on the assumption that all Fe is Fe^{2+}, the analyses are all very reasonable in terms of normal amphibole crystal chemical relations, and demonstrate the near-probe accuracy of the average formulae.

4.9.2.3   Chlorite and white micas in shales

The ability of AEM to provide accurate analyses and to differentiate between compositions of different populations of the same mineral are illustrated by a recent study of Li et al. (in prep.) of a shale sample from central Wales. The shale had been subjected to anchizonal metamorphic conditions; i.e., it had been weakly metamorphosed to a Kubler crystallinity index of 0.35°2θ, but not to the point where slaty cleavage had developed, as had occurred in equivalent higher grade rocks.

The sample has three populations of phyllosilicates: (1) Grains with a detrital shape up to tens of microns in diameter, of two types: one is referred to as "stacks" and consists of alternating packets of chlorite and white mica; the other consists only of white mica. (2) Fine-grained, intergrown, matrix illite and chlorite occurring in two populations: one has {001} subparallel to bedding and is authigenic in origin; the other is oriented preferentially parallel to a developing slaty cleavage direction and was derived from the first kind by dissolution and crystallization. (3) Irregular fracture-fillings of mica, cross-cutting stacks.

Table 4-2. Average amphibole compositions from AEM for five-amphibole assemblage from Tillotson Pond area, northern Vermont. (After Smelik and Veblen, 1989)

		Unexsolved	Glaucophane Cum. host	Win. host	Cummingtonite	Winchite	Actinolite	Hornblende
Tetrahedral	Si	7.908	7.934	7.891	7.906	7.629	7.542	6.938
	Al	0.092	0.057	0.109	0.094	0.371	0.458	1.062
Total	T site	8.000	8.000	8.000	8.000	8.000	8.000	8.000
Octahedral	Al	1.592	1.452	1.613	0.149	1.162	0.465	0.713
M1, M2, M3	Ti	0.003	0.001	0.007	0.006	0.001	0.004	0.007
	Mg	2.132	2.252	2.201	3.732	2.733	2.949	2.380
	Fe^{2+}	1.272	1.294	1.179	1.112	1.104	1.582	1.900
	Mn	0.000	0.000	0.000	0.000	0.000	0.000	0.000
Total	M1-M3	5.000	5.000	5.000	5.000	5.000	5.000	5.000
Octahedral	Ca	0.234	0.232	0.311	0.590	0.742	1.678	1.640
M4	Na	1.678	1.603	1.604	0.232	1.021	0.275	0.269
	Fe^{2+}	0.085	0.163	0.081	1.153	0.235	0.042	0.086
	Mn	0.002	0.001	0.003	0.025	0.002	0.006	0.006
	Mg	0.000	0.000	0.000	0.000	0.000	0.000	0.000
Total	M4	2.000	2.000	2.000	2.000	2.000	2.000	2.000
A site	Na	0.160	0.197	0.079	0.152	0.205	0.242	0.571
	K	0.012	0.009	0.007	0.013	0.023	0.016	0.033
Total	A site	0.172	0.206	0.086	0.164	0.227	0.259	0.604
No. of analyses		12	9	10	14	6	6	10

Note: Formulae were normalized to 23 O atoms basis, and all Fe was assumed to be FeO. All analyses are from sample no. VTPl.

Table 4-3. Formulae derived from AEM and EMPA[1] analyses for chlorite and white mica in shale, central Wales. (Courtesy G. Li)

	Chlorite						White Mica					
	Detrital in stacks[2]		Matrix bedding-parallel	Matrix cleavage-parallel	Detrital in stacks		Detrital homogeneous		Matrix bedding-parallel	Matrix cleavage-parallel	Fracture filling	
	AEM	EMPA	AEM	AEM	AEM	EMPA	AEM	EMPA	AEM	AEM	EMPA	
Si	2.56	2.59	2.57	2.56	6.22	6.16	6.01	5.99	6.02	6.13	6.05	
Al^{IV}	1.44	1.41	1.43	1.44	1.78	1.84	1.99	2.01	1.98	1.87	1.95	
Al^{VI}	1.53	1.61	1.61	1.53	3.59	3.62	3.84	3.86	3.77	3.61	3.86	
Ti	n.d.[3]	n.d.	n.d.	n.d.	0.07	0.03	0.03	0.01	0.01	0.04	0.01	
Fe^{2+}	3.22	3.23	2.97	3.20	0.14	0.19	0.05	0.07	0.06	0.13	0.06	
Mg	1.07	1.05	1.32	1.19	0.20	0.16	0.08	0.06	0.16	0.22	0.07	
Mn	0.03	0.02	0.05	0.03	n.d.	n.d.	n.d.	n.d.	n.d.	n.d.	n.d.	
Ca	n.d.	n.d.	n.d.	n.d.	n.d.	0.01	n.d.	n.d.	n.d.	n.d.	0.01	
Na	n.d.	n.d.	n.d.	n.d.	n.d.	0.10	0.32	0.44	0.52	0.73	0.26	
K	n.d.	n.d.	n.d.	n.d.	1.83	1.75	1.58	1.30	1.25	1.08	1.49	

[1]All EMPA analyses are averages of 5 to 7 analyses.
[2]Stacks are grains with a detrital shape, consisting of parallel pockets of chlorite and mica.
[3]n.d. = not detected

Table 4-3 lists typical compositions of each of the five types of white mica and three kinds of chlorite, with compositions as determined by EMPA where grains were large enough for comparison. The AEM-determined compositions were obtained from single analyses obtained with a Philips CM12 instrument by rastering the beam over selected areas. Chlorite and white mica in "stacks" with detrital shapes were analyzed by both EMPA and AEM, providing direct comparison of the methods, although analyses were obtained from different grains for each method.

The chlorite in stacks is Fe-rich and interpreted to have formed through replacement of detrital biotite. Bedding-parallel chlorite, of lowest Fe content, was derived through authigenesis of detrital clays, and the cleavage-parallel chlorite developed during tectonic deformation through dissolution of bedding-parallel chlorite.

The mica in stacks is phengitic with significant Fe and Mg, and was inferred to have formed through crystallization in fissures produced by layer separations of biotite/chlorite during stress. Detrital mica has low Fe and Mg contents, but high values of Na in interlayer sites. Bedding-parallel mica has low phengite components, but with a high Na content inferred to represent metastable solid solution. The cleavage-parallel

140

mica consists of interlayered paragonite and muscovite that can not be resolved even by AEM, but is on average more phengitic than bedding-parallel micas. The mica that cross-cuts stacks was analyzed only by EMPA, and has relatively high Na and intermediate phengite contents.

4.10 CONCLUDING REMARKS

The techniques of analytical electron microscopy have been perfected over the last few years to the extent that compositions of rock-forming minerals can be routinely determined with accuracy and sensitivity that approach those of electron microprobe analyses. Quantitative analyses can be obtained with a spatial resolution that approaches 10-20 nm; the compositions of much smaller phases can be qualitatively determined. Under normal conditions, data for standards need not be obtained during a given session, so that analyses of individual minerals can be rapidly obtained. Lastly, in contrast to older instruments, modern computer-controlled TEM/STEMs offer a variety of easily-interpreted imaging systems with easy transfer from TEM to STEM mode. Compositions can be directly obtained from areas that are separately characterized by diffraction and high resolution microscopy. As was true of the process of improving EMPAs in the 1960s, AEM analytical techniques have now been improved to the point where they can routinely be used in the solution of petrological problems, especially in those cases where the resolving power of EMPA is inadequate.

ACKNOWLEDGMENTS

I am grateful to Peter Buseck for his efforts in making this *Reviews in Mineralogy* possible, and for the constructive reviews of R.L. Freed and two anonymous reviewers. I especially want to express my gratitude and appreciation to all of the many graduate students and colleagues who have contributed to the chemical analytical techniques of our laboratory, starting with Larry Allard, Dave Blake, Will Bigelow, and Drew Isaacs, and continuing with Jung Ho Ahn, Jung Lee, Lancy Yau, Yen Hong Shau, Wei-Teh Jiang, Gejing Li, and many others. They have been the real sources of expertise and creativity with respect to the AEM capabilities of our laboratory, and to them I will always be grateful. This project was supported by NSF Grant EAR-9104565.

REFERENCES

Cliff, G. and Lorimer, G.W. (1975) The quantitative analysis of thin specimens. J. Microscopy 103, 203-207.
Goldstein, J.I., Williams, D.B. and Cliff, G. (1986) Quantitative x-ray analysis. In Principles of Analytical Electron Microscopy, D.C. Joy, A.D. Romig Jr., and J.I. Goldstein, eds., Plenum Press, New York, 155-218.
Jiang, W.-T., Essene, E.J. and Peacor, D.R. (1990b) Transmission electron microscopic study of coexisting pyrophyllite and muscovite: Direct evidence for the metastability of illite. Clays and Clay Minerals 38, 225-240.
Mackinnon, I.D.R. (1990) Low-temperature analyses in the analytical electron microscope. In I.D.R. Mackinnon and F.A. Mumpton eds., CMS Workshop Lectures, vol. 2, Electron-Optical Methods in Clay Science. Clay Minerals Society, Evergreen, Colorado, 90-106.
Potts, P.J. (1987) Energy dispersive x-ray spectrometry. In A Handbook of Silicate Rock Analysis, P.J. Potts, ed., Chapman and Hall, New York, 286-325.
Smelik, E.A. and Veblen, D.R. (1989) A five-amphibole assemblage from blueschists in northern Vermont. Am. Mineral. 74, 960-964.
Williams, D.B. (1987) Practical Analytical Electron Microscopy in Materials Science. Philips Electronic Instruments, Inc. Electron Optics Publishing Group, Mahwah, New Jersy, 153 p.
Williams, D.B., Goldstein, J.I. and Fiori, C.E. (1986) Principles of x-ray energy-dispersive spectrometry in the analytical electron microscope. In Principles of Analytical Electron Microscopy, D.C. Joy, A.D. Romig, Jr., and J.I. Goldstein, eds., Plenum Press, New York, 123-154.

Chapter 5. ELECTRON ENERGY-LOSS SPECTROSCOPY (EELS) AND ELECTRON CHANNELLING (ALCHEMI)

Peter R. Buseck Departments of Geology and Chemistry,
Arizona State University, Tempe, Arizona 85287, U.S.A.
Peter Self CSIRO, Private Bag No. 2, Glen Osmond,
South Australia 5064, Australia

5.1   INTRODUCTION

When a specimen is irradiated with an electron beam, interactions occur between the specimen and the beam electrons. These can be either elastic (i.e., when the energies of the beam electrons remain unchanged) or inelastic (i.e., when energy is lost by the beam electrons). Elastic interactions are observed as diffraction effects and are described in Chapters 1 to 3. The energy lost by beam electrons during inelastic interactions is converted into a variety of forms and can be observed in the electron microscope as various secondary signals. These secondary signals include x-radiation (as considered in Chapter 4), cathodoluminescence, and Auger electrons.

In addition to the production of secondary signals, the changes in energies of the beam electrons also provide information about both the types of atoms in the specimen and about their chemical states (bonding, valence, coordination). The measurement of the energy distribution of electrons that have passed through a specimen is called *electron energy-loss spectroscopy*, or EELS for short. The charm of that name is such that it has been widely used, both as defined above and for *electron energy loss*, so that one encounters expressions such as EELS spectra or EELS spectroscopy. In this chapter we accept that usage in spite of its obvious redundancy.

The energies of the transmitted electrons encountered using EELS fall into three ranges. Those that experience only elastic interactions exit the specimen with the same energy they had when they entered. They produce what is called the *zero-loss peak*, which in thin specimens is narrow and intense. The zero-loss peak also contains electrons that suffered minimum angular deviation. Next comes a range that spans energies between roughly 5 and 50 eV and that is called the *low energy-loss region*. It is largely produced by both plasmon and valence-electron interactions and is not generally of major utility in the study of minerals. Finally, there is the *high energy-loss region*, which results from ionization through release of inner-shell electrons and extends above roughly 50 eV. It is this region that is of greatest use for the elemental and chemical analysis of minerals*.

The effects of inelastic scattering are evident in the electron-diffraction pattern, although not generally in a readily quantifiable way. The main effect is to produce a diffuse background between the diffraction spots. The 000 spot, produced by the direct beam, lies in the center of the diffraction pattern and represents the zero-loss electrons as well as very weakly scattered electrons. The elastically scattered electrons produce the centers of the other diffraction spots. The inelastically scattered electrons from inner-shell ionizations and plasmon losses experience small angular deviations (Egerton, 1976), and so they produce

---

* In this chapter, the term "elemental analysis" is used to refer to the determination of atomic concentrations. The term "chemical analysis" is used to refer to the determination of crystal-chemical properties such as valence states.

intensity at the outer parts of the various diffraction spots. Phonon-loss electrons produce the diffuse intensity lying between diffraction spots, i.e., the diffuse background of diffraction patterns (Williams, 1984; Cowley, 1981).

Much has been written about EELS, and there are excellent review papers and books. Much of this chapter is based on material in Egerton (1986), Rez (1991), and Williams (1984); the reader should consult these works for more detail. Egerton is probably the most thorough and comprehensive treatment available. A useful guide to the form of EELS spectra is given by the "EELS Atlas" (Ahn and Krivanek, 1983).

Table 5-1. List of acronyms

		Section where cited
ALCHEMI	Atom location by channelling enhanced microanalysis.	5.3.1
CBED	Convergent beam electron diffraction.	Chapter 2
EDS	Energy dispersive x-ray spectroscopy.	Chapter 4; §5.2.3
EELS	Electron energy-loss spectroscopy.	5.1
ELNES	Electron energy-loss near-edge structure.	5.2.7
EXAFS	Extended x-ray absorption fine structure.	5.2.7.1
EXELFS	Extended energy-loss fine-structure.	5.2.7
FWHM	Full width at half maximum.	5.2.2.1
HRTEM	High-resolution TEM.	1.1
PEELS	Parallel EELS.	5.2.4.2
RDF	Radial distribution function.	5.2.7.1
SEM	Scanning electron microscope (microscopy).	
STEM	Scanning transmission electron microscope (microscopy)	1.2.2
TEM	Transmission electron microscope (microscopy).	
WDS	Wavelength dispersive spectroscopy.	Chapter 4
XAS	X-ray absorption spectroscopy.	5.2.7.1
ZAF	Z (atomic number), absorption and fluorescence corrections	Chapter 4

5.2   ELECTRON ENERGY-LOSS SPECTROSCOPY (EELS)

5.2.1   Electron energy-loss processes

There are four kinds of inelastic energy losses in solids. They are produced by (1) phonons, (2) plasmons, (3) valence electrons, and (4) inner-shell electrons. The latter two are collectively known as single-electron excitations. There is a minimum energy for a given type of excitation or transition to occur. Thus, each type of excitation has a threshold energy, and no energy losses occur below the threshold value for that particular transition. The four kinds of inelastic energy losses are considered below in the sequence of the magnitude of the energy losses they produce.

5.2.1.1   Phonons

Phonons are thermal vibrations, i.e., collective, quantized oscillations of the atoms throughout the crystal. In transmission electron microscopy, these vibrations are generated as a result of the conversion of kinetic energy from the electron beam into heat in the specimen. Phonons are the coupled vibrations of the atoms in a solid. The strength of the coupling between atoms is a measure of the bonding strength of the atoms and is directly related to the elastic constants of the solid. These coupled vibrations can be treated like the motion of beads on a vibrating string extending throughout the crystal. In the one-dimensional case this is readily visualized, but they are more difficult to visualize in higher dimensions such as in a 3-dimensional crystal (Chapters 4 and 5 in Kittel, 1986).

The number of interactions between beam electrons and thermal vibrations for a given energy loss is determined by the density of vibration states in the specimen, for which there are two main models. The first, proposed by Einstein, is a classical model and assumes that so many phonon states are excited that each atom can be considered as vibrating independently of all others. The second, proposed by Debye, takes the opposite approach by representing the solid as an elastic continuum (i.e., like a rubber band) and therefore assuming that the atomic vibrations are highly correlated. Both models can be used to give reasonable approximations to the density of vibration states over a reasonable range of temperatures: the Einstein model for high temperatures and the Debye model for low temperatures.

The reality is that the density of states is a much more complicated function than either the Einstein or Debye models predicts (Kittel, 1986). If the density of states were determined by EELS, then it could be used to give information about the atomic forces in the specimen. However, the energy of phonons is around 0.02 eV. Thus, in order to obtain useful information of energy losses caused by the generation of phonons, the energy spread of the high-keV electron source would have to be decreased greatly and the resolution of EELS spectrometers would have to be increased by an order of magnitude over the current value. Consequently, phonon measurements have not been used for mineralogical studies, and it is unlikely that they will be so used in the near future.

5.2.1.2   *Plasmons*

In metals and semiconductors, outer (valence) electrons can move freely throughout the crystal, whereas inner-shell electrons are bound to the atomic nucleus. A plasmon is a collective vibration of the entire valence-electron distribution throughout the crystal (Chapter 10 in Kittel, 1986). In energy-loss spectra from metals, the effect of plasmon interactions is seen as relatively sharp peaks, second in intensity only to the zero-loss peak. The minimum energy loss involved in plasmon excitation is between 10 and 20 eV. Plasmon interactions with beam electrons are important in metals and dielectric films, and the peaks they cause in EELS spectra can be interpreted to yield quantitative information regarding dielectric functions and electron densities in several energy bands for the special cases where there are strong signals (Rez, 1991). The number of electrons that lose energy to plasmons can also be used to determine specimen thickness (Joy, 1979; Egerton, 1986). However, plasmon-related effects are neither prominent nor easily interpretable in most minerals and thus are of little use for mineralogy.

5.2.1.3   *Single-electron excitations*

A single-electron excitation involves the beam electron knocking either an inner-shell electron or a valence electron into an empty state. These empty states may be those of an unfilled inner shell, the valence band, or the conduction band. Crystal electrons may also be ejected into the vacuum with an arbitrarily high energy (Fig. 5-1). For mineralogy, the most important process monitored in EELS measurements is the ionization caused by ejection of inner-shell electrons because it is this ionization that gives information about the elemental composition of a sample.

144

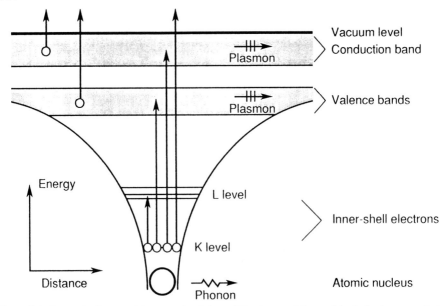

Figure 5-1. Schematic diagram of the energy levels in a solid and some of the possible single-electron excitation processes. As shown by the vertical arrows, electrons may be ejected to vacuum or into unfilled atomic states. Phonons and plasmons can also be excited (see text). [The symbols used for phonons and plasmons are those of Kittel (1986)]. The electrostatic potential of the atom is represented schematically by the two curved lines emanating from the atomic nucleus (bottom).

5.2.1.3.1 Valence electrons. Valence electrons (or conduction electrons in metals) are those involved in the bonding of atoms to form a solid. These electrons are shared between atoms, and their energies are a characteristic of the entire solid rather than of individual atoms. The energy states of valence electrons are not discrete but rather form an energy band (Fig. 5-1). Consequently, energy-loss processes involving valence electrons do not give rise to sharp features in EELS spectra but rather give a broad, continuously varying profile. Although valence electrons are very sensitive to crystal chemistry, measurements of their energies by EELS contains little directly interpretable elemental or chemical information. EELS signals from valence electrons are thus of only secondary importance for mineralogical purposes.

5.2.1.3.2 Inner-shell electrons. Inner-shell electrons are tightly bound to the atom. The minimum energy that a beam electron must lose to excite an inner-shell electron equals the energy needed to promote that crystal electron into the lowest-energy, empty electron state of the crystal. For inner-shell excitations, this minimum energy is known as the onset of ionization. In principle, the beam electron may lose any amount of energy larger than this minimum energy. Thus, the excitation of inner-shell electrons is recorded as a sharp rise in intensity in an energy-loss spectrum at the energy corresponding to the onset of ionization (Fig. 5-2).

The threshold energies for inner-shell ionizations are approximately equal to the corresponding absorption edges seen in soft x-ray absorption spectroscopy (e.g., using a synchrotron) and occur at slightly higher energies than the corresponding soft x-ray emission energies. The lower energy of emitted x-rays is because some energy is redistributed during the process of repopulating the inner-shell electron state. This x-ray emission process forms the basis of the EDS technique (Chapter 4).

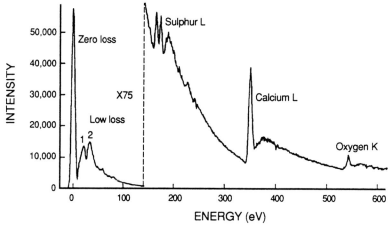

Figure 5-2. An example of an EELS spectrum (for anhydrite, $CaSO_4$). The spectrum shows the form of the zero-loss peak, low energy-loss region, and high energy-loss region. The effects of inner-shell excitations can be seen in the high energy-loss region. The peaks in the low energy-loss region, marked 1 and 2, are at energies of approximately 20 and 35 eV respectively. The latter peak can be attributed to Ca*M* excitations, whereas the former presumably results from a plasmon excitation.

5.2.2 The electron energy-loss spectrum

The energy-loss spectrum observed in a TEM results from a combination of all four inelastic energy-loss processes. By common convention, the electron energy-loss spectrum is divided into three energy regions, each of which corresponds to certain types of excitations. These three regions are the zero-loss peak, a low energy-loss region, and a high energy-loss region (Fig. 5-2).

5.2.2.1 The zero-loss peak

The zero-loss peak is produced by electrons that pass through the specimen with little or no angular deviation and that are relatively unaffected by energy-loss processes. In thin specimens the bulk of the electrons contribute to the zero-loss peak. These electrons exit the specimen with either the same energy with which they entered (having experienced only elastic interactions) or with a slightly decreased energy because of energy losses to thermal vibrations (i.e., phonons) in the specimen. For phonon interactions, a typical energy loss is 0.02 to 0.03 eV. Since the energy resolution in EELS is limited to about 1 eV by the spread in incident-beam energy and the inherent limits of spectrometer design, it is not possible to resolve individual phonon-loss peaks. Hence, both elastically scattered electrons and electrons undergoing only phonon-induced energy losses contribute to the zero-loss peak.

The utility of the zero-loss peak is two fold. Firstly, the full-width at half maximum (FWHM) of the zero-loss peak gives a direct measure of the performance of the microscope and the spectrometer. Secondly, the zero-loss peak can be used to measure specimen thickness. When specimen thickness increases, a greater number of electrons are scattered, and so the fraction of the signal in the zero-loss peak is reduced. Specimen thickness is important in quantitative EELS because the quantitative analysis techniques work best when electrons experience, on average, only a single inner-shell, energy-loss event. Thus, for optimum results in quantitative EELS analysis, it is important that the zero-loss peak is five to ten times more intense than the peaks in the adjoining low energy-loss region of the spectrum.

5.2.2.2   The low energy-loss region

The low energy-loss region is considered to extend from 5 eV, the upper limit of the zero-loss peak, to 50 eV. The energy losses measured in this region arise from inelastic scattering events involving plasmons or valence electrons. For metals and semiconductors, this region contains one or more well-defined, Gaussian-shaped plasmon peaks (e.g., Williams, 1984). The intensity of plasmon peaks relative to the zero-loss peak increases with specimen thickness. Furthermore, the number of plasmon peaks detectable above background increases with specimen thickness. These additional peaks are caused by beam electrons experiencing multiple, plasmon-induced, energy-loss events.

If a specimen does not contain free electrons, as in the case of most ionic and covalently bonded materials and so most minerals, the low energy-loss region consists of broad peaks with little structure. This is the situation for the spectrum of anhydrite (Fig. 5-2). The low energy-loss region for such a specimen is a complex mixture of scattering events involving outer-shell electrons, electrons in molecular orbitals (Isaacson, 1972), as well as volume and surface plasmons (Egerton, 1986). In the case of Figure 5-2, the peak at 35 eV can be attributed to Ca M excitations, while the peak at 20 eV presumably results from interactions with bonding (i.e., molecular orbital) electrons.

The region of an EELS spectrum below 50 eV does contain a considerable amount of information and can be used to determine physical properties of a specimen. For example, it is possible to determine the dielectric function of a specimen by applying Kramers-Kronig analysis (Egerton, 1986). In order to apply this form of analysis, the low energy-loss region of the spectrum must be recorded at the maximum possible resolution. However, in general the complex nature of the low energy-loss region of the spectrum means that this region is not widely used for analytical purposes.

5.2.2.3   The high energy-loss region

High energy-loss interactions have losses in excess of roughly 50 eV. At such energies, the intensity in EELS spectra falls away rapidly, so that the high energy-loss region of a spectrum comprises only a small part of the total intensity. This region of an EELS spectrum contains effects caused by the high-energy ejection of valence electrons from the specimen and the excitation of inner-shell (or "core") electrons. As inner-shell excitations are characteristic of each element, it is these that are of primary interest in analytical electron microscopy and mineralogical studies.

During ejection of an inner-shell electron, a beam electron can lose any energy above that required to knock the inner-shell electron into the lowest vacant energy state of the atom. Thus, at this critical energy, there is a rapid increase in the intensity of the EELS spectrum. This rapid increase is referred to as an "edge." The variations in the spectrum intensity at energies around the edge are called electron energy-loss near-edge structure (ELNES) and reflect effects caused by the excitation of inner-shell electrons to unfilled atomic-core and valence states. In addition to varying with specimen thickness, the exact form of the near-edge structure varies considerably from element to element (compare the signals for S, Ca, and O in Fig. 5-2) and crystal structure to structure. At energies greater than the critical energy for inner-shell electron excitation, the spectrum intensity gradually falls to background levels. There are, however, minor fluctuations that reflect other critical features of the target atom, such as bonding and nearest-neighbor interactions. These minor fluctuations are called extended energy-loss fine structure (EXELFS). Both EXELFS and ELNES are of great interest and are discussed in §5.2.7.

The useful energy range of an EELS spectrometer extends to around 2 keV. Above this energy the number of energy losses becomes so small that it is impractical to measure them. Hence, the energies of inner-shell excitations measured in an EELS spectrometer and used in analysis depend on the atomic number of the elements in the specimen. For Li to Si (Z = 3 to 14) K excitations are used, for Al to Sr (Z = 13 to 38) L_{23} excitations are used, and for Rb to Os (Z = 37 to 76) M_{45} excitations are used. The quality of EELS analyses decrease roughly proportionally to atomic number, so that EELS is not normally used for the heavier elements. In distinction to measurements in the low-loss region, those in the high-loss region encompass a spread of energies, and so high-energy resolution is not as critical.

The rapid increase in the spectrum intensity at the edge makes it fairly easy to distinguish among the elements. However, the gradual decrease in intensity following the edge means there is commonly significant overlap from the trailing tail arising from the excitation of inner-shell electrons of one element to the edge of another element. Such overlaps greatly complicate quantification and consequently limit the sensitivity and accuracy of chemical analysis using the EELS method.

Background intensity results from random losses in energy of the beam electrons as well as multiple scattering effects. The background decreases steadily with energy across the EELS spectrum, and it increases with sample thickness. Background intensity, and correction for its effects, are major factors in EELS spectra in the high-loss region (cf. §5.2.6.1).

5.2.3 Comparison of EELS to x-ray emission, energy-dispersive spectrometry (EDS)

As described in Chapter 4, x-ray fluorescence analysis by energy-dispersive spectroscopy (EDS) is the major application and use of analytical transmission electron microscopy. Together with wavelength-dispersive spectrometry, EDS is the primary method whereby analyses are performed with the electron microprobe and the analytical scanning electron microscope (see, for example, Goldstein et al., 1981). EDS is thus a convenient point of comparison for considering the pros and cons of EELS. The two methods are complementary; EELS is better suited for light elements, which are particularly difficult to measure by routine EDS analyses, whereas EDS is better for the measurement of heavier elements. Unless special thin-window or windowless detectors are used, Na is the lightest element that can routinely be measured by EDS, whereas elements as light as Li can be measured using EELS. A good review is provided by Goldstein and Williams (1992).

EELS reflects a primary event of energy change upon emission of an electron from the orbitals of an atom during an inelastic interaction between a beam electron and an atomic electron. Since most electrons that cause ionizations are scattered through relatively small angles, they can be collected by the EELS spectrometer and recorded, resulting in a high detection efficiency. According to Joy (1981), between 30 and 100% of the EELS electrons with energies between 10 and 1000 eV can be collected in the spectrometer, the exact value depending on the collection angle.

In contrast, EDS reflects a secondary event in the sense that it measures the energy of radiation released upon relaxation during the return of the atom to its ground state. A result is that EDS is less efficient because only a small fraction (the fluorescence yield) of the relaxation events result in the production of an x-ray. For example, for C, only 1 in 400 K-shell ionizations produces a characteristic x-ray, and it is still only 1 in 40 for Na. The remaining energy goes to produce Auger electrons (Goldstein et al., 1986; Williams, 1984). The low yield of emitted x-rays is further compromised by the relatively poor collection and detection efficiencies of x-ray detectors. As a consequence, only a small fraction of events that produce x-rays are detected. The differences in efficiencies mean that the counting time per energy channel can be shorter for EELS than for EDS.

Peak overlaps occur and potentially present serious problems in both EDS and EELS. With EDS, the peaks of the light elements are close to one another and are not always totally resolved by the detector. If light elements occur in widely different concentrations (e.g., a small amount of Mg in an aluminate), then the large peak of the major element can overlap and partly obscure the minority peak (Al and Mg, respectively, in this case). A more serious problem arises when there are mixtures of elements having widely different atomic numbers, in which instance an *L* or *M* line of the heavy element can occur at the same energy as a *K* line of the light element (*M/L* overlaps are also possible). Examples of such element interferences are S/Ag/Pb for galena, and Ba/Ti for benitoite and priderite. The narrow, well-defined character of EDS peaks gives EDS a considerable advantage when resolving peak overlaps. In EELS the long trailing tails after the edges means that, with rare exceptions, element interferences present a more severe problem with EELS than with EDS.

As with peak overlaps, background subtraction is more difficult for EELS than EDS. In EDS spectra, the background varies relatively slowly and predictably. The narrowness of the EDS peaks means that the interpolation required to calculate the background under a peak is not mathematically strenuous. For EELS, the long tails after the edge mean that background fitting is an extrapolation rather than an interpolation and hence is more prone to error.

Other important differences between the two techniques are in regard to standards and spatial resolution. EDS requires standards and elaborate but well-known correction procedures (i.e., ZAFs; see Chapter 4), whereas EELS is standardless and the correction procedures are less complex, although they are also considerably less accurate. The spatial resolution of EDS measurements are limited by the beam-broadening effects that occur in the sample during x-ray production. EELS, on the other hand, is mainly limited by the size of the electron probe. Since probes as small as a fraction of a nm can be produced with a dedicated STEM (§ 1.2.2), exceptionally high spatial resolutions are possible. The major limitation in this case is statistical, that is, the relatively small number of energy-loss events that are produced from extremely small areas within a finite time.

Spurious effects in the spectrum can be a problem with both EDS and EELS analyses. However, in general, spurious effects are smaller in EELS than in EDS measurements, primarily because only electrons from the region of interest enter the spectrometer. In contrast, in EDS there is an appreciable excited volume from which x-rays are generated (§4.4.1, 4.8.5), and the geometry of the EDS detector is such that it is possible for x-rays generated away from the illuminated area of the specimen, such as those caused by beam-limiting apertures, to be collected. As a result of these differences, fewer experimental precautions need to be taken with EELS to obtain simple measurements.

Finally, there are instrumental differences. For EELS, the signal is detected below the sample, so only modifications below the image plane of the TEM are required. Furthermore, the post-specimen lenses can be used to transfer the electrons to the entrance aperture of the spectrometer. Hence, critical lens parameters are unaffected by an EELS spectrometer, and so it is relatively straightforward to add one to a TEM.

The case for EDS differs because parts of the TEM must be physically modified to permit EDS measurements. The emitted x-rays must be detected above the sample and, to increase efficiency, the detector must be placed as close as possible to the specimen. To achieve this proximity, either the gap between the pole pieces of the objective lens must be wide enough to allow a clear path between the specimen and the detector, or there must be an exit port placed through the upper pole piece (§1.5.1). In the former case, widening the pole-piece gap will reduce the potential spatial resolution for imaging. In the latter case, it is difficult to maintain a uniform magnetic field within a pierced pole piece, again negatively influencing image resolution. In addition, it is commonly necessary to tilt the sample in order to make good EDS measurements, and there is a limited tilt range for which the measurements can be taken. There is no restriction on tilt, *per se*, for EELS.

In comparing EELS and EDS it is probably fair to say that EDS using a windowless or ultra-thin window detector is the preferable method of obtaining elemental analysis for all but the lightest elements. However, except in one or two cases (e.g., ALCHEMI), EDS can only give elemental information. EELS, on the other hand, can not only give elemental information but can also be used to obtain information on chemical and physical properties by the use of ELNES, EXELFS, and the low energy-loss region of the spectrum.

5.2.4   Instrumentation

5.2.4.1   The spectrometer

To record an EELS spectrum it is necessary to disperse the beam electrons according to their energy. Many instruments are capable of achieving this dispersion (Egerton, 1986), but the most prevalent is the magnetic-prism (sector) spectrometer. This form of spectrometer works on the principle that the radius of curvature of the path of electrons in a uniform magnetic field is dependent on the speed (i.e., energy) of the electrons. Electrons with a higher energy will not be deflected as much as electrons with a lower energy. A schematic diagram of a magnetic-prism spectrometer is shown in Figure 5-3a. The magnetic sector used to disperse the electrons extends for a little less than a quadrant, and the overall deflection is usually around -90°

If the magnetic-prism had dispersion properties alone, it would be necessary to collimate the electron beam so that the trajectories of electrons entering the spectrometer at different points do not overlap. However, in addition to its energy-dispersive properties, the magnetic-prism spectrometer also acts as a lens. The focussing properties of the spectrometer can be seen by considering the two entrance points **x** and **y** shown in Figure 5-3a. Although electrons with the same energy travel on circles with the same radius in the uniform magnetic field of the spectrometer, electrons entering at point **x** will have a longer path length in the field, and therefore a greater total deflection, than those entering at point **y**. Thus there will be an image plane on the exit side of the spectrometer conjugate with a plane on the entrance side.

The standard practice is to make the plane of the projector-lens crossover the plane imaged by the spectrometer. Then, because all electrons entering the spectrometer are focused to a point at the projector-lens crossover, electrons with the same energy will be focused to a point in the spectrometer image plane. For conventional EELS, electrons with a given energy loss are collected by placing a narrow slit (labeled the exit aperture in Fig. 5-3a) in the image plane of the spectrometer. The electrons are counted by a scintillator and photo-multiplier system behind the slit. The strength of the magnetic field of the spectrometer can be changed simply by altering the current passing through the magnet windings. Thus, electrons of different energies can be scanned sequentially through the slit and into

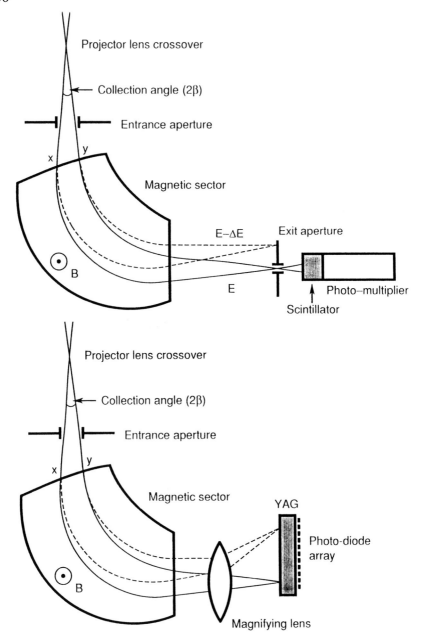

Figure 5-3. Schematic diagram of the magnetic-prism EELS spectrometer. (a) Spectrometer with photo-multiplier detection system and therefore serial data collection. (b) Spectrometer with a parallel data collection system using a photodiode array detector. Both (a) and (b) show the trajectories for electrons with two energies, **E**, and the slightly lower energy **E-ΔE**. Electrons entering the magnetic sector at point **x** have a longer trajectory in the magnetic field (B) of the spectrometer than those entering at point **y**. Hence, electrons entering at **x** are deflected through a greater angle than those entering at **y**, leading to a focussing effect.

the counting system. By accumulating counts for finite periods, a complete energy spectrum can be amassed. The procedure is analogous to wavelength-dispersive spectrometry in x-ray emission spectroscopy, in which counting also occurs via a multichannel, serial detection system. Normal EELS counting times are between 100 and 300 sec to obtain a complete spectrum.

The most important characteristic of the spectrometer is resolution, which is the ability to resolve peaks that are close in energy to one another (1 eV is commonly viewed as good resolution). Resolutions between 1.1 and 1.5 eV can be achieved with LaB_6 guns. Field-emission sources routinely do better than 1 eV and have produced resolutions as low as 0.3 eV (Batson, 1986). The resolution of the instrument can be varied by adjusting the width of the exit slit. When collecting a spectrum, it is pointless to use a resolution better than the channel width (i.e., the step size of the energy scan). The size of the optimum exit slit will depend on the dispersion of the spectrometer, i.e., the ability of a spectrometer to spatially separate electrons with different energies. Current spectrometers typically have dispersion values of a few microns per eV, and so exit slit sizes range from 1 to 5 µm. The optimum size of the exit slit for a given set of experimental conditions is set by obtaining the maximum height for the minimum width of the zero-loss peak.

Both the collection efficiency and the resolution of the system are greatly enhanced by using the lens system of the TEM and the imaging properties of the spectrometer. The collection angle is limited by an entrance aperture at the top of the spectrometer (Fig. 5-3). It is important that this aperture is aligned with the optic axis of the microscope in order to maximize the coupling between the projector lens and the spectrometer. The sizes of the objective and entrance apertures affect the spectrometer resolution and, in addition, edge-to-background ratios can be improved by decreasing the collection angles into the EELS spectrometer. However, a smaller entrance aperture causes a decrease in the count rate, and so an acceptable balance between aperture size, resolution, and background levels must be reached. The entrance-aperture size can usually be varied in the same way as the selected-area and diffraction apertures of the microscope. Most spectrometer entrance apertures are a few mm in diameter, and the effective collection semi-angle ranges from 5 to 100 mrad.

As with all lenses, the magnetic-prism spectrometer has aberrations; great care is taken in the manufacture of the spectrometer to minimize these aberrations. The shapes of the entrance and exit faces of the magnetic elements of the spectrometer are stringently designed and fabricated in order to achieve optimum focussing conditions. The spectrometer usually has stigmator and field-compensation coils that are adjusted to give maximum resolution. The setting of these coils is done at the time of installation, and the operator need only make minor adjustments, usually by observing the width of the zero-loss peak.

5.2.4.2   PEELS

At any one time when a slit detection system is used, 99.9% of the total spectrum is not being recorded. Clearly, the efficiency of such a counting system is far lower than if the entire spectrum, rather than just a small energy fraction, could be accumulated simultaneously, as is done with EDS x-ray emission spectroscopy. Parallel EELS ("PEELS") does just that. In this case, the serial detecting system is replaced by a magnifying quadrupole, YAG scintillator, and photodiode array, shown schematically in Figure 5-3b. The photodiode array simultaneously accumulates counts of different energies, thereby greatly improving the counting efficiency and reducing counting times relative to serial detection systems by a factor of ten or more. A magnifying quadrupole lens is required because the spacing of diodes in photodiode arrays is around 10 µm. This spacing would correspond roughly to a 5 eV resolution if the magnetic sector alone were used. The magnification of the quadrupole lens is variable, and resolutions can be adjusted to between 0.5 and 5 eV. The extra components used in PEELS units means that they are more expensive than serial-detection spectrometers.

A disadvantage of PEELS units is the non-uniform gain of the detection system. This non-uniformity arises from slight differences among the diodes in the photodiode array and from variations in the photo-response of the YAG scintillator. The former effect is generally random and can be corrected by obtaining two or more spectra with slightly different offsets, so that in each spectrum the intensity for a given energy loss is recorded by a distinct photodiode. By digitally realigning the zero-loss peaks and averaging each spectrum, the random gain effects can be averaged out.

Variations in the photo-sensitivity of the YAG scintillator are generally caused by thickness variations in the YAG crystal. These variations are systematic rather than random and are therefore more difficult to correct. One method of calibrating the spectrometer is to scan the zero-loss peak across the entire photodiode array and then make the appropriate digital corrections so that each detection element gives the same response.

5.2.4.3   The microscope

Almost any TEM or STEM has the potential to be used for EELS measurements. As indicated in §5.2.3, the only required modifications occur below the image plane (Fig. 1-1) and thus are relatively simple. The microscope can be used in virtually any mode when recording an EELS spectrum. In imaging mode the sample area to be analyzed is defined by the magnification of the microscope and the size of the entrance aperture of the spectrometer. Since the entrance aperture is much smaller than the viewing screen, the irradiated area is typically far larger than the portion of the image entering the spectrometer. Therefore, in imaging mode, high spatial resolution can be achieved, but the count rate will necessarily be low.

Large areas of specimen can be analyzed by operating the microscope in selected-area diffraction mode. Normally the undiffracted beam is centered on the entrance aperture of the spectrometer. This mode of operation gives high count rates, with a spatial resolu-tion limited by the size of the selected-area aperture. The effective maximum scattering angle of electrons entering the spectrometer can be altered by changing the camera length of the microscope.

In all operating modes the critical parameters are beam diameter (illuminated area) and beam current (current density). STEMs typically can produce narrower beams than conventional TEMs. When operated with the beam held fixed as in CBED mode, STEMs are capable of giving high spatial resolution at high current densities. For both TEM or STEM, a field-emission gun can greatly increase current density as well as decrease spot size. Field-emission guns also have the least spread in energy and therefore allow the highest possible energy resolution.

5.2.5   Elemental analysis by ionization-loss spectrometry

Elemental and chemical analysis using ionization-loss spectrometry is probably the major application of EELS for mineralogy, both in the past and for the foreseeable future. When recording and processing EELS spectra for quantitative analysis, the aim is to extract as accurately as possible the number of events caused by inner-shell excitations. The quality of EELS spectra is affected by many parameters, among which are spatial resolution and counting statistics. To achieve the best possible quantitative analysis, each of these parameters must be optimized.

5.2.5.1 Sample considerations

No special sample preparation is necessary for EELS analysis; the same sample can be used for EELS as for EDS analysis or for high-resolution imaging. It is important that the sample is as clean as possible and free of surface contaminants. The most common surface contaminant in electron microscopes is carbon, and such contamination by carbon and to a lesser extent oxygen is especially hard to avoid for samples such as metals. Carbon and oxygen have K edges at 284 and 532 eV respectively, and therefore significant amounts of such contamination can lead to serious problems for background subtraction.

Specimen effects can have important influences on EELS spectra. Variations in thickness strongly affect count intensities, as do specimen instabilities and orientation effects. The latter can give rise to electron channelling, such as is used to advantage for ALCHEMI (see § 5.3), with its consequent changes in EELS intensities. Changing the specimen tilt and noting the resulting changes in the EELS spectrum is a way of checking for the effects of channelling. Avoiding orientations along major zone axes minimizes channelling effects, as does the use of thin specimens and convergent beams.

5.2.5.1.1 Specimen thickness. The number of ionization events increases in proportion to thickness. Therefore, one might imagine that maximizing thickness would optimize peak-to-background ratios and thereby improve EELS analysis. However, the number of multiple-scattering events (§1.3.2) also increases with specimen thickness. These events remove intensity from the zero- and low-loss regions of a spectrum into the high-loss region, thereby increasing the background for high-energy losses. Unfortunately, the background intensity increases more rapidly than peak intensities, with the result that samples that are too thick are not useful for EELS. On the other hand, samples that are too thin have too few atoms to produce a sufficient number of ionization events to produce statistical reliability. Clearly, the challenge is to find the optimal thickness between samples that do not produce enough counts and those in which the background effects are too severe. The best balance is generally determined experimentally.

A common test for proper sample thickness is to examine the ratio of counts of the zero-loss peak, I_0, to those of the first plasmon peak, I_p. Various values have been proposed for the maximum acceptable I_p/I_0 ratio, ranging from 1 (Isaacson, 1978) to between 0.2 and 0.1 for certain compounds containing light elements (Zaluzec, 1980). The range of acceptable values is specimen-dependent, and it is not possible to give a reliable general figure for an acceptable I_p/I_0 ratio. However, it is evident that a strong zero-loss peak is an important test for a usable specimen thickness. Thicknesses between 15 and 60 nm work well for many minerals at an accelerating voltage of 100 kV.

The maximum usable specimen thickness can be increased by going to a higher accelerating voltage or by removing multiple-scattering effects from the spectrum. The removal of such plural scattering events involves the deconvolution of the appropriately scaled low energy-loss region of the spectrum from the high energy-loss region (Egerton, 1986; Schattschneider and Sölkner, 1984). Deconvolution improves the signal to background ratio and sharpens the fine-structure in the spectrum, although it can also introduce artifacts. However, deconvolution does not improve the signal to noise ratio, and so it is preferable to use specimens thin enough to be free of plural scattering wherever possible.

5.2.5.1.2 Spatial resolution. Spatial resolution is an area where EELS has a marked advantage relative to EDS analyses. Spatial resolution in EDS depends strongly on the excited volume (§4.4.1, 4.8.5) and thus is determined by the spread of the beam within the specimen as well as the specimen thickness. These effects limit the spatial resolution to greater than approximately 10 nm, even in the best of cases. In contrast, spatial resolution in EELS depends on the localization of the interaction between the electron beam and specimen. There are no appreciable beam-spreading effects since the electrons scattered at high angles are excluded by the entrance aperture. Therefore thickness is not an appreciable limitation on spatial resolution. However, for a given beam energy there is an inverse relation between localization and the energy of the interaction, such that spatial resolution is improved somewhat for high-energy losses (Colliex, 1982).

The resolution is controlled differently in TEM and STEM mode. The EELS spectrometer in TEM mode is focussed on an image of the specimen at the projector-lens crossover; in this case, spatial resolution is defined by the illuminated area in diffraction mode (Williams, 1984). In STEM mode the resolution is defined by the width of the STEM beam in the image plane.

Probe diameters in STEM mode can be as small as 0.5 nm, and so that is an inherent limit but, given its small size, not a highly significant limitation of the EELS spatial resolution. However, the number of ionization events and thus counting statistics for such small regions are low for reliable results in reasonable counting times. This limitation can be modified by extending the duration of measurement, but then one runs into instrumental problems such as specimen stability, instrument drift, and contamination. The result is that optimal values of spatial resolution, while still of the order of a few nanometers, are nevertheless considerably better than the resolution limit of EDS.

5.2.5.2   Spectrum accumulation

On average, approximately 95% of the intensity is in the zero- and low-loss regions of the spectrum (Williams, 1984). Therefore, the spectrometer must be able to record intensities over a wide dynamic range in order to sense the ionization-loss region, which is the most important part of the spectrum for our purposes. [The number of counts per channel in the high energy-loss region can be as much as four orders of magnitude lower than in the zero-loss peak.]

To record the spectrum, a large (10^2 to 10^3) gain change in amplification is used between the low- and ionization-loss regions of the spectrum. This gain change is evident in the spectral EELS plots by the vertical rise in counts near 100 eV (e.g., Figs. 5-2, 5-4). It is possible to achieve the same result by counting for longer times (i.e., higher dwell times) in the ionization-loss region, and in most cases both amplification and increased dwell times are used to optimize signal intensities. It is advantageous to scan the zero and low energy-loss regions of the spectrum rapidly so that the detector is not harmed by the high intensity in these regions. Furthermore, the large electron doses in this spectral region can overload the detector and cause "afterglow," which is a residual signal that can distort subsequent measurements. An interesting effect is that measured intensities will be lower if scanning from high energies down to the zero-loss region rather than the reverse, because such scanning avoids the residual effects on the spectrometer of the high intensity of the zero-loss peak. When using PEELS, "ghost peaks" can appear unless the beam is attenuated or deflected off the YAG detector.

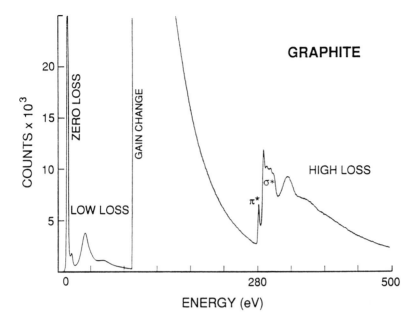

Figure 5-4. EELS spectrum of graphite showing the π^* peak used for calibration. The figure shows additional details such as the zero-loss peak and weak plasmon peaks in the low energy-loss region and a σ^* peak in the high energy-loss region. (Courtesy of B. Miner.)

5.2.5.3 Calibration of the spectrum

As in most spectrometric measure-ments, it is important to check the spectrometer for spurious signals and to calibrate the measurements. Spectrometer checks are commonly done by obtaining a spectrum without having a specimen in place, analogous to the "hole count" of EDS spectrometry. Such a spectrum should only show the zero-loss peak and a background having a rapidly decreasing intensity. The rate of decrease in background is a measure of the "afterglow" effect mentioned above. Any edges that might occur will be from electrons scattered from the TEM or spectrometer and so will indicate a need for realignment.

Graphite is a commonly used sample for spectrum calibration in EELS. The energies of the zero-loss peak and carbon K ionization edges are known (0 and 284 eV, respectively), and these are then used for the energy calibration. There is a problem, however, in that K-edge onset is not discrete, commonly extending over a range of several eV. Therefore it is conventional to assign the center of the sloping edge at the low-energy side of the edge with the energy of the K edge. The graphite spectrum (Fig. 5-4) has a pre-edge peak, called the π^* peak, that can be detected in high-resolution spectra. Where observed, this peak can be used as a calibration point for the 284 eV position.

Spectrometer alignment and operation can be set using the so-called "jump ratio," which is the ratio of the point of maximum intensity to the background channel that precedes the edge; for a well aligned spectrometer, the ratio should lie between seven and ten (Williams, 1984).

5.2.6   Quantitative analysis

With EDS it is possible to estimate abundances qualitatively by noting relative peak heights, and such estimation can be both rapid and convenient. However, such a procedure is not possible with EELS because of the broad shapes of the EELS edges, the decreasing probabilities of ionization with atomic number (decreasing efficiency of ionization), and the high background, which can also contain the cumulative signals of several lighter elements. In general, quantification of EELS spectra is more difficult and the results less accurate (with accuracies that are ±10 to 15%, relative) than for EDS.

5.2.6.1   Background subtraction

The first and probably most important step in the processing of EELS data for quantitative elemental analysis is removal of the background. Such background stripping in EELS spectra is normally done for one element at a time. Stripping can be difficult because of the cumulative effects of other elements having the "tails" of their edges overlap the edge of the element of interest.

Unlike EDS, where it is possible to model the background rather closely, the interactions contributing to EELS background are more complex and less well understood quantitatively. Hence, background-subtraction corrections for EELS are empirical. An inverse power-law relation of type $I = A \cdot E^{-r}$ is commonly used, where I is the intensity of the spectrum at a given energy, E. The parameters A, and r are empirical constants specific to the energy range of interest. The values for r are commonly between 2 and 5 (Williams, 1984). Where multiple edges overlap (e.g., Fig. 5-2), the effects of the spectral tails of the several elements must be considered, with appropriate adjustments of the empirical constants for the different energy regions of fit. The fit to background for each edge is made to the as-recorded spectrum. That is, background subtraction for a particular edge should never be carried out on data in which the background for another edge has already been corrected.

The empirical constants are commonly chosen by fitting a spectral region 50 to 100 eV prior to the edge, and then the curve is extrapolated under the edge of interest for a similar energy range. The number of counts between this curve as a base and the maxima of the edge then produces the integrated peak counts for the element of interest. Figure 5-5 shows an example of background fits and integration windows for a spectrum of anhydrite (cf. Fig. 5-2).

When a spectral fit is extrapolated to give the background under an edge, it is vital that the data before the edge is of high quality and that the energy window used for fitting the spectrum is sufficient. Obviously, where two edges are close together, determination of suitable constants and extrapolated curves may be difficult and can lead to errors in quantification. A good fitting window should have a width such that $E_{finish}/E_{start} < 1.5$ (Joy and Maher, 1981). The rapidly decreasing nature of the intensities of the background curve means that reliable curve fitting (and consequent extrapolation) can be done over narrower energy ranges at low energies than at high energies, where different empirical constants apply.

Another complication is that edge shapes vary significantly in EELS (Fig. 5-5), so that one cannot safely assume simple forms such as the Gaussians in EDS, and integrated measurements must be made for all energies under the edge. However, even the edge width is subject to interpretation because of the way it gradually tails off. Generally the width used for the fitting curve, described above, is used for peak integration.

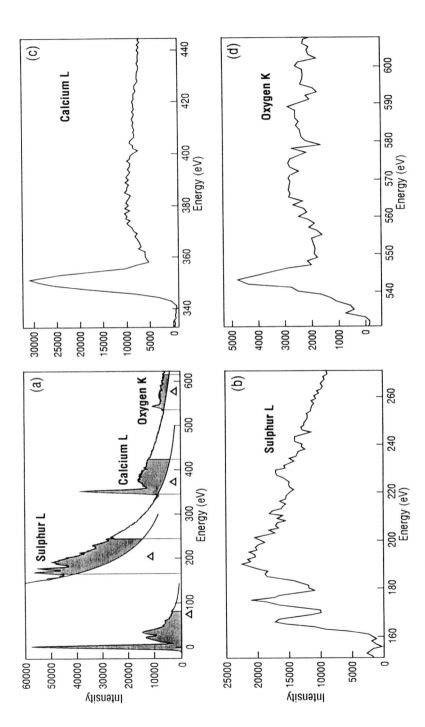

Figure 5-5. Empirical fits to background and background-subtracted intensities for anhydrite (as shown in Fig. 5-2). Integration windows suitable for quantitative analysis are shown in (a) and are indicated by width Δ. Expanded views of of the background-corrected S, Ca, and O edges are shown in (b), (c), and (d), respectively.

5.2.6.2 Determination of compositions

5.2.6.2.1 Complications. It would be agreeable if compositional determinations using EELS were straightforward and accurate, but there are several complications beyond those mentioned above. In theory it should be possible to count all electrons having a specified energy loss and thereby obtain absolute quantification. However, such counting has not been possible in practice. Moreover, ionization efficiencies vary from one element to another and also among different shells for a given element.

The details of background subtraction produce considerable uncertainty. The zero- and low-loss regions are used as standards during quantification. Obtaining the ratio of the counts under the zero- and low-loss regions to those from the region of interest helps correct for electron doses and spectrometer efficiencies. There are also uncertainties in properly defining the width of the energy window over which counts are to be obtained. Since the number of counts in the zero- and low-loss regions are orders of magnitude greater than for the region of interest for the element being studied, large errors can be generated. For this reason, a ratio method (described below) is commonly used.

Another difficulty in quantification is the large variation in edge shapes (Fig. 5-5), such that some edges are rather flat (especially those from L and M shells). In such cases, the peak-to-background ratios are low, and it is difficult to determine their exact dimensions. Peak overlaps add to the complications in multi-element samples. In the spectrum for anhydrite (Fig. 5-2) peak overlap is not a problem but for many other minerals, such as the Ca titanate, kassite, it is simply not possible to obtain accurate (better than 15%) analyses. In the particular case of kassite (Self and Buseck, 1991), the proximity of the TiL and the OK edges (Fig. 5-6) does not allow suitable windows for the integration of the Ti peak or the fitting of background for the O peak to be defined. Similar problems occur with EDS, but there the peaks have better defined shapes and commonly have greater peak-to-background ratios. Furthermore, the larger useable energy range in EDS means that elements produce multiple excitations (i.e., K, L, and M), and so in some instances peaks other than the overlapping ones can be used.

5.2.6.2.2 Procedures. Quantitative analysis is simply the conversion of the measured number of inner-shell excitation events to an atomic concentration. On face value this would seem straightforward. The measured intensity, I_s, caused by ionization of a particular shell (i.e., K, L, M, etc.) is given by

$$I_s = P_s I \qquad (5.1)$$

where I is the incident electron-beam current, and P_s is the probability that an incident electron ejects an electron from shell s (variables are summarized in Table 5-2). The probability, P_s, is the product of the total ionization cross section per atom, σ_s, and the number of atoms of the element present per unit area of the specimen, N. Determination of the value of N is the aim of quantitative analysis, and from Equation (5.1)

$$N = I_s /(I \sigma_s) \qquad (5.2)$$

Unfortunately, real life is not so straightforward. Equation (5.2) only applies if all ionization events are measured. In EELS it is only possible to collect scattered electrons over a restricted angle range and to integrate the intensity beyond an edge over a limited energy range. The collection angle and the energy range are determined by β, the collection semi-angle of the EELS spectrometer. (Fig. 5-3), and Δ, the width of the energy window used for peak integration (Fig. 5-5a). Under these restrictions Equation (5.2) becomes:

Table 5-2. Names of variables

Symbol	Definition	Section ref.
a, b	crystallographic sites	5.3.2
A, B	elements located entirely on sites **a** and **b**, respectively	5.3.2
X	element that has an unknown distribution between crystallographic sites, **a** and **b**	5.3.2
'	measurement taken at a channelling orientation	5.3.2
g	reciprocal lattice vector	5.3.1.1
I	incident electron-beam current	5.2.6.2.2
$\mathbf{I_s}$	measured intensity caused by electron ejection from the **s** shell	5.2.6.2.2
$\mathbf{I_o(\beta, \Delta)}$	measured incident intensity using specified values of β and Δ	5.2.6.2.2
$\mathbf{I_s(\beta, \Delta)}$	measured intensity arising from the ionization of the **s** shell	5.2.6.2.2
K	parameter that describes element- and instrument-dependent coefficients	5.3.2
N	number of atoms present per unit area of the specimen	5.2.6.2.2
$\mathbf{N_y}$	number of counts (intensity) of the measured signal for each element indicated by the subscript	5.3.2
$\mathbf{P_s}$	probability that an incident electron ejects an electron from shell **s**	5.2.6.2.2
s	atomic shell (e.g., K, L, M)	5.2.6.2.2
s'	subscript indicating atomic shell different from shell **s**	5.2.6.2.2
β	the collection semi-angle of the EELS spectrometer	5.2.6.2.2
Δ	width of the energy window used for peak integration	5.2.6.2.2
σ_s	total ionization cross section per atom	5.2.6.2.2
$\sigma_s(\beta, \Delta)$	partial ionization cross section	5.2.6.2.2
ξ_g	extinction distance	5.3.1.2

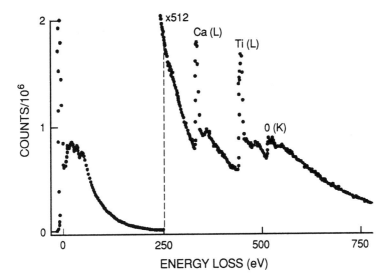

Figure 5-6. EELS spectrum of kassite [$CaTi_2O_4(OH)_2$]. The TiL and the OK signals overlap to such an extent that it is not possible to define suitable windows for background fitting and intensity integration.

$$N = \frac{I_s(\beta,\Delta)}{I_o(\beta,\Delta)} \cdot \frac{1}{\sigma_s(\beta,\Delta)} \tag{5.3}$$

where $\sigma_s(\beta,\Delta)$ is the partial ionization cross section, $I_o(\beta,\Delta))$ is the measured incident intensity, and $I_s(\beta,\Delta))$ is the measured intensity arising from the ejection of an electron from the s shell (Egerton, 1978; Egerton, 1981a)

The value of $I_o(\beta,\Delta)$ can be estimated by measuring the integrated intensity of the zero-loss and the low energy-loss regions of the spectrum up to the energy Δ, as shown in Figure 5-5a. The partial ionization cross sections can be obtained from tabulated values (e.g., Egerton, 1981b) or directly calculated (e.g., Egerton, 1986, App. B).

Equation (5.3) gives an absolute value for N but, because of uncertainties in $I_o(\beta,\Delta)$ and $\sigma_s(\beta,\Delta)$, the results of EELS quantification are not highly satisfactory. Relative errors of 10% are not unusual, and the uncertainties can be significantly greater. This problem can be minimized by determining the ratio of the concentrations for two elements rather than determining absolute concentrations. By using Equation (5.3), the concentration ratio of two elements A and B is given by

$$\frac{N_A}{N_B} = \frac{I_{As}(\beta,\Delta)}{I_{Bs'}(\beta,\Delta)} \cdot \frac{\sigma_{Bs'}(\beta,\Delta)}{\sigma_{As}(\beta,\Delta)} \tag{5.4}$$

where the parameters relating to the different elements are indicated by the subscripts A and B. The subscript s' for element B indicates that the intensities for the two elements do not have to be measured from the same atomic shell.

Equation (5.4) eliminates the need to measure the low energy-loss region of the spectrum (other than to check that a suitable specimen is being used). The ratio of the partial cross sections can be treated in much the same way as the "k-factors" in EDS analysis (see Chapter 4) and, just as in EDS analysis, can be determined from standard specimens (i.e., empirically) or by a standardless approach (i.e., calculated). When standards are used, it is imperative to ensure that the microscope and spectrometer operating conditions used to measure spectra for the standards and unknowns are identical. It is also necessary to ensure that the thickness of the standard specimens conform to the criteria discussed in §5.2.5.1.1. The advantage of the standards approach is that it eliminates the need to measure β. The ratio method (both with and without standards) is capable of giving relative concentrations to accuracies of around 5%.

5.2.6.3   Detection limits

Some of the factors limiting sensitivity and the detection limits are described in §5.2.6.2. However, even under optimal conditions, there is a minimum detectable concentration of one element in a matrix consisting of another element. The minimum detectable concentration is highly variable and ranges from a few tenths of a percent to a few percent. This minimum value is particularly important when searching for minor elements; when an element is not detected, the detection limit places a maximum value on the concentration of the element in the sample.

The criterion for detectability is that ionization events for the element of interest is discernible above the noise of the background signal. In general, the 3s limit is used to define this criterion so that an ionization signal, I, is classed as being detectable above a background signal, I_b, if

$$I \geq 3 (I + h\, I_b)^{1/2} \tag{5.5}$$

The parameter **h** is a factor representing the accuracy of the background extrapolation and has values in the range 5 to 15 (Egerton, 1986). As **I** is very much less than I_b, the minimum detectable signal, I_{min}, is given by

$$I_{min} = 3 (h\, I_b)^{1/2} \tag{5.6}$$

Using Equation (5.3), the minimum detectable atomic concentration of an element **X** in a matrix consisting of element **Y** is

$$\frac{N_{X\,min}}{N_Y} = \frac{3}{I_Y} \cdot \frac{\sigma_Y}{\sigma_X} (h\, I_b)^{1/2} \tag{5.7}$$

where σ_X and σ_Y are the partial cross sections for elements **X** and **Y** respectively, and I_Y is the integrated peak intensity for element **Y** (Rez, 1983; Egerton, 1986). The parameters β and Δ are implicit in Equation (5.7), and the same value of Δ should be used when evaluating I_Y and I_b.

The values for I_Y and I_b are determined experimentally. For example, using the spectrum shown in Figure 5-4 and tabulated values for the cross sections, the minimal detectable amount of Ti in C is approximately 0.7% while for N in C it is approximately 2%. Hence, for the particular sample used to record Figure 5-4, there is at most 2 at % N and 0.7 at % Ti. Other examples of detection limits are given by Egerton (1986) and Williams (1984).

5.2.7 Crystal chemistry and structural information

EELS spectra contain far more information than is obtained by integrating ionization events. The use of the low energy-loss region of an EELS spectrum to determine the dielectric function of a sample has already been mentioned (§ 5.2.2.2). The dielectric function is a sensitive indicator of the electronic band structure of a crystal (Kittel, 1986) which, in turn, is indicative of the crystal chemistry. In mineralogy, however, the major interest in crystal chemistry is not with the electronic band structure but rather the valence states and local environments of atomic species. These properties can be determined from EELS spectra by using the techniques of EXELFS and ELNES, details of which are reviewed by Spence (1988).

5.2.7.1 EXELFS: extended energy-loss fine structure

Fluctuations exist in the EELS edge at energies greater than edge onset. This fluctuation, or "fine-structure," occurs about 40 eV or more beyond onset and contains information about the bonding environment of the atom in question. The fine-structure is a combination of weak, sinusoidal variations comparable to extended x-ray absorption fine structure (EXAFS), used in x-ray absorption spectroscopy (XAS), commonly in conjunction with synchrotron radiation. The modulation arises from interference between the ejected electron and backscattered electrons from nearby atoms (Rez, 1991) and therefore contains information about local atomic environments.

EXELFS has great potential, but significant problems arise because of the relatively low number of counts plus the even lower intensities of the fluctuations; these effects combine to produce poor counting statistics. According to Williams (1984), the modulation is only approximately 5% of the edge intensity, and the wavelength of the fluctuation in the spectrum is in the range of 20 to 50 eV. Beyond the low intensities, limitations of EXELFS include sensitivity to irregularities in specimen thickness and the effects of multiple electron scattering events (dynamical effects).

Since EXAFS is probably more familiar than EXELFS to most mineralogists, it is useful to use it as a point of comparison. The information available from EXELFS and EXAFS is comparable, except that EXELFS has far greater spatial resolution because of the small size of the electron probe, whereas there is far greater intensity in EXAFS peaks and so it is more sensitive and easier to quantify. The two procedures are complementary and, ideally, they could be used in tandem.

The procedure for EXELFS data reduction is analogous to what is done with EXAFS, as reviewed in a previous *Reviews in Mineralogy* volume (Brown et al., 1988). Basically, the method is, in sequence, to subtract background, convert the intensity distribution to an electron wave function in reciprocal space, and calculate the Fourier transform (e.g., Disko et al., 1982). The result is a series of radial distribution functions (RDF) (one for each element in the sample) that reflect not only the average distance of the nearest neighbors from the ionized atom but also the number of nearest neighbors surrounding the ionized atom. A problem is that measurements of 300 to 400 eV of fine structure after an edge onset are needed for good quantification (Rez, 1991). Interferences from other elements commonly do not allow such long tails to be measured and, in any case, the weak fluctuations mean that long counting times are required to accumulate a sufficient number of counts for statistical reliability. According to Leapman et al. (1981), the RDF accuracies from EXELFS are of poor quality, indicating that only approximate answers are possible. However, in instances where such answers are desirable from localized regions of a sample (such as in disordered crystals or in glassy materials), EXELFS can provide unique information. An advantage of EXELFS is that directionally dependent information about nearest neighbor distances can be obtained by positioning the spectrometer entrance aperture so as to judiciously define the scattering angle experienced by a beam electron during an inelastic scattering event (Disko et al., 1982).

5.2.7.2 *Chemical shifts and ELNES: energy-loss near-edge fine structure*

Chemical shift and ELNES effects occur in the first 10 to 40 eV after the ionization-edge onset. Chemical shifts are small changes (approximately 2 eV) in the edge energy and are caused by differences in valence state. Consequently, accurate measurement of the edge energy can be used to determine the valence states of elements in a specimen. This technique has been applied successfully to a variety of minerals (e.g., Otten et al., 1985; Otten and Buseck, 1987a; Rask et al., 1987). When combined with ALCHEMI (see §5.3), the chemical shift can be used to give site-specific information about structures containing atoms of mixed valence state (Taftø and Krivanek, 1982a). An example for Mn, which occurs in several oxidation states in the geological environment, is shown in Figure 5-7. The changes in edge shape are clear. It is difficult if not impossible to obtain such information at such fine spatial resolution by any other technique. However, while it is easy to see changes in oxidation states, a limitation is that it is difficult to quantify the fractions of Mn, Mn^{2+}, Mn^{3+}, or Mn^{4+} that are present in any given sample.

ELNES reflects the electronic band structure of the specimen. ELNES modulations are stronger than those for EXELFS, but they are more difficult to interpret. They can, however, be used qualitatively to determine the atomic coordination of the elements in a sample. Just as the crystal band structure is changed by the coordination of the atoms in the structure, so too is the ELNES. By calibrating the form of the ELNES with samples containing elements in known coordination, the coordination of the same elements in

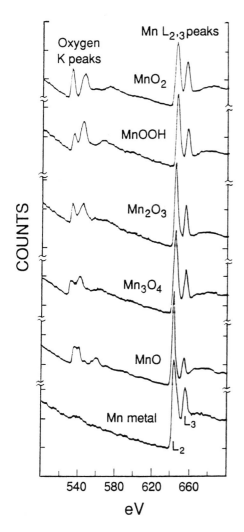

Figure 5-7. Comparison of EELS spectra of different Mn oxide minerals (in sequence from the top: pyrolusite, manganite, bixbyite, hausmannite, manganosite) and Mn metal. The changes in Mn and O peak position and shape with oxidation state is evident. After Rask et al. (1987).

poorly known samples can be specified (Taftø and Zhu, 1982). As with quantitative elemental analysis, the determination of atomic coordination by ELNES is only possible for thin specimens and fixed data-collection conditions.

5.2.8 Energy-filtered imaging

Energy-filtered images can be obtained by operating the microscope in STEM mode and using the counts in a predefined energy window as the imaged signal. For example, by choosing an energy window around the zero-loss peak, an image free of inelastic scattering effects (except for phonon events) can be obtained. By placing the window just above the characteristic edge of a particular element, a map of elemental concentration can be recorded. The quality of concentration maps can be greatly improved if compensation for spectral background (and therefore specimen thickness variation, etc.) is made by using the difference in counts between windows just above and just below the edge as the signal. In this situation, PEELS has considerable advantage over conventional EELS using an exit aperture. The improved images with PEELS arises because windows can be defined by combining the signals from a number of photo-diode elements, and the outputs for all the required energy windows can be collected simultaneously. For conventional EELS, the maximum energy-window size is limited by the maximum size of the exit aperture and, to obtain a background-corrected concentration map, two or more images with different energy windows must be collected and subtracted.

Background-corrected elemental maps obtained using EELS are of comparable quality to those obtained by EDS. EELS has the added advantages of high spatial resolution (especially when compared to SEM maps) and the ability to obtain maps for light elements (e.g., Li). With PEELS one also gets many more counts per second than with EDS.

In general, the STEM mode of a conventional TEM has a resolution some five to ten times worse that the resolution obtainable in high-resolution imaging mode. Furthermore, a STEM image takes considerably longer to obtain than a conventional TEM image. Instead of using STEM mode to obtain energy-filtered images, it is possible to place specially designed energy filters in-line with the microscope optics to obtain energy-filtered images directly on the viewing screen of the TEM. These filters are usually placed just below the objective lens, where the magnified image is still sufficiently small to fit entirely within the entrance aperture of the filter. The two most commonly used filtering devices are the prism-mirror system and the Ω (omega) filter.

The prism-mirror system, as designed by Castaing and Henry (1962), consists of two back-to-back magnetic-sector spectrometers and an electrostatic mirror (Rossouw et al., 1976; Egerton, 1986). The electrons entering the filter are dispersed and deflected by 90° in the first magnetic-sector spectrometer, reflected by the electrostatic mirror into the second spectrometer, which also disperses and deflects the electrons through 90°. The overall effect is that, depending on the settings of the magnetic-sector spectrometers, the path of electrons of one particular energy will be realigned with the optic axis of the microscope. The paths of electrons having other energies will not be aligned with the optic axis and can be eliminated using an aperture placed below the energy filter.

The Ω filter (Zanchi et al., 1977; Egerton, 1986) works on a similar principle to the prism-mirror system but replaces the electrostatic mirror with two magnetic-sector spectrometers. The path of the electrons in such a filter is similar to the Greek letter Ω and hence the naming of this device. The advantage of the Ω filter is that, because it does not have an electrostatic mirror, it does not require connection to a high-voltage power supply. (To reflect an electron, the voltage on the plates of the electrostatic mirror needs to be slightly more negative than the accelerating voltage used in the electron gun.)

By using these energy-filtering systems, high-resolution, energy-filtered images and diffraction patterns can be obtained. These are of use when it is desirable to obtain TEM signals that are free of inelastic signals. An example is for the quantitative interpretation of electron-diffraction data, especially those obtained with convergent beams. Another application is for imaging samples where chromatic aberrations limit image quality, such as can occur in specimens that are thicker than normally ideal for HRTEM imaging. In addition to the energy-filtering properties, adjustment of the post-filter lenses allows fully dispersed EELS spectra to be viewed on the microscope screen.

Microscopes that use energy-filtered imaging systems are now commercially available (e.g., the Zeiss EM912) and, although such specialized microscopes are new and not widely available, it is likely that they will be more available in the future. When combined with a cooled, charge-coupled device (CCD) camera (see §1.5.3) it is possible to achieve the ultimate in quantitative electron diffraction and imaging with essentially no background. The alternative approach for high-resolution is to scan the image across the entrance aperture of the spectrometer. For a 1024×1024 pixel image, the recording time will be about 10^6 times greater than with the Ω filter.

5.3   ELECTRON CHANNELLING (ALCHEMI)

5.3.1   Principles

5.3.1.1   General

Methods for quantitative elemental analysis, as described in Chapter 4 and the preceding sections of this chapter, assume that the electron-beam current is uniform throughout the specimen. For example, with the ratio method of quantitative EELS analysis [as described by Equation (5.4)] it is explicitly assumed that atoms of each atomic species receive exactly the same electron dose. This assumption is not warranted in all cases. In particular, when there is strong diffraction to low-order reflections, the electron-beam current in a crystal varies considerably from one point in the unit cell to the next.

As explained in Chapter 1, the diffraction pattern is the Fourier transform of the electron wavefunction at the exit surface of the specimen. Consequently, if there is strong diffraction, the exit-surface wavefunction is highly modulated. Given that the geometric positions of diffraction spots (i.e., the diffraction angles) are related to the unit-cell dimensions of the crystal, the modulation period of the electron current in a crystal will coincide with the unit-cell size. The exact form of the modulation depends on the amplitudes and phases of the diffracted beams (Chapters 1 and 3). Hence for a crystal oriented to a strongly diffracting condition, the current density in some parts of the unit cell is high while in other parts the current density is low.

To illustrate the variations of electron-beam current within a crystal, the semiconductor GaAs will be used as an example. This choice is made because GaAs has a relatively simple structure that can be used to demonstrate both the general as well as the more subtle points of beam-current variation. In addition, good crystals of GaAs are easy to obtain and prepare for electron microscopic study. The points demonstrated in the following discussion for GaAs apply equally well to minerals such as sphalerite (α-ZnS) that have the same crystallographic structure, and the principles are perfectly general. In terms of the following discussion and for the particular case of sphalerite, the S atoms can be considered as equivalent to Ga while Zn atoms can be considered as equivalent to As.

As an example of the form of beam-current modulations, consider the (00ℓ) atomic planes of GaAs (Fig. 5-8a). The Ga and As atoms lie on regularly-spaced, alternate planes. Using the techniques described in Chapter 3, it is possible to calculate the current density for various crystal orientations and thicknesses. To fully discuss the variation of electron-beam current modulations as a function of crystal orientation, it is necessary to define the terms used to describe the crystal tilt with respect to the incident electron beam. When a crystal is oriented so that the incident electron beam is at the exact Bragg angle for a given set of crystallographic planes (represented here by the reciprocal lattice vector **g**), the crystal is said to be at the "Bragg orientation" for **g**. At this orientation, the Ewald sphere (Fig. 1-5) passes through the reciprocal lattice vector **g**. A related orientation is when the crystal is tilted so that the incident beam is exactly perpendicular to the reciprocal lattice vector **g**. At this orientation the **g** and -**g** reflections are equidistant from the Ewald sphere

166

(i.e., equally excited), and the crystal is said to be oriented to the "symmetrical condition." Crystal orientations between the symmetrical condition and the Bragg orientation are said to be "inside" the Bragg angle. Similarly, crystal orientations where the angle between the incident beam and the crystallographic planes is greater than the Bragg angle are said to be "outside" the Bragg angle.

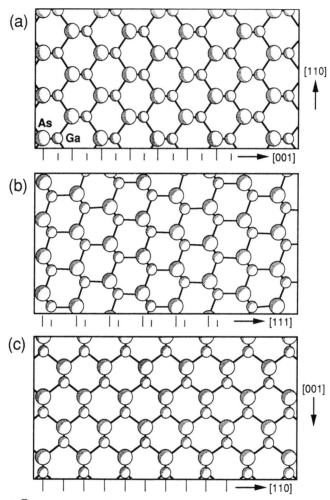

Figure 5-8. The [1 1̄ 0] zone of GaAs, which is isostructural with sphalerite, rotated so as to highlight different atomic planes. The (001), (111), and (110) planes are aligned vertically in (a), (b), and (c) respectively.

The calculated current densities for a crystal tilted just inside and just outside the Bragg angle for the 002 reflection [the 001 reflection is forbidden] are shown in Figure 5-9. The extent of the current variation for these examples is quite marked. For the condition where the crystal is tilted just inside the Bragg angle for the 002 reflection, the largest peaks in the current density fall on planes containing As atoms. For the condition where the crystal is tilted just outside the Bragg angle, the largest peaks fall on the Ga-containing planes. In general, the beam current is largest on planes with the highest electron density (i.e., containing the heaviest elements) for orientations inside the first-order Bragg angle.

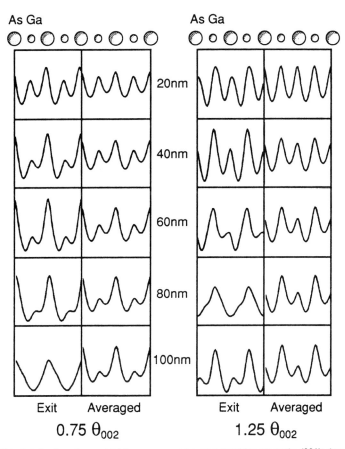

Figure 5-9. Calculated exit-surface and thickness-averaged current densities across the (001) planes of GaAs for various crystal thicknesses. The columns on the left are for a crystal tilted to 0.75 of the Bragg angle for the 002 reflection (i.e., tilted inside the Bragg angle). The columns on the right are for a crystal tilt of 1.25 θ_{002} (i.e., tilted outside the Bragg angle). The crystal thicknesses are shown down the center of the figure. The current densities at the crystal exit surfaces are shown in the columns marked "Exit" while the thickness-averaged current densities are shown in the columns marked "Averaged." The positions of Ga and As atoms are indicated at the top of the figure. The accelerating voltage is 100 kV.

Given that the number of atom-related, inelastic-scattering events (such as inner-shell excitations) is proportional to the electron current falling on the atom, the relative x-ray emission for As is higher than for Ga for tilts inside the Bragg angle and lower for tilts outside the Bragg angle. Inelastic events occur throughout the crystal and not just at the exit surface, and thus the number of inelastic events is represented more accurately by the thickness-averaged current density, as shown in Figure 5-9 for each of several thicknesses. The calculated ratio of electron current on Ga sites to the current on As sites as a function of crystal tilt with respect to (001) planes is shown in Figure 5-10. This ratio represents the number of inner-shell ionization events as measured by EELS and the ratio of characteristic x-ray emissions as detected in EDS. From Figure 5-10 it can be seen that the variations in the current ratio cease and the value of the ratio goes to unity at an orientation of around three Bragg angles for the first-order reflection. Thus, for the particular case of the 00ℓ reflections of GaAs, orientations outside the Bragg angle for the 006 reflection are free of channelling effects and suitable for quantitative analysis because each atomic species receives the same electron dose, i.e., the same electron current.

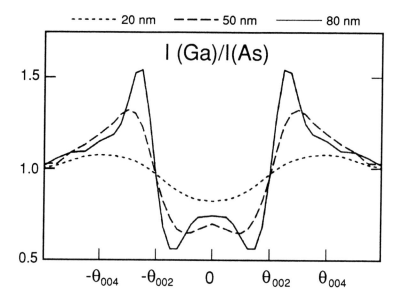

Figure 5-10. Calculated ratios of x-ray emissions from Ga and As as a function of crystal tilt around (001) planes. The horizontal axis shows the tilt in terms of Bragg angles; for example, at the point marked θ_{002}, the crystal is tilted to the Bragg angle for the 002 reflection. The accelerating voltage is 100 kV, and the ratios are shown for three crystal thicknesses.

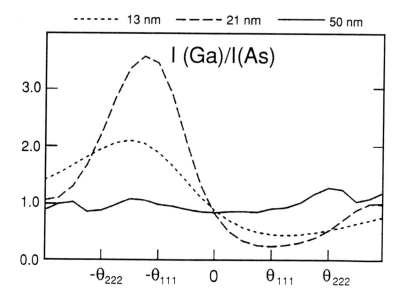

Figure 5-11. Calculated ratios of x-ray emissions from Ga and As as a function of crystal tilt around (111) planes. The horizontal axis shows the tilt in terms of Bragg angles (cf. Fig. 5-9). The accelerating voltage is 100 kV, and the ratios are shown for three crystal thicknesses. The lack of symmetry reflects the non-centro-symmetric character of GaAs (and a similar effect would obviously occur for sphalerite).

Figure 5-9 shows the current density for only two tilt conditions. There are of course many other tilt conditions, including those where the electron current is modulated so that most of the current passes between the atoms in the crystal. In such instances, the inelastic scattering of the incident-electron radiation by the atoms in the crystal is minimized, and so the absorption of this radiation by the crystal is reduced. The incident-electron radiation is then said to *channel* through the crystal, and the diffracting conditions for which channelling occurs are described as *channelling conditions*. As shown above for GaAs, there are also corresponding crystallographic orientations for which the intensity of the incident radiation is a maximum at the positions of the atoms (e.g., Fig. 5-9), in which case the intensity of inelastic scattering is maximized. In this discussion, the term *channelling* describes all conditions where there is either an anomalous increase or decrease in the absorption of the incident radiation. ALCHEMI (atom location by channelling-enhanced microanalysis) utilizes the non-uniform distribution of incident-electron radiation within a crystal for chemical and structural analysis.

An example of the utility of channelling for structural analysis is given by considering the (111) GaAs planes. The structure of GaAs is non-centrosymmetric, as demonstrated by the spacing between atoms along the [111] direction (Fig. 5-8b). This non-centrosymmetric nature is reflected in the current density of high-energy (beam) electrons passing through a GaAs crystal. Using calculations similar to those used for the (001) planes of GaAs as described above, the ratio of electron current at the Ga atomic sites relative to that at the As atomic sites can be plotted as a function of crystallographic orientation. The results of such calculations are shown in Figure 5-11. When the crystal is tilted to the Bragg angle for the $\bar{1}\,\bar{1}\,\bar{1}$ reflection (using directions as defined in Fig. 5-9) the electron current on the Ga atoms is high relative to the current on the As atoms. The situation is reversed when the crystal is tilted to the Bragg angle for the 111 reflection. The differences in the current density at different crystal tilts are mirrored in experimentally measured x-ray emissions (Taftø, 1983), as shown in Figure 5-12. Similar results could be achieved with EELS using the Ga and AsL edges. The variation in characteristic x-ray emissions in this case gives an absolute determination of crystal polarity and is therefore of benefit in HRTEM and diffraction analysis.

5.3.1.2 Experimental restrictions on the observation of channelling effects

The current ratios shown in Figures 5-10 and 5-11 highlight another feature of channelling effects in x-ray emissions; they are thickness dependent. For very thin crystals, where diffracted amplitudes are relatively small, the electron current variations are weak and so channelling effects are minor. With increasing crystal thickness, and therefore increased diffraction, the variation in current density becomes stronger. This increase in variation is apparent for the crystal thicknesses shown in Figure 5-10. In Figure 5-11, however, the magnitude of the channelling effects initially builds up with crystal thickness but then falls away and becomes virtually nonexistent at a thickness of 50 nm.

This falling off in intensity is caused by the changes in the relative phase of the undiffracted and diffracted beams. At a thickness of half the extinction distance (ξ_g) [see Equation (3.28) in §3.3.4], there is a $\pi/2$ change in the relative phases of the undiffracted and diffracted beams. In HRTEM, this effect appears as a contrast reversal in the image at a crystal thickness of $\xi_g/2$. Hence, at crystal thicknesses greater that $\xi_g/2$, the modula-tions in electron-beam current are shifted relative to the modulations at thicknesses less than $\xi_g/2$, and so channelling effects diminish. (There will, of course, be further phase changes at $3\xi_g/2$, $5\xi_g/2$, etc.) Thus, for the (111) planes of GaAs, where the extinction distance is approximately 56 nm, channelling effects cannot be observed for a crystal thickness greater than 50 nm. For the (001) planes of GaAs, the extinction distance is

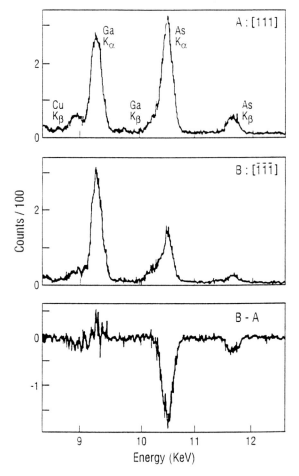

Figure 5-12. Measured characteristic x-ray intensities from a single crystal of GaAs. In the upper plot the crystal is oriented at the Bragg angle for the 111 reflection (with the [111] direction as defined in Fig. 5-8). In the middle plot the crystal is oriented to the Bragg angle for the $\bar{1}\,\bar{1}\,\bar{1}$ reflection. There is a marked difference in the relative x-ray emissions for Ga and As, as highlighted by the difference in intensities shown in the lower plot. The variations are in accord with the calculated results shown in Figure 5-11. The measurements were made using a chemically thinned, single crystal of GaAs. The accelerating voltage was 100 kV. The Cu peak arises from the grid used to mount the sample.

much larger (approximately 540 nm), and so the channelling effects can be observed at much greater thicknesses than for the (111) planes.

Channelling effects are also diminished by beam absorption and the general smearing of the beam current by inelastic scattering events. Thus, in general, suitable crystal thicknesses for observing channelling effects are between $\xi_g/4$ and $3\xi_g/4$. For most crystals this range corresponds to thicknesses between 10 and 40 nm. The crystal thicknesses used to measure channelling effects are normally those considered satisfactory for quantitative thin-film analysis as used by EDS and EELS.

Another factor that diminishes channelling effects is beam divergence. The calculations shown in Figures 5-9 and 5-10 assume that there is no beam divergence; in a realistic situation the results shown in these figures should be averaged over a tilt range corresponding to the beam divergence. For the (002) planes of GaAs, a beam divergence of half a Bragg angle (i.e., $\theta_{002}/2$) would virtually wipe out any channelling effects, even at the optimum thickness. Thus, in order to observe channelling effects, beam divergence should be kept to below 1/5 of a Bragg angle for the first-order reflection. This condition is achieved by using a condenser-lens aperture of such a size that, when the microscope is operated in diffraction mode (see Chapter 1), the diameter of the diffraction spots is less than 1/5 of the distance between the undiffracted beam and the lowest-order reflection.

5.3.1.3   Axial channelling

So far, discussion has been limited to crystallographic planes. However, this is an unnecessary restriction, and the discussion can be applied equally well to low-order zone-axis orientations. The two cases are usually distinguished as planar channelling and axial channelling. In low-order zone-axis orientations, the atoms do not line up in planes but, instead, line up in columns. As there are generally a large number of diffracted reflections at zone-axis orientations (and therefore a large number of Fourier coefficients contributing to the electron current), the current variations are considerably greater for axial orientations than for planar diffraction (Pennycook and Narayan, 1985; Rossouw and Maslen, 1987). In fact, for the [110] zone of GaAs (Fig. 5-8) the ratio of the electron-beam current on the As and Ga sites can be as much as ten times greater than shown in Figure 5-11 for the planar case.

In a manner similar to the planar case, the electron-beam current falls on columns of high electrostatic potential for orientations inside the Bragg angles of the first-order reflections. At these orientations, most of the electron-beam current is channelled along the columns of atoms and, consequently, the amount of inelastic scattering is high and the beam transmission through the specimen is poor. Anyone who has tried to orient a thick specimen for HRTEM imaging will be aware of this effect. Once the crystal is oriented inside the Bragg angle for the first-order reflection, the transmitted intensity all but vanishes. This effect, of course, makes axial orientations highly favorable for observing channelling effects in characteristic x-ray emissions and inner-shell ionization events. The disadvantage of axial channelling is that variations in the electron-beam current are very sensitive to beam tilt and commonly do not behave in a readily predictable manner.

5.3.2   Site-occupancy determination

The dominant use of channelling effects in characteristic x-ray emissions and inner-shell ionization events is for the quantitative determination of atomic-site occupancies. For example, if there were impurity atoms, say In or P, in a crystal of GaAs, the variation in characteristic x-ray emissions or inner-shell ionization events could be used to determine on which sites (either the Ga or As sites) the impurity atoms were located. (The analogy for the isostructural sphalerite might be the site distribution of minor In and Sb.) One way of quantifying the distribution would be to compare experimental results for various crystal tilts with results predicted from calculations, such as those shown in Figures 5-10 and 5-11. However, this method is not practical. The best achievable experimental variation in the ratio of intensities from As and Ga x-ray emissions was a factor of two (Fig. 5-12); the theoretical result predicts a maximum variation of about eight (Fig. 5-11). This difference in experimental and theoretical ratios arises because of a number of factors including beam divergence and variations in crystal thickness and tilt over the irradiated area of the specimen.

A break-through in experimental determination of atomic-site occupancies came with the realization that atoms lying in known crystallographic positions in the crystal could be used as internal calibrations for the electron-beam current on specific atomic sites (Spence and Taftø, 1983). Thus, for a crystal of GaAs containing In and P impurities, the intensity of characteristic x-ray emissions or the number of inner-shell ionization events for Ga and As can be used as calibrations; the signals from In and P, once standardized against Ga and As, can then be converted to site-occupancy values.

The standard method of determining site occupancies requires measurements at just two crystal orientations: one where there is no channelling (i.e., high tilt) and the other where there is strong channelling. The strong channelling condition can be either planar or axial. There is no specific requirement on the channelling orientation other than that the variation in the measured signal (either x-ray emissions or inner-shell-ionization events) is as large as possible.

To see how site occupancies can be quantified, let X be an element that has an unknown distribution between two crystallographic sites, a and b. Furthermore, let element A be located entirely on site a and let element B be located entirely on site b; these are called the reference elements. At a non-channelling crystal orientation, the thickness-averaged beam current (I) is equal on both sites, and so the measured signal intensity for each element is

$$N_A = K_A I \tag{5.8}$$

$$N_B = K_B I \tag{5.9}$$

$$N_X = K_X I \tag{5.10}$$

where N represents the number of counts (intensity) of the measured signal for each element as indicated by the subscript, and K represents parameters that describe element- and instrument-dependent coefficients such as atomic cross sections, absolute elemental concentrations, specimen thickness, and detector efficiency. These factors are closely related to the k-factors used in quantitative elemental analysis.

At a channelling orientation, the thickness-averaged, electron-beam current on site a (I_a) differs from the thickness-averaged current on site b (I_b). Thus the signal intensities for elements A and B are given by

$$N'_A = K_A I_a \tag{5.11}$$

$$N'_B = K_B I_b \tag{5.12}$$

respectively. The prime is used to indicate a measurement taken at a channelling orientation. If element X has a fractional occupancy of $f_X(a)$ on site a [and therefore $1 - f_X(a)$ on site b], then the signal intensity from element X is given by

$$N'_X = f_X(a) K_X I_a + [1 - f_X(a)] K_X I_b \tag{5.13}$$

These six Equations [(5.8) to (5.13)] can be solved for $f_X(a)$, thereby giving

$$f_X(a) = \frac{N'_X/N_X - N'_B/N_B}{N'_A/N_A - N'_B/N_B} \tag{5.14}$$

Equation (5.14) gives a quantitative expression for the site partitioning of element X between the a and b crystallographic sites. The equation involves only measured intensities, and so the use of Equation (5.14) does not require a knowledge of k-factors, specimen thickness, or even absolute elemental concentrations.

Although Equation (5.14) only requires measurements at two orientations, statistics are greatly improved by evaluating $f_X(a)$ for several pairs of channelling and non-channelling orientations. In this way $f_X(a)$ can be measured to an accuracy of around 5%. To obtain reliable results it is critical that the non-channelling spectrum is of high quality. This determination is made in two ways. The first is by comparing spectra obtained at two or more presumed (on the basis of few or no observed diffraction spots) non-channelling crystal orientations. For example, non-channelling could be checked by comparing spectra taken for the crystal tilted sequentially in opposite directions from a known strong reflection. No channelling should occur for either of these orientations, and so there should be no difference in the relative intensities of the characteristic emissions. A second check is to compare an experimental non-channelling spectrum with that expected from the known composition of the crystal; the relative intensities of the characteristic emissions of the non-channelling spectrum should agree with those predicted from the chemical composition of the crystal.

Not all crystallographic zones or planes are suitable for ALCHEMI. For example, the (110) planes of GaAs (Fig. 5-8c) contain equal amounts of Ga and As, and so the individual Ga and As sites will not be differentiated. [Channelling effects will be observable for the (110) planes, but signals from Ga and As will vary in unison.] Suitable zones or planes for ALCHEMI can be chosen either by studying the crystal structure in the form of maps, as shown in Figure 5-8, or by randomly searching zones and planes in the TEM until channelling effects are observed. The manipulation of a crystal to both channel-ling and non-channelling conditions can be achieved by observing the positions of Kikuchi lines (Chapter 2) in the diffraction patterns of the crystal (e.g., Taftø and Buseck, 1983).

The first structure used for the determination of site occupancies using an equation of the form of (5.14) was gahnite (Taftø and Spence, 1982). Gahnite is a $ZnAl_2O_4$ spinel in which the Zn and Al atoms are tetrahedrally and octahedrally coordinated by O respectively. Along the [100] direction, the Zn and Al atoms (i.e., the tetrahedral and octahedral sites) are arranged on alternate planes (Fig. 5-13). Hence, Zn and Al can be used as reference elements in the determination of site occupancies for impurity elements such as Fe and Mn. For example, Taftø and Spence (1982) found that $62\pm2\%$ of Fe atoms lay in tetrahedral sites for a $ZnAl_2O_4$ spinel containing 3% Fe.

Equations (5.8) to (5.14) can be readily modified to allow for other situations. For example, Otten (1987) shows how to modify the equations if one of the reference atoms is spread with a known distribution over the two sites. Forsteritic olivine provides a good example; Si provides a unique reference for one site, and Mg is assumed to be uniformly distributed over the two sites of interest (Smyth and Taftø, 1982; McCormick et al., 1987).

The structure of olivine (Fig. 5-14) is interesting for ALCHEMI. A common question is the distribution of metal cations on the M1 and M2 sites. Perpendicular to [001], these sites lie on alternate planes, and Si lies in the same plane as the M2 sites. Hence, Si provides a reference for the M2 sites even though there is no substitution at the Si sites. ALCHEMI was used together with the favorable alignment of sites in olivine to resolve the problem of the location of minority P in phosphoran olivine (Self and Buseck, 1983b; Buseck and Clark, 1984). The Si acts as a reference only for planar channelling conditions using the 002 reflections (the 001 reflections are forbidden). Axial ALCHEMI is not possible in this case.

The ALCHEMI equations can be modified to take into account situations in which there is only one reference element. In this case, Equations (5.8), (5.9), (5.11), and (5.12) are reduced to just two equations (either element A or element B is not present). In order to

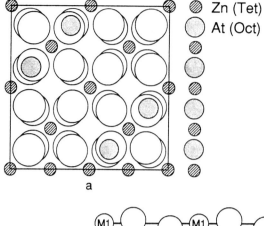

Figure 5-13. [001] projection of gahnite, $ZnAl_2O_4$ spinel. In this projection, Zn and Al atoms lie on alternate planes (as shown at the side of the figure) and thereby provide a reference for impurity atoms. In this drawing many Al atoms are obscured by the oxygen atoms (large, open circles).

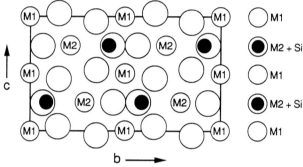

Figure 5-14. [100] projection of olivine. The M1 and M2 sites lie on alternate (00ℓ) planes, as shown at the side of the figure. The Si atoms (small solid circles), although not coincident with either the M1 or M2 sites, provide a reference for the M2 sites. Oxygen atoms are shown as large open circles.

evaluate $f_X(\mathbf{a})$, extra chemical information must be supplied. This information can take the form of a detailed chemical analysis or known information about atomic ratios. For example, Taftø and Buseck (1983) used the known atomic ratio of Al and Si to determine the distribution of Al and Si in an orthoclase feldspar where K is the only reference element. McCormick (1986) combined x-ray structure-refinement constraints with ALCHEMI to locate cation vacancies in M2 in chemically complex mantle omphacite pyroxenes. Otten (1987) suggested an iterative approach to evaluating $f_X(\mathbf{a})$ when there is insufficient information for the direct solution of the ALCHEMI equations. Smyth and McCormick (1988) reviewed mineralogical applications of this technique.

5.3.2.1 Statistical approach

The simplicity of Equation (5.14) belies an inherent weakness in the equation; the denominator is the difference of two ratios. When the difference of the intensity ratios from the reference atoms is small, any statistical or systematic experimental errors in the determination of the ratios are magnified in the value calculated for $f_X(\mathbf{a})$. This effect is dramatically demonstrated by Otten and Buseck (1987b) in their study of garnet. Their work shows that in order to obtain accurate site occupancies there should be a variation of at least 25% in the ratio of signal intensities from the reference atoms between the channelling and non-channelling orientations. In this respect, axial channelling is better than planar channelling because of the much stronger orientation effects at axial orientations.

The inherent weakness in ALCHEMI using Equation (5.14) led Turner et al. (1991) to propose a new method of analyzing ALCHEMI data. These workers realized that in an ALCHEMI experiment the signal from an element of unknown site distribution (i.e., X in the above) was being linearly correlated with the signals from the reference elements (i.e., A and B). Thus, instead of using results from just two orientations at a time [cf. Equation (5.14)], a statistically more rigorous approach is to evaluate the linear-correlation coefficients of the measured signals using multivariate analysis on a number of experimental results obtained at different (random) orientations, both channelling and non-channelling.

Following Turner et al. (1991), if there is a set of atoms, p (where p = 1 to P), having unknown distributions on sites occupied (not necessarily uniquely) by a set of M reference atoms, i (where i = 1 to M), then for a given orientation (i.e., a single x-ray or EELS spectrum)

$$N_p(s) = \sum_i \alpha_{ip} N_i(s) \tag{5.15}$$

where N is the measured signal for the element indicated by the subscript and α_{ip} is the linear-correlation coefficient for the signal from element p with respect to the signal from element i. Parameter s is used to indicate the results are from a single channelling orientation.

Procedural details for measuring suitable spectra are given by Turner and White (1992a). They recommend recording at least 12 spectra taken at different channelling orientations around a low-order zone axis. Furthermore, the spectral intensities at one orientation (usually the exact zone-axis condition) should be measured repeatedly to ensure that the specimen is not being contaminated or experiencing significant radiation damage while the spectra are being collected.

The key parameters in Equation (5.15) are the correlation coefficients, α_{ip}. Having determined these coefficients by multivariate analysis on all collected spectra, the overall concentration of element p (c_p) can be computed using the formula

$$c_p = \sum_i \frac{\alpha_{ip} n_i / k_{ip}}{1 + \sum_q \alpha_{iq} / k_{iq}}, \tag{5.16}$$

where n_i is the number of reference atoms of type i per unit cell and k_{iq} is the k-factor ratio, k_i/k_p, for atoms of type i and p (Chapter 4). The fractional concentration of atoms of type p on sites containing element i is

$$f_{ip} = \frac{\alpha_{ip} n_i / k_{ip}}{c_p \left(1 + \sum_q \alpha_{iq} / k_{iq} \right)} \tag{5.17}$$

According to Turner and White (1992a), site occupancies can be determined to accuracies of around 5% for variations in spectral intensities of only 5% between channelling and non-channelling orientations. Accuracies improve for larger variations in spectral intensities. The disadvantage of the technique is that it requires knowledge of k-factors as used in analytical electron microscopy. The accuracy of the site-occupancy determinations therefore reflects the accuracy of the k-factors used in Equations (5.16) and (5.17). This restriction led Turner and White (1992b) to propose a more accurate method of experimentally evaluating k-factors.

5.3.2.2   Localization

In discussing channelling effects, we stated that the intensity of characteristic emissions is proportional to the thickness-averaged beam current at the atomic sites. This observation is correct, but it is necessary to define exactly what is meant by the "atomic sites." Instead of sitting stationary at one point in a crystal, thermal energy causes atoms to vibrate constantly. Thermal vibration requires the introduction of the Debye-Waller factor in the calculation of diffracted intensities (Chapter 3). The amplitudes of thermal vibrations vary from element to element and from structure to structure but, in general, are of the order of 0.01 to 0.05 nm.

In addition to the smearing of the atomic sites by thermal vibrations, a high-energy electron does not have to impact head-on with an atom to excite an inner-shell electron or to undergo any other inelastic scattering event. The interaction range (distance) for inelastic scattering is given by the impact parameter, $\langle \mathbf{b} \rangle$. Classically, the impact parameter defines the maximum distance at which a high-energy electron can be from an atom in order to undergo an inelastic scattering event. This parameter has the form

$$\langle \mathbf{b} \rangle = \frac{h\mathbf{v}}{4\pi\Delta E} \bullet \mathbf{r} \tag{5.18}$$

where h is Planck's constant, $\mathbf{v}$ is the speed of the beam electron, and ΔE is the energy lost by the electron-beam for EELS and the energy of the characteristic x-ray for EDS. The parameter $\mathbf{r}$ is a constant that is dependent on the scattering-angle range of a electron beam (Bourdillon et al., 1981). For characteristic x-ray emissions, the beam electrons are scattered to all angles, and in this case

$$\mathbf{r} \quad = \quad [\, ln\,(\,4\,E\,/\,\Delta E\,)\,]^{-1/2} \tag{5.19}$$

where E is the incident-beam energy (Pennycook and Narayan, 1985). For EELS, the value of $\mathbf{r}$ is set by the maximum scattering angle as defined by the acceptance angle of the spectrometer. For most instrumental configurations used for EELS, $\mathbf{r}$ will be approximately 1. Thus for 120-kV electrons and an energy loss of 1.74 keV (i.e., the energy of SiK_α x-rays), the impact parameter for x-ray emissions is 0.01 nm and for EELS is approximately 0.03 nm.

The combination of thermal vibrations and the electron-interaction range effectively delocalizes the atom positions in a structure. Phenomena related to these effects are collectively known as *localization* effects. To allow for localization effects, the electron-beam current should not only be thickness averaged but also averaged over the distance corresponding to the sum of the thermal-vibration amplitude and the impact parameter. Obviously, if the sum of the thermal vibration amplitude and the impact parameter is greater than the interplanar spacing, channelling effects will not be observed. This condition places a lower limit on the energy loss for which channelling effects can be observed (e.g., Self and Buseck, 1983a). In general, total delocalization only occurs for energy losses of well below 1 keV.

More importantly for ALCHEMI, if the elements in a structure have widely differing delocalizations (i.e., markedly different thermal vibration amplitudes or characteristic emission energies), the ALCHEMI equations [i.e., Equations (5.8) to 5.17)] will not be applicable. In general localization effects are less severe for planar channelling than for axial channelling. This difference arises because for planar channelling the electron current varies fairly slowly and smoothly across the unit cell (e.g., Fig. 5-9) while for axial channelling the beam current is strongly peaked, with peak half widths comparable to thermal vibration amplitudes (Pennycook and Narayan, 1985).

Localization effects can be incorporated into the ALCHEMI equations by the introduction of correction factors (Pennycook and Narayan, 1985; Otten and Buseck, 1987b). These correction factors usually result in a change of between 5 and 10% in the calculated value of $f_X(a)$.

5.3.2 EELS ALCHEMI

Although the above discussion on channelling applies equally well to EELS as to EDS, EELS ALCHEMI is considerably more difficult to achieve experimentally than EDS ALCHEMI. In addition to the difficulties of background subtraction, peak overlap, and choice of integration windows as described in §5.2.5 and §5.2.6, localization effects limit the lower energy at which channelling effects can be observed. The lower energy limit for the observation of channelling effects in EELS is around 300 eV (Self and Buseck, 1983a). The low-energy limit for quantitative site-occupancy determination is somewhat higher.

Localization effects are dependent on scattering angle, as indicated by the parameter **r** in the expression for the impact parameter [Equation (5.18)]. As a general rule, the larger the scattering angle the greater the localization (i.e., the smaller the impact parameter). Correspondingly, Taftø and Krivanek (1982b) show how the entrance aperture of the EELS spectrometer can be positioned to increase interaction localization and enhance channelling effects for planar diffraction conditions.

EELS ALCHEMI is of considerable value for the determination of crystallographic-site occupancies for atomic species of different valence states. Taftø and Krivanek (1982a) combined channelling effects with the chemical shifts in the edge energies for Fe^{2+} and Fe^{3+} L-shell ionizations to show that in a chromite spinel all Fe^{2+} atoms lay on tetrahedral sites and all Fe^{3+} atoms lay on octahedral sites. The combination of EELS and EDS ALCHEMI therefore has the potential of being used quantitatively to determine a complete chemical picture of many mineral structures.

5.4 EELS and ALCHEMI: CLOSING THOUGHTS

In this chapter we discuss two techniques that in some ways are opposite and in other ways similar. ALCHEMI and analytical analysis by EELS are similar in that both present somewhat specialized aspects of electron microscopy, best suited for solving problems that are not well addressed by other methods. EELS represents technological excellence; modern EELS spectrometers are instruments requiring machining tolerances of the order of microns. ALCHEMI represents a practical application of electron diffraction theory. In Chapter 3, the question "Are Bloch waves real?" was considered; anyone who has seen channeling effects in EDS or EELS spectra is left with little doubt as to the answer.

EELS equipment has been around for a relatively long time but until the 1980s, because of the high manufacturing precision required, spectrometers were only available in laboratories that had access to extensive engineering facilities. This situation changed with the advent of compact EELS spectrometers (Krivanek and Swann, 1981), which were then priced within the budgets of many laboratories. Consequently, when these spectrometers became available there was a considerable boost in EELS use and a renewed interest in analytical EELS techniques. However, the availability of reliable, windowless detectors for EDS has resulted in a decrease in the use of EELS. The windowless EDS detector offers the possibility of using one detector to quantify almost all elements. It also provides experimental (you only have to learn one technique) and budgetary (you only have to buy one instrument) advantages. Similarly, ALCHEMI has recently received less attention than in the high-activities days of the early 1980s.

Perhaps the single most important factor in the relatively limited use of EELS and ALCHEMI is that neither technique is easy, nor are they readily suited to a "black box" approach. Each specimen must be considered individually, and the techniques can be challenging. For example, the EELS spectrum shown in Figure 5.6 is one of many taken to determine the oxygen content of kassite (Self and Buseck, 1991). The purpose of the EELS experiments was to determine if the chemical formula of kassite is $CaTi_2O_5$ or $CaTi_2O_4(OH)_2$. The result is that the O to Ca atomic ratio is 5.6 ± 0.5. Although this result favors the second of the two formulas listed above, it is hardly conclusive. The reason for this uncertainty is the overlap of the Ti and O edges. Similarly for ALCHEMI, the spectra shown in Figure 5.12 demonstrating channelling effects is the best of a series of spectra from samples prepared by a variety of techniques; many of the spectra taken showed only small variations in the relative heights of the Ga and As peaks.

In both EELS and ALCHEMI, experimental parameters conspire to mask the effect one is trying to measure. Many first-time users of ALCHEMI have been discouraged because they either saw no effect (specimen too thick, beam convergence too high, wrong crystallographic zone) or obtained answers that were obviously incorrect (i.e., the measured channelling effect was not sufficiently large to give reliable results from equation 5.14). Similarly, overlap and other problems in EELS can be frustrating, especially when compared to the relative simplicity of light-element analysis using windowless EDS detectors.

The secret to obtaining usable results from both EELS and ALCHEMI is to be persistent. The statistical approach to ALCHEMI looks as if it will remove some of the problems, paving the way for ALCHEMI to find greater application in mineralogy and geochemistry. The future for EELS for light-element analysis seems less certain, given the upsurge in EDS analysis using windowless detectors. It may well be that EELS will find its niche in applications arising from the new generation of dedicated, energy-filtering TEMs. In any case, EELS will always find application in crystal-chemistry studies through the use of ELNES and chemical shifts for the determinations of chemical states of elements, where it provides data that are essentially unavailable by any other technique.

ACKNOWLEDGMENTS

We thank Ms. S. Selkirk at Arizona State University and the staff of the CSIRO Division of Soils Publications Unit, in particular Ms. D. Smith and Mr. G. Rinder, for the preparation of figures. Dr. B. Miner kindly made Figure 5-4 available. The helpful comments of Drs. J.C.H. Spence, A.R. Wilson, and A.E.C. Spargo are acknowledged, as are reviews by Drs. J.R. Smyth and D.B. Williams. The work was supported in part by NSF grant EAR-8708529.

REFERENCES

Ahn, C.C. and Krivanek, O.L. (1983) EELS Atlas. Arizona State University HREM Facility and Gatan Inc., Tempe, Arizona, 166 pp.

Batson, P.E. (1986) High-energy resolution electron spectrometer for 1-nm spatial analysis. Rev. Sci. Instrum. 57, 43-48.

Bourdillon, A.J., Self, P.G. and Stobbs, W.M. (1981) Crystallographic orientation effects in energy dispersive x-ray analysis. Phil. Mag. A 44, 1335-1350.

Brown G.E., Jr., Calas, G., Waychunas, G.A. and Petiau, J. (1988) X-ray absorption spectroscopy: applications in mineralogy and geochemistry. In Spectroscopic Methods in Mineralogy and Geology, F.C. Hawthorne, ed., Rev. Mineral. 18, 431-512.

Buseck, P.R. and Clark, J. (1984) Zaisho—A pallasite containing pyroxene and phosphoran olivine. Mineral. Mag. 48, 229-235.

Castaing, R. and Henry, L. (1962) Filtrage magnetique des vitesses en microscopie électronique. C.R. Acad. Sci. Paris B255, 76-78.

Colliex, C. (1982) Electron energy-loss analysis in materials science. In Electron Microscopy, 1982, 10th Int'l Cong., Deutsche Ges.Elektronenmikroskopie 1, 159-166.

Cowley, J.M. (1981 Diffraction Physics, 2nd ed. North Holland, Amsterdam.

Disko, M.M., Krivanek, O.L. and Rez, P. (1982) Orientation-dependent extended fine structure in electron-energy-loss spectra. Phys. Rev. B 25, 4252-4255.

Egerton, R.F. (1976) Inelastic scattering and energy filtering in the transmission electron microscope. Phil. Mag. 34, 49-65.

Egerton, R.F. (1978) Formulae for light-element microanalysis by electron energy-loss spectrometry. Ultramicrosc. 3, 243-251.

Egerton, R.F. (1981a) The range of validity of EELS microanalysis formulae. Ultramicrosc. 6, 297-300.

Egerton, R.F. (1981b) Values of K-shell partial cross-section for electron energy-loss spectrometry. J. Microsc. 123, 333-337.

Egerton, R.F. (1986) Electron Energy Loss Spectroscopy. Plenum, New York, 410 pp.

Goldstein, J.I., Newbury, D.E., Echlin, P., Joy, D.C., Fiori, C. and Lifshin, E. (1981) Scanning Electron Microscopy and X-ray Microanalysis. Plenum Press, New York, 673 pp.

Goldstein, J.I. and Williams, D.B. (1992) Analytical electron microscopy—current state and future directions. Microbeam Anal. J., in press.

Goldstein, J.I., Williams, D.B. and Cliff, G. (1986) Quantitative x-ray analysis. In Principles of Analytical Electron Microscopy, D.C. Joy, A.D. Romig, Jr. and J.I. Goldstein, eds. Plenum Press, New York, 155-217.

Isaacson, M. (1972) Interaction of 25 keV electrons with the nucleic acid bases, adenine, thymine and uracil. (II) Inner-shell excitation and inelastic cross sections. J. Chem. Phys. 56, 1813-1818.

Isaacson, M. (1978) All you might want to know about ELS (but are afraid to ask): a tutorial. In Scanning Electron Microscopy, 1978, O. Johari, ed. SEM Inc., AMF O'Hare, Chicago, Illinois, 763-776.

Joy, D.C. (1979) The basic principles of electron energy loss spectroscopy. In Introduction to Analytical Electron Microscopy, eds. J.J. Hren, J.I. Goldstein and D.C. Joy, Plenum Press, New York, 223-244.

Joy, D.C. (1981) Electron energy loss spectroscopy. In Quantitative Microanalysis with High Spatial Resolution, G.W. Lorimer, M.H. Jacobs and P. Doig, eds. The Metals Society, London, 127-135.

Joy, D.C. and Maher, D.M. (1981) The quantitation of electron energy loss spectra. J. Microsc. 124, 37-48.

Kittel, C. (1986) Introduction to Solid State Physics, 6th ed., Wiley, New York. 646 pp.

Krivanek, O.L and Swan, P.R. (1981) An advanced electron energy loss spectrometer. In Quantitative Microanalysis with High Spatial Resolution. Proc. The Metals Society Conf., umist/Manchester, 25-27 March 1981, The Metals Society, London.

Leapman, R.D., Grunes, L.A., Fejes, P.L. and Silcox, J. (1981) Extended core-edge fine structure in electron energy-loss spectra. In EXAFS Spectroscopy, B.K. Teo and C.D. Joy, eds. Plenum Press, New York, 217-239.

McCormick, T.C. (1986) Crystal chemical aspects of non-stoichiometric pyroxenes. Am. Mineral. 71, 1434-1440.

McCormick, T.C., Smyth, J.R. and Lofgren, G.E. (1987) Site occupancies of minor elements in synthetic olivines as determined by channeling-enhanced x-ray emission. Phys. Chem. Minerals 14, 368-372.

Otten, M.T. (1987) A practical guide to ALCHEMI. Philips Electron Optics Bull. 126, 21-28.

Otten, M.T. and Buseck, P.R. (1987a) The oxidation state of Ti in hornblende and biotite determined by electron energy-loss spectroscopy with inferences regarding the Ti substitution. Phys. Chem. Minerals 14, 45-51.

Otten, M.T. and Buseck, P.R. (1987b) The determination of site occupancies in garnet by planar and axial ALCHEMI. Ultramicrosc. 23, 151-158.

Otten, M.T., Miner, B.A., Rask, J.H. and Buseck, P.R. (1985) The determination of Ti, Mn and Fe oxidation states in minerals by electron energy-loss spectroscopy. Ultramicrosc. 18, 285-290.

Pennycook, S.J. and Narayan, J. (1985) Atom location by axial-electron-channeling analysis. Phys. Rev. Letts. 14, 1543-1546.

Rask, J.H., Miner, B.A. and Buseck, P.R. (1987) Determination of manganese oxidation states in solids by electron energy-loss spectroscopy. Ultramicrosc. 21, 321-326.

Rez, P. (1983) Detection limits and error analysis in energy-loss spectrometry. In Microbeam Analysis, 1983, R. Gooley, ed. San Francisco Press, San Francisco, CA, 153-155.

Rez, P. (1991) Microanalysis by electron energy loss spectroscopy and energy dispersive x-ray analysis. In Physical Methods of Chemistry IV, B.W. Rossiter and J.F. Hamilton, eds. 2nd ed., Chap. 6.

Rossouw, C.J., Egerton, R.F. and Whelan, M.J. (1976) Applications of energy analysis in a transmission electron microscope. Vacuum 26, 427-432.

Rossouw, C.J. and Maslen, V.W. (1987) Localization and ALCHEMI for zone axis orientations. Ultramicrosc. 21, 277-288.

Schattschneider, P. and Sölkner, G. (1984) Generalization of a new approach to the retrieval of angle-resolved single-loss profiles. Phil. Mag. B50, 53-62.

Self, P.G. and Buseck, P.R. (1983a) Low-energy limit to channelling effects in the inelastic scattering of fast electrons. Phil. Mag. A48, L21-L26.

Self, P.G. and Buseck, P.R. (1983b) High-resolution structure determination by ALCHEMI. 41st Ann. Proc. Elect. Micros. Soc. Am., 178-179.

Self, P.G. and Buseck, P.R. (1991) Structure model for kassite, $CaTi_2O_4(OH)_2$. Am. Mineral. 76, 238-287 (1991).

Smyth, J.R. and McCormick, T.C. (1988) Earth science applications of ALCHEMI. Ultramicrosc. 26, 77-86.

Smyth, J.R. and Taftø, J. (1982) Major and minor element site occupancies in heated natural forsterite. Geophys. Res. Letts. 9, 1113-1116.

Spence, J.C.H. (1988) Techniques closely related to high-resolution electron microscopy. In High-Resolution Transmission Electron Microscopy, P.R. Buseck, J.M. Cowley and L. Eyring, eds. Oxford University Press, Oxford, 190-243.

Spence, J.C.H and Taftø, J. (1983) ALCHEMI: a new technique for locating atoms in small crystals. J. Microsc. 130, 147-154.

Taftø, J. (1983) Structure-factor phase information from two-beam electron diffraction. Phys. Rev. Letters 51, 654-657.

Taftø, J. and Buseck, P.R. (1983) Quantitative study of Al-Si ordering in an orthoclase feldspar using an analytical transmission electron microscope. Am. Mineral. 68, 944-950.

Taftø, J. and Krivanek, O.L. (1982a) Site-specific valence determination by electron energy-loss spectros-copy. Phys. Rev. Letters 48, 560-563.

Taftø, J. and Krivanek, O.L. (1982b) Characteristic energy losses from channeled 100 keV electrons. Nuclear Instr. Methods 194, 153-158.

Taftø, J. and Spence, J.C.H. (1982) Atomic site determination using the channeling effect in electron-induced x-ray emission. Ultramicrosc. 9, 243-248.

Taftø, J. and Zhu, J. (1982) Electron energy-loss near edge structure (ELNES), a potential technique in the studies of local atomic arrangements. Ultramicrosc. 9, 349-354.

Turner, P.S., White, T.J., O'Connor, A.J. and Rossouw, C.J. (1991) Advances in ALCHEMI analysis. J. Microsc. 162, 369-378.

Turner, P.S. and White, T.J. (1992a) Experimental procedures for the precise determination of elemental partitioning coefficients in complex structures by ALCHEMI. J. Computer-Assisted Microsc., in press.

Turner, P.S. and White, T.J. (1992b) On the determination of precise k-factors for AEM. J. Microsc. 166, RP1-RP2.

Williams, D.B. (1984) Practical analytical electron microscopy in materials science. Philips Electronic Instruments Inc., Electron Optics Publishing Group, New Jersey, 153 pp.

Zaluzec, N.J. (1980) The influence of specimen thickness in quantitative electron energy loss spectroscopy. Proc. 38th Ann. Mtg., Elect. Microsc. Soc. Am., Claitor's Pub. Co., Baton Rouge, LA, 112-113.

Zanchi, G., Sevely, J. and Jouffrey, B. (1977) An energy filter for high voltage electron microscopy. J. Microsc. Spectrosc. Electron. 2, 95-104.

Chapter 6. ELECTRON MICROSCOPY APPLIED TO NONSTOICHIOMETRY, POLYSOMATISM, AND REPLACEMENT REACTIONS IN MINERALS

David R. Veblen   Department of Earth and Planetary Sciences, The Johns Hopkins University, Baltimore, Maryland 21218, U.S.A.

6.1 INTRODUCTION: STRUCTURAL BASIS FOR NONSTOICHIOMETRY AND REACTIONS IN MINERALS

One of the primary applications of selected-area electron diffraction (SAED), transmission electron microscopy (TEM), and especially high-resolution TEM (HRTEM) is to detect and characterize defects that cause deviations from ideal stoichiometry in minerals and to show how such defects are involved in mineral reactions. This chapter (1) describes the types of defects that cause nonstoichiometry; (2) considers the relatively limited information that can be gained from TEM studies of disordered point defects in crystals; (3) discusses minerals in which nonstoichiometry results from partially ordered arrays of point defects that *can* be imaged with the TEM; (4) explores examples from TEM studies of minerals in which nonstoichiometry results from polysomatic defects; and (5) discusses the role of such polysomatic defects in mineral reactions.

A great deal has been written about the relationships between crystal defects and nonstoichiometry, as well as other chemical, physical, and transport properties. Reviews that address the structural basis for nonstoichiometry include those by Wadsley (1964), Greenwood (1970), Hannay (1973), Kröger (1974), Lasaga (1980), Schmalzried (1981), Tilley (1987), Sørenson (1981), and Veblen (1985a, 1991, 1992).

6.1.1 What is nonstoichiometry?

Some minerals seldom, if ever, deviate far from a fixed chemical composition and structural formula, for example quartz, SiO_2. Similarly, any student of mineralogy is familiar with solid solutions having a structural formula that accounts for chemical variations, such as the isomorphous substitution of Fe^{2+} for Mg on the M1 and M2 sites in olivine: $(Mg_{2-x}Fe_x)_{\Sigma 2}SiO_4$. Although the chemical composition is variable in this case, the coefficients are all integral, if the structural formula is written in terms of crystallographic sites: $(M1)_1(M2)_1SiO_4$. We consider such "normal" olivine to be *stoichiometric*, and for the present discussion we define as stoichiometric any mineral that can be described by a structural formula with integral coefficients when written in terms of the occupancies of crystallographic sites.

In contrast, many minerals, at least in some occurrences, have structural formulae that *cannot* be written with integral coefficients. An example is disordered laihunite, an olivine-derivative mineral with a formula that can be written as

$$(Fe^{2+}_{2-1.5x}Fe^{3+}_x)_{\Sigma 2-x/2}SiO_4.$$

In this case, the coefficient giving the total occupancy of the M sites is not 2.0, but rather 2.0 - x/2, where x is variable; if we were to write a more complex formula in terms of M1 and M2, we would find that both of these sites have non-integral coefficients. In the present paper, we refer to crystalline compounds with such nonintegral coefficients in the structural formula as *nonstoichiometric*.

Table 6-1. Examples of nonstoichiometry resulting from point defects

	Ordered?	Disordered?	Reference
Cation Vacancies			
Pyrrhotite, $Fe_{1-x}S$	Yes	Yes	Nakazawa et al., 1975
Bornite-digenite, Cu_5FeS_4-Cu_9S_5	Yes	Yes	Pierce & Buseck, 1978
Laihunite,	Yes	Yes	Kitamura et al., 1984
$(Fe^{2+}_{2-1.5x}Fe^{3+}_{x})_{\Sigma2-x/2}SiO_4$			
Olivine,	Yes	Yes	Banfield et al., 1990
$(Mg_{2-y}Fe^{2+}_{y-1.5x}Fe^{3+}_{x})_{\Sigma2-x/2}SiO_4$			
Spinels, $\square_{x/3}A^{2+}_{1-x}B^{3+}_{2+2x/3}O_4$	Yes	Yes	Viertel & Seifert, 1979;
			Gleitzer & Goodenough, 1985
Wüstite, $Fe_{1-x}O$	Yes	Yes	Hazen & Jeanloz, 1984
Pyroxenes, $Ca_{0.5}\square_{0.5}AlSi_2O_6$	Yes	No	McCormick, 1986
component			
Sheet silicates (interlayer & oct. sites)	Yes	No	Veblen, 1992
Framework silicates (alkali site)	Yes	No	Smith & Brown, 1988
Cation and Molecular Interstitials			
Manganese oxides (tunnel	Yes	Yes	Wadsley, 1964
or interlayer sites)			
Smectites (interlayer sites)	Yes	No	Veblen, 1992
Zeolites (cavity sites)	Yes	Yes	Barrer, 1978
Anion Vacancies			
Mullite, $Al_2(Al_{2+2x}Si_{2-2x})O_{10-x}\square_x$	Yes	Yes	Schryvers et al., 1988
Perovskites & derivatives	Yes	Yes	Smyth, 1989
Anion Interstitials			
Fluorite structures, e.g., $UO_{2\pm x}$	Yes	Yes	Manes & Benedict, 1985

Many students of mineralogy think of nonstoichiometry as a relatively uncommon phenomenon, restricted to minerals such as pyrrhotite ($Fe_{1-x}S$) and wüstite ($Fe_{1-x}O$), for which the chemical formulae typically have been written with a variable coefficient such as 1-x. This is an incorrect perception. In fact, substantial deviations from ideal stoichiometry occur commonly, affecting the majority of common rock-forming silicates and oxides, at least under some conditions of crystallization or subsequent alteration. Table 1 lists some examples of common nonstoichiometric minerals in which the nonstoichiometry is caused by various point-defect mechanisms, as discussed below.

6.1.2 Point defects

Point defects in crystals are nonperiodic features, or mistakes, that are localized around a point (although any defect obviously has finite dimensions and therefore is not strictly a point). The simplest types of point defects are *vacancies,* which are empty crystallographic sites that are normally occupied by an atom, and *interstitial defects,* which are atoms occupying sites that are normally vacant. These simple defects can involve either cations or anions and are illustrated schematically in Figure 6-1a,b,c.

Some types of point defects, such as vacancies and interstitials, can alter the stoichiometry of a mineral. For example, pyrrhotite exhibits a range of compositions between FeS and approximately Fe_7S_8 as a result of large numbers of vacancies on Fe sites that are fully occupied in the parent stoichiometric mineral troilite (which crystallizes in the NiAs structure type). On the other hand, these simple point defects can

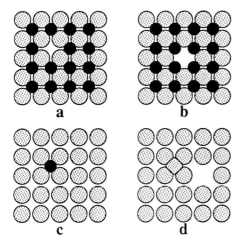

Figure 6-1. Schematic diagram of simple point defects. (a) Cation vacancy. (b) Anion vacancy. (c) Cation interstitial defect. (d) Anion Frenkel defect.

couple to form combination defects, which may not change the stoichiometry. For example, an anion vacancy coupled with a cation vacancy of the same valence (called a Schottky defect) maintains charge balance in the crystal, as does a cation vacancy coupled with a cation interstitial or an anion vacancy coupled with an anion interstitial (Frenkel defects) (Fig. 6-1d). In a compound with the simple stoichiometry AX, Schottky defects do not alter the overall stoichiometry of the crystal, and Frenkel defects do not cause nonstoichiometry in any case.

6.1.3 Extended defects

Dislocations (one-dimensional, or line defects) and several different types of two-dimensional defects collectively are called extended defects. In some cases, dislocations can influence crystal composition, but such chemical variations are appreciable only at extremely high dislocation densities (Buseck and Veblen, 1978; Wriedt and Darken, 1965a,b). Hence, dislocations will not be considered further in this chapter.

Two-dimensional defects are often called planar defects, even though they commonly have complex morphology rather than being strictly planar. Some types of planar defects, such as stacking faults, simple twins, and antiphase boundaries, generally do not have large effects on crystal stoichiometry, although certain elements may be segregated along such defects, such as Ca along antiphase boundaries in pigeonite (e.g., Carpenter, 1978). The reader is referred to Chapters 7, 11, and 12 of this volume for discussions of such defects.

In contrast, several other types of planar defects can cause profound deviations from ideal stoichiometry. In the field of solid-state chemistry, one type of chemical defect is called a *crystallographic shear plane*, which involves the collapse or dilation of the ideal structure across a planar boundary (Tilley, 1987). The composition and structure in the affected region of the crystal deviate from those of the parent structure, resulting in nonstoichiometry when a disordered array of shear planes is present. (If the shear planes are periodic, the crystal can be described as a distinct, ordered, stoichiometric structure with a discrete though often complex structural formula.) *Chemical twins* are similar to crystallographic shear planes, but the operation across the boundary is a point-group operation not included in the symmetry of the parent structure, rather than a translational operation. Like crystallographic shear planes, disordered chemical twinning results in nonstoichiometry, whereas ordered arrays of chemical twin planes result in discrete derivative compounds. To describe planar features that do not fit the crystallographic shear or chemical twin categories, solid-state chemists use a third description, called *intergrowth*, in which two types of slabs having distinct structure and chemical composition are interleaved.

In fact, as discussed by Veblen (1991), all of these types of planar defects can be described as intergrowth structures, using the formalism of *polysomatism*. As defined in a classic paper by Thompson (1978), a *polysome* is "a crystal ... that can be regarded as made of chemically distinct layer modules." A *polysomatic series* is a group of crystalline compounds (*e.g.*, minerals) that possess the same types of modules in different ratios or sequences, and the general term for this type of structure mixing is polysomatism. Polysomes also can be regarded as hybrid structures or, more fancifully, as "mineralogical mules" (another term for which we have J.B. Thompson, Jr., to thank).

The principle of polysomatism is illustrated schematically in Figure 6-2. The pure end-member minerals (A) and (B) are made entirely from A slabs and B slabs respectively (Fig. 6-2a, b), where the A slabs and B slabs differ in both structure and chemistry. The polysome (AB) is made by rigorously alternating the A and B slabs (Fig. 6-2c), whereas the polysome (ABB) contains the slab sequence ...ABBABBABBABB... (Fig. 6-2d). Clearly, the stoichiometries of (A), (AB), (ABB), (B), and any other polysomes containing only A and B slabs lie along a mixing line and are collinear.

Electron microscopy has shown us in recent years that polysomatic minerals commonly contain defects that can be described as mistakes in the ideal, ordered sequence of structural slabs. A simple example of such a defect is shown in Figure 6-3a, where an extra A slab is inserted in (or a B slab is omitted from) an otherwise ordered crystal of the polysome (AB). Such a defect would render its host crystal nonstoichiometric, with the deviation from stoichiometry being toward the stoichiometry of A. The structural formula might be represented as $A_{1+x}B_1$, rather than the ideal formula A_1B_1 (here A and B represent the structural formulae of the individual A and B slabs). Furthermore, "pathologically disordered" crystals with highly disorganized or even random sequences of slabs can occur, as shown in Figure 6-3b (the sequence of A and B slabs in this illustration was assembled through a series of coin flips). Although the stoichiometry of such a disordered material is constrained to lie between those of the end-member structures, it can vary widely between the two extremes. The stoichiometry of the region shown in Figure 6-3b, for example, is $A_{16}B_{21}$, where A and B again refer to the end-member stoichiometries. In a macroscopic crystal, such polysomatic disorder can cause the stoichiometry to vary almost smoothly, as a function of the ratio A/(A+B).

6.1.4 Polysomatic reactions

Reactions can be classified in a number of different ways. One broad category of reactions of interest to geoscientists can be referred to as solid-state reactions, which involve solids as both reactant(s) and product(s). By definition (Schmalzried, 1981), solid-state reactions also can involve fluids. However, with the exception of TEM studies of fluid inclusions (Guthrie et al., 1991; Guthrie and Veblen, 1992), it is typically only the solids that are investigated with electron microscopy. Many reactions occur by simple dissolution and precipitation mechanisms, in which the products bear no orientation relationship to the reactant minerals, and the reactant and product minerals are even commonly separated spatially in a rock. Although defects may play a role in such

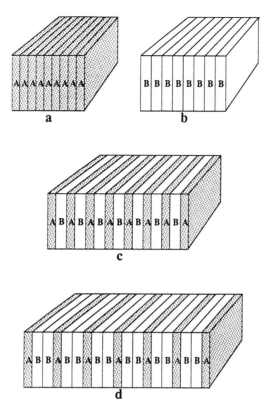

Figure 6-2. Schematic representation of polysomatism. (a) The polysome (A), made entirely of A slabs. (b) The polysome (B), containing only B slabs. (c) The polysome (AB). (d) The polysome (ABB). All these polysomes are stoichiometrically collinear, lying on a mixing line between the stoichiometries of the A slabs and the B slabs.

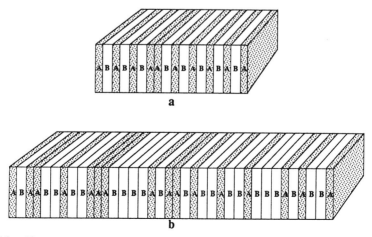

Figure 6-3. Nonperiodic polysomes. (a) A simple polysomatic defect consisting of an extra A slab (alternatively described as a missing B slab) in the otherwise ordered polysome (AB). (b) A highly disordered structure consisting of a random array of A and B slabs.

186

reactions by altering mineral solubilities, dissolution rates, and growth rates, they do not play an intimate structural role.

In solid-state reactions, however, one mineral directly replaces another, and the orientation of the product can be strongly controlled by the reactant crystal structure. Such replacement reactions are sometimes called topotactic or topotaxic to denote the three-dimensional orientation relationship. In cases where both the reactant and product are members of the same polysomatic series, such reactions commonly occur by a mechanism that involves nucleation and growth of narrow slabs of the product mineral. Thus, a slab of the reactant structure is replaced by a slab having a different sequence of modules. The geometry of such a process, in which a single module of the polysome (A) is replaced by a B module, is illustrated in Figure 6-4. The structural mechanism by

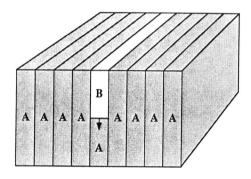

Figure 6-4. Schematic diagram of a simple, volume-preserving polysomatic reaction. A single module of the polysome (A) is replaced by a B slab, which terminates at a line defect that advances through the crystal as the reaction proceeds.

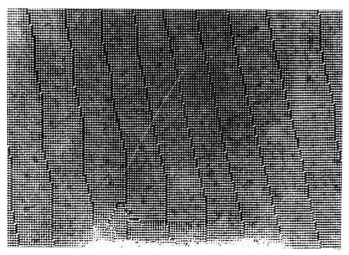

Figure 6-5. HRTEM image of Nb-doped W oxide, $(W, Nb)O_{2.93}$. The approximately vertical features are planar defects (crystallographic shear planes), and the dark blotches between them are interpreted as small aggregates of interstitial point defects. [Used by permission of the publisher of *Ultramicroscopy*, from Bursiil, 1985, Fig. 6, p. 7.]

which such reactions occur can be strongly influenced by strain between the parent structure and the modules that replace it. For example, due to strain considerations, the thermodynamically stable phase may be precluded from forming, leading to the growth of metastable minerals. It is also critical to note that such "solid-state" reactions in geological systems may involve not only the solids (minerals), but they typically are also mediated by fluids that act as catalysts and transport media. Indeed, many geololgical reactions involving replacement of one mineral by another simply would not occur at all if it were not for fluids that pervade the environment where the reaction occurs (see, for example, reactions involving clays, as discussed by Peacor, this volume).

6.2 IMAGING POINT DEFECTS

6.2.1 Disordered and small clusters of point defects

Crystalline specimens used for electron microscopy typically are at least several unit cells thick. For this reason, isolated point defects in minerals generally do not produce interpretable contrast in electron micrographs, because they are swamped by the signal from normal regions of the crystal and by noise that arises from irregular sample surfaces or amorphous material coating the sample (due either to sample preparation or contamination in the electron microscope). Single, isolated atoms can, however, be imaged with modern electron microscopes (Isaacson et al., 1979), and, in theory, it should be possible to image individual point defects within an ideal specimen (Gribelyuk and Zakharov, 1987). Indeed, there are cases in which point defects form clusters that can be imaged readily with TEM methods (Veblen and Cowley, 1992).

Several examples of point-defect contrast are presented by Bursill (1985). For both interstitial and vacancy clusters, maximum contrast occurs in bright-field, phase-contrast images near Scherzer focus; an objective aperture is selected that includes the central beam and diffuse intensity around it, but that excludes all of the Bragg diffraction beams. In some cases, point-defect contrast also occurs in normal HRTEM images, as seen between the disordered planar defects in Figure 6-5 for nonstoichiometric, niobium-doped, tungsten oxide, $(W,Nb)O_{2.93}$. In this image, the dark blotches have been interpreted through image simulations as small aggregates of reconstructed interstitial defects (Bursill, 1985). Variations in image density between nominally identical sites may also occur due to different cation occupancies, as in the image of tourmaline shown in Figure 6-6 (Iijima et al., 1973). These variations can best be seen by viewing this figure at a distance or at a low angle.

6.2.2 Ordered point defects

If point defects are ordered into linear or planar arrays, they can produce strong contrast in HRTEM images, since a row of crystallographic sites rich in vacancies obviously will have lower scattering power than a row of fully occupied sites (and a row of interstitial defects will have greater scattering power than a row of unoccupied interstitial sites). For example, in pyrrhotite, $Fe_{1-x}S$, nonstoichiometry is accommodated by vacancies on the metal sites of the ideal NiAs structure type. At low temperatures, these vacancies order into columns that can be imaged using either dark-field or bright-field HRTEM (i.e., either excluding or including the central beam in the set of beams that are used to form the image) (Pierce and Buseck, 1974; Nakazawa et al., 1975). In dark-field images, the contrast is approximately reversed compared to bright-field HRTEM images: in the dark-field image shown in Figure 6-7a, bright white spots correspond to columns of Fe sites with higher Fe occupancy than the dimmer white spots,

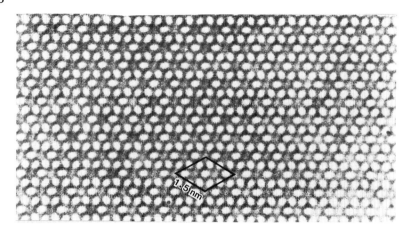

Figure 6-6. Variations in image density for the nominally identical crystallographic sites located at the centers of the "flowers" in a c-axis HRTEM image of tourmaline. The variations can be seen more clearly by viewing at a distance or at a low angle. [Used by permission of the publisher of *Mineralogy and Petrology* [formerly *Tschermaks Mineralogische und PetrographischeMitteilungen*, from Iijima et al., 1973, Fig. 1a, p. 219.]

which correspond to columns rich in vacancies. Several studies (Dódony and Pósfai, 1990; Nakazawa et al., 1975; Pierce and Buseck, 1974) have shown that twinning and antiphase boundaries involving the vacancy distribution are common in pyrrhotite. In addition, pyrrhotite specimens possessing incommensurate superstructures (known as NC pyrrhotite) locally contain ordered regions that are separated by a disordered sequence of antiphase boundaries.

Other examples of minerals that contain ordered or partially ordered arrangements of vacancies that have been imaged with HRTEM include the following: (1) carlosturanite, as shown in Figure 6-7b, apparently contains slabs of vacant Si sites that are occupied in the related serpentine mineral lizardite (Mellini et al., 1985); (2) laihunite (Kitamura et al., 1984) and olivine-laihunite intergrowths (Banfield et al., 1990), as shown in Figure 6-7c; (3) maghemite as in Figure 6-7d (Banfield, 1990); (4) low-symmetry vesuvianite, which can possess small twin domains within which vacancies and occupied cation sites alternate (Veblen and Wiechmann, 1991); and (5) sulfides in the bornite-digenite system (Pierce and Buseck, 1978).

6.2.3 Short-range ordered point defects

Intermediate between the cases of disordered and ordered vacancy distributions is the situation in which order exists, but the size of the ordered domains is so small (perhaps 10 nm or smaller) that they do not produce the diffraction effects that are usually associated with ordered structures, such as reasonably sharp superstructure diffraction spots. This case is commonly referred to as "short-range order," or SRO. Instead of superstructure spots, SRO typically produces an organized pattern of diffuse intensity that appears between the Bragg spots in SAED or X-ray diffraction patterns (Clapp and Moss, 1968; Cowley, 1950). Images produced by imaging the diffuse intensity in dark field can exhibit a fine-scale mottling related to the domain structure (Howie, 1988), although this speckling cannot be interpreted directly as the SRO domains themselves, due to dynamical scattering effects and the overlapping of individual domains (Cowley, 1973).

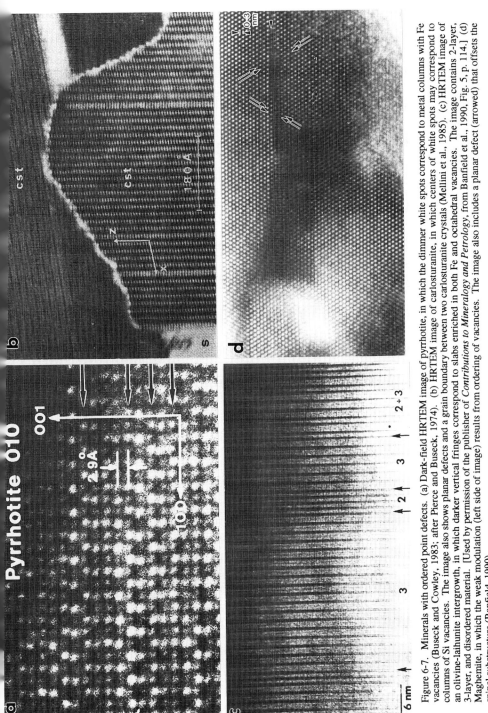

Figure 6-7. Minerals with ordered point defects. (a) Dark-field HRTEM image of pyrrhotite, in which the dimmer white spots correspond to metal columns with Fe vacancies (Buseck and Cowley, 1983; after Pierce and Buseck, 1974). (b) HRTEM image of carlosturanite, in which centers of white spots may correspond to columns of Si vacancies. The image also shows planar defects and a grain boundary between two carlosturanite crystals (Mellini et al., 1985). (c) HRTEM image of an olivine-laihunite intergrowth, in which darker vertical fringes correspond to slabs enriched in both Fe and octahedral vacancies. The image contains 2-layer, 3-layer, and disordered material. [Used by permission of the publisher of *Contributions to Mineralogy and Petrology*, from Banfield et al., 1990, Fig. 5, p. 114.] (d) Maghemite, in which the weak modulation (left side of image) results from ordering of vacancies. The image also includes a planar defect (arrowed) that offsets the spinel substructure (Banfield, 1990).

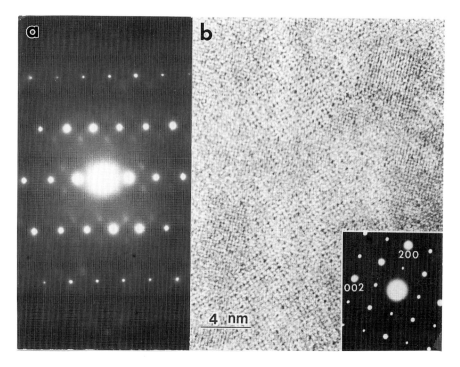

Figure 6-8. Manifestations of short-range order. (a) SAED pattern from phase E, showing organized diffuse intensity between Bragg spots (Kudoh et al., 1992). (b) HRTEM image of a natural perovskite mineral (loparite), showing an irregular distribution of cations (dark spots) in clusters, indicative of SRO (Hu et al., 1992).

Such images do, however, provide a semi-quantitative view of the scale of the domain structure (Tanaka and Cowley, 1987).

Although most TEM studies of SRO have been restricted to metals and other important materials, it has been shown that the high-pressure, hydrous magnesium silicate known as phase E produces SAED patterns with diffuse intensity (Fig. 6-8a) and speckled dark-field images characteristic of SRO (Kudoh et al., 1992). Hu et al. (1992) showed similar diffuse intensity from naturally occurring perovskites and also produced impressive HRTEM images suggesting an irregular cation distribution with short-range-ordered clusters (Fig. 6-8b).

6.3   IMAGING POLYSOMATIC DEFECTS

In this section, I will discuss examples of using HRTEM to image polysomatic disorder in minerals. The discussions are not exhaustive, but rather are intended to illustrate the types of information that can be derived from such investigations. Not all of the polysomatic mineral groups are discussed, and the reader is directed to several papers for more information: Veblen (1991, 1992), Ferraris (1986), Makovicky (1989), and references within those papers.

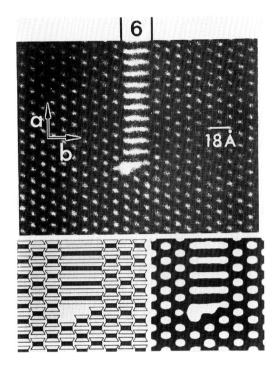

Figure 6-9. A sextuple-chain defect terminating in the amphibole anthophyllite. Top: Experimental image. Bottom: Structural model in I-beam representation and com-puter-simulated image (for the JEOL JEM 100B instrument).
(After Veblen and Buseck, 1980.)

6.3.1 Why do we use HRTEM to image polysomatic minerals?

Conventional amplitude-contrast imaging (CTEM), such as the single-beam dark-field technique, is ideally suited to many problems involving the determination of fault vectors or operations for defects such as antiphase boundaries, stacking faults, twin boundaries, and dislocations (see other chapters in this volume). HRTEM, on the other hand, is the method of choice for imaging polysomatic defects, as well as ordered polysomatic structures. This is in part because polysomatic defects almost by definition have appreciable width (the width of the anomalous polysomatic slab or slabs), and HRTEM typically can be used to resolve these slabs or even to resolve details within individual slabs. Furthermore, in a polysomatic series where the two types of slabs have the same width, there are commonly certain defects that produce virtually no net displacement of the surrounding structure on either side of the fault. For example, the termination of a sextuple-chain slab in amphibole is coherent, in the sense that the surrounding amphibole is distorted little or not at all (Fig. 6-9). Since there is no net displacement, conventional bright-field and dark-field experiments produce only minimal contrast, whereas the defect in the HRTEM image of Figure 6-9 is quite obvious.

In fact, when two blocks of amphibole are separated by a polysomatic defect, there are three possible displacement vectors that relate the amphibole on the two sides of the fault (Veblen and Buseck, 1981), in addition to the null vector $0\mathbf{a} + 0\mathbf{b} + 0\mathbf{c}$. Thus, in a process such as the random nucleation of amphibole during the replacement of pyroxene, when the amphibole blocks resulting from different nucleation events grow together, 75% of the interfaces will be polysomatic defects with the displacement vectors $(1/4\mathbf{b} + 1/2\mathbf{c})$, $(1/2\mathbf{b})$, or $(-1/4\mathbf{b} + 1/2\mathbf{c})$. Appropriate dark-field images of these faults will exhibit strong

192

contrast. The remaining 25% of the interfaces are not detectable, however, because the two adjoining domains of amphibole have displacements of zero (i.e., there actually is no fault, because the two domains are related to each other by the null vector, and the amphibole grew together in registry). Alternatively, the two domains may be separated by a slab of material with a displacement that is equivalent to zero by addition or subtraction of a lattice vector (e.g., a defect with sextuple-chain structure) (Veblen and Bish, 1988).

These possibilities are illustrated in Figure 6-10. One-fourth of the interfaces will produce little or no contrast in amplitude-contrast experiments. Similarly, if one were to

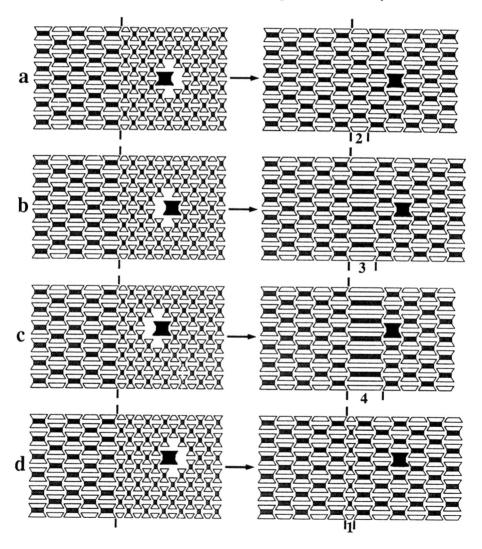

Figure 6-10. Schematic diagram showing the four possible relationships between amphibole domains formed by separate nucleation events during the topotactic replacement of pyroxene. See text for explanation. (After Veblen and Buseck, 1981.)

use conventional amplitude-contrast methods to image a random distribution of chain-width errors in amphibole (single-chain, triple-chain, quadruple-chain, etc.), or in pyroxene (double-chain, triple-chain, quadruple-chain, etc.), at least some of the defects would be effectively invisible, regardless of the imaging conditions. However, all of these defects are readily observed and easily interpreted with a single HRTEM image obtained with the electron beam parallel to **c**.

Another problem with conventional amplitude-contrast imaging is that the image interpretation can be ambiguous. For example, as shown in Figure 6-11, the displacement vector that can be determined from CTEM does not serve to identify the type of fault uniquely. Thus, a quintuple-chain defect and a single-chain defect produce the same displacement of the amphibole structure on either side of the fault (Fig. 6-11a). As another example, replacement of pyroxene in some cases produces quadruple-chain defects (Veblen and Buseck, 1981), whereas in other occurrences the defect with this fault vector consists of a pair of single-chain slabs (Veblen and Bish, 1988) (Fig. 6-11b). Again, HRTEM can distinguish easily among these defects, whereas CTEM cannot. For all of the above reasons, HRTEM is the method of choice for studying structural disorder in polysomatic structures, as well as in ordered polysomes that occur on too fine a scale to be examined with x-ray diffraction methods.

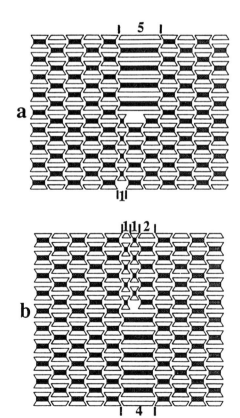

Figure 6-11. (a) Schematic diagram showing that the structural displacement associated with a quintuple-chain defect in amphibole is equivalent to that of a single-chain defect. (b) Similar, showing that the displacement from a quadruple-chain defect is the same as that from a pair of single-chain defects.

6.3.2 Biopyriboles

6.3.2.1 Introduction to biopyribole nomenclature and HRTEM

The biopyriboles include pyroxenes, amphiboles, 2:1 sheet silicates, and other chain silicates containing 2-repeat silicate chains, such as jimthompsonite (triple chains), chesterite (alternating double and triple chains), and other ordered and disordered materials (Thompson, 1981). The term *biopyribole* was first used by Johannsen (1911) as a field geologist's word for unidentified black minerals. Johannsen also coined the term *pyribole* to indicate chain silicate members of the biopyribole group. Thompson (1978) extended the usage of these terms to include the recently discovered, rarer members of the group, such as clinojimthompsonite. It should be noted that these terms are commonly misused to denote only these rarer structures with triple and wider chains; Thompson's definition, however, clearly includes amphiboles and pyroxenes as pyriboles and biopyriboles.

The biopyriboles undoubtedly are the polysomatic minerals most studied by HRTEM. The polysomatic defect structures and reaction mechanisms involving biopyriboles have been reviewed previously (Buseck et al., 1980; Buseck and Veblen, 1988; Veblen, 1981; Veblen, 1991), and the reader is directed to those discussions for background. Rather than repeating the early work, I will use selected biopyribole examples to illustrate various aspects of HRTEM studies on polysomatic minerals in general, as well as to provide discussion of some of the recent work on biopyriboles.

6.3.2.2 Interpretation of biopyribole HRTEM images

Essential to any HRTEM study is the interpretation of images. Historically, image interpretation was purely intuitive, based on the microscopist's best guess of the structures represented by the experimental images. This method was prone to error. For example, the unit cell in the vesuvianite image from the classic paper by Buseck and Iijima (1974, their Fig. 8) is displaced from its correct position by one-half of a unit-cell translation. In the old days, there was no shame in such an error based on intuitive interpretation! We now know that the interpretation was incorrect, because subsequent computer image simulations by the same authors revealed the correct placement of the unit cell (O'Keefe et al., 1978). In fact, subsequent calculations have shown that a substantial majority of the intuitive interpretations from the early days of HRTEM *are* correct, including most of the other interpretations presented by Buseck and Iijima (1974).

Today, many papers that utilize HRTEM images include computer simulations of the images, which make the interpretation much more rigorous, especially if the experimental and simulated images can be matched at several different focuses and/or specimen thicknesses. The important topic of simulations is covered in detail by Self in Chapter 3 of this volume. However, the example of biopyribole simulations serves to point out the importance of computer simulation in the interpretation of images from polysomatic structures.

Figure 6-12 shows early simulations for anthophyllite, jimthompsonite, and chesterite (Veblen and Buseck, 1979), compared to experimental images obtained under the optimum defocus condition. It is obvious to the eye that there is good agreement between calculation and experiment. There should, of course, be agreement, for these structures are known in detail from x-ray structure refinements, and we learn nothing new about these ordered structures from this exercise. This is not so for images of defects, which are nonperiodic features and hence not amenable to detailed study by x-ray diffraction. Here, electron microscopy is typically the only method available for determining the structure of these features, and computer simulation is the only accepted way of showing one's interpretation to be reasonable.

Unfortunately, most algorithms for computing images perform calculations at

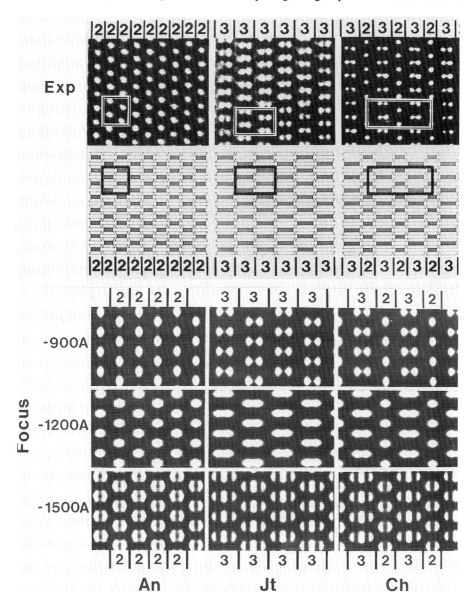

Figure 6-12. Computer-simulated, **c**-axis images of anthophyllite ("An"), jimthompsonite ("Jt"), and chesterite ("Ch"), compared to experimental images ("Exp") obtained near the optimum defocus of -900 Å for a JEOL JEM 100B instrument. Crystal thickness for the simulations was 53 Å, and focus values for the simulations are given in Å. (After Veblen and Buseck, 1979.)

discrete, periodic points in reciprocal space, and therefore they require the simulated structure to be periodic in real space. The programs for calculating images therefore must be "tricked" into believing that the defect is part of a periodic structure. This is done by a method known as periodic continuation, in which the defect is embedded in a relatively large block of normal structure. The resulting unit cell is generally quite large, and, as a result, periodic continuation simulations are considerably more lengthy than simulations for ideal, nondefective crystals. Figure 6-9 shows an experimental image of a sextuple-chain defect and its termination in amphibole, along with a structural model for the defect in I-beam representation and its computed image. The artificially large unit cell chosen for this simulation contained more than 2900 atoms.

6.3.2.3 Recent HRTEM studies of amphiboles and other hydrous pyriboles

As indicated previously, HRTEM studies of structural disorder in amphiboles and other pyriboles have been reviewed previously in the *Reviews in Mineralogy* series (Veblen, 1981), and it is not my intention to rehash earlier results. In this section, however, I review some of the more recent results of HRTEM studies on pyriboles.

6.3.2.3.1 Further observations of chain-width disorder in pyriboles. Several papers have elucidated the occurrence of chain-width disorder in amphiboles, of structurally disordered pyriboles, and of ordered wide-chain pyriboles. Akai (1982) showed that metasomatic alteration of clinopyroxene in the Akatani ore deposit, Japan, produced mixtures of amphibole and triple-chain silicate. Furthermore, AEM indicated that a Ca-rich analog of clinojimthompsonite (pure triple chains) occurs in blocks up to 100 nm wide. This is presumably a mineral in which the M5 sites contain Ca, rather than (Fe,Mg) as in clinojimthompsonite, and it is likely that the chemical formula is close to $Ca_2(Mg,Fe)_8Si_{12}O_{32}(OH)_4$, i.e., it is probably the triple-chain analog of augite and actinolite. Although this mineral has not been named, it is likely that it is very widespread in small amounts: though not analyzed chemically, similar material is commonly formed by weathering and low-temperature alteration of augite (Nakajima and Ribbe, 1980; Veblen and Buseck, 1981).

Yau (1986) demonstrated that sediments of the Salton Sea geothermal area contain calcic amphibole and pyroxene with coherent, random intergrowths of triple and, more rarely, quadruple and quintuple chains. The euhedral crystals apparently are produced by direct growth from a fluid, rather than by reaction from a primary pyroxene or ordered amphibole. Thus, they are analogous to disordered double/triple-chain silicates that have been directly synthesized in the Na/Mg pyribole system (Drits et al., 1976). Yau et al. give a temperature for this occurrence of 310 to 330°C, in the biotite zone.

Schumacher and Czank (1987) provided a thorough examination of anthophyllite from the classic locality at Orijärvi, Finland. They found pervasive intergrowth of jimthompsonite, chesterite, and disordered, mixed-chain silicates with the amphibole, and they presented a detailed discussion of the metamorphic petrology of this occurrence.

HRTEM studies of garnet-pyroxene coronas found in anorthositic gabbros from western Norway have shown that the pyroxenes contain fine-scale intergrowths of amphibole (Fig. 6-13), as well as replacement of orthopyroxene by sheet silicates (Griffin et al., 1985; Mellini et al., 1983). These results, combined with other data, suggest low but finite P_{H_2O} during granulite-facies metamorphism. The TEM results also show that retrograde hydration is much more extensive than indicated by petrographic observations. Amphibole alteration products finely intergrown in slowly cooled pyroxenes have also been observed recently in orthopyroxene megacrysts from a Labrador anorthosite (Veblen

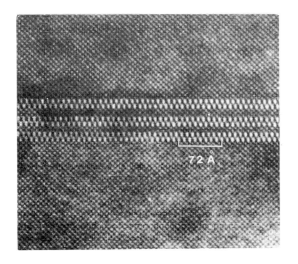

Figure 6-13. Slabs of amphibole that have partially replaced clinopyroxene from a garnet-pyroxene corona in anorthositic gabbro (viewed down c-axis). Each amphibole slab is two chains wide. [Used by permission of the publisher of *Contributions to Mineralogy and Petrology*, from Griffin et al., 1985, Fig. 6, p. 335.]

and Bish, 1988) and in augite from the Balmuccia Massif in northeastern Italy (Skrotzki et al., 1991).

In a HRTEM investigation of experimentally heated talc specimens, Konishi and Akai (1991) found that the most abundant intermediate products of the talc → enstatite reaction are highly disordered sequences of single, double, and wider chains, mixed with remnant talc. Large patches of amphibole structure or the polysome (MPP), which consists of alternating single and double silicate chains, were not observed. Akai et al. (1985) also observed disordered (010) pyroxene slabs in natural oxyhornblende, which they suggested were produced by dehydroxylation of primary amphibole.

6.3.2.3.2 Amphibole asbestos.

HRTEM observations have been used previously to infer that amphibole asbestos disaggregates into fibers along grain boundaries, and perhaps secondarily along planar defects (Alario Franco et al., 1977; Veblen, 1980). It also has been suggested by several authors that chainwidth defects are more abundant in asbestiform amphiboles than they are in massive amphibole (e.g., Veblen, 1981; and see the fine review by Chisholm, 1983). These issues were systematically addressed in an excellent study by Dorling and Zussman (1985, 1987), who examined 53 samples of massive, prismatic, acicular, and asbestiform tremolite and actinolite. Their results clearly show that finescale multiple twinning on {100} is restricted to asbestiform samples and that, on average, asbestiform and massive (nephrite jade) samples contain more chain-width defects than prismatic and acicular samples. However, prismatic and acicular amphiboles still can contain considerable numbers of wide-chain defects. Asbestiform tremolite appears to be restricted to very Al-poor compositions.

The earlier results are also underscored by Cressey et al. (1982), who studied asbestiform grunerite and anthophyllite, and by Ahn and Buseck (1991), who used an ultrahigh-resolution TEM to produce impressive images of crocidolite (riebeckite) asbestos. Figure 6-14 shows several fibrils that contain chain-width defects and that are separated from each other by grain boundaries. Ahn and Buseck noted that these are commonly low-angle boundaries, i.e., that fibrils are commonly only slightly rotated with respect to their neighbors. Furthermore, planar defects containing edge dislocation components suggest that some fibrils developed during growth by fragmentation of larger

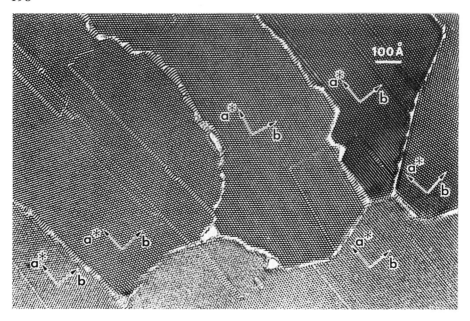

Figure 6-14. HRTEM image of crocidolite (riebeckite asbestos), showing fibrils, which contain chain-width errors, separated by low-angle grain boundaries (Ahn and Buseck, 1991).

crystals; this process may have been assisted by tectonic stresses present during the growth of the asbestos.

6.3.2.3.3 Synthetic amphiboles and triple-chain silicates.

It has been known for some time that synthetic amphiboles and triple-chain silicates can contain abundant chain-width defects (Drits et al., 1976; Veblen et al., 1977), and the past decade has seen further activity in this area. Maresch and Czank (1983a, 1988) explored the degree of structural disorder in synthetic $Mn_xMg_{7-x}Si_8O_{22}(OH)_2$ amphiboles as a function of composition and experimental conditions, finding that chain-width disorder is widespread in such synthetic products. Maresch and Czank (1983b) also presented an extensive catalog of HRTEM images showing the types of extended defects that can occur in synthetic amphibole. Systematic changes in chain-width defect density as a function of composition also were demonstrated in the anthophyllite-gedrite system (M. Czank, personal comm., 1991). Gribelyuk et al. (1988) further showed that chain-width disorder is important in synthetic triple-chain silicates containing Na, Ni, and Co.

The degree of chain-width disorder has also been evaluated for some synthetic calcic amphiboles. Maresch and Czank (1988) reported that tremolite synthesized at 700°C and 5 kbar contained only about 70% undisturbed amphibole, whereas Skogby and Ferrow (1989) indicated that Fe-rich tremolite with Fe/Mg between 0.07 and 0.14 contains a much lower defect density. Ahn et al. (1991) systematically investigated synthetic hydroxyl tremolite, fluor-tremolite, and magnesio-hornblende and found that the degree of chain-width disorder varies considerably from sample to sample. An important question for future studies is whether or not chain-width disorder affects the thermodynamic stability of amphiboles and other chain silicates. Jenkins and Clare (1990) noted that there is a 40°C difference in the thermal stability limits of natural and synthetic tremolite, but they concluded that substitution of F in the natural sample may account for the difference. This does not, however, demonstrate conclusively that chain width errors did not contribute to the lower thermal stability of the synthetic, pure hydroxyl tremolite.

Welch et al. (1992) examined intergrowths of the triple-chain silicate $Na_4Mg_8Si_{12}O_{32}(OH)_4$ and its amphibole polymorph, $Na_{2.67}Mg_{5.33}Si_8O_{21.33}(OH)_{2.67}$, with a variety of methods, including HRTEM and ^{29}Si magic-angle spinning NMR. They demonstrated that NMR is a very promising technique for obtaining bulk measurements of the amounts of double and triple chains, at least in synthetic samples containing no Fe. If future studies bear out this conclusion, NMR obviously will become an important technique for investigating intergrown biopyriboles, since HRTEM is inherently a statistically poor technique for determining bulk properties of a sample.

6.3.2.3.4 Short-range ordered polysomes. HRTEM is useful not only for studies of disordered materials, but it is also of value for determining the structures of ordered minerals that occur only in small amounts and hence are not suitable for investigation by x-ray diffraction methods. One application of this sort is the structure determination of ordered, short-range polysomes that may occur in blocks only a few unit cells wide. As shown in Table 2, a number of such polysomes were reported by Veblen and Buseck

Table 6-2. Observed short-range polysomes with large unit cells

Sequence	Repeats	Probability (p)*	Repeat Length (nm)	Reference**
(24)	40	3.67×10^{-23}	2.7	2
(233)	11	3.33×10^{-8}	3.6	1,2
(234)	49	3.45×10^{-67}	4.05	2
(2111)	46	1.62×10^{-43}	2.25	2
(2233)	15	1.31×10^{-16}	4.5	1
(2333)	45	1.62×10^{-42}	4.95	1
(2234)	4	5.41×10^{-5}	4.95	1
(2343)	6	2.15×10^{-8}	5.4	2
(322212)	5	1.93×10^{-8}	5.4	2
(232233)	4	2.06×10^{-5}	6.75	1
(222333)	3	1.08×10^{-3}	6.75	1
(233224)	3	7.21×10^{-5}	7.2	2
(433323)	5	1.94×10^{-8}	8.1	1
(2332323)	3	3.33×10^{-4}	8.1	1
(222222333333)	7	2.25×10^{-21}	13.5	2
(32422232422222)	3	4.60×10^{-9}	15.3	2
(43332343332423)	4	5.82×10^{-17}	18.9	1
(42423322222222-324222222222222222)	3	3.29×10^{-16}	32.8	2
(32324233222222-323232222222222222)	3	1.70×10^{-17}	32.8	2

* The probability p as defined by Veblen and Buseck (1979).
** Reference 1 is Veblen and Buseck (1979); reference 2 is Grobety (1992).

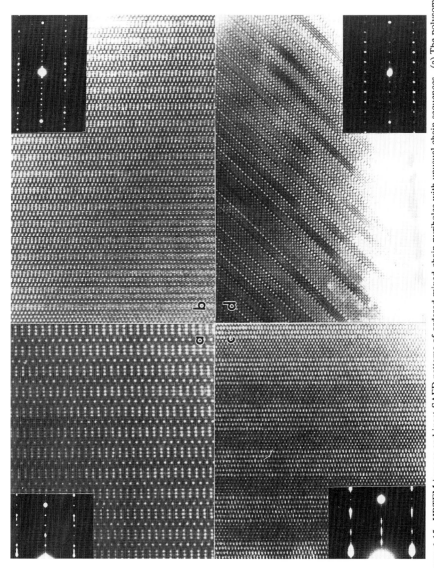

Figure 6-15. HRTEM images and inset SAED patterns of ordered mixed-chain pyriboles with unusual chain sequences. (a) The polysome (24). (b) The polysome (234). (c) The polysome (222222333333). (d) A region of the polysome (2111) with two defects that contain extra P slabs. (Grobety, 1992.)

(1979), and several more recently have been observed by Grobety (1992). The former are from Chester, Vermont (Veblen and Burnham, 1978), and the latter are from ultramafic lenses in the Maggia nappe in the Lepontine Alps, Upper Valle Maggia, Ticino, Switzerland, for which temperature and pressure were estimated to be 650±50°C, 4 kbar (Grobety, 1992; see also Nissen et al., 1980). All of the polysomes listed in Table 2 occur in amounts that suggest they form as the result of some non-random growth or reaction mechanism, according to the statistical test of Veblen and Buseck (1979). In other words, they apparently did not form simply as the result of the random combination of different structural elements (such as double- and triple-chain slabs).

Figure 6-15 shows some examples of ordered biopyriboles that occur only in small amounts (Grobety, 1992). Once one understands c-axis images of other ordered pyriboles, from computer simulations and experimental images of more abundant ordered polysomes (*i.e.*, pyroxene, amphibole, chesterite, jimthompsonite), it is possible to determine the structure of such short-range ordered materials by inspection. Note that the stoichiometry of such ordered structures (and disordered biopyriboles as well) can be determined simply by counting the numbers of M and P slabs and multiplying these numbers by the chemical formulae of such slabs (Veblen and Buseck, 1979; see also §6.4.1). It is not known why these short-range polysomes occur in nature, but Grobety (1992) points out that some of the simpler ones are minimum-energy structures according to the polysomatic energy model of Price and Yeomans (1986). Note also that Table 2 lists the first long-period ordered polysomes containing *single* silicate chains: the structure (2111), which Grobety observed in several different places, with a maximum of 46 adjacent unit cells, and the structure (322212).

6.3.3 Pyroxenoids

Whereas pyroxene chains have a repeat unit of two tetrahedra in the **c** direction, pyroxenoids are single-chain silicates that ideally possess chain periodicities of 3, 5, 7, or 9, as shown schematically in Figure 6-16. In several papers, it has been shown that the pyroxenoids can be treated as modular structures (Koto et al., 1976; Narita et al., 1977; Takéuchi and Koto, 1977), and Thompson (1978) formalized these ideas by showing that there exists a polysomatic series between the clinopyroxene structure and the wollastonite structure. Within this series, pyroxene is the polysome (P), wollastonite consists purely of W slabs, (W), whereas the 5-repeat rhodonite is (PW), 7-repeat pyroxmangite and pyroxferroite are (PPW), and 9-repeat ferrosilite-III is (PPPW).

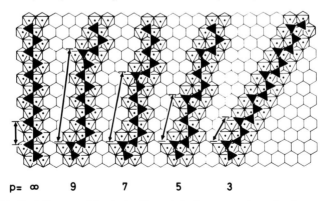

p= ∞   9   7   5   3

Figure 6-16. Idealized diagrams of the single silicate chains in pyroxene (p = ∞) and pyroxenoids with chain periodicities of 3, 5, 7, and 9. [Used by permission of the publisher of *Physics and Chemistry of Minerals,* from Czank and Liebau, 1980, Fig. 1, p. 86.]

As in all other known polysomatic series, members of the pyroxene-pyroxenoid series exhibit structural disorder that results from either missing or extra P slabs, i.e., mistakes occur in the ideally periodic sequences of P's and W's. This type of disorder was experimentally first recognized independently in at least four different laboratories and reported in papers by Czank and Liebau (1979, 1980), Ried and Korekawa (1980), Jefferson et al. (1980), and Buseck et al. (1980). Following Czank and Liebau (1980), such faults are commonly called "chain periodicity faults," because they upset the ideally perfect periodicity of the silicate chains parallel to c. In addition, HRTEM studies have shown that structures in the P-W polysomatic series are also prone to polytypic, or stacking, disorder. HRTEM applications to polytypism in other structures are discussed by Baronnet elsewhere in this volume and therefore will not be discussed here in detail, but I note here that such disorder occurs in pyroxenes (Livi and Veblen, 1989, and references therein); in wollastonite (Palomino et al., 1977; Wenk et al., 1976); in rhodonite (Jefferson and Pugh, 1981; Jefferson et al., 1980); in pyroxmangite and pyroxferroite (Czank, 1984; Reiche et al., 1978); and in ferrosilite-III (Czank and Simons, 1983). It should also be noted that interpretation of the rather complex HRTEM images for minerals in this structural system has been assisted by computer simulations undertaken by Jefferson et al. (1980) and Smith et al. (1981).

Naturally occurring pyroxenoids typically have a rather low density of chain periodicity faults. Czank and Liebau (1980) found such defects in several well-crystallized natural rhodonite and pyroxmangite samples and noted that typically $\Delta p = \pm 2$, where Δp is the chain periodicity of the host structure, minus the "periodicity" of the fault. For example, faults in pyroxmangite ($p = 7$) generally have the 5repeat rhodonite structure ($7 - 2$) or the 9-repeat ferrosilite-III structure ($7 + 2$), but not the 3-repeat wollastonite or bustamite structure ($7 - 4$). This relationship does not, however, hold for synthetic pyroxenoids or for all natural samples. For example, naturally occurring lunar pyroxferroite ($p = 7$) contains 11- and 13-faults, in addition to the more common 5- and 9-faults (Czank and Liebau, 1983). Other studies in which chain periodicity faults were found in naturally occurring rhodonite or pyroxmangite include those by Ried and Korekawa (1980), Alario Franco et al. (1980), Buseck et al. (1980), and Veblen (1985b). Ried and Korekawa (1980) and Czank (1981) also found chain periodicity faults in natural babingtonite, which has a crystal structure closely related to that of rhodonite. Examples of chain periodicity faults in pyroxmangite and rhodonite are shown in Figure 6-17 (Czank and Liebau, 1980).

Many, but not all, synthetic pyroxenoids and pyroxenes have higher chain-periodicity defect densities than natural samples, except for those natural occurrences in which reactions from one structure to another were arrested due to kinetic factors (Veblen, 1985b). In synthetic samples, the defect density has been shown to decrease with increased annealing time (Jefferson and Pugh, 1981). Compositions that have been investigated include $MnFeSi_2O_6$ (Catlow et al., 1982; Jefferson and Pugh, 1981); $En_{50}Fs_{30}Wo_{20}$, $En_{94}Wo_{06}$, and $En_{80}Wo_{20}$ (Ried, 1984); $Mg_{0.04}Ca_{0.21}Fe_{0.75}SiO_3$ (Ried and Korekawa, 1980); $FeSiO_3$ (Czank and Simons, 1983); $Mn(Si,Ge)O_3$ (Czank, 1984); $MnSiO_3$ and $(Mn,Zn)SiO_3$ (Simons and Czank, 1986). Examples of chain periodicity and stacking disorder in synthetic pyroxenoids are shown in Figures 18 and 19 (Czank and Simons, 1983; Jefferson and Pugh, 1981).

A particularly interesting observation is that long-period polysomes occur in some synthetic pyroxenoids, similar to the observation of such long-period structures in biopyriboles (see discussion above). Czank (1984) observed a repeating arrangement of 9-repeat chain periodicity faults within $Mn(Si,Ge)O_3$ ($p = 7$) that gave rise to a

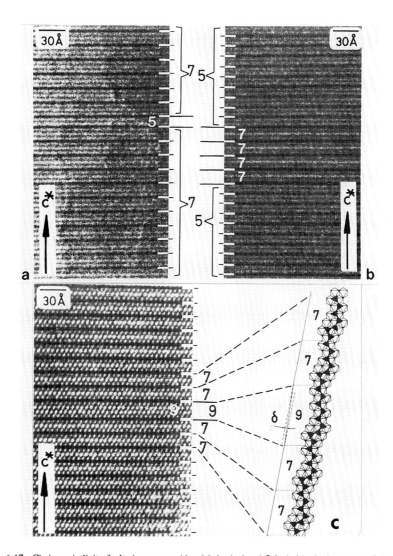

Figure 6-17. Chain periodicity faults in pyroxenoids. (a) An isolated 5 fault (rhodonite structure) in pyroxmangite (p = 7). (b) Four pyroxmangite units in rhodonite. (c) A 9 fault (ferrosilite III-type structure) in pyroxmangite. [Used by permission of the publisher of *Physics and Chemistry of Minerals*, from Czank and Liebau, 1980, Figs. 2 and 3, p. 89.]

superstructure with 400 tetrahedra, or 91 nm, in the repeating chain unit. Ried and Korekawa (1980) reported a 10.5-nm superstructure with each unit cell containing the sequence 555555557. It is quite possible that such long-period structures result from spiral growth about screw dislocations, although there is no direct evidence for this growth mechanism. Similar superstructures are very rare in natural pyroxenoids, although Veblen (1985b) observed the polysome (557) to repeat six times in one part and eight times in another part of the same disordered "crystal." He also observed five unit cells of the polysome (57557). Although these structures pass the test for statistical significance of Veblen and Buseck (1979), it is not inconceivable that they result simply from random combination of 5 and 7 units.

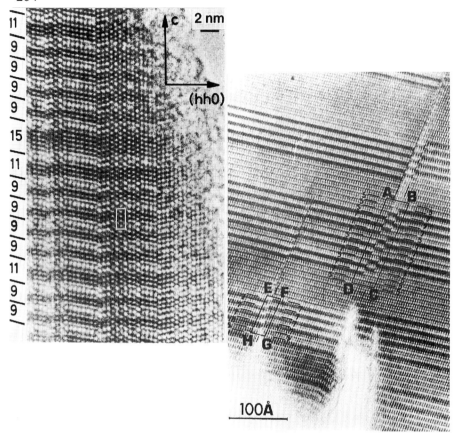

Figure 6-18 (left). Chain periodicity faults (11 and 15) in synthetic ferrosilite III. Stacking disorder (stacking faults and twinning) causes the (001) fringes to deviate from linearity. [Used by permission of the publisher of *Physics and Chemistry of Minerals*, from Czank and Simons, 1983, Fig. 8, p. 233.]

Figure 6-19. Chain periodicity faults (7) in synthetic rhodonite (p = 5). The chain periodicity faults are offset by stacking faults, as seen in boxes ABCD and EFGH. [Reproduced with permission of *Acta Crystallographica*, Jefferson and Pugh, 1981.]

There have been several studies of reactions in which a pyroxene or pyroxenoid structure is transformed into a different pyroxenoid. These investigations will be discussed in the §6.4.3 on polysomatic reactions.

6.3.4 Sheet silicates

The sheet silicates are probably the most defective group among the major rock-forming silicates, very commonly suffering from the twin humiliations of polytypic disorder, polysomatic disorder, or both. HRTEM studies of polytypism in sheet silicates are discussed by Baronnet (this volume, Chapter 7), and the polysomatic phenomenon of mixed layering that occurs in clay minerals is discussed by Peacor (this volume). In this section, I will explore two other facets of polysomatism in sheet silicates: (1) mixed layering in higher-temperature, more coarsely crystalline silicates with flat sheets and (2) polysomatic phenomena in modulated layer silicates.

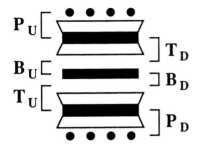

Figure 6-20. Polysomatic modules for sheet silicates similar to those used by Thompson (1978). U and D indicate up and down modules; P = phlogopite, T = talc, and B = brucite (Guthrie and Veblen, 1990b).

6.3.4.1 Polysomatic basis for mixed layering in sheet silicates

Thompson (1978) showed that the structures of sheet silicates can be rationalized in terms of a polysomatic model. For this treatment, he used modules similar to those shown in Figure 6-20. By fitting these different slabs together in regularly repeating patterns, it is possible to construct the structures of the common ordered sheet structures, such as those of the mica phlogopite (P_DP_U), the lizardite form of serpentine (B_DT_U), the hydroxide brucite (B_DB_U), talc (T_DT_U), or pure trioctahedral chlorite ($T_DT_UB_DB_U$). Other modules obviously can be used for dioctahedral structures and other structural and chemical variants. Similarly, more complicated sequences, in which there is more than one type of layer, do occur. These are commonly referred to as mixed-layer structures.

There is no law, however, that Thompson's modules must be assembled in a perfectly repeating sequence, and indeed nature very commonly builds aperiodic sequences and thus creates what I will refer to as mixed-layering disorder. An example would be the following sequence embedded within an otherwise perfect crystal of phlogopite:

$$...P_DP_UP_DP_UP_DP_UP_DP_UB_DB_UP_DP_UP_DP_UP_DP_UP_DP_U....$$

The pair of B slabs creates an isolated hydroxide sheet within the phlogopite, and if considered together with the surrounding 2:1 structure, this fault could also be interpreted as a single unit cell of chloritelike material embedded in the mica. Indeed, defects of just this sort have been reported in 2:1 sheet silicates by numerous workers employing HRTEM, as have extra or missing 2:1 layers in chlorite (Amouric et al., 1988; Banfield et al., 1989; Eggleton and Banfield, 1985; Ferrow and Roots, 1990; Ferrow et al., 1990; Le Gleuher et al., 1990; Lee and Peacor, 1985; Lee et al., 1986; Lo and Onstott, 1989; Maresch et al., 1985; Olives Baños, 1985; Olives Baños and Amouric, 1984; Olives Baños et al., 1983; Ristich and Huyck, 1989; Veblen, 1980, 1983; Veblen and Buseck, 1980, 1981; Veblen and Ferry, 1983; Yau et al., 1984). As we shall see shortly, mixed-layering defects that can be represented as non-periodic sequences of modules like those in Figure 6-20 also have been reported from chlorite, structures of the serpentine group, and other sheet silicates.

6.3.4.2 Interpretation of HRTEM images of mixed-layer sheet silicates

As with many other mineral groups, image interpretation in the early studies on sheet silicates was largely intuitive. Iijima and Buseck (1978) and Amouric et al. (1981) provided the first image simulations for micas and showed that at optimum defocus the layers should be dark, and the interlayer regions relatively lighter. The primary motivation for these studies, however, was to enable analysis of the two-dimensional image details to permit the extraction of detailed polytype information from HRTEM images. The importance of these contributions is discussed by Baronnet in Chapter 7, this volume.

206

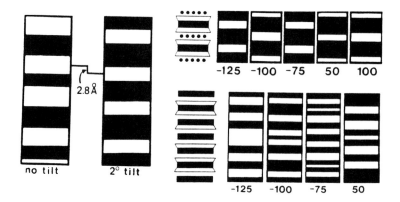

Figure 6-21 (left). Computer-simulated images (Philips 420T instrument) for R1 illite/smectite with electron beam perfectly parallel to the layers (left) and with a misorientation of 2° (right). The apparent position of the layers shifts appreciably for the misoriented case. [Used by permission of the Clay Minerals Society, from Guthrie and Veblen, 1989, Fig. 7, p. 7.]

Figure 6-22 (right, top). Simulated images (Philips 420T) for phlogopite at different defocus values. The contrast reverses even for relatively small, 25-nm changes in focus on either side of the Scherzer focus (-100 nm). (Guthrie and Veblen, 1990b.)

Figure 6-23 (right, bottom). Simulated images (Philips 420T) for a mixed-layer structure containing hydroxide, 1:1, and 2:1 layers. Only the perfect Scherzer-focus image (-100 nm) is directly interpretable in terms of the layer thicknesses. (Guthrie and Veblen, 1990b.)

Unlike the two-dimensional experimental images and image simulations used to explore polytypism in sheet silicates, most of the studies of mixed layering have involved the interpretation of one-dimensional images produced by imaging the 00 1reciprocal lattice row. For this type of image, computer simulations have been presented only recently by Guthrie and Veblen (1989, 1990a,b). The simulations suggest that many interpretations in the literature are probably correct, but the potential exists for major errors in interpretation. The potential for error is exacerbated by the fact that most sheet silicates undergo rather rapid beam damage, and hence the microscopist must work fast to obtain images. Under this condition, and especially when working with small or deformed crystals, it is likely that errors will be made in orientation of the specimen and in setting the proper microscope defocus; it is clear from experience and simulation studies that interpretation of images obtained under inappropriate conditions can lead to incorrect conclusions.

Several types of imaging artifacts are possible. Even when imaging conditions are perfect, the bright fringes of 1:1 structures may not exactly overlie the interlayer regions, as one might suppose. This effect is even more pronounced and occurs for all sheet silicate structures when the sample is not in perfect orientation (i.e., when the electron beam is not exactly parallel to the layers of the structure). Figure 6-21 shows that the apparent layer positions can shift considerably, even for relatively small orientation errors. Thus, unless the microscopist knows that the crystal was truly in perfect orientation, it is incorrect to interpret specific fringes as representing a specific part of the crystal structure.

For some structures, image contrast does not change rapidly with small deviations of focus from the Scherzer focus. For others, however, such as the phlogopite shown in

Figure 6-22, small focusing errors can lead to complete contrast reversal, so that interpretation of dark fringes as corresponding to the structural layers would be incorrect. Important changes in image contrast can occur not only for the simple ordered structures, such as micas, but also for mixed-layer structures. Figure 6-23 shows how images of an assembly of hydroxide, 1:1, and 2:1 layers change as a function of defocus. The image at $\Delta f = -100$ nm could be interpreted correctly, since the widths of the dark fringes are simply related to the thicknesses of the layers they represent. However, deviations in focus of only 25 nm to either side of this condition produce images that are not readily interpretable. It is essential, therefore, that direct interpretations of such mixed-layer structures be made only when it is certain that the microscope focus and sample orientation were very closely controlled.

Simulations also show that a number of other imaging artifacts are possible with sheet silicates. For a detailed discussion relating to specific structures, and for guidelines for obtaining and interpreting such images of sheet silicates, see Guthrie and Veblen (1990b).

6.3.4.3 HRTEM observations of mixed layering

Many different types of mixed layering phenomena have been observed in sheet silicates with HRTEM. Here I consider only the actual structures observed; their relevance to reactions in which one sheet silicate replaces another are discussed in §6.4.2.

As noted above, intergrowth of hydroxide sheets with 2:1 sheet silicates, especially biotite and phlogopite, has been reported numerous times from a variety of igneous and metamorphic rock types. Such defects can be interpreted as isolated chlorite layers intergrown with the mica. Extra 2:1 layers within chlorite also have been observed commonly; these can be interpreted as layers of mica or talc intergrown in a chlorite matrix. Similarly, missing 2:1 layers in chlorite locally give rise to brucitelike configurations. It is also common to observe larger packets of normal biotite occurring with packets of normal chlorite (e.g., Yau et al., 1984; Ferrow et al., 1990). HRTEM studies also have demonstrated the existence of highly disordered intergrowths of mica and chlorite that might truly be considered as mixed crystals of the two structures, rather than as isolated defects within more normal mica or chlorite (Fig. 6-24).

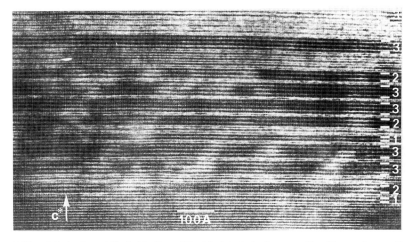

Figure 6-24. A "mixed crystal" of chlorite and the trioctahedral mica wonesite. The numbers on the right indicate how many adjacent layers there are in the mica packets (Veblen, 1983).

Perhaps the most interesting intergrowths involving the mica and chlorite structures are ordered materials with the layer sequence 2:1-2:1-hydroxide (Maresch et al., 1985; Olives Baños, 1985). These can be interpreted as mixed-layer sheet silicates consisting of an ordered 1:1 mixture of biotite and chlorite, with a d_{001} value of approximately 2.4 nm. Stoichiometrically, this ordered structure lies midway between biotite and chlorite, and unpublished AEM results indicate that the chemical composition is collinear with those of coexisting K-biotite and chlorite (Huifang Xu, personal comm., 1992). The structure of this material is similar to that of kulkeite, an ordered 1:1 mineral consisting of alternating chloritelike layers and layers of Na-rich talc (Schreyer et al., 1982).

Although mixed-layering phenomena involving the trioctahedral mica and chlorite structures are the most commonly reported, other types of mixed layering have also been observed in relatively coarse-grained sheet silicates with HRTEM. A partial list follows. (1) Kaolinite/biotite, which forms as an intermediate product of the reaction of biotite to kaolinite (Ahn and Peacor, 1987). (2) Phengite/biotite, which forms by reaction of phengite to biotite (Worden et al., 1992). (3) Vermiculite/biotite, which forms by weathering of biotite (Banfield and Eggleton, 1988). (4) Hydrobiotite/biotite, which forms during biotite weathering (hydrobiotite is a regularly alternating interstratification of vermiculite and biotite or phlogopite layers) (Ilton and Veblen, 1988). (5) 1:1 sheet silicate/biotite/chlorite (i.e., berthierine or other serpentinelike structure with biotite or chlorite) (Ahn and Peacor, 1985; Amouric et al., 1988; Banfield et al., 1989). (6) Biotite/muscovite (Ferrow et al., 1990; Iijima and Zhu, 1982). (7) Illite/chlorite (Ahn et al., 1988; Lee and Peacor, 1985). (8) Muscovite/chlorite (Lee et al., 1986). (9) Phengite/paragonite (Ahn et al., 1985). (10) Chlorite/berthierine/paragonite/stilpnomelane/chloritoid (Banfield et al., 1989). (11) Phlogopite/serpentine (Livi and Veblen, 1987). (12) Serpentine/chlorite, which forms during weathering of enstatite (Le Gleuher et al., 1990).

Taken together, these and other studies demonstrate clearly that even relatively coarse-grained sheet silicates commonly display mixed-layering structural disorder on scales down to the unit-cell level. Given that sheet silicates are also prone to polytypic disorder (Baronnet, Chapter 7, this volume), it might be argued that the sheet silicates form the most structurally disordered group of common rock-forming minerals. A last point is that sheet silicates exhibiting polysomatic disorder are inherently non-stoichiometric. For example, a "biotite" crystal that contains intergrown layers of chlorite structure possesses neither biotite nor chlorite stoichiometry, but rather lies somewhere between these end members. Petrologists who analyze sheet silicates should be aware that polysomatic intergrowths can occur at scales well below the resolution of the light microscope or electron microprobe. A "bad" analysis may simply be an analysis from a nonstoichiometric region of the sample.

6.3.4.4 Modulated layer silicates

Modulated layer silicates are in many ways similar to normal 1:1 and 2:1 layer silicates, but in general they have larger unit-cell parameters parallel to the sheets, as a result of structural perturbations involving the silicate and octahedral sheets. The structures and crystal chemistry of modulated layer silicates are described in two excellent review papers by Guggenheim and Eggleton (1987, 1988). The modulated layer silicates may be classified into two groups, one in which tetrahedral sheets are connected to octahedral sheets in an islandlike arrangement, and the other in which these tetrahedral-octahedral linkages form strips. Some members of the latter group can be treated as polysomes. For example, antigorite can be represented as two types of slabs in which the apical oxygens of the tetrahedral sheets reverse direction, separated by a

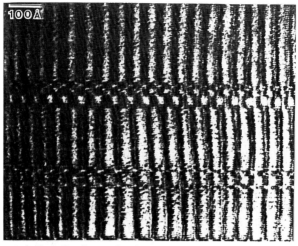

|10|10|13|13|15|15|18|18|21|15|15|13|15|13|13|13|15|15|

Figure 6-25.   HRTEM image of antigorite, showing variation in modulation periodicity due to different numbers of lizardite modules per unit cell (indicated by numbers at bottom). (From Spinnler, 1985.)

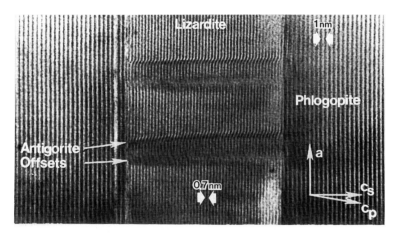

Figure 6-26.   A lamella of serpentine enclosed in phlogopite. The serpentine consists of nonperiodic layer reversals characteristic of antigorite, intergrown in the planar structure of lizardite. (Livi and Veblen, 1987.)

number of slabs having the lizardite structure (Spinnler, 1985; Livi and Veblen, 1987). By varying the numbers of lizardite slabs between the reversal slabs, the modulation period of antigorite can be varied, as shown in Figure 6-25, resulting in structural disorder when more than one period occurs in a crystal. This type of structural disorder is common in antigorite (Spinnler, 1985; Spinnler et al., 1983), and Mellini et al. (1987) showed that the disorder varies systematically with metamorphic grade. Low-temperature antigorite tends to be highly disordered, displaying a complex range of defects, whereas high-temperature antigorite is generally much better ordered. In addition, the reversal slabs can intergrow within serpentine that is dominantly lizardite (Fig. 6-26), leading to disordered structures intermediate between lizardite and normal antigorite (Livi and Veblen, 1987).

Carlosturanite also can be represented as a polysomatic structure (Fig. 6-27), and Mellini et al. (1985) have shown that disorder exists that can be interpreted as missing or extra S slabs, which are slabs of lizardite structure (Fig. 6-7b). Guggenheim and Eggleton (1986) used HRTEM to show that a similar type of structural disorder occurs in minnesotaite. In this case, however, the disorder can involve individual strips of

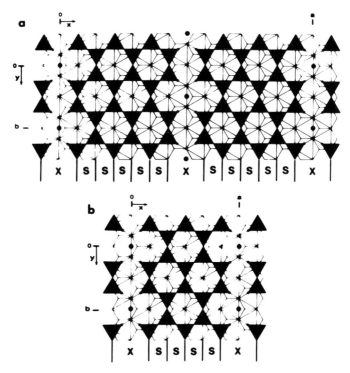

Figure 6-27. (a) Polysomatic model for carlosturanite, (SSSSSX), which consists of serpentine (S) slabs intergrown with slabs containing tetrahedral vacancies (X). (b) The polysome (SSSSX), which has not been observed in ordered form but occurs as defects in carlosturanite. (Mellini et al., 1985.)

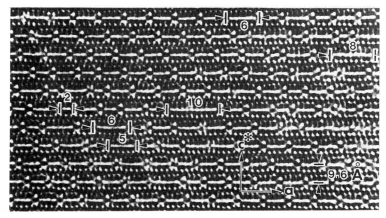

Figure 6-28. HRTEM image of minnesotaite showing strip-width disorder. Most of the tetrahedral strips are 3 or 4 tetrahedra wide, but anomalous strips (labeled) occur with widths of 2, 5, 6, 8, and 10 (Ahn and Buseck, 1989.)

structure, rather than entire slabs. These results were confirmed by Ahn and Buseck (1989), who obtained images displaying this type of disorder with a very high-resolution instrument (Fig. 6-28).

Because of the difficulty in finding good single crystals for x-ray diffraction studies, HRTEM and electron diffraction also have made major contributions to the study of the ideal structures of modulated layer silicates (Guggenheim et al., 1982). It is likely that these methods will continue to make important contributions to our understanding of this structurally complex group of minerals.

6.3.5 Olivine and the humite group

Thompson (1978) showed that olivine and the humite group of minerals (norbergite, chondrodite, humite, and clinohumite) comprise a polysomatic series. This series contains the ordered structures (N), (NO), (NOO), (NOOO), and (O), where N is a slab of norbergite structure, and O is a slab of olivine (Fig. 6-29). This raises the possibility that these structures might be subject to structural disorder based on errors in the sequence of N and O slabs.

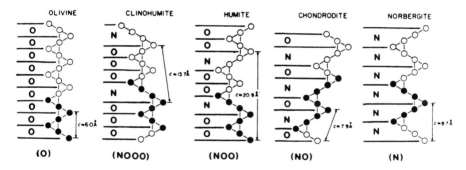

Figure 6-29. Schematic diagram of ordered structures in the olivine-humite group polysomatic series, shown as polysomes containing olivine O slabs and norbergite N slabs. [Veblen, 1990; after Papike and Cameron, 1976, with permission of the American Geophysical Union.]

6.3.5.1 Structural disorder in olivine

Although many olivine samples are devoid of planar defects (White and Hyde, 1982a), several different types of planar defects are at least possible. Intergrowths with the spinel (γ-phase) or β-phase polymorphs may occur, although these are not poly-somatic intergrowths and hence do not affect the stoichiometry of the olivine. However, from the polysomatic model for the olivine/humite group, one might expect to find N slabs intergrown with olivine under appropriate conditions, and such defects would render the olivine nonstoichiometric. In fact, Kitamura et al. (1987) have reported such defects in mantle-derived olivine from the Buell Park kimberlite, Arizona. Since the N slabs possess the nominal chemical formula $Mg(OH,F)_2 \cdot Mg_2SiO_4$, these defects result in the incorporation of hydrogen and fluorine into the structure. Indeed, Kitamura et al. found that this defective olivine produces OH infrared absorption peaks typical of the humite group. Similarly, Drury (1991) found that certain planar defects separating dissociated partial dislocations in mantle olivine are isostructural with the OH-rich layer in minerals of the humite group. Water in the mantle may thus be incorporated into olivine via planar defects, as well as into other nominally anhydrous minerals, such as pyroxene (Rossman, 1988; Skogby et al., 1990), with possible implications for deformation processes and reactions in upper-mantle rocks.

A different phenomenon that can afflict fayalite is oxidation to the olivinelike mineral laihunite, which contains layers of vacancy-rich and Fe^{3+}-rich structure (Shen et al., 1986; Tamada et al., 1983). These layers produce a superstructure with a c-axis that is a multiple of the olivine c-axis (e.g., a $2c$ or $3c$ superstructure). The superstructure was imaged in HRTEM studies by Kitamura et al. (1984) and Kondoh et al. (1985). Their results showed that the $2c$ and $3c$ structures can coexist in the same crystal and that there can be considerable disorder resulting from the intergrowth of these and other superstructures.

In an experimental study of olivine oxidation, Iishi et al. (1989) produced laihunite-type materials from olivines having a number of different chemical compositions, such as $MgFeSiO_4$. Subsequently, laihunitelike superstructures have been found in volcanic olivines that are even more magnesian, with Mg/(Mg + Fe) ≤ 0.66 (Banfield et al., 1990). Figure 6-7c shows a region of such an oxidized olivine with the $3c$, $2c$, and disordered structures. Although the structural details of these materials are not known, the dark bands separated by lighter areas suggest that Fe may have segregated from Mg. If this is the case, these oxidized olivinelike materials may be represented as polysomatic intergrowths involving slabs of forsteritic olivine intergrown with Fe-rich slabs. More HRTEM work is clearly needed on naturally occurring olivines to determine how common such intergrowths are in olivines from different geological environments.

6.3.5.2 Structural disorder in the humite group

Just as intergrown N slabs produce defects in olivine, it might be expected from the polysomatic model that minerals of the humite group might contain planar defects resulting from anomalies in the ideally periodic sequence of N and O slabs. Indeed, Müller and Wenk (1978) showed that such defects based on missing O slabs do occur in clinohumite, locally producing chondrodite or humite structure. In an extensive study of the defect chemistry in 42 Mg-rich natural and synthetic samples of the olivine-humite series, White and Hyde (1982a) found that among natural samples clinohumite tended to have the highest polysomatic defect density, with humite being less defective. The defect densities varied widely, however, some samples being defective and others being perfectly periodic, and most of the observed defects corresponded to extra O slabs. No defects were observed in the natural chondrodite, norbergite, or olivine samples that were studied. The synthetic samples displayed the same trends, with some clinohumite and humite samples defective (Fig. 6-30) but chondrodite and norbergite well-ordered. In the synthetic samples, several long-period structures were observed, analogous to the long-period structures sometimes observed in Mg-rich pyriboles. These structures consisted of ordered intergrowths of NOOO (clinohumite structure) with slabs of NOOOOO.

White and Hyde (1982b) also examined 14 natural samples of the Mn-rich analogs of the olivine-humite group (tephroite, sonolite, manganhumite, and alleghanyite). Although no planar faults were observed in the Mn-olivine tephroite, the humite-group analogs ranged from perfect crystals to highly defective. The observed defects corresponded to both extra and missing O slabs. White and Hyde (1983) further used HRTEM and electron diffraction to clarify the structure of the related mineral leucophoenicite. They also showed that leucophoenicite contains many extended defects, some of which are analogous to the polysomatic faults found in the humite group.

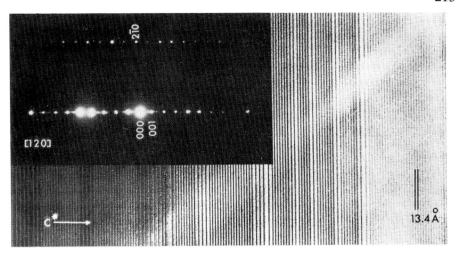

Figure 6-30. Polysomatic faulting in a specimen that is primarily clinohumite. The faults correspond to the insertion of extra O slabs. [Used by permission of the publisher of *Physics and Chemistry of Minerals*, from White and Hyde, 1982a, Fig. 7, p. 57.]

6.3.6 Other polysomatic mineral series

The polysomatic minerals described above are among the petrologically most important, and their defect structures are therefore of interest to the mineralogist and petrologist alike. There are, however, many other important rock-forming and ore-forming minerals, as well as related synthetic compounds, that can be represented as polysomes and that belong to polysomatic series. A few of these are listed here, and the reader is urged to read the references to these groups.

1. Oxides with the ReO_3 structure type and the perovskite structure type (stuffed derivatives of ReO_3) (Tilley, 1980).

2. Oxides with the rutile and related structure types (Tilley, 1980).

3. Numerous sulfosalts (Makovicky, 1989; Takéuchi, 1978).

4. The oxide-oxysulfide series schafarzikite, versiliaite, and apuanite (Ferraris et al., 1986).

5. Enstatite IV in the $Mg_2Si_2O_6$-$LiScSi_2O_6$ system (Takéuchi, 1978).

6. Rhönite (Bonaccorsi et al., 1990).

7. The oxyborates ludwigite, orthopinakiolite, pinakiolite, and takéuchiite (Bovin et al., 1981).

8. Tunnel and sheet Mn-oxides (Turner and Buseck, 1979; Veblen, 1991).

9. Oxide/silicates of the bixbyite-braunite group (de Villiers and Buseck, 1989).

10. Carbonates of the bastnaesite-synchisite series (Van Landuyt et al., 1975).

All of these polysomatic groups share an important characteristic with the groups discussed in earlier sections: in at least some samples, structural disorder occurs that can be described as deviations from the ideally periodic sequence of structural slabs. The occurrence of such disorder is the topic of the next section.

6.3.7 How common is polysomatic disorder in minerals?

Imagine the following scientific paper:

Introduction

Defect structures in minerals are important, and therefore we should know if there are extended defects in mineral X.

Experimental

A transmission electron microscope was used to examine ion-milled samples of mineral X from locality Y.

Results

Other than a few dislocations, no extended defects were observed in mineral X.

Conclusions

Based on our data, we conclude that mineral X can occur in a relatively defect-free state. However, it is possible that samples formed under other conditions, or parts of our specimen we did not examine, might contain extended defects other than dislocations.

I would assert that hundreds of studies of this sort have been carried out by the electron microscopy community, but would anyone ever write a scientific paper of this sort? Would any journal ever publish it? Probably not. Yet, it can be argued that "negative" results of this sort are extremely important, that it is essential to know not only what minerals and occurrences are highly defective, but also to recognize those minerals that tend to be highly regular and under what conditions they crystallize with very few defects. But let's face it; if one is a high-resolution transmission electron microscopist, it's much more fun to work with and publish on pathologically disordered specimens than it is to work on agonizingly boring material with no defects. Thus, there is a built-in bias in this field, which has produced numerous papers on defective minerals but relatively few reporting on non-defective samples.

There have been a few studies, however, in which the lack of defects has been reported. Two of these notable exceptions are the thorough studies of Dorling and Zussman (1987), who examined 54 different samples of tremolite and actinolite, and White and Hyde (1982a,b; 1983), who worked on a total of 59 specimens from the humite group, Mn-humite group, and leucophoenicite. Beyond investigations such as these, most of the negative results never make it out of the lab, except perhaps as anecdotes that become part of the HRTEM folklore known to only a select few.

So, much of the remainder of this section is anecdotal, based on my own negative results and those of others I have known during my sixteen years in electron microscopy. This is not science and should not be taken as such. It represents my opinions, hunches, and interpretation of what has and has not been published on polysomatic disorder in a few select groups of rock-forming minerals. I should probably be shot for this, and some of my colleagues may see to it, but here goes.

6.3.7.1 Pyriboles

6.3.7.1.1 Pyroxenes. There are examples of pyroxenes that contain a great deal of submicroscopically intergrown amphibole and/or other hydrous biopyribole, such as the orthopyroxene megacryst of Veblen and Bish (1988). However, using IR spectroscopy, which is a much better bulk technique than HRTEM, Skogby et al. (1990) found that this particular sample was the most hydrogen-rich of all the pyroxenes they examined. In fact, HRTEM studies of many pyroxenes show them to have very low defect densities, at least in the crystal interiors. It is very common to find hydrous biopyriboles intergrown with pyroxenes near grain boundaries and fractures, where hydrothermal or weathering fluids had access, but away from these sites many igneous and metamorphic pyroxenes contain no polysomatic defects. At the same time, some pyroxenes are pervasively altered, without obvious indications at the light microscope level. In addition to polysomatic disorder, pyroxenes are prone to twinning, stacking faults, and antiphase domains, especially in ferromagnesian species.

6.3.7.1.2 Amphiboles. Chain-width defects apparently can occur in any amphibole, although ferromagnesian amphiboles appear to be particularly prone to this type of disorder. However, some anthophyllite and gedrite samples are almost devoid of these defects. From the extensive study of tremolite-actinolite by Dorling and Zussman (1987), it appears that polysomatic defects are abundant in nephrite, abundant in tremolite asbestos but not in actinolite asbestos, rare in acicular forms, and uncommon in most prismatic specimens, although one prismatic sample contained abundant chain-width errors. There are thus trends in the occurrence of polysomatic disorder in naturally occurring amphiboles, but no rigorous generalizations can be drawn. On average, synthetic amphiboles seem to contain more chain-width disorder than natural specimens, but again no firm generalizations can be made. End-member anthophyllite seems to be particularly prone to such disorder, but results of Maresch and Czank (1983, 1988) and Ahn et al. (1991) make it clear that polysomatic disorder is dependent on a number of variables, including composition, temperature, and run duration. Like pyroxenes, amphiboles are prone to {100} twinning and stacking faults, especially for ferromagnesian compositions and asbestiform varieties.

6.3.7.1.3 Other hydrous pyriboles. Natural specimens of silicates with triple chains (jimthompsonite, chesterite, et alia) typically display a variety of chain-width defects and associated displacive faults. Synthetic samples show a wide range of states from ordered to highly disordered (Drits et al., 1976; Welch et al., 1992), depending on temperature and presumably other variables.

6.3.7.2 Pyroxenoids

Many natural samples contain few chain periodicity faults (Czank and Liebau, 1980), but others contain a very high density of such defects (Veblen, 1985b). Synthetic pyroxenoids and pyroxenes appear to be much more disordered on average than natural samples, and the state of disorder is again dependent on run duration, composition, temperature, and presumably other variables. In addition to polysomatic variations, pyroxenoids are prone to stacking disorder.

6.3.7.3 Sheet silicates

Polysomatic disorder is so common in sheet silicates that the term "mixed layering" has long been used to describe the phenomenon, especially with respect to clay minerals. It is now clear (see the numerous references in §6.3.7.3 on sheet silicates) that more coarsely crystallized varieties also commonly show interstratification of one layer type within another. It seems that any biotite crystal that has experienced a bit of hydrothermal alteration or weathering will contain intercalated layers of chlorite, vermiculite, hydrobiotite, kaolinite, or another structure. In addition, low-temperature micas may contain other types of layers as growth defects (Amouric et al., 1988). Other sheet silicates are prone to their own variations on the theme of polysomatic disorder. In addition to rampant polysomatic disorder, stacking disorder is extremely common in sheet silicates. I do not believe I have ever bumped into a mica or chlorite crystal while using the electron microscope that did not have at least some stacking faults or twins. The modulated sheet silicates display an additional complication—defects involving the periodicities within individual sheets (i.e., structural variation in the **a-b** plane). Because of all these common complications, I like to say that the sheet silicates as a group suffer from pathological disorder.

6.3.7.4 Olivine and the humite group

Because of the importance for the rheology of the mantle, there have been numerous studies of dislocations in olivine. One would expect, therefore, that if olivine commonly possesses a high density of planar defects, this would have been noticed. In addition, direct attempts to observe planar faults in olivine have shown that at least some olivine specimens do not contain them, or possess planar defects at extremely low densities (White and Hyde, 1982a). Nonetheless, as discussed above, some olivine samples contain a high density of planar defects having the structure of a humite-group mineral (probably the result of alteration), whereas others contain large amounts of intergrown laihunite (probably the result of intermediate-temperature oxidation reactions).

Thanks to the studies of White and Hyde (1982a,b; 1983), we know that norbergite and chondrodite tend to be free of, or have a very low density of, polysomatic defects. Clinohumite and humite, on the other hand, can range from ordered to highly defective, with the defects corresponding to either extra or missing O slabs.

6.3.7.5 Other polysomatic structures

As for all the other polysomatic series in the mineral kingdom, we can make a statement similar to those for the above groups: in some occurrences, polysomatic disorder does exist, at least to a limited extent, but the degree of disorder tends to vary from occurrence to occurrence and/or even from one area of a crystal to the next. I know of no mineral belonging to a polysomatic series that has been studied thoroughly with HRTEM for which no polysomatic defects have been found.

6.3.7.6 Rock-forming minerals not belonging to polysomatic series

The silica minerals and feldspars are examples of important rock-forming mineral groups for which no polysomatic relationships have been noted, and, indeed, no defects of the type we have been describing have been observed in these groups. This is not to say that these minerals do not have defects that can be examined with HRTEM. Low-temperature quartz commonly contains abundant planar defects and very fine-scale intergrowths with moganite (Heaney and Post, 1992), and tridymite and cristobalite typically contain high densities of stacking defects (stacking faults and twins). Defects are rampant in the feldspars, too, and TEM studies have made major contributions to feldspar crystal chemistry. They do not, however, contain polysomatic defects, and hence they are beyond the scope of this paper.

6.3.7.7 So why do we need HRTEM?

Because they are capable of introducing new types of crystallographic sites into a mineral, or at least altering the ratios of different types of sites, polysomatic defects can affect the way trace or major elements are partitioned in a mineral with respect to other, coexisting phases (Buseck and Veblen, 1978; Veblen and Buseck, 1981). Therefore, polysomatic disorder can affect trace element composition, geothermobarometry, as well as mechanical properties and other physical and transport properties of a mineral. From the above review, it should be clear that minerals such as pyroxenes, amphiboles, micas, and olivine can occur devoid of such defects, but they are not free of them all of the time. Those using these minerals in geochemical investigations would therefore be wise to use electron microscopy to verify that their minerals are wholesome, rather than being shot through with defects. The same may well be true for those utilizing Ar and other isotopic dating methods: little is directly known about the effects of these and other defects on Ar retention.

The past 20 years have seen numerous HRTEM investigations on a wide variety of minerals, rock-forming and otherwise. Surely mineralogically oriented studies will continue, but it is likely that HRTEM more and more will find its way into the fields of trace element geochemistry, isotope geochemistry, and petrology. Surely there is room for more studies that include large suites of specimens, such as those of Dorling and Zussman (1987) on calcic amphiboles and White and Hyde (1982a,b) on the humite group. Perhaps it is even more important to pursue broad-based studies of petrologically coherent suites of samples, so that the defect types, densities, and distributions can begin to be placed in a field context and rigorously related to conditions of petrogenesis, alteration, and weathering. Studies of experimentally produced polysomes are also needed for the same reason. It is surprising how little we still understand about the precise formation mechanisms and conditions at which polysomatic defects form.

6.4 POLYSOMATIC REACTIONS

HRTEM, especially in combination with one of the electron diffraction methods, is an ideal tool for determining the geometrical characteristics of reactions in which a solid is replaced by one or more crystalline solids. The diffraction experiment can be used to determine the orientation relationships between the reactant and product minerals. A high-resolution imaging experiment can then be used to determine whether the reaction involves narrow polysomatic lamellae replacing the host material, as shown schematically in Figure 6-4 for a reaction in which a single polysomatic slab is replaced, or whether the reaction proceeds by bulk replacement along a broad, two-dimensional reaction front (which can be described alternatively as a grain boundary between reactant and product). These two geometries define what we can call lamellar reactions and bulk reactions. Both types of reaction geometry are known to occur in pyriboles (Veblen and Buseck, 1981), in pyroxenoids (Veblen, 1985b), and in sheet silicates (Veblen and Ferry, 1983; Yau et al., 1984).

It may be less obvious that HRTEM images commonly can be used to deduce the volume change for a reaction and thus allow the precise solid-state reaction to be written, either in terms of idealized end-member components, or, if the reactant and product are analyzed chemically, in terms of the actual mineral compositions. Knowing the portion of the reaction involving the solids further allows one to determine the mass balances between the minerals and their surrounding medium, for example a metamorphic fluid. Examples of this application of HRTEM are provided below.

6.4.1 Reactions with little or no volume change: biopyriboles

There are examples of reactions in which one pyribole replaces another via a mechanism involving considerable volume change, i.e., that can be described as involving a dislocation having a component of the Burgers vector parallel to the **b**-axis (Veblen and Buseck, 1981). However, most lamellar reactions of this type do not involve appreciable volume change, and the terminations of the advancing lamellae are coherent (i.e., they do not involve dislocations or appreciable strain). Even when they do involve a dislocation at the termination, the Burgers vector is commonly normal to **b** (Skrotzki et al., 1991; Veblen and Buseck, 1981). For example, Skrotzki et al. (1991) give the Burgers vector for dislocations at terminations of 0.9-nm {010} amphibole lamellae replacing pyroxene as 1/2[101].

We can use the simple example of a sextuple-chain slab replacing a 2.7-nm slab of the amphibole anthophyllite (Fig. 6-9) to see how we can deduce the stoichiometric change involved in a reaction (Veblen and Buseck, 1980). In this case, the slabs MPMPMP of amphibole are replaced by the slabs MMMMMP of sextuple-chain silicate (the actual sequences have one-half of a P slab at either end, with MPMPM and MMMMM respectively lying between the two half P slabs). The part of the reaction involving solids can then be written

$$MPMPMP \rightarrow MMMMMP$$

or
$$3M + 3P \rightarrow 5M + P.$$

We know that the M slabs, like talc, possess the stoichiometry $A_3T_4O_{10}(OH)_2$, where A are the octahedrally coordinated cations Mg and Fe (here not designated by the usual M, to avoid confusion with the mica slabs). T are the tetrahedral cations, in this case primarily Si. The P slabs have stoichiometry $A_4T_4O_{12}$ (i.e., twice the usual six-oxygen pyroxene formula). The reaction is then

$$3\ A_3T_4O_{10}(OH)_2\ +\ 3\ A_4T_4O_{12}\ \rightarrow\ 5\ A_3T_4O_{10}(OH)_2\ +\ A_4T_4O_{12}$$

or
$$A_{21}T_{24}O_{66}(OH)_6 \rightarrow A_{19}T_{24}O_{62}(OH)_{10}.$$

Obviously, this is not a balanced reaction, because it takes only the solid phases into account. In order to balance it, we also must include components that are taken from or donated to the altering fluid. We may not know the exact speciation of the components in the fluid, but one way of balancing this reaction is

$$A_{21}T_{24}O_{66}(OH)_6\ +\ 4\ H^+\ \rightarrow\ A_{19}T_{24}O_{62}(OH)_{10}\ +\ 2\ A^{2+}.$$

This treatment tells us that there is no need for tetrahedrally coordinated cations (Si) to be transported during the reaction, but hydrogen (in some form) must diffuse into the structure to the reaction site, and octahedrally coordinated cations must be transported from the reaction site in the crystal to the fluid.

The above treatment of the reaction of amphibole to form sextuple-chain silicate takes into account the stoichiometry of the reactants and products. However, it tells us nothing about the detailed behavior of Mg and Fe, for example. For a more detailed view of this reaction, it would be necessary to know not only the stoichiometries of amphibole and sextuple-chain silicate, but also their chemical compositions, specifically the ratios of the octahedral cations A and tetrahedral cations T, as discussed in the next section.

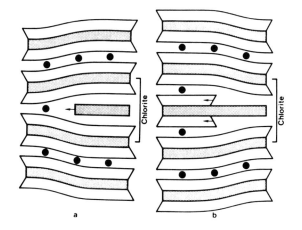

Figure 6-31. Two simple one-layer mechanisms by which biotite can be transformed to chlorite. Mechanism (a) involves a volume increase and requires the introduction of substantial numbers of octahedral cations. Mechanism (b) results in a volume decrease and the release of large numbers of tetrahedral cations. Both mechanisms result in release of interlayer cations to the altering fluid. (Veblen and Ferry, 1983.)

6.4.2 Reactions involving a change in volume: biotite-chlorite

As noted in §6.3.7.3 on sheet silicates, there have been numerous studies of 2:1 sheet silicates intergrown with chlorite on the unit-cell scale. Since such intergrowths commonly form during reactions (Chayes, 1955), several authors have used HRTEM observations to study the detailed reaction mechanisms and chemistry for biotite → chlorite (Eggleton and Banfield, 1985; Ferrow et al., 1990; Olives Baños and Amouric, 1984; Olives Baños et al., 1983; Ristich and Huyck, 1989; Veblen and Ferry, 1983; Yau et al., 1984), or for chlorite → biotite (Maresch et al., 1985). As pointed out by Veblen and Ferry (1983), there are two simple reaction mechanisms by which chlorite can replace biotite one layer at a time, as shown in Figure 6-31. (If run in reverse, these mechanisms represent biotite replacing chlorite.) Mechanism 1 (Fig. 6-31a), called a 1:1 reaction by Yau et al. (1984), involves brucitization of a mica interlayer and can be written 1Bi → 1 Chl. Mechanism 2 (Fig. 6-31b), called a 2:1 reaction by Yau et al. (1984), involves dissolution of the tetrahedral sheets of a biotite layer, leaving a remnant brucitic layer, and can be written 2Bi → 1Chl. As pointed out by Yau et al. (1984), there are other possible mechanisms, such as the approximately volumetric reaction 14Bi → 10Chl, but the two simple one-layer mechanisms seem to operate in many cases of biotite-chlorite replacement.

Simply by inspection, it is clear that mechanism 1 involves a volume decrease, releases K to the altering fluid, and requires introduction of octahedrally coordinated cations. Mechanism 2, however, involves an increase in volume, release of K, but also release of substantial amounts of tetrahedrally coordinated cations. These qualitative relationships can be quantified if the molar volumes are known (e.g., if the unit-cell parameters are measured) and if the biotite and chlorite compositions are analyzed. For example, in the case of Veblen and Ferry (1983), it was calculated that mechanism 1 results in a volume increase of 39.8%, whereas mechanism 2 yields a reduction in volume of 30.1%. Chemical reactions written using analyzed biotite and chlorite compositions were as follows:

<u>Mechanism 1</u>

$K_{0.95}Na_{0.01} \square_{0.23}Mg_{0.59}Fe_{1.65}Mn_{0.04}Ti_{0.18}Al_{1.63}Si_{2.72}O_{10}(OH)_2$ (Biotite)

$+ 6.04 H_2O + 0.59 Mg + 1.46 Fe + 0.05 Mn + 1.17 Al \rightarrow$

$\square_{0.11}Mg_{1.18}Fe_{3.11}Mn_{0.09}Al_{2.80}Si_{2.71}O_{10}(OH)_8$ (Chlorite)

$+ 0.01\ H_4SiO_4 + 0.18\ Ti + 6.03\ H^+ + 0.12\ \square + 0.95\ K + 0.01\ Na$

Mechanism 2

$K_{0.95}Na_{0.01}\ \square_{0.23}Mg_{0.59}Fe_{1.65}Mn_{0.04}Ti_{0.18}Al_{1.63}Si_{2.72}O_{10}(OH)_2$ (Biotite)

$+ 2.55\ H^+ + 2.45\ H_2O \rightarrow$

$1/2\ \square_{0.11}Mg_{1.18}Fe_{3.11}Mn_{0.09}Al_{2.80}Si_{2.71}O_{10}(OH)_8$ (Chlorite)

$+ 1.36\ H_4SiO_4 + 0.18\ Ti + 0.18\ \square + 0.09\ Fe + 0.23\ Al + 0.95\ K + 0.01\ Na$

These two reactions illustrate the point that different structural reaction mechanisms imply very different consequences for the composition of the altering fluid. As suspected from our simple inspection of Figure 6-31, both processes release substantial K, mechanism 1 requires substantial introductions of octahedral cations (Mg, Fe, Al), whereas mechanism 2 involves the release of substantial Si. Furthermore, reaction via mechanism 1 releases substantial H^+, whereas mechanism 2 requires the addition of substantial H^+, suggesting that the operating reaction mechanism may be controlled in part by fluid pH. In fact, whereas Veblen and Ferry (1983) observed terminations of biotite layers characteristic of mechanism 2, other authors have observed mechanism 1 in some cases (e.g., Olives Baños and Amouric, 1984).

The above treatment of reactions using actual compositions demonstrates an important point: the specific structural mechanism of a polysomatic reaction has major implications for the actual chemistry of the reaction, as well as for the volume change associated with the reaction. Thus, when a petrologist writes a supposed reaction, he or she may well be inadvertently making important assumptions about the structural mechanism of the reaction. Conversely, if the mechanism can be determined from HRTEM observations, then the petrologist should be able to write out the specific reaction based on observation, rather then whim, and thereby constrain the mass balance between solids and fluid.

6.4.3 Alternative descriptions and AEM of polysomatic reactions: pyroxenoids

Although we tacitly described the biotite-chlorite reactions in terms of the polysomatic model for their structures, we could have used an alternative description by noting that mechanism 1, for example, can be described in terms of an edge dislocation, in that the termination of the advancing brucitic layer is a linear feature with displacement normal to the line defect. Additionally, we could have noted the magnitude of the displacement with respect to the unit cell of the biotite and described the reaction in terms of movement through the structure of partial dislocations having Burgers vectors with edge components of approximately 0.42[001]. (Since the introduced hydroxide sheet presumably introduces a polytypic shift, there may also be components parallel to the sheets, but these are not observed in the usual one-dimensional HRTEM experiment.) Normally, partial dislocations mark the terminations of stacking faults, but the line defects in this description occur instead at the terminations of polysomatic faults.

Because the pyroxenoids with chain periodicities $p \geq 5$ can be described accurately as mixtures of clinopyroxenelike P slabs and wollastonitelike W slabs (Angel and Burnham, 1991; Veblen, 1991), transformation reactions in which pyroxenes are replaced by pyroxenoid, or one pyroxenoid is replaced by another, can be described adequately by noting the initial and final sequence of polysomatic slabs. Such a description tacitly implies certain structural displacements that are inherent in P or W slabs. Alternatively, the reactions can be described in terms of the structural displacements alone, as done by Angel et al. (1984) and Angel (1986). Several authors have noted that different pyroxenoids intergrow on (001), whereas pyroxenoids intergrown with clinopyroxene are parallel to $\{11\bar{1}\}$ of the pyroxene (Czank and Simons, 1983; Ried, 1984; Veblen, 1985b). Based both on theoretical grounds and TEM observations of synthetic intergrowths, Angel and coworkers noted that the pyroxene-pyroxenoid reactions can proceed by the motion of line defects along the pyroxene $\{111\}$ planes, and they discussed in detail the displacement vectors of these defects. They also noted that these line defects are not strictly partial dislocations, because their passage through the reactant structure involves rather complex displacements of atoms in the moving defect core.

There are thus two rather different ways to describe pyroxene-pyroxenoid reactions: a polysomatic description, in which the changing sequence of polysomatic slabs is emphasized, and a dislocation description, in which the resulting structural displacements are emphasized. Both descriptions are valid. The polysomatic description may be the easiest for the nonspecialist to comprehend and describes explicitly the initial and final structures. The line defect description, however, makes it clear that the formation of polysomatic faults involves considerable strain at the slab terminations in these structures, and it has the virtue of explicitly stating the orientations and magnitudes of the structural displacements. This dichotomy is also found in the description of humite-type faults in olivine. Whereas these defects can be described completely with the polysomatic model for these structures, they are also aptly described in terms of the displacements on the line defects at the terminations of the polysomatic slabs. Thus, both Kitamura et al. (1987) and Drury (1991) determined the line-defect displacements and used these to show that the planar faults they bounded were consistent with clinohumite structure intergrown in olivine. The polysomatic description is most likely to be used by microscopists who specialize in phase-contrast (HRTEM) experiments, whereas the dislocation description is more likely to be used by microscopists doing amplitude-contrast imaging. The most complete description of these defect structures and reactions would state both the rearrangement of polysomatic slabs and the specific displacements on the line defects found at the advancing tips of the polysomatic lamellae.

Polysomatism clearly provides a powerful structural description of replacement reaction mechanisms, and it additionally provides a simple method for assessing the changes in stoichiometry associated with these reactions (see §6.4.1, above). Although much can be learned about the specific chemical compositions of reactants and reaction products from electron microprobe and other bulk analytical methods, a thorough knowledge of changes in chemical composition during solid-state reactions requires a more detailed determination of the relationships between chemistry and structure, including the structure of intermediate, disordered regions of the sample. With present technology, it appears that the only way of observing these relationships is with a combination of HRTEM and AEM performed on the same specimen areas, as utilized by Livi and Veblen (1992). As shown schematically in Figure 6-32, analyses and HRTEM images are obtained of the same regions, including the unreacted mineral, final reaction products, and intermediate, disordered materials. Chemical variables then can be correlated with reaction progress variables, such as ratios of polysomatic slabs.

In the case described by Livi and Veblen (1992), the reaction involved replacement of the CaMn clinopyroxene, johannsenite, by the Mn-rich pyroxenoids, rhodonite and pyroxmangite, and the reaction progress was monitored by using the polysomatic slab ratio P/(P + W) (Fig. 6-33). These results showed that the chemical composition is not a simple function of the structure. In the initial stage of reaction, fine-scale lamellae of pyroxene and pyroxenoid form, and there is a linear relationship between structure and composition (see region between P and PPW in Fig. 6-33). This suggests that during this

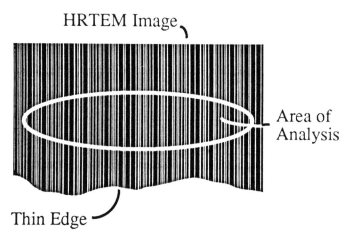

Figure 6-32. A schematic diagram showing how polysomatic variations can be correlated to chemistry. In this example for pyroxenoids, the numbers of P and W slabs were determined from HRTEM images, and the corresponding chemical compositions were determined by AEM analyses obtained from the same regions as the images. (Livi and Veblen, 1992.)

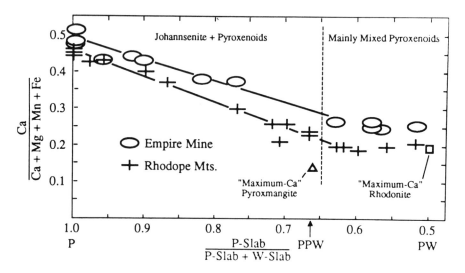

Figure 6-33. The relationship between structure, in terms of the ratio of P and W slabs, and Ca content for pyroxenoids from two localities. In both of these occurrences, johannsenite (Ca-Mn pyroxene) was caught in the act of reacting to disordered mixtures of pyroxene/pyroxenoid and pyroxenoid/pyroxenoid, as well as ordered rhodonite and pyroxmangite (see text). (Livi and Veblen, 1992.)

reaction, Ca and Mn are partitioned in johannsenite and pyroxenoid slabs in much the same way as in the bulk structures. However, the products of this stage in the reaction are primarily rhodonite, pyroxmangite, and disordered mixtures of pyroxenoid with p = 5, 7, and other periodicities (Veblen, 1985b). Among these varied products, it appears that composition is constant and not correlated with the ratio between P and W slabs. Indeed, it appears that the Ca contents of these pyroxenoids are considerably higher than those previously reported for rhodonite and pyroxmangite. In one of the localities studied, these high-Ca, presumably metastable pyroxenoids subsequently reacted to coarser-grained rhodonite with more normal Ca contents. The relationships between structural change and chemical change during these pyroxene-pyroxenoid reactions are thus quite complicated (Livi and Veblen, 1992). It is to be hoped that more studies of polysomatic reactions will involve direct observation of both structure and composition in the future.

ACKNOWLEDGMENTS

I thank Donald R. Peacor for a very constructive review of this paper. I also thank the many authors who permitted me to reproduce figures from their HRTEM studies. I am particularly indebted to Bernard Grobety for providing detailed information from his Ph.D. dissertation, prior to publication, and photographs of the long-period ordered polysomes he observed in Alpine ultramafic rocks (Fig. 6-15 and Table 2). Work described in the present paper was supported in part by NSF grant EAR-8903630 and DOE grant DE-FG02-89ER14074.

REFERENCES

Ahn, J.H. and Buseck, P.R. (1989) Microstructures and tetrahedral strip-width order and disorder in Fe-rich minnesotaites. Am. Mineral. 74, 384-393.

Ahn, J.H. and Buseck, P.R. (1991) Microstructures and fiber formation mechanisms of crocidolite asbestos. Am. Mineral. 76, 1467-1478.

Ahn, J.H., Cho, M., Jenkins, D.M. and Buseck, P.R. (1991) Structural defects in synthetic tremolitic amphiboles. Am. Mineral. 76, 1811-1923.

Ahn, H.H. and Peacor, D.R. (1985) Transmission electron microscopic study of diagenetic chlorite in Gulf Coast argillaceous sediments. Clays Clay Minerals 33, 228-236.

Ahn, J.H. and Peacor, D.R. (1987) Kaolinitization of biotite: TEM data and implications for an alteration mechanism. Am. Mineral. 72, 353-356.

Ahn, J.H., Peacor, D.R. and Coombs, D.S. (1988) Formation mechanisms of illite, chlorite and mixed-layer illite-chlorite in Triassic volcanogenic sediments from the Southland Syncline, New Zealand. Contrib. Mineral. Petrol. 99, 82-89.

Ahn, J.H., Peacor, D.R. and Essene, E.J. (1985) Coexisting paragonite-phengite in blueschist eclogite: A TEM study. Am. Mineral. 70, 1193-1204.

Akai, J. (1982) Polymerization process of biopyribole in metasomatism at the Akatani ore deposit, Japan. Contrib. Mineral. Petrol. 80, 117-131.

Akai, J., Tomita, K. and Yamaguchi, Y. (1985) High resolution electron microscopic observation of pyroxene lamellae in oxyhornblende. Earth Sci. 38, 334-345.

Alario Franco, M., Hutchison, J.L., Jefferson, D.A. and Thomas, J.M. (1977) Structural imperfection and morphology of crocidolite (blue asbestos). Nature 266, 520-521.

Alario Franco, M., Jefferson, D.A., Pugh, N.J. and Thomas, J.M. (1980) Lattice imaging of structural defects in a chain silicate: the pyroxenoid mineral rhodonite. Mat. Res. Bull. 15, 73-79.

Amouric, M., Gianetto, I. and Proust, D. (1988) 7, 10 and 14 Å mixed-layer phyllosilicates studied structurally by TEM in pelitic rocks of the Piemontese zone (Venezuela). Bull. Minéral. 111, 29-37.

Amouric, M., Mercuriot, G. and Baronnet, A. (1981) On computed and observed HRTEM images of perfect mica polytypes. Bull. Minéral. 104, 298-313.

Angel, R.J. (1986) Transformation mechanisms between single-chain silicates. Am. Mineral. 71, 1441-1454.

Angel, R.J. and Burnham, C.W. (1991) Pyroxene-pyroxenoid polysomatism revisited: A clarification. Am. Mineral. 76, 900-903.

Angel, R.J., Price, G.D. and Putnis, A. (1984) A mechanism for pyroxene-pyroxenoid and pyroxenoid-pyroxenoid transformations. Phys. Chem. Minerals 10, 236-243.

Banfield, J.F. (1990) HRTEM Studies of Subsolidus Alteration, Weathering, and Subsequent Diagenetic and Low-Grade Metamorphic Reactions. Ph.D. dissertation, The Johns Hopkins Univ., Baltimore, Maryland.

Banfield, J.F. and Eggleton, R.A. (1988) Transmission electron microscope study of biotite weathering. Clays Clay Minerals 36, 47-60.

Banfield, J.F., Karabinos, P. and Veblen, D.R. (1989) Transmission electron microscopy of chloritoid: Intergrowth with sheet silicates and reactions in metapelites. Am. Mineral. 74, 549-564.

Banfield, J.F., Veblen, D.R. and Jones, B.F. (1990) Transmission electron microscopy of subsolidus oxidation and weathering of olivine. Contrib. Mineral. Petrol. 106, 110-123.

Barrer, R.M. (1978) Zeolites and clay minerals as sorbents and molecular sieves. Academic Press, London.

Bonaccorsi, E., Merlino, S. and Passero, M. (1990) Rhönite: Structural and microstructural features, crystal chemistry and polysomatic relationships. Eur. J. Mineral. 2, 203-218.

Bovin, J.O., O'Keeffe, M. and O'Keefe, M.A. (1981) Electron microscopy of oxyborates. I. Defect structures in the minerals pinakiolite, ludwigite, orthopinakiolite, and takéuchiite. Acta Crystallogr. A37, 28-35.

Bursill, L.A. (1985) The interpretation of HREM images of crystals. Ultramicroscopy, 18, 1-10.

Buseck, P.R. and Cowley, J.M. (1983) Modulated and intergrowth structures in minerals and electron microscope methods for their study. Am. Mineral. 68, 18-40.

Buseck, P.R. and Iijima, S. (1974) High resolution electron microscopy of silicates. Am. Mineral. 59, 1-21.

Buseck, P.R., Nord, G.L., Jr. and Veblen, D.R. (1980) Subsolidus phenomena in pyroxenes. In C.T. Prewitt, ed. Pyroxenes, Rev. Mineral. 7, 117-211.

Buseck, P.R. and Veblen, D.R. (1978) Trace elements, crystal defects, and high resolution electron microscopy. Geochim. Cosmochim. Acta, 42, 669-678.

Buseck, P.R. and Veblen, D.R. (1988) Mineralogy. In P.R. Buseck, J.M. Cowley and L. Eyring, eds., High-Resolution Transmission Electron Microscopy, p. 308-377. Oxford University Press, Oxford.

Carpenter, M.A. (1978) Nucleation of augite at antiphase boundaries in pigeonite. Phys. Chem. Minerals 2, 237-251.

Catlow, C.R.A., Thomas, J.M., Parker, S.C. and Jefferson, D.A. (1982) Simulating silicate structures and the structural chemistry of pyroxenoids. Nature 295, 658-662.

Chayes, F. (1955) Potash feldspar as a by-product of the biotite-chlorite transformation. J. Geol. 63, 75-82.

Chisholm, J.E. (1983) Transmission electron microscopy of asbestos. In S.S. Chissick and R. Derricott, eds., Asbestos, p. 85-167. Wiley, New York.

Clapp, P.C. and Moss, S.C. (1968) Correlation functions of disordered binary alloys, II. Phys. Rev. 171, 754-763.

Cowley, J.M. (1950) X-ray measurement of order in single crystals of Cu_3Au. J. Appl. Phys. 21, 24-30.

Cowley, J.M. (1973) High-resolution dark-field electron microscopy. Acta Crystallogr. A29, 437-540.

Cressey, B.A., Whittaker, E.J.W. and Hutchison, J.L. (1982) Morphology and alteration of asbestiform grunerite and anthophyllite. Mineral. Mag. 46, 77-87.

Czank, M. (1981) Chain periodicity faults in babingtonite, $Ca_2Fe^{2+}Fe^{3+}H[Si_5O_{15}]$. Acta Crystallogr. A37, 617-620.

Czank, M. (1984) Structural properties of pyroxenoids. Electr. Micros. 1984, 1010-1012.

Czank, M. and Liebau, F. (1979) Periodizitätsfehler in Pyroxenoiden—eine neue Art von Baufehlern (elektronenmikroskopische Untersuchungen). Fort. Mineral. 57, 23-24.

Czank, M. and Liebau, F. (1980) Periodicity faults in chain silicates: A new type of planar lattice fault observed with high resolution electron microscopy. Phys. Chem. Minerals 6, 85-93.

Czank, M. and Liebau, F. (1983) Chain periodicity faults in pyroxferroite from lunar basalt 12021. Lunar Planet. Sci. XIV, p. 144-145. Lunar and Planetary Sci. Inst., Houston, Texas.

Czank, M. and Simons, B. (1983) High resolution electron microscopic studies on ferrosilite III. Phys. Chem. Minerals 9, 229-234.

de Villiers, J.P. and Buseck, P.R. (1989) Stacking variations and nonstoichiometry in the bixbyite-braunite polysomatic mineral group. Am. Mineral. 74, 1325-1336.

Dódony, I. and Pósfai, M. (1990) 19901Pyrrhotite superstructures. Part II: A TEM study of 4C and 5C structures. Eur. J. Mineral. 2, 529-535.

Dorling, M. and Zussman, J. (1985) An investigation of nephrite jade by electron microscopy. Mineral. Mag. 49, 31-36.

Dorling, M. and Zussman, J. (1987) Characteristics of asbestiform and non-asbestiform calcic amphiboles. Lithos, 20, 469-489.

Drits, V.A., Goncharov, V.A. and Khadzhi, I.P. (1976) Formation conditions and physicochemical constitution of triple chain silicate with (Si_6O_{16}) radical (in Russian) Izvestiya Akad. Nauk., Ser. Geol. 7, 32-41.

Drury, M.R. (1991) Hydration-induced climb dissociation of dislocations in naturally deformed mantle olivine. Phys. Chem. Minerals 18, 106-116.

Eggleton, R.A. and Banfield, J.A. (1985) The alteration of granitic biotite to chlorite. Am. Mineral. 70, 902-910.

Ferraris, G., Mellini, M. and Merlino, S. (1986) Polysomatism and the classification of minerals. Rend. Soc. Italiana Mineral. Petrol. 41, 181-192.

Ferrow, E. and Roots, M. (1990) A preparation technique for TEM specimens; application to synthetic chlorite. Eur. J. Mineral. 1, 815-819.

Ferrow, E.A., London, D., Goodman, K.S. and Veblen, D.R. (1990) Sheet silicates of the Lawler Peak granite, Arizona: chemistry, structural variations, and exsolution. Contrib. Mineral. Petrol. 105, 491-501.

Gleitzer, C. and Goodenough, J.B. (1985) Mixed-valence iron oxides. Struc. Bonding, 61, 1-76.

Greenwood, N.N. (1970) Ionic Crystals, Lattice Defects and Nonstoichiometry. Chemical Publishing Co., New York.

Gribelyuk, M.A. and Zakharov, N.D. (1987) Possibility of observation of point defects in high-resolution electron microscopy. Sov. Phys. Crystallogr. 32, 349-354.

Gribelyuk, M.A., Zakharov, N.D., Hillebrand, R. and Makarova, T.A. (1988) Microstructure of chain silicates investigated by HRTEM. Acta Crystallogr. B44, 135-142.

Griffin, W.L., Mellini, M., Oberti, R. and Rossi, G. (1985) Evolution of coronas in Norwegian anorthosites: re-evaluation based on crystal-chemistry and microstructures. Contrib. Mineral. Petrol. 91, 330-339.

Grobety, B. (1992) Electron Microscopy on Mineral Intergrowths in Metamorphic Rocks Ph.D. dissertation, ETH Zürich, Switzerland.

Guggenheim, S., Bailey, S.W., Eggleton, R.A. and Wilkes, P. (1982) Structural aspects of greenalite and related minerals. Canadian Mineral. 20, 1-18.

Guggenheim, S. and Eggleton, R.A. (1986) Structural modulations in iron-rich and magnesium-rich minnesotaite. Canadian Mineral. 24, 479-497.

Guggenheim, S. and Eggleton, R.A. (1987) Modulated 2:1 layer silicates: Review, systematics, and predictions. Am. Mineral. 72, 724-738.

Guggenheim, S. and Eggleton, R.A. (1988) Crystal chemistry, classification, and identification of modulated layer silicates. In S. W. Bailey, ed. Hydrous Phyllosilicates, Rev. Mineral. 19, 675-725.

Guthrie, G.D., Jr. and Veblen, D.R. (1989) High-resolution transmission electron microscopy of mixed-layer illite/smectite: Computer simulations. Clays Clay Minerals 37, 1-11.

Guthrie, G.D., Jr. and Veblen, D.R. (1990a) High-resolution transmission electron microscopy applied to clay minerals. In L.M. Coyne, S.W.S. McKeever and D.F. Blake, eds., Spectroscopic Characterization of Minerals and Their Surfaces, p. 75-93. Am.Chemical Soc., Washington, D.C.

Guthrie, G.D., Jr. and Veblen, D.R. (1990b) Interpreting one-dimensional high-resolution transmission electron micrographs of sheet silicates by computer simulation. Am. Mineral. 75, 276-288.

Guthrie, G.D., Jr., Veblen, D.R. and Navon, O. (1991) Submicrometer fluid inclusions in turbid-diamond coats. Earth Planet. Sci. Let. 105, 1-12.

Guthrie, G.D., Jr. and Veblen, D.R. (1992) Turbid alkali feldspars from the Isle of Skye, Northwest Scotland. Contrib. Mineral. Petrol. 108, 298-304.

Hannay, N.B., ed. (1973) The chemical structure of solids. Plenum Press, New York, 540 p.

Hazen, R.M. and Jeanloz, R. (1984) Wüstite $(Fe_{1-x}O)$: A review of its defect structure and physical properties. Rev. Geophys. Space Phys. 22, 37-46.

Heaney, P.J. and Post, J.E. (1992) The widespread distribution of a novel silica polymorph in microcrystalline quartz varieties. Science 255, 441-443.

Howie, A. (1988) Highly disordered materials. In P.R. Buseck, J.M. Cowley and L. Eyring, eds., High-Resolution Transmission Electron Microscopy, p. 607-632. Oxford University Press, Oxford.

Hu, M., Wenk, H.-R. and Sinitsyna, D. (1992) Microstructures in natural perovskites. Am. Mineral. 77, 359-373.

Iijima, S. and Buseck, P.R. (1978) Experimental study of disordered mica structures by high-resolution electron microscopy. Acta Crystallogr. A34, 709-719.

Iijima, S., Cowley, J.M. and Donnay, G. (1973) High resolution electron microscopy of tourmaline crystals. Tschermaks Mineral. Petrogr. Mitt. 20, 216-224.

Iijima, S. and Zhu, J. (1982) Electron microscopy of a muscovite–biotite interface. Am. Mineral. 67, 1195-1205.

Iishi, K., Okamoto, K. and Kadomi, M. (1989) Formation of laihunite from Fe-(Mg,Co,Mn,Ca) olivines. N. Jahrb. Mineral. Mh. 1989, 345-356.

Ilton, E.S. and Veblen, D.R. (1988) Copper inclusions in sheet silicates from porphyry Cu deposits. Nature 334, 516-518.

Isaacson, M., Ohtsuki, M. and Utlaut, M. (1979) Electron microscopy of individual atoms. In J.J. Hren, J.I. Goldstein, and D.C. Joy, eds., Introduction to Analytical Electron Microscopy, p. 343-368. Plenum Press, New York.

Jefferson, D.A. and Pugh, N.J. (1981) The ultrastructure of pyroxenoid chain silicates. III. Intersecting defects in a synthetic iron-manganese pyroxenoid. Acta Crystallogr. A37, 281-286.

Jefferson, D.A., Pugh, N.J., Alario-Franco, M., Mallinson, L.G., Millward, G.R. and Thomas, J.M. (1980) The ultrastructure of pyroxenoid chain silicates. I. Variation of the chain configuration in rhodonite. Acta Crystallogr. A36, 1058-1065.

Jenkins, D.M. and Clare, A.K. (1990) Comparison of the high-temperature and high-pressure stability limits of synthetic and natural tremolite. Am. Mineral. 75, 358-366.

Johannsen, A. (1911) Petrographic terms for field use. J. Geol. 19, 317-322.

Kitamura, M., Kondoh, S., Morimoto, N., Miller, G.H., Rossman, G.R. and Putnis, A. (1987) Planar OH-bearing defects in mantle olivine. Nature 328, 143-145.

Kitamura, M., Shen, B., Banno, S. and Morimoto, N. (1984) Fine textures of laihunite, a nonstoichiometric distorted olivine-type mineral. Am. Mineral. 69, 154-160.

Kondoh, S., Kitamura, M. and Morimoto, N. (1985) Synthetic laihunite ($\square_x Fe^{2+}_{2-3x} Fe^{3+}_{2x} SiO_4$), an oxidation product of olivine. Am. Mineral. 70, 737-746.

Konishi, H. and Akai, J. (1991) Depolymerized pyribole structures derived from talc by heating. Phys. Chem. Minerals. 17, 569-582.

Koto, K., Morimoto, N. and Narita, H. (1976) Crystallographic relationships of the pyroxenes and pyroxenoids. J. Japan Assoc. Mineral., Petrol., Econ. Geol. 71, 248-254.

Kröger, F.A. (1974) The Chemistry of Imperfect Crystals. North-Holland, Amsterdam.

Kudoh, Y., Finger, L.W., Hazen, R.M., Prewitt, C.T., Kanzaki, M. and Veblen, D.R. (1992) Phase E: a high-pressure hydrous silicate with unique crystal chemistry. Phys. Chem. Minerals, in press.

Lasaga, A.C. (1980) The atomistic basis of kinetics: Defects in minerals. In A.C. Lasaga and R.J. Kirkpatrick, eds., Kinetics of geochemical processes, Rev. Mineral. 8, 261-319.

Le Gleuher, M., Livi, K.J.T., Veblen, D.R., Noack, Y. and Amouric, M. (1990) Serpentinization of enstatite from Pernes, France: Reaction microstructures and the role of system openness. Am. Mineral.. 75, 813-824.

Lee, J.H. and Peacor, D.R. (1985) Ordered 1:1 interstratification of illite and chlorite: a transmission and analytical electron microscopy study. xxx 33, 463-467.

Lee, J.H., Peacor, D.R., Lewis, D.D. and Wintsch, R.P. (1986) Chlorite-illite/muscovite interlayered and interstratified crystals: a TEM/STEM study. Contrib. Mineral. Petrol. 88, 372-385.

Livi, K.J.T. and Veblen, D.R. (1987) "Eastonite" from Easton, Pennsylvania: A mixture of phlogopite and a new form of serpentine. Am. Mineral. 72, 113-125.

Livi, K.J.T. and Veblen, D.R. (1989) Transmission electron microscopy of interfaces and defects in intergrown pyroxenes. Am. Mineral. 74, 1070-1083.

Livi, K.J.T. and Veblen, D.R. (1992) An analytical electron microscopy study of pyroxene-to-pyroxenoid reactions. Am. Mineral. 77, 380-390.

Lo, C.-H. and Onstott, T.C. (1989) [39]Ar recoil artifacts in chloritized biotite. Geochim. Cosmochim. Acta, 53, 2697-2711.

Makovicky, E. (1989) Modular classification of sulphosalts--current status; Definition and application of homologous series. N. Jahrb. Mineral. Abh. 160, 269-297.

Manes, L. and Benedict, U. (1985) Structural and thermodynamic properties of actinide solids and their relation to bonding. Struc. Bonding, 59/60, 75-126.

Maresch, W.V. and Czank, M. (1983a) Phase characterization of synthetic amphiboles on the join $Mn_x^{2+} Mg_{7-x} [Si_8 O_{22}](OH)_2$. Am. Mineral. 68, 744-753.

Maresch, W.V. and Czank, M. (1983b) Problems of compositional and structural uncertainty in synthetic hydroxyl-amphiboles; with an annotated atlas of the realbau. Per. Mineral.--Roma, 52, 463-542.

Maresch, W.V. and Czank, M. (1988) Crystal-chemistry, growth-kinetics and phase-relationships of structurally disordered (Mn^{2+},Mg)-amphiboles. Fortschr. Mineral. 66, 69-121.

Maresch, W.V., Massone, H.-J. and Czank, M. (1985) Ordered and disordered chlorite/biotite inter-stratifications as alteration products of chlorite. N. Jahrb. Mineral. Abh. 152, 79-100.

McCormick, T.C. (1986) Crystal-chemical aspects of nonstoichiometric pyroxenes. Am. Mineral. 71, 1434-1440.

Mellini, M., Ferraris, G. and Compagnone, R. (1985) Carlosturanite: HRTEM evidence of a polysomatic series including serpentine. Am. Mineral. 70, 773–781.

Mellini, M., Oberti, R. and Rossi, G. (1983) Crystal-chemistry and microstructures of pyroxenes and amphiboles in the coronas of the Bergen arcs and of the Sognefjord region, western Norway. Per. Mineral. —Roma 52, 583-615.

Mellini, M., Trommsdorff, V. and Compagnoni, R. (1987) Antigorite polysomatism: behavior during progressive metamorphism. Contrib. Mineral. Petrol. 97, 147-155.

Müller, R.F. and Wenk, H.-R. (1978) Mixed-layer characteristics in real humite structures. Acta Crystallogr. A34, 607-609.

Nakajima, Y. and Ribbe, P.H. (1980) Alteration of pyroxenes from Hokkaido, Japan, to amphibole, clays and other biopyriboles. N. Jahrb. Mineral. Mh. 6, 258-268.

Nakazawa, H., Morimoto, N. and Watanabe, E. (1975) Direct observation of metal vacancies by high-resolution electron microscopy. Part I: $4C$ type pyrrhotite (Fe_7S_8). Am. Mineral. 60, 359-366.

Narita, H., Koto, K. and Morimoto, N. (1977) The crystal structures of $MnSiO_3$ polymorphs (rhodonite- and pyroxmangite-type). Mineral. J. 8, 329-342.

Nissen, H.-U., Wessicken, R., Wooensdregt, C.F. and Pfeifer, H.R. (1980) Disordered intermediates between jimthompsonite and anthophyllite from the Swiss Alps. Inst. Phys. Conf. Ser. 52, 99-100.

O'Keefe, M.A., Buseck, P.R. and Iijima, S. (1978) Computed crystal structure images for high resolution electron microscopy. Nature 274, 322-324.

Olives Baños, J. (1985) Biotites and chlorites as interlayered biotite–chlorite crystals. Bull. Minéral. 108, 635-641.

Olives Baños, J. and Amouric, M. (1984) Biotite chloritization by interlayer brucitization as seen by HRTEM. Am. Mineral. 69, 869-871.

Olives Baños, J., Amouric, M., De Fouquet, C. and Barronet, A. (1983) Interlayering and interlayer slip in biotite as seen by HRTEM. Am. Mineral. 68, 754-758.

Palomino, J.M., Jefferson, D.A., Hutchison, J.L. and Thomas, J.M. (1977) Linear and planar defects in wollastonite. J. Chem. Soc. 1977, 1834-1836.

Papike, J. J. and Cameron, M. (1976) Crystal chemistry of silicate minerals of geophysical interest. Revs. Geophys. Space Phys. 14, 37-80.

Pierce, L. and Buseck, P.R. (1974) Electron imaging of pyrrhotite superstructures. Science 186, 1209-1212.

Pierce. L. and Buseck, P.R. (1978) Superstructuring in the bornite-digenite series: A high-resolution electron microscopy study. Am. Mineral. 63, 1-16.

Reiche, M., Messerschmidt, A. and Bautsch, H.-J. (1978) Pyroxferroit in einem Anorthosit-Fragment der Luna 20-Probe. Z. Geol. Wiss. Berlin, 6, 709-718.

Ried, H. (1984) Intergrowth of pyroxene and pyroxenoid; chain periodicity faults in pyroxene. Phys. Chem. Minerals 10, 230-235.

Ried, H. and Korekawa, M. (1980) Transmission electron microscopy of synthetic and natural fünferketten and siebenerketten pyroxenoids. Phys. Chem. Minerals 5, 351-365.

Ristich, A.M. and Huyck, H.L.O. (1989) Chloritization of biotite associated with porphyry copper systems. Geol. Soc. Am. Abstr. Prog. 21.

Rossman, G.R. (1988) Vibrational spectroscopy of hydrous components. In F.C. Hawthorne, ed., Spectroscopic Methods in Mineralogy and Geology, Rev. Mineral. 18, 193-206.

Schmalzried, H. (1981) Solid State Reactions. Verlag Chemie, Weinheim.

Schreyer, W., Medenbach, O., Abraham, K., Gebert, W. and Müller, W.F. (1982) Kulkeite, a new metamorphic phyllosilicate mineral: Ordered 1:1 chlorite/talc mixed-layer. Contrib. Mineral. Petrol. 80, 103-109.

Schryvers, D., Srikrishna, K., O'Keefe, M. A. and Thomas, G. (1988) An electron microscopy study of the atomic structure of a mullite in a reaction-sintered composite. J. Mater. Res. 3, 1355-1361.

Schumacher, J.C. and Czank, M. (1987) Mineralogy of triple- and double-chain pyriboles from Orijärvi, southwest Finland. Am. Mineral. 72, 345-352.

Shen, B., Tamada, O., Kitamura, M. and Morimoto, N. (1986) Superstructure of laihunite-$3M$ ($\square_{0.40}Fe^{2+}_{0.80}Fe^{3+}_{0.80}SiO_4$). Am. Mineral. 71, 1455-1460.

Simons, B. and Czank, M. (1986) Chain periodicity faults in pyroxenoids. Physics of Minerals and Ore Microscopy—IMA 1982, p. 143-149. Int' Mineralogical Assoc.

Skogby, H., Bell, D.R. and Rossman, G.R. (1990) Hydroxide in pyroxene: variations in the natural environment. Am. Mineral. 75, 764-774.

Skogby, H. and Ferrow, E. (1989) Iron distribution and structural disorder in synthetic calcic amphiboles studied by Mössbauer spectroscopy and HRTEM. Am. Mineral. 74, 360-366.

Skrotzki, W. Müller, W.F. and Weber, K. (1991) Exsolution phenomena in pyroxenes from the Balmuccia Massif, NW-Italy. Eur. J. Mineral. 3, 39-61.

Smith, D.J., Jefferson, D.A. and Mallinson, L.G. (1981) The ultrastructure of pyroxenoid chain silicates. II. Direct structure imaging of the minerals rhodonite and wollastonite. Acta Crystallogr., A37, 273-280.

Smith, J.V., and Brown, W.L. (1988) Feldspar Minerals. Springer-Verlag, Berlin.

Smyth, D.M. (1989) Defect equilibria in perovskite oxides. In A. Navrotsky and D.J. Weidner, eds., Perovskite: A structure of great interest to geophysics and materials science, p. 99-103. American Geophysical Union, Washington, D.C.

Sørenson, O.T., Ed. (1981) Nonstoichiometric Oxides. Academic Press, New York.

228

Spinnler, G.E. (1985) HRTEM study of antigorite, pyroxene–serpentine reactions and chlorite. Ph.D. dissertation, Arizona State University, Tempe, Arizona.

Spinnler, G.E., Veblen, D.R. and Buseck, P.R. (1983) Microstructure and defects of antigorite. Proc. Electron Microscopy Soc. America, 41, 190-191.

Takéuchi, Y. (1978) "Tropochemical twinning": A mechanism of building complex structures. Recent Prog. Nat. Sci. Japan, 3, 153-181.

Takéuchi, Y. and Koto, K. (1977) A systematics of pyroxenoid structures. Mineral. J. 8, 272-285.

Tamada, O., Shen, B. and Morimoto, N. (1983) The crystal structure of laihunite ($\square_{0.40}Fe^{2+}_{0.80}Fe^{3+}_{0.80}SiO_4$). Mineral. J. 11, 382-391.

Tanaka, N. and Cowley, J.M. (1987) Electron-microscope imaging of short-range order in disordered alloys. Acta Crystallogr. A43, 337-346.

Thompson, J.B., Jr. (1978) Biopyriboles and polysomatic series. Am. Mineral. 63, 239-249.

Thompson, J.B., Jr. (1981) An introduction to the mineralogy and petrology of the biopyriboles. In D.R. Veblen, ed., Amphiboles and Other Hydrous Pyriboles--Mineralogy, Rev. Mineral. 9A, 141-188.

Tilley, R.J.D. (1980) Non-stoicheiometric crystals containing planar defects. Surf. Def. Prop. Solids, 8, 121-201.

Tilley, R.J.D. (1987) Defect Crystal Chemistry and Its Applications. Blackie & Son, Glasgow.

Turner, S. and Buseck, P.R. (1979) Manganese oxide tunnel structures and their intergrowths. Science 203, 456-458.

Van Landuyt, J. and Amelinckx, S. (1975) Multiple beam direct lattice imaging of new mixed-layer compounds of the bastnaesite-synchesite series. Am. Mineral. 60, 351-358.

Veblen, D.R. (1980) Anthophyllite asbestos: Microstructures, intergrown sheet silicates, and mechanisms of fiber formation. Am. Mineral. 65, 1075-1086.

Veblen, D.R. (1981) Non-classical pyriboles and polysomatic reactions in biopyriboles. In D.R. Veblen, ed., Amphiboles and Other Hydrous Pyriboles—Mineralogy. Rev. Mineral. 9A, 189-236.

Veblen, D.R. (1983) Microstructures and mixed layering in intergrown wonesite, chlorite, talc, biotite, and kaolinite. Am. Mineral. 68, 566-580.

Veblen, D.R. (1985a) Extended defects and vacancy non-stoichiometry in rock-forming minerals. In R.N. Schock, ed., Point defects in minerals, p. 122-131. American Geophys. Union, Washington, D.C.

Veblen, D.R. (1985b) TEM study of a pyroxene-to-pyroxenoid reaction. Am. Mineral. 70, 885-901.

Veblen, D.R. (1991) Polysomatism and polysomatic series: A review and applications. Am. Mineral. 76, 801-826.

Veblen, D.R. (1992) Structural types of non-stoichiometry in minerals. In A.S. Marfunin, ed., Higher Mineralogy, in press. Springer-Verlag, Berlin.

Veblen, D.R. and Bish, D.L. (1988) TEM and x-ray study of orthopyroxene megacrysts: Microstructures and crystal chemistry. Am. Mineral. 73, 677-691.

Veblen, D.R. and Buseck, P.R. (1979) Chain-width order and disorder in biopyriboles. Am. Mineral. 64, 687-700.

Veblen, D.R. and Buseck, P.R. (1980) Microstructures and reaction mechanisms in biopyriboles. Am. Mineral. 65, 599-623.

Veblen, D.R. and Buseck, P.R. (1981) Hydrous pyriboles and sheet silicates in pyroxenes and uralites: Intergrowth microstructures and reaction mechanisms. Am. Mineral. 66, 1107-1134.

Veblen, D.R., Buseck, P.R. and Burnham, C.W. (1977) Asbestiform chain silicates: new minerals and structural groups. Science 198, 359-365.

Veblen, D.R. and Cowley, J.M. (1992) Direct imaging of point defects by HRTEM. In A.S. Marfunin, ed., Higher Mineralogy, in press. Springer-Verlag, Heidelberg.

Veblen, D.R. and Ferry, J.M. (1983) A TEM study of the biotite–chlorite reaction and comparison with petrologic observations. Am. Mineral. 68, 1160-1168.

Veblen, D.R. and Wiechmann, M.J. (1991) Domain structure of low-symmetry vesuvianite from Crestmore, California. Am. Mineral. 76, 397-404.

Viertel, H. U. and Seifert, F. (1979) Physical properties of defect spinels in the system $MgAl_2O_4$-Al_2O_3. N. Jahrb. Mineral. Abh. 134, 167-182.

Wadsley, A.D. (1964) Inorganic non-stoichiometric compounds. In L. Mandelcorn, ed., Non-stoichiometric Compounds, p. 98-209. Academic Press, New York.

Welch, M.D., Rocha, J. and Klinowski, J. (1992) Characterization of polysomatism in biopyriboles: double-/triple-chain lamellar intergrowths. Phys. Chem. Minerals 18, 460-468.

Wenk, H.-R., Müller, W.F., Liddel, N.Y. and Phakey, P.P. (1976) Polytypism in wollastonite. In H.-R. Wenk, P. E. Champness, J. M. Christie, J. M. Cowley, A. H. Heuer, G. Thomas and N.J. Tighe, eds., Electron Microscopy in Mineralogy, p. 324-331. Springer-Verlag, Berlin.

White, T.J. and Hyde, B.G. (1982a) Electron microscope study of the humite minerals: I. Mg-rich specimens. Phys. Chem. Minerals 8, 55-63.

White, T.J. and Hyde, B.G. (1982b) Electron microscope study of the humite minerals: II. Mn-rich specimens. Phys. Chem. Minerals 8, 167-174.

White, T.J. and Hyde, B.G. (1983) An electron microscope study of leucophoenicite. Am. Mineral. 68, 1009-1021.

Worden, R.H., Droop, G.T.R. and Champness, P.E. (1992) The influence of crystallography and kinetics on phengite breakdown reactions in a low-pressure metamorphic aureole. Contrib. Mineral. Petrol. 110, 329-345.

Wriedt, H.A. and Darken, L.S. (1965a) Lattice defects and the solution of nitrogen in a deformed ferritic steel: Part I--Experimental data and thermodynamic analysis. Trans. Metal. Soc. AIME 233, 111-122.

Wriedt, H.A. and Darken, L.S. (1965b) Lattice defects and the solution of nitrogen in a deformed ferritic steel: Part II--Identification of defect sites and influence on composition. Trans. Metal. Soc. AIME 233, 122-130.

Yau, Y., Anovitz, L.M., Essene, E.J. and Peacor, D.R. (1984) Phlogopite-chlorite reaction mechanisms and physical conditions during retrograde reactions in the Marble Formation, Franklin, New Jersey. Contrib. Mineral. Petrol. 88, 299-306.

Yau, Y.-C., Peacor, D.R. and Essene, E.J. (1986) Occurrence of wide-chain Ca-pyriboles as primary crystals in the Salton Sea Geothermal Field, California, USA. Contrib. Mineral. Petrol. 94, 127-134.

Chapter 7.   POLYTYPISM AND STACKING DISORDER

Alain Baronnet   CRMC[2] Campus Luminy, Case 913,

13288 Marseille Cedex 09, France

7.1   INTRODUCTION

7.1.1   Historical background and chapter outline

Polytypism is a crystallographic property of a large number of inorganic crystals. Discovered in silicon carbide by Baumhauer (1912, 1915) using optical microscopy, this phenomenon was intensely studied between 1950 and 1970, during which time experimentalists combined optical microscopy with x-ray diffraction techniques. Many polytypes were determined structurally, with much effort focused on three prototype compounds: silicon carbide (SiC), zinc sulfide (ZnS), and cadmium iodide (CdI$_2$). Also, exciting ideas evolved regarding the possible origins of the phenomenon. These ideas developed by the close collaboration of solid state physicists, crystallographers, mineralogists, and crystal growers. Polytype research developed along three lines: structural, phase (at equilibrium), and kinetic (disequilibrium) approaches. The state of the art has been reviewed by Verma and Krishna (1966), Trigunayat and Chadha (1971), Verma and Trigunayat (1973), Baronnet (1978), Krishna (1983), and Trigunayat (1991).

The intimate link between polytypism and stacking of structural modules (Chapter 6) was recognized early. Crystals exhibiting polytypism may have random displacements of modules to produce disorder. X-ray diffraction patterns of polytypes often show reflections indicating ordering along certain reciprocal lattice rows, but a degree of concomitant random stacking disorder also may be evidenced as diffuse streaks connecting these reflections. During the last decade or so, understanding of the relationship between stacking faults and polytypism has been emphasized (e.g., Pandey and Krishna, 1982; Sebastian and Krishna, 1987).

The purpose of this chapter is first to define and delineate the concept of polytypism with respect to concepts such as polysomatism, polymorphism, and modulated structures. Examples of polytypic minerals are given to illustrate variations with respect to composition and single-layer structures. Subsequently, polytype microstructures are described in terms of stacking faults, which affect the so-called basic structures. This is followed by a short summary of current equilibrium and kinetic theories of polytypism. The lively debate concerning these theories will imply the need of descriptions of polytype microstructures at the atomic-scale as allowed by transmission electron microscope (TEM). Specimen preparation, diffraction and imaging techniques relevant to polytypism are detailed. The use of the TEM and its potential to define polytypic relatons is illustrated with layer silicates, primarily micas. Results relative to growth mechanisms, recrystallization and stacking mode definition are first discussed for small crystals. In large crystals, local microstructures and their variations are described. Based on the integrated use of diffraction, imaging, and microanalytical facilities of modern high-resolution electron microscopes, the chemical signature of biotite microstructures is discussed. Finally, examples of polytype parentage are used to suggest possible transformation mechanisms concerning layer silicates.

7.1.2 Definitions of polytype structures and their notations

There is little agreement on an exact definition of "polytypism." It varies from a narrow definition (e.g., Verma and Krishna, 1966) to a broader one (Bailey et al., 1977; Guinier et al., 1984). More recently, it was expanded considerably to encompass polysomatism as well (Angel, 1986; Zvyagin, 1988).

7.1.2.1 Definition by Verma and Krishna (1966)

Polytypism occurs for some structures composed of closest-packed atoms, ions or molecules and/or by the stacking of nearly structurally invariant layers or modules (layered crystals). Polytypes of one substance are found in an almost unlimited number of crystallographically distinct structures with identical, or nearly identical, chemical composition. In polytypic crystals, the first and invariant coordination polyhedron of specific atoms may be built up in different manners with the second- or even higher-coordinations being different.

7.1.2.2 Modifications by Bailey et al. (1977) and Guinier et al. (1984)

"An element or compound is polytypic if it occurs in several different structural modifications, each of which may be regarded as built up by stacking layers of nearly identical structure and composition, and if the modifications differ only in their stacking sequence." The main extension here is the word "nearly," which allows some small variability of the structure and composition of the layers.

7.1.2.3 Definition-broadening by Angel (1986) and Zvyagin (1988)

On the basis of the admittedly narrow range of structural and composition variability as given in the above definitions, Angel (1986) broadened the definition of polytypism to almost any modular structure. Under this definition, structural variation derived from similar structure fragments can be called polytypes. Such fragments can be blocks, rods, or layers (Zvyagin, 1988).

However, I believe that the narrow definition should be maintained, in agreement with Veblen (1991). He argues that the classical separation between polytypes and polysomes has been useful in practice and that the classification of polysomes as a subset of polytypes is confusing. This progressive broadening of the definition of polytypes has occurred in response to an increasing number of structure relations. However, this broadening may be detrimental with regard to specific properties of strict polytypes, as they are detailed in §7.1.5 below.

7.1.2.4 Layers and stacking sequences

Layers involved in polytypic stacking sequences may be of two distinct types. In one type, strong bonds within layers and weak bonds between layers allow layers to be considered as separate entities. In the other type, layers do not occur as coherent units since intralayer bond strengths are not significantly different from interlayer bond strengths. Sphalerite is a typical example, wherein ZnS layers occur parallel to four symmetrically related {111} planes.

Layers can be defined as lying parallel to flat (F) faces and are called "slices" (Hartman and Perdok, 1955). Such layers contain at least two non-colinear directions of strong bond chains. The physical consequence of such slices are apparent during crystal growth of F faces; specific layer-by-layer growth mechanisms may operate at small departures from equilibrium (Sunagawa, 1977).

Layered structures are those where layers are parallel to a unique plane in the structure. A densely-packed, strongly-bonded intralayer slab contrasts with a weakly bonded interlayer region. Examples include the layer silicates, graphite and molybdenite.

Building layers (BLs) are unit layers, i.e., the fundamental structural units involved in the building of a polytype. Two lattice translations, say **a** and **b**, define the invariant planar lattice of the BL. The distances a and b will be preserved regardless of the nature of stacking the BLs in a polytype.

In regularly periodic polytypes, the c lattice dimension may vary. The lattice vector **c** defines a slab that repeats some number of BLs to generate the structure. The way the BL within each slab is stacked defines the *layer stacking sequence*. The *periodicity* N of the periodic polytype is the number of BL involved along the repeat distance (Fig. 7-1). Any *periodic polytype* may be fully described by NX [layer stacking sequence symbolism], with X representing the crystal system of the polytype as: C = cubic, H = hexagonal, R = rhombohedral, T = trigonal, Or = orthorhombic, M = monoclinic and Tc = triclinic. Obviously, the knowledge of the layer stacking sequence implies knowledge of both N and X. This NX (e.g., 2H, 3C, 4H, 9R, etc.) is the Ramsdell notation (Ramsdell and Kohn, 1951). The definition of NX is obtained through the inspection of x-ray or electron diffraction patterns (see §7.5.3). N is the number of layers in the repeating unit, and X indicates the symmetry of this unit.. The larger the value of N, the greater the probability for one NX value to correspond to more than one stacking sequence. Therefore, the nomenclature NX_n is used (e.g., $2M_1$, $2M_2$, $3Tc_1$, etc., for mica polytypes), with the value of n defined in order of the discovery of the specific polytype.

The above notation cannot be applied to *disordered* or *semi-random polytypes* because of the lack of a layer-stacking-sequence repeat. In those cases, a one-layer notation is followed by a lower case d (for disordered) or r (for random). Example: $1M_d$ or $1M_r$ mica as a set of BLs with an individual monoclinic lattice each stacked in a disordered scheme.

Many notations are available to describe the respective positions of successive BLs along the repeat distance (Verma and Krishna, 1966; Zvyagin, 1967; Trigunayat, 1991). It is beyond the scope of this chapter to discuss all of them. Rather, the universal ABC notation is used for closest-packed substances and the more specialized RTW (Ross, Takeda and Wones, 1966) and Zvyagin (1967) notations are used for layered structures like those of the micas.

7.1.2.5 The ABC notation for closest-packing

The only way in which rigid spheres of equal radii may be arranged in a plane, here corresponding to a BL, in a closest-packed manner, is shown in Figure 7-2. Positions A, at the center of each sphere, form a planar hexagonal array of 6mm symmetry. Spheres in this plane are designated as having A positions. Such a plane contains two kinds of curved triangular interstices in equal number: the first one points upwards (Δ) and is labeled B whereas the second points downwards (∇) and is labeled C. The B and C positions are related to the A positions by [1/3, 2/3] and [2/3, 1/3] translations, respectively. Imagine a second identical closest-packed plane positioned over the former one as a start toward 3D-close packing. All spheres of the second plane can be positioned either over the B voids or

Figure 7-1. Polytype stacking of successive layers. Diagrammatic sketch of a three-layer polytype.

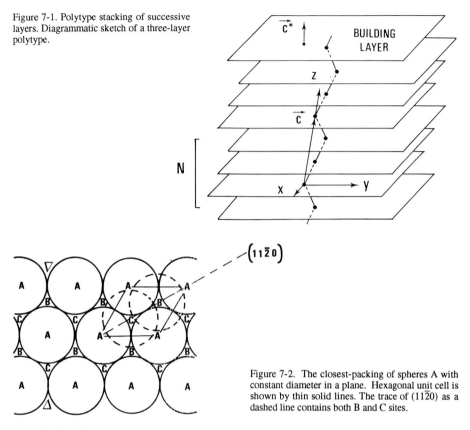

Figure 7-2. The closest-packing of spheres A with constant diameter in a plane. Hexagonal unit cell is shown by thin solid lines. The trace of $(11\bar{2}0)$ as a dashed line contains both B and C sites.

over the C voids, thus defining a B or C plane, respectively. Between two contiguous closest-packed planes, there are equal numbers of (small) tetrahedral and (large) octahedral voids. If we consider the stacking of many more layers, the entire 3D-structure can be described by a set of letters A, B, and C, with each letter referring to the sphere position in a given plane with respect to a standard A plane. Due to the closest packing requirement, AA, BB or CC sequences are not possible in a stacking sequence.

The periodicity N may range from 2 to ∞, although the most common closest-packed stacking sequences are:

2H (A B A B A B ...) designated hexagonal closest packing (h.c.p.)

3C (A B C A B C ...) designated cubic closest packing (c.c.p.) or face-centered cubic lattice (f.c.c.)

Theoritically an infinite number of different stacking sequences may occur although some are more common than others. For example, common sequences of the closest-packed structures of zinc sulfide and silicon carbide are:

4H   (IA B C B I A B C BI ....)

6H   (IA B C A C B I A B C A C B I ....)

15R   (IA B C A C B C A B A C A B C B I A B C A C B C A B A C A B C BI ...)

As shown in Figure 7-3, a full representation of the stacking sequences of any closest-packed structure is accessible in $\{11\bar{2}0\}$ planes, as indexed on an hexagonal cell. This plane contains centers of spheres of all layers. Note that the origin of any layer stacking sequence can be defined within any layer of the repeated unit, i.e.,

| A B C A C B| = | B C A C B A | = | C A C B A B | = | A C B A B C | = ...

Although it provides a complete description of the structure, the ABC notation may be cumbersome for polytypes with large stacking sequences. More compact notations, also based on the ABC concept, are those of Hägg (1943) and Zhdanov (1945). As seen in Figure 7-3, an upward shift from A to B, B to C, or C to A denotes a positive (+) shift of [+1/3, +2/3] in the BL plane whereas a shift from A to C, C to B or B to A involves a reverse shift (-) of [-1/3, -2/3]. Hägg accounted for the positive or negative translations by (+) or (-) signs, respectively. Accordingly,

2H stacking is represented as (+ - + - + - ....),   3C as (+ + + + ...),

4H as (+ + - - + + - - ...),   6H as (+ + + - - - + + + - - - ....)

and 15R as (+ + + - - + + + - - + + + - - ....).

To provide for a more compact notation, Zdhanov grouped the number of consecutive (+) and consecutive (-) signs as a symbol consisting of a pair of numbers between brackets. The Zdhanov notation for 2H is (11), of 4H (22), of 6H (33) and of 15R (323232) or $(32)_3$. As a special case, 3C is written as (∞), a notation that is widely used in the close-packed polytype literature.

7.1.2.6 Ross-Takeda-Wones and Zvyagin notations for micas

These notations apply to micas by virtue of its layer-like features. They apply to stacking modes of successive layers where no interlayer shift occurs, and instead only a layer rotation or intralayer shift occurs. The idealized structure of the mica building unit is represented in Figure 7-4. The mica layer is composed of two opposing tetrahedral (T) sheets with sixfold symmetry, with a single octahedral (O) sheet sandwiched between them. The tetrahedral sheets are formed by (Si,AlO_4) tetrahedra linked by bridging "basal" oxygen ions O_B. These tetrahedra form connected coplanar rings and each has a non-bridging apical oxygen O_A. All O_A are located on one side with respect to the O_B plane. The octahedral sheet is formed where two of these tetrahedral sheets are brought together with their O_A planes facing each other. The octahedral sheet is formed by the closest packing of contiguous O_A planes, with added OH^- and F^- ions completing them. The closest packing constraints require the two tetrahedral sheets to be shifted laterally with respect to each other. The symmetry of the tetrahedral-octahedral-tetrahedral (TOT) assembly is thus reduced from hexagonal to monoclinic. Two-thirds of the octahedral sites are filled (mainly by Al) in dioctahedral micas, whereas all of those sites are occupied by Mg, Fe^{2+}, mainly in trioctahedral micas. In micas, the negatively charged TOT silicate sheets are linked together by rather large interlayer, positively charged cations such as K^+, Na^+ or Ca^{2+}. They are arranged in a plane. These alkali- or alkaline-earth ions occupy the centers of hexagonal prisms between the two hexagonal O_B rings of adjacent TOT units. Because the weakest ionic bonds in the mica structure are the (K, Na, Ca)-O_B bonds, the BL of mica is defined as the TOT unit occuring between two successive interlayer cation planes.

The conventional unit cell of the BL (Fig. 7-5) is monoclinic and C-centered. Lattice parameters are:

236

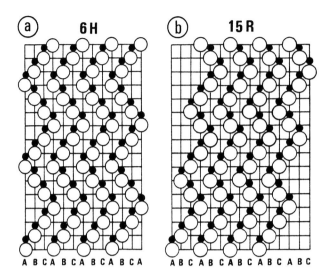

Figure 7-3. Zigzag sequences of the closest-packed stackings 6H (a) and 15R (b) of zinc sulphide in the $(11\overline{2}0)$ projection. S: large open circles; Zn: small solid circles. [From Verma and Krishna (1966), Figs. 20 and 21, p. 87.]

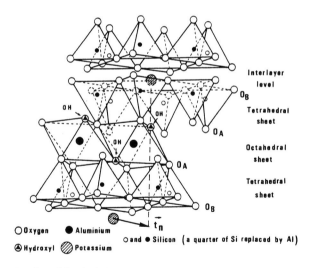

Figure 7-4. Perspective view of the single layer structure of the micas with the stacking vector t_n shown. [From Baronnet (1980), Fig. 3, p. 453.]

$$b = \sqrt{3}a \sim 9.0 \text{ Å}, c \sim 10 \text{ Å}, \beta = \cos^{-1} (-a/3c).$$

Accordingly, the C-centered unit cell is shifted by a *layer stagger* (or *stacking vector*) $t_n = -1/3 \, a_n$ from the BL numbered n to the next layer numbered n + 1. To preserve the coordination polyhedron of interlayer cations, t_{n+1} may adopt six possible orientations with respect to t_n so that the interlayer stacking angles obey the relation $(t_n \wedge t_{n+1}) = 0°$; 60°; 120°; 180°; 240°; 300°; i.e., 0 mod $\pi/3$. As required for polytypes, the first nearest neighbors of all atoms in the structure remain undisturbed.

The Ross-Takeda-Wones (RTW) notation ascribes a series of N number symbols written within square brackets that refer to the successive interlayer stacking angles ($t_n \wedge t_{n+1}$) involved in the periodic sequence, from bottom (Fig. 7-6) to top. 0, 1, $\bar{1}$, 2, $\bar{2}$ and 3 correspond to 0°, + 60°, - 60° (= + 300°), + 120°, - 120° (= + 240°) and 180°, respectively (Fig. 7-6a). Over a periodic sequence, these a_n symbols obey the relation (Takeda, 1971):

Figure 7-5. The unit-cell of the mica single layer as projected onto (001). [From Baronnet (1980) Fig. 4, p. 453.]

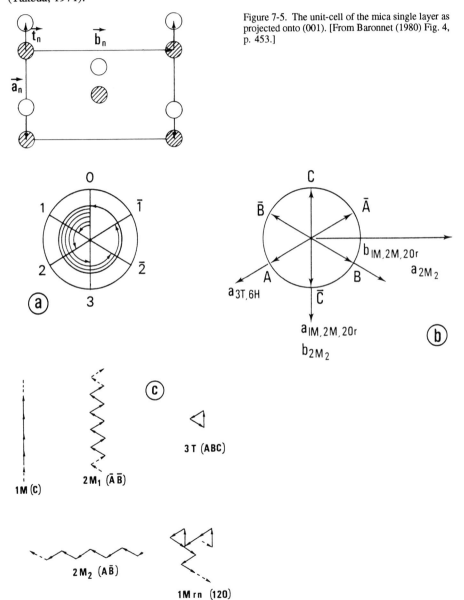

Figure 7-6. Notations for the stacking of mica layers. (a) The rotational RTW notation, (b) the positional Zvyagin's notation, (c) stacking vector representation of the common basic structures of the micas.

$$\sum_{n=1}^{N} a_n = 0 \bmod 6$$

A cycling permutation over the series of symbols does not change the sequence.

The common mica polytypes may be written using the combined Ramsdell and RTW symbolisms as:

$$1M[0], 2M_1[2\bar{2}], 2M_2[1\bar{1}], 3T[222] \text{ or } [\bar{2}\ \bar{2}\ \bar{2}]$$

and are represented by stacking vectors in Figure 7-6c.

The RTW notation refers to the relative rotations between successive BLs and consequently is "position-independent." Location of the position of an individual BL in a long-period or a disordered stacking sequence may be advantageous. This information is available from a high-resolution TEM image (see §7.5.5). The Zvyagin notation involves the azimutal position of each successive BL around c^*, the normal to the layer (Fig. 7-6b). A series of N block letters can define the orientation of any BL with respect to a standard layer position C. For any monoclinic or orthorhombic polytype, this C position is taken as opposite to the shortest (001) plane lattice vector. Again, the common mica polytypes may be written using the Ramsdell and Zvyagin symbolisms as: 1M(C), $2M_1(\bar{A}\ B)$, $2M_2(A\ \bar{B})$, 3T(ABC) or (ACB). As in the case of the ABC notation for close packed structures, the RTW and Zvyagin notations are well suited for describing finite and/or aperiodic stacking sequences and stacking faults in mica structures.

In the mica literature, 1Mr n (120) and 1Mr n (60) have been used for statistically disordered stacking structures (Ross, Takeda and Wones, 1966). 1Mrn (120) refers to successive stacking angles of k.120° with k chosen randomly (r) between 0, 1, and 2. 1 Mrn (60) corresponds to stacking angles of k.60° with random choice of k among 0, 1, 2, 3, 4, or 5. The turbostratic stacking (any angle of rotation of the layers) is not considered in our definition of polytypes, because nearest-neighbor coordination of interlayer atoms is not preserved.

7.1.3   Polytypism versus polysomatism

Polysomatism in a crystal structure (Thomson, 1978; Veblen, 1991; §6.1.3) may be portrayed as a mixture of at least two types of slabs, say M and N. Separately, all M slabs, as well as all N slabs, stacked together, form distinct crystal structures with different stoichiometries. Thus, M and N are end-member structures of a polysomatic series. *Polysomes* are polysomatic structures constructed with different combinations of those end-members: the periodic structures (M), (MMN), (MN), (MNN) and (N) form a polysomatic series but disordered combinations are also accepted as such. An obvious consequence of polysomatism is that all members of a polysomatic series are stoichiometrically colinear i.e., the chemical composition of any one of them will be a linear combination of the stoichiometries of M and N (Veblen, 1991). Perhaps the most common example of polysomatic series among minerals is provided by the biopyriboles. They include the pyroxenes (P), amphiboles, and talc or mica (M). Amphiboles (PM) and jimthomsonite (PMM) are intermediate ordered members. After properly scaling the chemical compositions of (P) as $(Mg,Fe)_4Si_4O_{12}$ and (M) = talc as $(Mg,Fe)_3Si_4O_{10}(OH)_2$, the composition of the ferromagnesian pyriboles may be calculated as

$$n_P(Mg,Fe)_4Si_4O_{12} + n_M(Mg,Fe)_3Si_4O_{10}(OH)_4.$$

Symbols n_P and n_M stand for the numbers of P and M slabs in the polysomatic sequence, respectively. The number of cations in a polysome may be accurately predicted considering a linear combination of those amounts in end-member structures (Veblen, 1991).

Crystallographic constraints on the two polysome end-members are: (1) structural continuity should exist between adjoining M and N slabs, and (2) the matching surfaces of M and N slabs must have the same plane-group symmetry. Constraint (1) implies that the interface free energy between M and N is low, i.e., the surface free energies of the contact plane are almost equal for both M and N. In biopyriboles, such a contact plane is normal to the plane of tetrahedral chains (P), ribbons (MP), and sheets (M). In mixed layer clay minerals—smectite, illite, vermiculite, kaolinite, etc. (Reynolds, 1980) or mixed layer micas—chlorites, talc, serpentines of higher metamorphic grade (e.g., Reynolds, 1988; Veblen, 1983; Amouric et al., 1988; Baronnet and Onrubia, 1988; Eggleton and Banfield, 1985; Olives Banos, 1985; Olives Banos et al., 1983), (001) is the contact plane between layer components. This results from the high compatibility between the structures and from similar lattice dimensions of the basal planes of all sheet silicates (very similar O,OH basal frameworks due to closest-packing). Similarly, biopyriboles and many silicates and aluminosilicates have similar O,OH arrangements owing to anion packing relations. Constraint (2) regarding the plane-group symmetry of matching surfaces of M and N is not rigorous. Dimensions of planar unit cells of contact planes may be nearly identical or some integral multiple of each other. Thus, polytypism appears to be a restricted case of polysomatism. As a modular concept, polytypism involves component slabs with both the same structure and the same composition. Based on the strict definition of polytypism, it follows that:
- polytypes have identical chemical compositions;
- the varying repeat unit of polytypes along the normal to the layer modules is equal to, or an integral multiple of, the layer module thickness;
- if the surface structures of adjoining layers are identical, then different stacking modes of consecutive layers are possible for module surfaces of high symmetry; there must be several ways to stack layers without distorting the first nearest neighbor atoms at the surface of the modules;
- interface free energy is not significant. Instead, a stacking-fault energy is defined for comparison of different stacking modes (see §7.3.5). Because the first nearest neighbor configurations of all atom are preserved, differences in stacking fault energies are vanishingly small for polytypes.

Polysomatic and polytypic relations may be simultaneously involved. For example, coarse mixed layer silicates (e.g., talc-serpentine as in Fig. 7-7a; mica-chlorite as in Fig. 7-7b) occur when end-member structures segregate as thick packets of slabs each (e.g., ... M M M N N N N M M M ...).

7.1.4 Polytypism and modulated structures arising from site ordering/positional displacements

Many minerals may exhibit compositional or displacive *modulations* that arise from cation site ordering or from slight displacements, or from both (Morimoto, 1978; Buseck and Cowley, 1983). Modulations may produce either strictly periodic or statistically periodic arrangements. In reciprocal space, modulations are detected as weaker satellite spots around the stronger diffractions of the substructure. If the spacing between satellite spots is an integral submultiple of the substructure reflection spacing, then the modulation is said to be *commensurate*. In real space, the modulation forms a superstructure involving an integral number of subunits. An *incommensurate* modulation is indicated by a non-integral submultiple of the substructure reflection spacing. Polytypes that are strictly

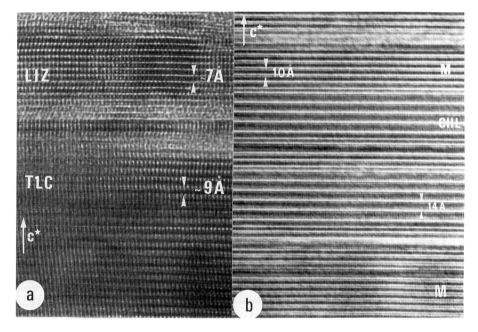

Figure 7-7. Segregated mixed-layering or polytypism in polysomes: (a) talc-lizardite intergrowth; (b) biotite-chlorite intergrowth.

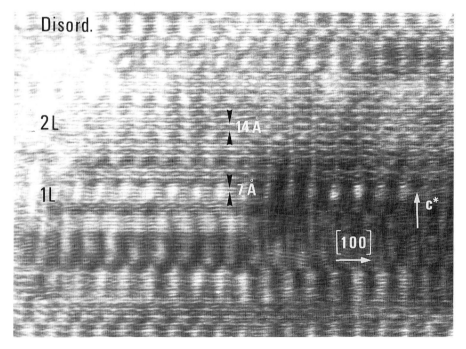

Figure 7-8. HRTEM image of antigorite seen along [010]. Commensurate modulations extend along [100]. Ordered (one-layer and two-layer) polytypes pile up along [001]*, normal to the mean plane of layers.

periodic are one-dimensionally modulated structures along the normal to the layers. Other characteristics of modulated N-layered polytypes are: (1) a commensurate polytype forms when the BL corresponds to the substructure and an integral number of BLs determines the repeat, (2) displacive modulations involve discrete positional displacements, usually transverse in nature, (3) cation ordering as a cause of modulations may be of secondary importance, as implied by the definition of polytypism.

In Figure 7-8, a HRTEM image of antigorite viewed along [010] shows sine-wave displacive modulations along [100] and various polytype repeats along [001]* similar to those reported by Spinnler et al. (1982) and Mellini et al. (1987).

7.1.5   Polytypism *vs.* polymorphism

Crystalline substances adopting more than one atomic structure while maintaining a fixed chemical composition (Verma and Krishna, 1966) are known as *polymorphs*. There is a finite number of these polymorphic modifications, with each having a well-defined stability field in P,T space. A polymorph, in its stability field, has a minimum Gibbs free energy G. When compared to any other possible polymorph, we can write:

$$\Delta G = \Delta H - T\Delta S \; = \Delta E + P\Delta V - T\Delta S \; , \qquad (7.1)$$

with ΔG, ΔH, ΔS, ΔE, and ΔV as the changes of free energy, enthalpy, total entropy, internal energy, and molar volume, respectively. P and T are the total pressure and the absolute temperature of interest. For polymorphs, the slope of phase boundaries in P,T space may be determined from the Clausius-Clapeyron equation:

$$dP/dT = \Delta S/\Delta V \; . \qquad (7.2)$$

This relation is possible because of the non-zero values of both ΔS and ΔV.

For polytypes, ΔE and ΔV are almost equal to zero. This similarity results from preserved first nearest neighbors in the structures. ΔS, as the sum of the configurational entropy difference and of the vibrational entropy difference, is usually small. Thus ΔG, in Equation (7.1), hardly differs from zero, and dP/dT is indeterminate—see Equation (7.2). Only at high temperature the $T\Delta S$ term may stabilize some polytypes (see Jagodzinski's theory discussed in §7.4.1.1).

Stabiliby fields for polytypic modifications generally cannot be defined in a P,T diagram. Furthermore, the coexistence of several modifications in an apparent "single crystal" (syntactic coalescence) suggests a violation of the Gibbs phase rule, assuming equilibrium. The density, specific heat, melting point, and other properties are nearly constant among polytypes. However, electronic and optical properties are usually sensitive to the polytype modification.

Structurally, polytypism may be regarded as a special case of polymorphism wherein the structure varies in only one dimension (Schneer, 1955). However, some transformations between polytypes may occur through annealing or cooling (e.g., ZnS – Steinberger, 1983; SiC – Jepps and Page, 1983). Therefore, for some materials, there may be intermediate conditions between polymorph and polytype. The later is illustrated by basic structures as discussed in §7.3.1.

7.2   MINERAL EXAMPLES OF POLYTYPISM

Polytype structures are found among minerals from simple (monoelemental) to very complicated chemistries (complex silicates). Structure types are commonly based on close-

242

Table 7-1. - Some minerals of various chemical groups which exhibit polytypism. n : natural, s : synthetic. Numbers between square brackets correspond to references below.

COMPOSITION	STRUCTURE TYPE	POLYTYPES	REFERENCES
Elements			
C	diamond type	diamond	[1], [2]
		lonsdaleite 2H	
	graphite (n + s)	2H	[3], [4]
		3R	[5], [6]
Oxides			
SiO_2	tridymite		[7]
$SiO_2.nH_2O$	opal (colloïdal crystal)	3C, 9R	[8]
Hydroxides			
$Al(OH)_3$	Layered	bayerite, gibbsite, norstrandite	[9-11]
Sulphides			
ZnS	close-packed	n: zinchlende 3C,	[12-14]
		würtzite 2H, 4H, 6H, 15R	
		8H, 10H	[15]
		9R, 12R, 21R	[16]
		s: more than 200 polytypes	[17]
MoS_2	layered and close-packed	n : molybdenite	[18-20]
		$2H_1$, 3R	
Sulfates			
$Fe_2O_3 \text{-} 3SO_3 \text{-} 9H_2O$		n : coquimbite (H)	[21-22]
		paracoquimbite (R)	
Carbonates			
$(CeF)_2Ca(CO_3)_3$		n : α-parisite (H_1)	[23]
		β-parisite (R)	
		γ-parisite (H_2)	
Silicates			
	chain-silicates		
$Mg_2Si_2O_6$		clinoenstatite (1M)	[24]
		orthoenstatite (2Or)	
$Ca_3Si_3O_9$		wollastonite (A A)	[25]
		parawollastonite (A A B B)	
	sheet-silicates		
	1 : 1 type		
$Mg_3Si_2O_5(OH)_4$		lizardite (1T, $2H_1$)	[26]
$Al_2Si_2O_5(OH)_4$		kaolins : kaolinite 1Tc	[27]
		dickite 2M	
		nacrite 2M	
	2 : 1 type		
$Mg_3Si_4O_{10}(OH)_2$		talc (1Tc, 2M ?)	[28], [29]
$K(Mg,Fe)_3AlSi_3O_{10}(OH,F)_2$ (biotite)		micas (1M, $2M_1$, $2M_2$, 3T, 23 complex polytypes)	[30], [31]
$K Al_2 AlSi_3O_{10}(OH)_2$ (Muscovite)			
	2 : 1 - 0 : 1 type		
$(Mg,Al)_3 AlSi_3O_{10}(OH)_2$ $Mg_3(OH)_6$ (clinochlore)		chlorite (12 possible polytypes of 1 L types)	[32]

[1] Frondel and Marvin, 1967
[2] Ergun and Alexander, 1962
[3] Bernal, 1924
[4] Hassel and Mark, 1924
[5] Debye and Scherrer, 1916
[6] Lipson and Stokes, 1942
[7] Flörke, 1955
[8] Monroe, Sass and Cole, 1969
[9] Bragg and Claringbull, 1965
[10] Lippens, 1961
[11] Saalfeld and Jarchow, 1968
[12] Frondel and Palache, 1948
[13] Frondel and Palache, 1950
[14] Seaman and Hamilton, 1950
[15] Evans and McKnight, 1959
[16] Haussühl and Müller, 1963
[17] Steinberger, 1983
[18] Frondel and Wickman, 1970
[19] Newberry, 1979 a
[20] Newberry, 1979 b
[21] Ungemach, 1935
[22] Bandy, 1938
[23] Ungemach, 1935
[24] Buseck and Iijima, 1975
[25] Wenk et al, 1976
[26] Bailey, 1988 a
[27] Bailey, 1963
[28] Rayner and Brown, 1966
[29] Stemple and Brindley, 1960
[30] Bailey, 1984
[31] Baronnet, 1980
[32] Bailey, 1988 b

periodic are one-dimensionally modulated structures along the normal to the layers. Other characteristics of modulated N-layered polytypes are: (1) a commensurate polytype forms when the BL corresponds to the substructure and an integral number of BLs determines the repeat, (2) displacive modulations involve discrete positional displacements, usually transverse in nature, (3) cation ordering as a cause of modulations may be of secondary importance, as implied by the definition of polytypism.

In Figure 7-8, a HRTEM image of antigorite viewed along [010] shows sine-wave displacive modulations along [100] and various polytype repeats along [001]* similar to those reported by Spinnler et al. (1982) and Mellini et al. (1987).

7.1.5 Polytypism *vs.* polymorphism

Crystalline substances adopting more than one atomic structure while maintaining a fixed chemical composition (Verma and Krishna, 1966) are known as *polymorphs*. There is a finite number of these polymorphic modifications, with each having a well-defined stability field in P,T space. A polymorph, in its stability field, has a minimum Gibbs free energy G. When compared to any other possible polymorph, we can write:

$$\Delta G = \Delta H - T\Delta S = \Delta E + P\Delta V - T\Delta S , \qquad (7.1)$$

with ΔG, ΔH, ΔS, ΔE, and ΔV as the changes of free energy, enthalpy, total entropy, internal energy, and molar volume, respectively. P and T are the total pressure and the absolute temperature of interest. For polymorphs, the slope of phase boundaries in P,T space may be determined from the Clausius-Clapeyron equation:

$$dP/dT = \Delta S/\Delta V . \qquad (7.2)$$

This relation is possible because of the non-zero values of both ΔS and ΔV.

For polytypes, ΔE and ΔV are almost equal to zero. This similarity results from preserved first nearest neighbors in the structures. ΔS, as the sum of the configurational entropy difference and of the vibrational entropy difference, is usually small. Thus ΔG, in Equation (7.1), hardly differs from zero, and dP/dT is indeterminate—see Equation (7.2). Only at high temperature the TΔS term may stabilize some polytypes (see Jagodzinski's theory discussed in §7.4.1.1).

Stabiliby fields for polytypic modifications generally cannot be defined in a P,T diagram. Furthermore, the coexistence of several modifications in an apparent "single crystal" (syntactic coalescence) suggests a violation of the Gibbs phase rule, assuming equilibrium. The density, specific heat, melting point, and other properties are nearly constant among polytypes. However, electronic and optical properties are usually sensitive to the polytype modification.

Structurally, polytypism may be regarded as a special case of polymorphism wherein the structure varies in only one dimension (Schneer, 1955). However, some transformations between polytypes may occur through annealing or cooling (e.g., ZnS – Steinberger, 1983; SiC – Jepps and Page, 1983). Therefore, for some materials, there may be intermediate conditions between polymorph and polytype. The later is illustrated by basic structures as discussed in §7.3.1.

7.2 MINERAL EXAMPLES OF POLYTYPISM

Polytype structures are found among minerals from simple (monoelemental) to very complicated chemistries (complex silicates). Structure types are commonly based on close-packing or on layering. It is beyond the scope of this chapter to exhaustively list all polytypic minerals. Instead, Table 7-1 gives selected examples of minerals to illustrate the large range of chemistries and structures. Bonding between building layers may consist of covalent (diamond type), van der Waals (graphite, talc), hydrogen (hydroxides, chlorite, lizardite, kaolins) or essentially ionic (enstatite, micas) bonds. Disordered and/or semi-random polytypes are ubiquitous in all substances in Table 7-1.

7.3 TOPOLOGY OF LAYER STACKING SEQUENCES

7.3.1 Basic structures

Some polytypes of a given substance are more abundant than others. These "basic structures" (BS) are usually short-period polytypes. Except for rhombohedral structures where the repeat of the unit cell is three times that of the stacking sequence repeat, the number of layers in the periodic stacking sequence of basic structures is usually smaller than or equal to 6. Examples of basic structures are 2H, 3C, 4H, 6H, and 15R for zinc sulphide, 1M, $2M_1$, $2M_2$, 3T for micas, etc.

Most basic structures satisfy the "homogeneity condition" (Zvyagin,1988). There is "equivalence of the relative position of each layer (building module) among the others and equivalence of transition from the preceding layer to the next one for all layers." This is the case for 2H, 3C, 1M, $2M_1$, $2M_2$, and 3T. "If layer positions are characterized by their azimuthal orientations and relative displacements, homogeneity implies either identical consecutive rotations and displacements of the successive layers" (as in 3C, 1M, and 3T) "or a regular alternation of rotations and/or displacements having equal absolute values but opposite signs" (as in 2H, $2M_1$, and $2M_2$). Smith and Yoder (1956) deduced all theoretically possible short-period polytypes of the micas on this basis.

The phase relationships of basic structures are transitional between polymorphs and polytypes (Baronnet, 1989). 3C-ZnS (zincblende at low T, ß-ZnS) and 2H-ZnS (wurtzite at high T, α-ZnS) have stability fields with a distinct α-ß conversion temperature of 1020 ± 5°C at atmospheric pressure. The transition is reversible and very sluggish (Steinberger, 1983). Thus, zincblende and wurtzite behave as true polymorphic phases whereas 4H, 6H, and 15R do not.

Among the micas, 1M and $2M_1$ appear to be the dominant, stable basic structures of trioctahedral and dioctahedral species, respectively (Baronnet, 1980). Evidence for this is shown experimentally by monotropic transformations of any other polytype to a unique modification: see, for example, the 1 Mrn(120) → 1 M conversion for trioctahedral phlogopite $KMg_3AlSi_3O_{10}(OH)_2$ (Yoder and Eugster, 1954), the 1Mrn(120) → 1M → $2M_1$ conversion for dioctahedral muscovite $KAl_2AlSi_3O_{10}(OH)_2$ (Yoder and Eugster, 1955; Velde, 1965; Muklamet-Galeyev et al., 1985) at constant annealing temperature. Different degrees of stability are indicated by the varying ability of certain basic structures to transform by dissolution/crystallization or solid-state transformation processes (Baronnet, 1981; Pandey, 1981).

7.3.2 Disorder and complex polytypes

In polytype-rich substances, basic structures exhibit variations in the density of stacking faults (as defined in §7.3.4), with distributions ranging from disordered to fully ordered. First disordered and then semi-ordered distributions frequently occur during interpolytypic solid-state transformations (Pandey, 1981; Pandey and Krishna, 1982;

Sebastian and Krishna, 1987). Fully ordered stacking faults result in superstructures called *long-period polytypes* or *complex polytypes*, the stacking sequences of which may extend over thousands of Ångströms. The great majority of observed complex polytypes have layer stacking sequences that are usually based on a single structure, but sometimes on two basic structures (Pandey and Krishna, 1983).

7.3.3 Syntactic coalescence

Most "single crystals" consisting of multiple polytypes are generally made of two or more basic structures coexisting as coherent intergrowths aligned parallel to the stacking axis; this is called syntactic coalescence. X-ray diffraction shows that syntactic coalescence is indicated by the combination of diffraction spots of the polytype components on "polytype-sensitive" diffraction rows.

Such data about the coexistence of these intergrowths and the possible transformations between polytypes are controversial and produce endless discussions about the equilibrium or metastable nature of polytype structures.

7.3.4 The concept of stacking faults

A stacking fault is a break in the normal stacking sequence of a basic structure, i.e., the stacking rule is locally violated. The concept of the stacking fault is connected intimately to that of polytypism, because the "mistake" is only possible if there are several acceptable positions for adjacent layers. In Figure 7-9a, a unique position exists for any layer of chrysotile cylinders next to a neighboring one in order to produce closest packing. Consequently, there is no possibility of either polytypism or stacking faults for cylinders of equal diameters. In Figure 7-9b, where closest-packing of equal-diameter spheres of amorphous silica (opal) is shown, polytypism and stacking faults are possible because a choice exists between two possible positions (see §7.1.2).

The nomenclature for stacking faults is given below for closest-packed substances but this may also be used for layered structures. Frank (1951) considers two *geometrical types* of faults.

7.3.4.1 Intrinsic faults

A perfect stacking sequence occurs up to a single composition (or contact) plane; i.e.,

(1)   2H   .... A B A B C B C ....

(2)   3C   .... A B C A C A B C ....

Sequence (2) is illustrated in Figure 7-10a and corresponds to the removal of one close-packed plane (B) from the normal stacking sequence. The composition plane is either an atomic plane or a non-atomic plane.

7.3.4.2 Extrinsic faults

For such faults, the composition plane does not belong to either adjacent structure, i.e.:

(3)   2H   ...A B A B C A B A B ...

(4)   3C   ...A B C A C B C A B ...

246

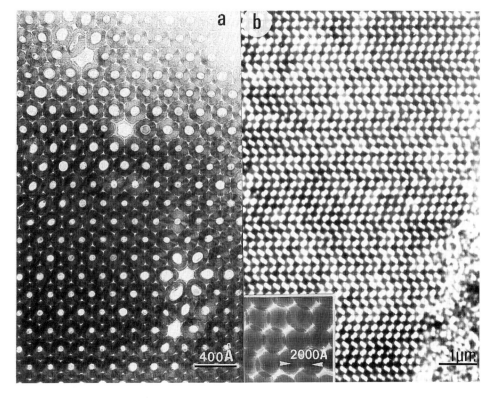

Figure 7-9. Two types of stackings in colloidal crystals: (a) 2D closest-packing of isodiameter rolls of chrysotile; (b) 3D closest-packing of isodiameter spheres of hydrous and amorphous silica-forming opal. [Courtesy of J.P. Gauthier.]

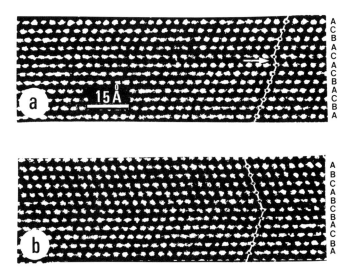

Figure 7-10. 2D structure images of stacking faults in cubic regions of SiC single crystals: (a) an intrinsic stacking fault of type (2); (b) the common twin fault of type (5). [Both from Schwamm et al. (1991) Fig. 10, p. 71.]

In both examples, the fault corresponds to the insertion of an extra atomic plane into the normal sequence (a C layer), and the composition plane is always an atomic plane.

Another *genetic classification*, partly in common with the above geometrical classification, involves stacking faults that are distinguished by physical processes that presumably cause them (Sebastian and Krishna, 1987).

7.3.4.3   Growth faults

Such faults result from the incorrect addition of a new layer during the layer-by-layer growth of a crystal, followed immediately by subsequent layers that obey the original stacking rule.

Sequence (1) can be caused by growth. A common example for c.c.p. is:

(5)   faulted 3C .... A B C A B C B A C B A....

which also can be considered to be a twin fault (Fig. 7-10b).

7.3.4.4   Deformation faults

A deformation fault is produced when two parts of the crystal slip past each other along a basal lattice plane through the partial slip vectors $s_i = \pm \mathbf{a}\,/3\;\langle 10\bar{1}0\rangle$ (hexagonal setting):

$$2H \;\;.... \; A\,B\,A\,B\,A\,B\,A\,B \;....$$
$$\downarrow\;\downarrow\;\downarrow\;\downarrow$$
$$2H \;\;.... \; A\,B\,A\,B\,C\,A\,C\,A \;.... \qquad\qquad (6)$$

is a typical deformation fault in a 2H matrix as

$$3C \;\;.... \; A\,B\,C\,A\,B\,C\,A\,B\,C \;....$$
$$\downarrow\;\downarrow\;\downarrow\;\downarrow\;\downarrow$$
$$\text{faulted}\;\; 3C \;\;.... \; A\,B\,C\,A\,C\,A\,B\,C\,A \;.... \qquad\qquad (7)$$

is in a cubic structure.

7.3.4.5   Layer displacement faults

These faults involve changing the position of a single layer, or a pair of adjacent layers, and leaving the rest of the stacking unaffected.

$$2H \;\;.... \; A\,B\,A\,B\,A\,B\,A\,B \;....$$
$$\downarrow$$
$$\text{faulted}\;\; 2H \;\;.... \; A\,B\,A\,B\,C\,B\,A\,B \;.... \qquad\qquad (8)$$

For the cubic phase, a single layer displacement is forbidden since two successive layers have the same position and therefore would violate the close-packing requirements. However, the transposition of a block of two layers is possible:

$$3C \;\;.... \; A\,B\,C\,A\,B\,C\,A\,B\,C \;....$$
$$\downarrow\;\uparrow$$
$$\text{faulted}\;\; 3C \;\;.... \; A\,B\,C\,A\,C\,B\,A\,B\,C \;.... \qquad\qquad (9)$$

As noted by Jagodzinski (1954), such faults do not affect the long-range correlation of the layers in the crystal. Regardless of the type of fault, such a planar defect locally introduces a cubic sequence in a hexagonal matrix, or conversely, a locally hexagonal sequence in a cubic matrix. Although widely used in polytype literature, this system of genetic classification is confusing because one fault configuration can result from several physical processes (Pandey and Krishna, 1982). The geometrical type of classification can be applied to layered structures like layer silicates (Baronnet et al., 1981; Pandey et al., 1982).

7.3.5 Energetics of stacking faults

Stacking fault energies are small in polytypic crystals (Tiwari et al., 1974), because first nearest neighbor bonding configurations are the same at the fault and in the perfect structure. Regardless, stacking fault energies may be used effectively in predicting the occurrence probabilities of polytypes and faulted structures (e.g., Tiwari et al., 1975). Such stacking fault energy increases mainly the crystal energy and thus the free energy of a polytype stacking. Accordingly, the smaller the stacking fault energy, the higher the occurrence probability of a polytype stacking. The calculation of such energies may thus serve to predict most probable faults as well as most probable polytypes. Qualitative estimates are possible only because the accuracy of calculated bond strengths is poor for the small stacking-fault energy differences between models. Hirth and Lothe (1968) proposed two calculation models based on: (1) the estimation of the distortional energy using the interaction of faulted layers, and (2) the direct use of bond energies involved in a central force model. A short outline of the first method is given below.

7.3.5.1 Empirical stacking fault energy by Hirth and Lothe (1968)

Hirth and Lothe (1968) wrote the stacking fault energy (SFE) as:

$$SFE = \sum_n r_n \phi_n$$

where ϕ_n is the distortion energy per faulted pair of layers with n-layer separation and r_n is the number of such faulted pairs. An estimate of the SFE up to ϕ_2 terms is usually sufficient to rank stacking faults.

For close-packed polytypic structures, $\phi_1 = 0$. For example, consider the faulted 2H sequence (8):

.....A B A B C B A B A ...
$$\phi_2 \quad \phi_2$$

$\phi_1 = 0$. For two-layer separation, the same letters appear in the normal 2H sequence. We evaluate the number of ϕ_2 terms by enumerating the number of unlike pairs by passing step-by-step across the fault. Clearly, the empirical stacking fault energy is $SFE(8) = 2\,\phi_2$.

Although the SFE calculation method was developed for closest-packed structures, it may also be applied to layered structures. Baronnet et al. (1981) and Pandey et al. (1982) determined SFE's for all possible intrinsic and extrinsic stacking faults of the 1M, 2M$_1$, and 3T basic structures of the micas. The calculation is more involved than for close-packed structures because ϕ_1 terms differ from zero and take different values for each basic structure. ϕ_2 terms may be more numerous but if we assume that $\phi_2 \ll \phi_1$, then all ϕ_2

terms contribute the same energy. Let us consider a layer-displacement fault in the 1M matrix:

$$\phi_1\phi_1$$

$$\phi_2 \quad \phi_2$$

Zvyagin   .... C C C C A C C C ....
RTW   .... 0 0 0 2 $\bar{2}$ 0 0 ....

The use of RTW symbols to label the ϕ_n terms allows identification of the various n components and the evaluation of those which are equivalent by symmetry. Here we have

$\phi_1(0) = 0$; $\phi_1(2) = \phi_1(\bar{2})$; and we assume $\phi_2(02) = \phi_2(20) \simeq \phi_2(2\bar{2})$.

The SFE is thus: SFE $= 2\,\phi_1 + 2\,\phi_2$.

Empirical stacking fault energy for a polytype. As seen above, long-period or complex polytypes of period N may be described as a superstructure where there is a periodic repetition of one or more sets of stacking faults in a given basic structure. Thus, the stacking fault energy of a complex polytype may be estimated with respect to the basic structure by the summation of the SFE's over the layer periodicity N of the polytype. For example, consider the complex polytype of mica based on $2M_1$ with N = 9:

.... $\bar{A}$ $\bar{B}$ $\bar{A}$ I $\bar{A}$ $\bar{B}$ $\bar{A}$ $\bar{B}$ $\bar{A}$ $\bar{C}$ $\bar{A}$ $\bar{B}$ $\bar{A}$ I $\bar{A}$ $\bar{B}$ $\bar{A}$ ....
.... 2 $\bar{2}$ 0   2 $\bar{2}$ 2 $\bar{2}$ $\bar{2}$ 2 2 $\bar{2}$ 0 2 $\bar{2}$ ....

As $\phi_1(2) = \phi_1(\bar{2}) = 0$, $\phi_1(0) \neq 0$, $\phi_2(2\bar{2}) = \phi_2(22) = 0$, $\phi_2(22) = \phi_2(\overline{22}) \simeq \phi_2(02) = \phi_2(0\bar{2}) = \phi_2(20) = \phi_2(\bar{2}0)$, the estimated SFE of this hypothetical 9 Tc polytype of micas is:

$$\text{SFE (9Tc)} = 1\phi_1(0) + 4\phi_2 \ .$$

The SFE is not normalized to the layer repeat N, and thus the SFE serves to compare excess energies of polytypes with the same repeats but with different layer stacking sequences (Pandey and Krishna, 1975). Such stacking fault energy estimation was used to evaluate the most probable stacking faults and complex polytypes based on the minimum SFE's [closest-packed structures: SiC (Pandey and Krishna, 1975a, 1976); layer structures: CdI_2 (Pandey and Krishna, 1975b) and micas (Baronnet et al., 1981; Pandey et al., 1982)]. Special emphasis is given here to qualitative energy comparisons because these data are easily extracted from high-resolution TEM images of layer stacking sequences—see §7.6.4.

7.4   CURRENT THEORIES OF POLYTYPISM

Over the previous four decades, more than a dozen theories have been proposed to explain the formation of polytypes (see reviews by Baronnet, 1978; Trigunayat and Chadha, 1971; Trigunayat, 1991). The purpose of this section is to review the background of those theories that can be tested by TEM observations. They can be broadly classified into categories: (1) equilibrium-based theories called "equilibrium" or "thermodynamic" theories and (2) "growth" or "kinetic" theories.

Equilibrium or thermodynamic theories:

- Jagodzinski's disorder theory (Jagodzinski, 1954)
- The Axial Next Nearest Neighbor Ising (ANNNI) model (Price and Yeomans, 1984; Angel et al., 1985)
- Landau phase-transition theory-based model of Salje et al. (1987).

Growth or kinetic theories:

- The perfect-matrix model of spiral growth (Frank, 1951b; Baronnet, 1975a)
- The faulted-matrix model of spiral growth (Pandey and Krishna, 1975a,b; 1976; Baronnet et al., 1981; Pandey et al., 1982)
- The epitaxial model by Vand and Hanoka (1967)
- The periodic slip mechanism (Mardix et al., 1968).

Of these, the disorder theory, the ANNNI model, and kinetic models based on the role of screw dislocations during and after growth are discussed briefly.

7.4.1   Equilibrium theories

7.4.1.1   Jagodzinski's disorder theory

Jagodzinski (1954) noted that polytypic crystals are commonly characterized by the existence of one-dimensional disorder superimposed on a periodic structure. Such disorder arises from the random distribution of stacking faults into an otherwise ordered structure. By considering the way such stacking faults could originate, he argued that the Helmholtz concept of minimum free energy determines the assumed stability of a polytypic system. At constant temperature, the minimum value of $G = H - TS$ is controlled by the maximum entropy, S, since the internal energy H is nearly constant for any layer stacking sequence. At first sight, this implies maximum disorder in the system if only configura-tional entropy is considered. However, Jagodzinski argued that S is the sum of the configurational entropy S_c and the vibrational entropy S_v. Since S_v displays a reverse dependence on the state of order, then $S_v + S_c$ is expected to show two maxima as a function of the degree of disorder α (Fig. 7-11). One maximum occurs at $\alpha = 0$ (perfect order) and the second at $\alpha \simeq 10\%$ (roughly 10% of the single layers displaced from their normal position in the periodic structure). Jagodzinski experimentally verified that the frequency distribution of polytypes in a batch of SiC crystals was a function of their state of order. In good agreement with theory, he found values of α near 10%.

The theory, however, is unable to account for the existence of well-ordered long-period polytypes, which should be partly disordered. In addition, the vibrational energy differ-ences between polytypes of SiC or ZnS were shown (Weltner, 1969) to be negligibly small, so these differences cannot contribute significantly towards stabilization.

7.4.1.2   The ANNNI model

This model also assumes that polytypism is at least partly an equilibrium phenomenon. Introduced by Elliott (1961) to describe the magnetic ordering in rare earth compounds, the Axial Next Nearest Neighbor Ising (ANNNI) model was applied to polytypic systems by Ramesesha (1984), Smith et al. (1984), and Price and Yeomans (1984).

The ANNNI model considers Ising spin variables $S_i = \pm 1$ associated with each site of a

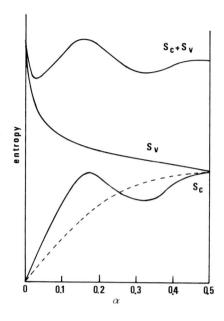

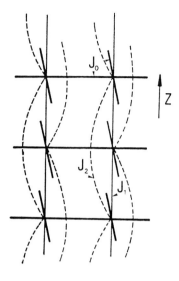

Figure 7-11. Configurational (S_c) and vibrational (S_v) entropies of a polytype system, and their sum, as a function of the degree of order α. [From Verma and Krishna (1966) Fig. 6, p. 265.]

Figure 7-12. Scheme of the intralayer (J_0) and interlayer (J_1 and J_2) interaction parameters in a layered system.

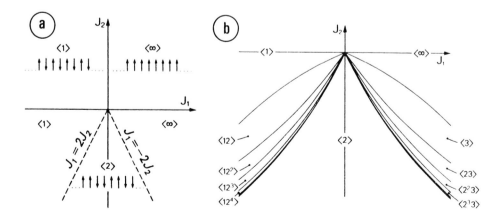

Figure 7-13. Phase diagrams for the ANNNI model in the J_1,J_2 space: (a) ground-state, (b) at finite temperature. [From Price and Yeomans (1984) Figs. 2 and 3, p. 451.]

cubic lattice. The interaction between spins is described by the Hamiltonian:

$$H = -J_0 \sum^{\perp Z} S_i S_j - J_1 \sum^{//Z} S_i S_j - J_2 \sum^{//Z} S_i S_j$$

The first term refers to a ferromagnetic interaction ($J_0 > 0$) between nearest neighbor spins lying in a plane perpendicular ($\perp Z$) to a specified axial direction Z. The second and third terms account for interactions between first and second neighbors spins along Z (//Z) (Fig. 7-12). The properties of the model depend on the relative signs of J_1 and J_2. The application to layered polytypic structures involves spins that are ferromagnetically aligned within each layer and, therefore, it is sufficient to consider ordering schemes along the axial direction. In a slab of consecutive layers the two possible positions of a layer may be described by $\uparrow$ (spin up) or $\downarrow$ (spin down). This description may be applied to binary choice in the position of adjacent close-packed layers (e.g., B or C position relative to A, Zhdanov's notation or h (hexagonal)-k(cubic) first nearest neighbor environment of layers, etc).

The ground state of the ANNNI model is shown in Figure 7-13a in J_1, J_2 space. The ferromagnetic (..$\uparrow\uparrow\uparrow\uparrow$..) ($J_1,J_2>0$), antiferromagnetic (..$\uparrow\downarrow\uparrow\downarrow$..) ($J_1<0$, $J_2>0$) states share a boundary with a phase made of alternating couples of like spins (..$\uparrow\uparrow\downarrow\downarrow\uparrow\uparrow\downarrow\downarrow$..) ($|J_1/J_2|<2$). For $J_2<0$ there is competition between the first- and second-neighbor interactions so that for $|J_1/J_2|>2$ the first neighbor dominates, and the ferromagnetic and antiferromagnetic phases persist to the above boundary. Along a boundary, all states made by mixtures of spin sequences of neighboring states have equal energy. Boundaries on which the ground state is infinitely degenerate are multiphase lines (the case for $J_2 < 0$ in Fig. 7-13a). At a finite temperature near each of the multiphase lines, there is an infinite number of stable phases (Fig. 7-13b). These latter phases contain periodic spin sequences involving the arrangement of spins found in surrounding ground states in varying proportions. Phases are separated by first-order transitions. Increasing temperature results in the appearance of new stable sequences while some of the existing ones become unstable and disappear (Bak and Boehm, 1980; Selke and Duxbury, 1984). The conversion of spin positions to layer positions as described by the Hägg's notation (see §7.7.2) is as follows: spin up ($\uparrow$) = (+), spin down ($\downarrow$) = (-). This allows one to transpose spin sequences to polytypic sequences. Considering Figure 7-3, we have:

... $\uparrow\uparrow\uparrow\uparrow$ ...	= .. + + + + + ...	= 3 C (∞)
... $\uparrow\downarrow\uparrow\downarrow$ ...	= .. + - + - + ...	= 2 H (11)
... $\uparrow\uparrow\downarrow\downarrow\uparrow\uparrow\downarrow\downarrow$ ...	= .. + + - - + + - - ...	= 4 H (22)
... $\uparrow\uparrow\uparrow\downarrow\downarrow\downarrow$ ...	= .. + + + - - - ...	= 6 H (33)
... $\uparrow\uparrow\uparrow\downarrow\downarrow\uparrow\uparrow\uparrow\downarrow\downarrow$	= .. + + + - - + + + - - ...	= 15 R $(32)_3$
... $\uparrow\downarrow\downarrow\uparrow\uparrow\downarrow\downarrow$ ...	= .. + - - + + - - ...	= 21 R $(1222)_3$

The Zdhanov spins are considered to interact according to the Hamiltonian. Accordingly, the ANNNI model predicts a phase diagram for polytypes. Most of the basic structures (short-period polytypes) are present, but it is most interesting that complex polytypes (long-period polytypes) are also present in transition regions between BS stability fields. As J_1 and J_2 are expected to vary with applied external conditions (temperature, pressure,...) the variation of such intensive conditions is believed to result in a trajectory through J_1, J_2 parameter space, i.e., to induce polytype phase transitions.

The ANNNI model is most successful in showing that short range competing interactions (up to the third neighbor) between component layers can result in long-period stacking sequences. It succeeds in describing broad features of polytypism, although it also has severe limitations. It fails to predict many of the observed polytypic structures, particularly those with extremely long periods. The model introduces the interaction parameters phenomenologically, but does not consider the crystal chemical properties of the BS. Therefore, the specific behavior of various polytypic substances cannot be described by this model in its present form. This model is also restricted mainly to closest-packed structures: for instance, micas for which three stacking modes are possible for a new layer would require a three-state model, the Potts model.

7.4.1.3   The model based on Landau phase-transition theory

Salje et al. (1987) described a detailed study of the reversible polytype transformation 2H → 12R in PbI$_2$. They developed another phenomenological approach to the formation of polytypes. They use the Landau parameter ξ, a long-range order parameter. ξ decreases continuously with temperature and vanishes at the transition temperature. The free energy of the system is expanded in terms of ξ. Hence, the differences in the free energies of the various polytypes of a substance can themselves be expressed as a function of an order parameter. This can crudely be done with a pseudo-spin model in which the orientations of the pseudo-spins are based on the layer positions. Thus, 2H appears as a ferro phase and 12R as an antiferro phase, i.e., basic structures. A fully disordered spin system (called a para-phase) may also be described. All topotactic phase transitions take place between these three phases. Accordingly, the succession of phases during cooling would be

$$\text{disordered} \rightarrow \text{antiferro(12R)} \rightarrow \text{ferro(2H)} ,$$

as observed experimentally. Long-period polytypes are also considered here as intermediate, stabilized phases. Although a promising approach, this model suffers from its "ad hoc" character, and applies to PbI$_2$ phase transformations only.

7.4.2   Kinetic theories

These theories involve the presence of screw dislocations in polytypic crystals. These dislocations may be operative during growth as in the perfect- and faulted-matrix models and the Vand Hanoka's model, or after growth as in the periodic slip model.

7.4.2.1   The perfect matrix model of spiral growth

Soon after Burton et al. (1951) published their famous screw dislocation theory of crystal growth by which the faces of a crystal grow at a significant rate even at extremely low supersaturations, Frank (1951) described the role such linear defects may play in ordering the layer stacking sequences of polytypes. The link between screw dislocations and polytypism in SiC was inferred by the frequent observation of growth spirals on the {0001} faces by phase-contrast optical microscopy (e.g., Verma, 1951, 1953).

A growth spiral occurs on a crystal face where a screw dislocation emerges from the surface (Fig. 7-14). This exposes a perpetual growth step at the end point of the dislocation line. The fault-vector (or Burgers vector) $\vec{b}$ of the dislocation has a component $\vec{b}_n$ normal to the face such as $|\vec{b}_n| = N_s t$; N_s is an integer and t is the single layer thickness. This requirement allows the bottom layer of the exposed ledge to align correctly with the helical surface of the basal face. The exposed ledge may be monolayered ($N_s = 1$) or multilayered ($N_s > 1$). During spiral growth, the face advances parallel to itself by the lateral extension of the exposed ledge without nucleation of new layers. The fundamental property of growth spirals with regard to polytypism is that any new turn of the exposed ledge of N_s layers around the dislocation core will stack the packet of N layers over itself. Thereafter, the block of N_s layers is repeated with N_s-layer periodicity along the screw axis. The stacking sequence of these N_s layers is repeated accordingly. Frank (1951) postulated that, before any screw dislocation forms, the initial platelet grew at high supersaturation in accord with the surface nucleation mechanism of "birth and spread" of new layers onto the basal faces. During that stage, the platelet could adopt any one of the possible basic structures of the compound, ordered or disordered.

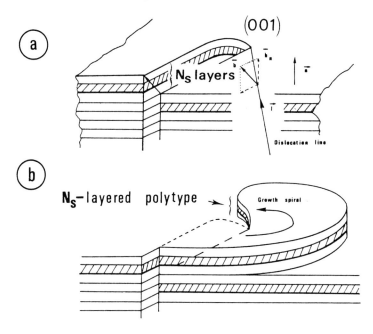

Figure 7-14. From a screw dis-location outcropping from a flat face (a) to the growth spiral which provides a structure memory (b).

Table 7-2. Polytypes generated by screw dislocations of increasing strength on the perfect $2M_1$ basic structure of mica.

$N_s =$	Exposed ledge structure	Resulting structure
1	$\mid \overline{A} \mid$ or $\mid \overline{B} \mid$	$1M\,(\overline{A})$ or $(\overline{B})$
2	$\mid \overline{A}\ \overline{B} \mid$ or $\mid \overline{B}\ \overline{A} \mid$	$2M_1\,(\overline{A}\ \overline{B}) = 2M_1\,(\overline{B}\ \overline{A})$
3	$\mid \overline{A}\ \overline{B}\ \overline{A} \mid$ or $\mid \overline{B}\ \overline{A}\ \overline{B} \mid$	$3Tc\,(\overline{A}\ \overline{B}\ \overline{A})$ or $3Tc\,(\overline{B}\ \overline{A}\ \overline{B})$
4	$\mid \overline{A}\ \overline{B}\ \overline{A}\ \overline{B} \mid$ or $\mid \overline{B}\ \overline{A}\ \overline{B}\ \overline{A} \mid$	$2M_1\,(\overline{A}\ \overline{B}) = 2M_1\,(\overline{B}\ \overline{A})$
5	$\mid \overline{A}\ \overline{B}\overline{A}\ \overline{B}\ \overline{A} \mid$ or $\mid \overline{B}\ \overline{A}\ \overline{B}\ \overline{A}\ \overline{B} \mid$	$5Tc\,((\overline{A}\ \overline{B})_2\,\overline{A})$ or $((\overline{B}\ \overline{A})_2\,\overline{B})$
2n	$\mid (\overline{A}\ \overline{B})_n \mid$ or $\mid (\overline{B}\ \overline{A})_n \mid$	$2M_1\,(\overline{A}\ \overline{B}) \to\, = 2M_1\,(\overline{B}\ \overline{A})$
2n + 1	$\mid (\overline{A}\ \overline{B})_n\overline{A} \mid$ or $\mid (\overline{B}\ \overline{A})_n\overline{B} \mid$	$(2n+1)\,Tc\,((\overline{A}\ \overline{B})_n\,\overline{A})$ or $((\overline{B}\ \overline{A})_n\,\overline{B})$

For an ordered BS, there are two possible relations for a Burgers vector after the screw dislocation forms:

(1) $|\vec{b_n}|$ is an integral multiple of the BS repeat distance. In this case, the BS structure continues to expand and no new polytype is formed.

(2) $|\vec{b_n}|$ is a non-integral multiple of the height of the BS unit cell. Such an imperfect dislocation will produce a new polytype structure with repeat distance equal to $|b_n|$, the height of the exposed ledge. For rhombohedral structures, the repeat is tripled and has value $3|b_n|$.

Frank (1951), Mitchell (1957) and Krishna and Verma (1965) derived in this way all the expected polytypes of SiC created from spiral growth on the 6H, 4H, and 15R BSs. By varying N_s, the position of the last layer, and the BS, they accounted for many of the observed polytype structures. Later, Baronnet (1975) extended the model to micas as an application to layered structures. For instance, Table 7-2 (above) lists the polytype structures generated by spiral growth on the $2M_1$ ($\overline{A}\ \overline{B}$) basic structure. For even values of N_s, the $2M_1$ BS grows further by the spiral, whereas for N_s odd, a triclinic polytype series develops epitaxially onto the BS. By also considering 3T, 2Or, $2M_2$ and 1Mrn (120) BS's, Baronnet (1975) rationalized roughly one-third of the known long-period polytypes of the micas.

The rather limited success of the model applied to a stacking-fault-free BS (or perfect matrix) inspired the following faulted-matrix model.

7.4.2.2   The faulted-matrix model of spiral growth

Vand (1951) suggested that a stacking fault could be present within the exposed ledge and therefore, involved in the final stacking sequence of the spirally grown polytype. Pandey and Krishna (1975a,b; 1976) reexamined the derivation of theoretical polytypes of SiC and CdI_2 by spiral growth but after insertion of a stacking-fault near the crystal surface before the screw dislocation originates. The general procedure is as follows:

• choose a basic structure,
• determine all the intrinsic and extrinsic stacking faults possible in the BS,
• among the stacking faults, select those expected to occur most frequently, i.e., those with the least stacking-fault energies (SFE) as calculated in §7.3.5,
• deduce all possible polytype structures that result from the extension of the exposed ledge of the screw dislocation. Such a ledge may contain successively one of each type of low-SFE stacking faults at different layer positions,
• vary Ns,
• estimate the SFE of the so-deduced polytypes as mentioned in §7.3.5 and retain those of lowest SFE among those with the same repeat distance.

This deductive process has been successful in explaining all polytype structures of SiC and CdI_2. In particular, the modified Frank model accounted for "polytype families", i.e., groups of structures with the same repeat but different layer stacking sequences. Also, the occurrence of complex polytypes with repeats of an integral multiple of the BS unit cell could be explained. The above model was applied to mica polytypes by Baronnet et al. (1981) and Pandey et al. (1982). As shown in Table 7-3, 14 of the 20 complex mica polytypes are predicted by the model.

The spiral growth models received experimental support by the observation of spirals on the basal faces of polytypic crystals. A one-to-one correspondance between the spiral

Table 7-3. List of the x-ray solved complex polytypes of the micas and performances of the perfect-matrix model (PMM) and the faulted-matrix model (FMM). + : explained, - : non-explained.

#	Ramsdell's notation	Stacking sequence (RTW)	PMM	FMM
1	$3Tc$	$[02\bar{2}]$	-	+
2	$3M$	$[1\bar{2}1]$	-	-
3	$4Tc_1$	$[0132]$	-	-
4	$4Tc_2$	$[(0)_2 2\bar{2}]$	-	+
5	$4M_1$	$[22\bar{2}0]$	+	+
6	$4M_2$	$[22\bar{2}\bar{2}]$	-	+
7	$5Tc$	$[(2\bar{2})_2 0]$	+	+
8	$5M$	$[2222\bar{2}]$	+	+
9	$6M$	$[1\bar{2}1\bar{1}2\bar{1}]$	-	-
10	$8Tc_1$	$[(2\bar{2})_3\bar{2}2]$	-	+
11	$8Tc_2$	$[(0)_6 2\bar{2}]$	-	+
12	$8Tc_3$	$[(0)_3 2\bar{2}\bar{2}202]$	-	-
13	$8M$	$[(222)_2 2\bar{2}]$	+	+
14	$9Tc$	$[(0)_7 2\bar{2}]$	-	+
15	$10Tc$	$[(2)_5\bar{2}2200]$	-	-
16	$11M$	$[(222)_3 2\bar{2}]$	+	+
17	$14Tc$	$[(0)_{12} 2\bar{2}]$	-	+
18	$14M$	$[(222)_4 2\bar{2}]$	+	+
19	$18Tc$	$[(0)_{n1}2(0)_{n2}(0)_{n3}2]$ $n1+n2+n3=15$	-	-
20	$23Tc$	$[(0)_{21} 2\bar{2}]$	-	+

step height and the polytype repeat distance or its multiple has been verified by combining optical interferometric methods and x-ray diffraction techniques on SiC (Verma, 1951), CdI_2 (Forty, 1952) and mica (Amelinckx, 1952; Amelinckx and Dekeyser, 1953). Recently, however, the relationship between spiral growth and polytypism was disputed for CdI_2 (Sarna and Trigunayat, 1989; Trigunayat, 1991).

7.4.2.3   *The epitaxial model by Vand and Honoka (1967)*

This model is also a variation of the Frank model. Vand and Honoka (1967) assumed that a screw dislocation may form in a crystal by epitaxial growth on a foreign nucleus instead of from "buckling and slip" of the initial platelet. This epitaxial mode of growth is believed to cause the exposed ledge to be faulted in a more flexible way than by any homogeneous nucleation process of the platelet. Long-period polytypes do not require a genetic basic structure to form. Also, the syntactic coalescence of several polytypes can be explained by this model. However, it remains rather speculative since a foreign nucleus within polytype crystals has not been demonstrated systematically.

7.4.2.4   *The periodic slip mechanism*

Mardix et al. (1968) proposed an ingenious mechanism to explain the formation of synthetic ZnS polytypes grown from the vapor phase. Such polytypes form in the solid-state by nucleation and subsequent expansion of stacking faults along helical atomic planes of the crystal after growth.

At high temperature ($T \simeq 1250°$) well above the 3C-2H transition temperature, ZnS forms as wurtzite (2H) needles and then platelets. As the crystals cool to room

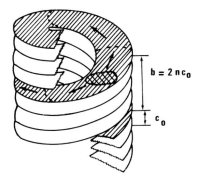

temperature, the 2H structure converts into polytype domains of faulted 2H, 3C, 4H, 6H, and more complex polytypes. They form parallel strips perpendicular to the **c** axis as a dislocation line running parallel to **c**, as observed by x-ray topography (Kiflawi, 1972). This dislocation is of screw character and develops the parent needle. The screw has a pitch equal to $2nc_0$, with n as an integer and c_0 as the interlayer spacing along the **c** axis ($c_0 = c_{2H}/2$).

The basal (00.1) layers form a set of interleaved helical ramps around the dislocation line (Fig. 7-15). Every second layer is a part of the same lattice plane. Now, if an intrinsic stacking fault with a Schockley partial as a borderline, nucleates within one layer, it will tend to expand along the spiral ramp. Accordingly, any set of stacking faults will occur at distances along the dislocation line in relation to the pitch of the screw; ordered polytypes will result. The number, distribution and nature of propagation of the stacking faults control the stacking sequences generated. Moderate mechanical stresses and even prolonged storage near room temperature also contribute to the expansion of the faults. The elementary slip associated with the propagation of the fault along one turn repeats on the 2nth, 4nth, 6nth ...etc, so that periodic slip accompanies the polytype change.

The transformation of the h.c.p. to the f.c.c. structure is illustrated:

$$
\begin{array}{ll}
\dots A\ B|\ A\ B\ A\ B\ A\ B \dots & 2H \\
\qquad\quad \downarrow \\
\qquad C\ A|\ C\ A\ C\ A \\
\qquad\qquad \downarrow \\
\qquad\qquad B\ C|\ B\ C \\
\qquad\qquad\quad \downarrow \\
\qquad\qquad\qquad A\ B \\
\hline
\dots A\ B\ C\ A\ \ B\ C\ A\ B \dots & 3C
\end{array}
$$

Planes of slip are indicated by vertical lines and blocks of faulted layers are on the right side. The most frequent modulus of the Burgers vector, $2c_0$, for the dislocation in the 2H structure is working here.

7.4.3 Discussion

Each of the above phenomenological and mechanistic models applies to polytype relations of specific substances and, taken together, they encompass many aspects of

258

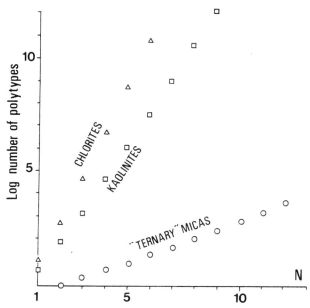

Figure 7-16. The number of non-equivalent layer-stacking sequences for some layer silicates as a function of the number N of layers in the repeat. [Drawn from data by McLarnan (1981).]

polytypism. However, none are universal, probably because polytypism involves thermo-dynamic, growth, and post-growth developments, which contribute together or separately to the process of polytype formation, depending on the structure and composition of the material. However, the role of stacking faults is universal, either related to growth or post-growth processes. Further understanding of polytypism requires more data on local stacking microstructures and their variations in crystals. Transmission electron microscopy (TEM) is an important tool for this purpose and, at times, the only one available to study polytypism and growth mechanisms of small crystals.

7.5 TEM TECHNIQUES RELEVANT TO POLYTYPISM

7.5.1 X-ray data *vs*.TEM data

X-ray diffraction techniques provide structural data averaged over an order of magnitude of 10^{15} unit cells. Such techniques give an idealized picture of a crystal. This lack of spatial resolution is not detrimental for nearly perfect crystals (e.g., beryl, tourmaline, etc.). However, complex microstructures of minerals (exsolution textures, microtwinning, planar defects, intergrowths) disturb intrinsic properties of the (assumed) perfect structure. With regard to polytypism, many excellent single-crystal studies have allowed the determination of the repeat period, symmetry, and the layer stacking sequences (Verma and Krishna, 1966; Farkas-Jahnke, 1983). For periodic polytypes, the stacking structure can be solved provided that a coherent domain is sufficiently large, there is no overlap of diffraction patterns of several polytypes (syntactic coalescence), and the layer repeat N is not too large. Long layer repeats involve a huge number of possible stacking structures with the same periodicity (Fig. 7-16). The structure solution is often conducted from a "trial and error" process consisting of trying to match observed diffracted intensities along polytype sensitive diffraction rows with calculated diffracted intensities from trial stacking sequences. In doing so, the uniqueness of the solution may not be obvious for large N, i.e., for long-period polytypes.

In an ordered structure, randomly distributed stacking faults contribute to diffuse intensity outside the Bragg reflection positions (e.g., Plançon, 1981; Reynolds, 1980). Thus, it is difficult to extract data about the nature and density of faults. Finally, semi-random or totally random stacking of layers cannot be studied by x-ray diffraction.

Transmission electron microscopy provides us with data complementary to x-ray diffraction. TEM is able to provide localized microstructural data both in the diffraction mode (reciprocal space) by selected area diffraction and in the imaging mode (direct space). Modern high resolution TEM's have point-to-point resolution between $\simeq$ 1.5 and 3 Å so that the layered module of polytypic minerals may often be resolved. Therefore, projections of polytype sequences can be observed when imaged properly (e.g., Van Landuyt et al.., 1983). Also, randomly distributed stacking faults may be directly imaged. The local analysis of partial order in disordered stacking sequences is also possible by HRTEM.

However, in contrast to x-ray diffraction techniques, TEM does not provide precise data about mean atom positions, in part because the electron diffracted beams are commonly subject to multiple diffraction effects (P.R. Buseck, J. Steeds, P. Self, this volume). Thus, the kinematic theory of diffraction cannot be used for thick specimens. Another drawback of TEM is that many minerals become amorphous when exposed to the intense electron beam required for HRTEM. On the other hand, conventional transmission electron microscopy (CTEM) allows imaging of surfaces with replica techniques, and correlations can be made between crystal morphology, growth mechanisms of basal faces, and polytypism (Baronnet, 1976, 1978, 1980).

7.5.2 Specimen preparation techniques

7.5.2.1 Surface studies—replica techniques

The platinum-carbon replica technique and the gold decoration technique may be used to determine the surface microtopography of polytype crystal platelets. Platinum-carbon (Pt-C) shadowing of basal faces is carried out by inclined condensa-tion of a 100-150 Å thick Pt-C mixed film under 10^{-4}-10^{-5} torr residual pressure (Skatulla and Horn, 1960). The microcrystals are dispersed over a thin glass plate (Baronnet, 1972, 1973) or a mica flake (Nadeau, 1985). After shadowing, the layer coating the crystals and the substrate are floated on diluted HF (for silicates). After total dissolution of the crystals and separation of the substrate, the Pt-C film is washed in water, placed on electron microscope grids, and examined by TEM in transmission mode. The particle morphology, including growth or deformation steps, is imaged mainly by contrast in the Pt density (Fig. 7-17a).

The gold-decoration technique (Basset, 1958; Sella et al., 1958) was successfully applied to clay minerals by Gritsaenko and Samotoyin (1966) and then used to characterize microsteps on synthetic micas (Baronnet, 1972, 1976) and natural clays (Sunagawa and Koshino, 1975; Tomura et al., 1979; Yonebayashi and Hattori, 1979; Kitagawa et al., 1983). After dispersion of the crystallite population in distilled water and collection on a cover glass, the specimen is dried. The cover glass (with attached crystals) is then heated to 300-550°C for 1-2 hours under a vacuum of 10^{-5}-10^{-6} torr, to degas the contamination layer. Cooling to 100-150°C is followed by flash evaporation of gold and subsequent 100-Å-thick carbon coating. Procedures for HF dissolution, water cleaning, and transfer to the microscope are identical to those of the Pt-C shadowing procedure. When successful,

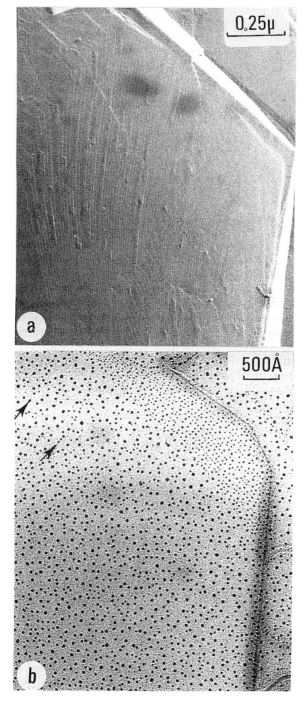

Figure 7-17. High resolution surface replica techniques usable with a conventional TEM. Trains of elementary steps (10 Å in height) on micas as seen by (a) Pt-C shadowing, (b) Au-decoration. The regular train of steps forms a vicinal face at the border of the mica habit.

flashed gold particles as small as 20-30 Å in size are observed by TEM. These particles are scattered over the decorated surface but are aligned preferentially along microsteps (Fig. 7-17b). The essential characteristics of the microtopographic methods are the ability to faithfully image step features including small (vertical "resolution") and narrow (horizontal "resolution") steps. Comparisons of several techniques are given in Table 7-4.

Table 7-4. Comparisons of the performances of the surface visualization techniques. TEM: transmission electron microscopy; SEM: scanning electron microscopy; OM: optical microscopy.

Technique	step height	step separation	comments	
			capabilities	drawbacks
<u>TEM</u> Pt-C shadowing	≤ 7 Å	~ 7 Å	relief, respective altitude seen ; step height measurements	directional contrast
Au-decoration	< 2 Å	~ 30 Å	exceptional resolutions	no altitude data, no step height measurement
<u>SEM</u> secondary electron and back-scattered electron imaging	50-100 Å	> 100 Å	not restricted to platy crystals very easy use	total loss of elementary steps
<u>OM</u> phase-contrast imaging	< 5 Å	> 1 μm	exceptional vertical resolution use friendly	very anisotropic resolutions large and flat surfaces needed

7.5.2.2   Bulk studies

Preparation techniques for crystalline substances to be examined by TEM are given by Buseck in Chapter 1. We discuss here only their applications to polytypic crystals. Depending on their mechanical properties, crushed-grain mounts may or may not be appropriate. This sample preparation technique works for brittle crystals without perfect cleavage (e.g., SiC), but destroys or at least modifies the layer stacking sequences for soft crystals or those affected by plastic deformation (e.g., chlorite, clay minerals, ZnS- see §7.4.2).

Slicing resin-impregnated materials with an ultramicrotome equipped with a diamond knife is the main preparation technique for separated, micron-sized, natural or synthetic, crystal populations for TEM examination along the layers (e.g., Amouric, 1981; Amouric and Baronnet, 1983). For larger crystals, only portions are preserved from mechanical damage during the slicing (e.g., Iijima and Buseck, 1978).

Ion milling is probably the least-damaging technique for polytypes, provided that mechanical damage involved in the initial preparation of the thin section is eliminated. However, the moderate heating (≈ 200°C) of the specimen during the ion-milling process may cause polytype transformations (e.g., ZnS): an ion mill with a cooling stage is recommended. A resin technique is now available for ion thinning (Ferrow and Roots, 1989).

7.5.3   Diffraction techniques

Because of the heterogeneity of polytype microstructures and their fine-scale inter-growths, the diffraction techniques for polytypes have recently evolved from x-ray sources to electron sources. With increasing resolution, these techniques involve normal x-ray beams, x-ray microbeams, reflection high-energy electron diffraction (RHEED), parallel beam selected-area electron diffraction (SAED), and convergent-beam electron diffraction (CBED; ess Chapter 2). Among electron-diffraction techniques especially suitable for polytype studies, SAED is dominant relative to RHEED, and CBED is less commonly used. RHEED provides good ($\simeq$ 100 kV) diffraction patterns from the lateral faces of polytype crystals with a resolution of $\simeq$ 2.5 μm and allows easy scanning of polytype structures along the normal to the layers. Tedious preparation techniques are not required. Accordingly, RHEED has provided significant data on parallel intergrowth of polytypes over con-siderable distances (cm range). Gauthier and colleagues have studied syntactic coalescence in SiC (Gauthier and Michel, 1977) and ZnS (Gauthier, 1980).

7.5.3.1   General diffraction features

The most useful data about a polytype structure occurs along reciprocal lattice rows normal to the layer plane. Consequently, the incident beam of electrons must be parallel to the layers. Thus, systematic 00ℓ reflections always appear, with the possible addition of lateral hkℓ rows, depending on the azimuthal position of the crystal around the normal to the layers (c* axis). For kinematic diffraction, certain hkℓ rows are insensitive to the superstructure because of systematic extinctions due to the structure of the BL. No super-lattice information is inherent for 00ℓ data. More generally, extinctions occur along hkℓ diffraction rows such as (h-k) = 0 (mod 3) for closest-packed substances like SiC, ZnS, and CdI$_2$, and k = 0 (mod 3) = 3n for most layer silicate polytytpes.

Any other hkℓ row has information about the stacking sequence. The periodicity N of the polytype is determined from the reciprocal distance d^*_N between successive diffraction spots versus d^*_1 the reciprocal distance between single layer (subcell) spots as: N = d^*_1/d^*_N. A similar result occurs by adding one to the number of equispaced superstructure spots found between consecutive substructure spots (see Figs. 7-18a and b).

As N increases with long-period polytypes, d^*_N becomes so small that it is difficult to distinguish a row of discrete, closely spaced reflections from continuous streaking along such a row. Accordingly, a regular long-period polytype is easily confused with a disordered stacking sequence. Nonetheless, from sharp electron-diffraction patterns, Van Landuyt et al. (1983) were able to distinguish 1300 Å-repeat SiC polytypes and Bell and Wilson (1977) and Bigi (1980) showed complex mica polytypes, beyond 11 layers in repeat. The symmetry class of the polytype is deduced from the geometry of its reciprocal unit cell.

7.5.3.2   SAED and CBED

SAED patterns are commonly used in TEM work because of the spacial resolution of the aperture on the crystal and the high resolution of spots on the diffraction photograph. Space resolution is achieved by the selected-area aperture being inserted in one of the image planes of the electron microscope such that only a small circular field of the sample contributes to the diffraction pattern (Chapter 1). On 100-200 kV commercial instruments, the smallest selected diameter on the specimen is $\simeq$ 400 nm. SAED is required to separate polytype components across an intergrowth, since such a topotactic coherent intergrowth

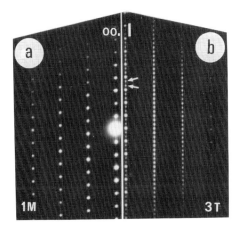

Figure 7-18. Zone-axis SAED pattern of ordered mica polytypes seen along the layers, (a) one-layer (1M) repeat or substructure, and (b) three-layer (3T) repeat. Superstructure spots are marked by arrows.

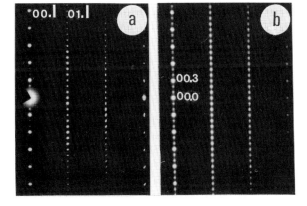

Figure 7-19. The modification of relative intensities of diffraction spots of a 3T polytype of mica as a function of the specimen thickness.

(a) Thinner than 100 Å.
(b) Thicker than 500 Å.

causes overlap of the reciprocal rows of the various components. The sharpness of diffraction spots is optimized by the use of a roughly parallel electron beam and by using a selected aperture that is not too small. As mentioned above, a crystal structure cannot be easily solved from SAED patterns because of dynamical interactions between diffracted beams. However, a high-quality SAED pattern is a prerequisite to forming high-resolution images of the polytypes as shown below.

The thickness of the sample can greatly affect the SAED pattern. For example (Figs. 7-19a,b): (1) forbidden reflections may appear if stacking-sensitive reciprocal rows are simultaneously excited, (2) diffraction spots may gain intensity with respect to the transmitted beam, (3) diffraction spots loose their sharpness due to the increase in inelastic electrons contributing to a halo. Convergent-beam electron diffraction (CBED) is of little use in the study of polytypism because diffraction disks overlap chronically along spot-crowded rows and because of the high electron dose received by the specimen (producing specimen damage for many polytypic compounds).

7.5.4   Imaging techniques

As imaging procedures with a TEM (and their interpretations) are detailed in Chapters 1 and 3 of this volume, imaging techniques are only briefly discussed here.

7.5.4.1   *Conventional transmission electron microscopy (CTEM)*

CTEM refers to the use of a TEM at low resolutions, generally using amplitude contrast (§1.2.4; §1.4.4). CTEM may be used to image the c lattice dimension if this period is greater than the lattice fringe resolution of the microscope (5-10 Å even for low-resolution microscopes). Since long-period polytypes are made by the periodic repetition of a small set of stacking faults, a bright-field technique consists of enhancing the contrast of these faults with respect to the surrounding matrix. This effect is achieved by slightly tilting the crystal away from the exact Bragg position (zone axis orientation). Thus, the zero'th order Laue zone of the SAED is brought into diffracting position so that a set of closely spaced spots is intensified along polytype-sensitive rows. In thick regions, the 00ℓ row then exhibits forbidden reflections close to the transmitted beam as a result of dynamical diffraction, so that the polytype contrast modulation along $\mathbf{c}^*$ may be imaged (Fig. 7-20a) using a medium-sized objective aperture centered on the transmitted beam 000. Dark-field imaging may be used under similar conditions by centering the objective aperture on the set of polytype reflections. This procedure also works for diffuse streaks in order to image stacking disorder (Fig. 7-20b).

7.5.4.2   *High resolution transmission electron microscopy (HRTEM)*

The high resolution TEM of today will soon be the CTEM of tomorrow! Electron microscopes with a point-to-point resolution of less than about 3 Å are commonly considered high-resolution instruments.

1-D lattice imaging. To obtain data regarding N and to count the number of single layers involved in the contrast sequence of sublattice fringes, the lattice-fringe resolution must be greater than the layer thickness (say 2.5 Å for SiC and ZnS, $\simeq$ 7.3 Å for serpentines, $\simeq$ 10 Å for micas, etc.). One-dimensional (1-D) lattice images will suffice for this purpose.

The procedure for bright-field 1-D imaging is as follows:

- choose a thick portion of the polytype region,
- find the 00ℓ row in SAED mode,
- tilt the crystal with a double-tilt specimen stage until "polytype-sensitive" diffraction rows appear. If necessary, tilt slightly around the normal to the row. Forbidden reflections along 00ℓ should be visible, if N > 1,
- center the objective aperture on 000 so that only some 00ℓ reflections pass through it,
- take a through-focus series (§1.3.6) starting at underfocused conditions to bracket the Scherzer focus [$\Delta f_S \simeq -1.2\ (C_S\lambda)^{1/2}$, with C_S as the spherical aberation coefficient of the objective lens and λ the electron wavelength] (§1.2.6; 1.3.5).

Warning: the 1-D images of polytypes are conclusive for documenting N only if the above procedure is followed (e.g., Rule et al., 1987). In case only lateral diffraction rows containing forbidden reflections appear (e.g., $h0\ell$ and h, 3h, ℓ for micas) in addition to 00ℓ, the 1-D lattice images always show a one-layer repeat contrast (Fig. 7-21). However

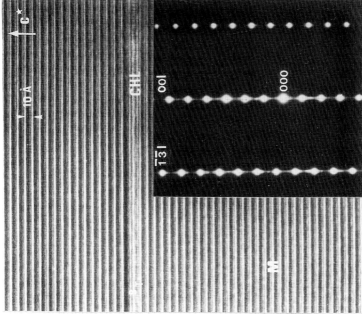

265

Figure 7-20 (left). Conventional TEM images of biotite stacking sequences. (a) Bright-field image of intergrown complex polytypes. [From Baronnet and Kang (1989) Fig. 1, p. 481.] (b) Dark-field image of stacking disorder.

Figure 7-21 (above). <310> zone axis, 1D image of the biotite substructure and of a few intergrown chlorite layers. The one-layer contrast of (001) lattice fringes (10 Å apart) shows up irrespective of the actual layer stacking sequence.

this repeat is inconclusive regarding polytypic repeat and disorder in the crystal. Similarly, results are inconclusive if 00ℓ spots only are excited in the SAED. This has important consequences about order-disorder in layer stacking deduced from the "blind" recording of 1D-lattice fringes of very small (along $\mathbf{c}^*$) crystallites of natural (Amouric and Parron, 1985) or synthetic (Amouric et al., 1978) clay minerals. Synthetic clay minerals cannot easily produce a recordable, single-grain diffraction pattern. The general rules are:

(1) any crystallite showing a one-layer period of contrast is not necessarily indicative of a one-layer polytype,

(2) any other type of contrast modulation (two-layer, three layer, ..., disordered) is representative of the stacking repeat assuming that mixed-layering of structure types is not present; e.g., R = 1 illite/smectite.

Guthrie and Veblen (1989) simulated (001) lattice fringes of $2M_1$ muscovite. They showed that the apparent single-layer thickness can deviate from 10 Å and therefore mimic an ordered 1/1 mixed-layering.

2D lattice imaging. Though more difficult to obtain, the two-dimensionally modulated lattice images display much information (N, symmetry class, stacking sequence) about the polytype, but as a projection parallel to the incident beam. Now, in addition to single layer resolution, the shortest lateral periodicity along the layers may also be resolved ($\simeq 2.7$ Å for ZnS and SiC, $\simeq 4.4$ Å for most layer silicates). Only axial-illumination, bright-field, high-resolution images are considered because image calculations can be safely made for them with the multislice method (O'Keefe et al., 1978) of image calculation.

The main experimental steps for obtaining high-resolution 2D-images of polytypes are:

- Choose a thin (<100-200 Å) wedge, transparent to electrons over a sufficiently large area.
- Use a double-tilt specimen stage and orient the crystal in SAED mode so that a principal direction (zone axis) is parallel to the electron beam. This is achieved when a densely-populated reciprocal-lattice plane appears. It is critical to obtain an exact (centered) orientation. Proper orientations are obtained when all hkl and $\bar{h}\,\bar{k}\,\bar{l}$ reflections of the zero-order Laue zone are symmetric and/or when the first-order Laue zone on the transmitted beam in slight convergent-beam mode is centered. The selective aperture should be as small as possible to minimize the averaging effect of the hour-glass curvature of the ultra-thin crystal about its orientation (case of easily bent minerals).
- Choose an objective aperture that omits the space frequencies just beyond the first zero of the transfer function (see Chapters 2 and 3), i.e., the reciprocal of the point-to-point resolution of the microscope. Center the aperture carefully around the transmitted beam so that no asymmetry is introduced to the diffracted beams hkl and $\bar{h}\,\bar{k}\,\bar{l}$ passing near the aperture cut-off.
- Switch to the real-space image mode and correct for astigmatism at the appropriate magnification on nearby amorphous carbon film or on a contaminated / amorphous edge near the crystal.
- Check the exact focus near the zone of interest by minimizing the contrast of an amorphous edge.
- Shift to this zone; wait for mechanical stabilization.
- Take a series of underfocused, through-focus images that bracket the Scherzer focus...
- and hope that there was no specimen drift and only minimal electron damage of the sample!

7.5.5 From the HR image to the layer stacking sequence

7.5.5.1 Structure image and image simulations

As with any type of microstructure, the polytype images vary greatly with changes of focus, specimen thickness and misorientation, and other operating characteristics of the microscope (accelerating voltage, beam divergence, spherical and chromatic aberration coefficients, objective aperture size, etc.). Thus, it is essential to obtain images as close as possible to the projected charge density of the actual structure in the viewing direction. This condition is theoretically met at Scherzer focus for a very thin sample (<50 Å) and this image is referred to as the "structure image". However, the conditions may depart significantly from the ideal as shown by 2-D image simulation studies of mica polytypes at 100 kV (Amouric et al., 1981) or 200 kV acceleration voltages (Ducellier, 1991) for which -1200 to -1400 Å defocus is better than the ≈ -900 Å Scherzer defocus. Furthermore, the full set of high resolution simulated images as a function of both thickness and defocus proved invaluable in avoiding erroneous interpretation of the stacking sequence projections (Figs. 7-22a and b) deduced from non-ideal experimental images.

7.5.5.2 The 2D to 3D problem

For closest-packed substances, projected stacking sequences on $\{11\bar{2}0\}$ determine unequivocally the 3-D stacking sequence (see §7.1.2). This is not the case for polytypic substances having non-closest-packing configurations for which the projected structure may correspond to several possible stacking sequences. For example, for micas, when viewed along densest [100] and ‹110› directions in the mica layer, the c vector of each layer may project vertically (0) or inclined to the right (+) or to the left (-) from one layer to the next (see Fig. 7-22a and those of §7.6.3 and §7.6.4). If Zvyagin's C layer position is set parallel to the electron beam (Amouric and Baronnet, 1983), we have the correspondence:

$$+ \Rightarrow \bar{A} \text{ or } B; - \Rightarrow \bar{B} \text{ or } A; 0 \Rightarrow C \text{ or } \bar{C}$$

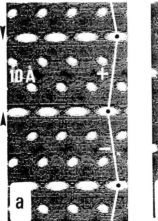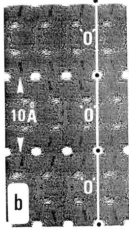

Figure 7-22. Two simulated HRTEM images of a 2M₁ biotite seen along [100] using the multislice method. Operating conditions are: acceleration voltage: 200kV, semi-angle of beam convergence: 0.30 mrad, objective aperture radius: 0.35 Å⁻¹, depth of focus: 90 Å, spherical aberration coefficient: 3.5 mm, (a) defocus: Δf = -1200 Å, specimen thickness: e = 50 Å. Brighter spots mark correctly the ... + - + - ... sequence of shifts. (b) Δf = -800 Å, e = 105 Å. Brighter spots indicate erroneously a wrong ... 000 ... sequence of shifts which may be misleading for polytype identification (seemingly 1M seen along [100] or [1̄00]). (After Ducellier, 1991).

so that two distinct layer positions may be inferred from one projection of a layer. Starting from a projected sequence of N layers, there are 2^{N-1} possible stacking sequences including symmetry-equivalent solutions. For mica sequences with six possible stacking angles $(t_n \wedge t_{n+1}) = 0 \mod \pi/3$ (see §7.1.2), Iijima and Buseck (1978) found that three distinct projections would be necessary to determine the 3D structure. Two projections are sufficient for "ternary" micas with $(t_n \wedge t_{n+1}) = 0 \mod 2\pi/3$, such as biotite and muscovite. In the latter case, Amouric and Baronnet (1983) further deduced that one projection provides two solutions only, which are enantiomorphs (mirror images). Because the idealized mica layer has a (010) symmetry plane, these two solutions are energetically equivalent. This relation will be used below to interpret HRTEM images of micas.

The situation, however, is more complex for chlorite (Fig. 7-27, below), the single layer of which consists of two sheets: a talc-like sheet (T) of general composition $(Mg,Fe,Al)_3(Si,Al)_4O_{10}(OH)_2$, and a brucite-like sheet (B), $(Fe,Mg,Al)_3(OH)_6$. Due to the numerous ways of stacking (B) on (T) and then (T) on (B) there are 12 non-equivalent chlorite polytypes with a one-layer repeat (Bailey, 1988b). Accordingly, many different structures have the same projection and the interpretation of HRTEM images must be constrained by x-ray diffraction (Spinnler et al., 1984; Bons and Schryvers, 1989).

Two relations should be considered when interpreting 3D stacking variations: (1) any periodicity of the projection of a polytype may not be indicative of that of the 3D-structure, and (2) any aperiodicity of the projection of a polytype is identical to that of the 3D structure.

7.6 LAYER SILICATE POLYTYPISM: TEM RESULTS

Examples of TEM examination of layer silicates are presented below. This presentation addresses current theories of polytypism and/or suggests further research. Furthermore, examples are given in which polytypism is used as a tool to constrain replacement mechanisms of layer silicates.

7.6.1 Micas as prototype layer silicates for growth features

Mica minerals demonstrate the relationship between growth features and polytypism because: (1) micas display a simple case of polytypism described only by layer rotations. This contrasts with 1:1 (e.g., lizardite, kaolinite) and other 2:1 (talc, pyrophyllite, chlorite, vermiculite) layer silicates where interlayer shifts are also involved. (2) Polytype transformations do not occur in mica when they are annealed (Takéuchi and Haga, 1977) or deformed. Other layer silicates are subject to interlayer glides so that stacking may change under thermal and/or mechanical stress. Consequently, synthetic and natural micas are good candidates for TEM analysis of polytype microstructures with interpretation strictly as growth features.

7.6.2 Growth mechanisms of mica basal faces

Micas from quenched hydrothermal synthesis experiments were studied by micro-topography of {001} faces by replicas made from inclined Pt-C shadowing and gold decoration techniques (Baronnet, 1972; 1975b; 1976, 1980; Baronnet et al., 1976). The examination of samples with different duration times show the following features:

• Growth starts with nucleation of very thin platelets exhibiting smooth {001} faces, with or without shallow growth islands (Fig. 7-23a). These patterns are

Figure 7-23. Growth features on synthetic micas as imaged by Pt-C shadowing. (a) Growth islands on lepidolite (see arrows). (b) Interlaced growth spiral on phengite-smaller steps marked by arrows are elementary (10 Å in height) while others are of double height.

typical of a 2D nucleation-and-growth process of new layers taking place under high supersaturation rates.

• With increasing run durations, growth continues for some crystals by means of growth spirals (Fig. 7-23b), whereas no increase in thickness occurs for those with flat faces. The BCF mechanism (Burton et al., 1951) is the only mechanism that insures basal face growth under lower supersaturation rates. By observing samples that were transitional between 2D-nucleation and spiral growth of basal faces, the origin of screw dislocations (Baronnet, 1973) was identified as: (i) recombination of dendritic arms of single crystallites, (ii) edgewise merging of separate mica platelets.

• The shape of the single mica layer is anisotropic (trioctahedral micas) to very anisotropic (dioctahedral micas) with elongation along [100] and minimum growth rate along [010] (single layer 1M setting) in agreement with Au-decoration results on natural sericite (Tomura, 1985). Such anisotropic shapes do not depend on the way the underlying layer is oriented. Growth step patterns as indicated by spirals can be used to determine layer stacking sequences from the screw-dislocation mechanism of growth. These spirals commonly cause the 1M and $2M_1$ basic structures (BS's) to grow (Baronnet and Amouric, 1986) with a Burgers vector equal to, or an integral multiple of, the BS repeat distance. Much less frequently, complex polytypes are generated, with layer stacking sequences of

$$3Tc[2\bar{2}0](\bar{A}\ \bar{B}\ \bar{A}) \quad \text{and} \quad 5Tc[(2\bar{2})_20](\bar{A}\ \bar{B}\ \bar{A}\ \bar{B}\ \bar{A}),$$

i.e., based on the $2M_1$ BS for muscovite, paragonite and phengite. These results are predicted from the perfect-matrix model of spiral growth (see Table 7-2). For synthetic phlogopite, complex polytypes are based on a more or less faulted 1M BS, i.e., in keeping with the faulted-matrix model of spiral growth.

These experiments clearly show that the BS ordering is present in mica platelets before growth spiral mechanisms operate. Consequently, the process of BS ordering is related to early stages of crystal growth, i.e., to the 3D-nucleation and 2D-nucleation growth stages. In other words, spiral growth has nothing to do with the generation of sequences of ordered basic structures. A strong dependence of the growth habit upon the stacking sequence of synthetic dioctahedral micas is also observed. 1M muscovite, paragonite or phengite develop as platy elongated laths with well developed {001}, {010} plus minor {100} and {110}. The $2M_1$ habit is typically diamond-shaped with major {001} and {110} and minor {010}, whereas the 3T form is pseudohexagonal: {00.1} and {10.0}. Hence, the slowest growing faces are those which contain the slowest growing ledges of the single layers. X-ray powder diffraction combined with CTEM observation of natural illite appears to produce a similar conclusion (Inoue et al., 1988) regarding the shapes of 1M-laths and $2M_1$ diamond-to-hexagonal-platelets.

7.6.3 Polytypism and small crystal size

7.6.3.1 In situ *nucleation stacking*

To confirm the differences between the initially formed BS and the complete crystal whose growth is controlled by screw dislocation, Amouric and Baronnet (1983) conducted short isothermal, isobaric nucleation experiments of muscovite, at P_{H2O} = 1 kbar, and at three nucleation temperatures T_N = 355°, 502° and 630°C. For each run, the products were microtomed and the layer-stacking sequences of individual, dislocation free, crystallites were determined through HRTEM 2D-images (Fig. 7-24). The detailed analysis of crystallites a few layers in thickness produced the following conclusions:

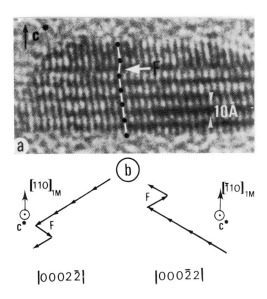

Figure 7-24. Short stacking sequence of a microtomed crystallite of synthetic muscovite. (a) 2D-HRTEM image slightly misoriented. (b) Stacking vector representation of the two enantiomorphous possible stackings and RTW symbols. [From Amouric and Baronnet (1983) Figs. 10a,b, p. 153.]

- At 355°C: disordered stacking 1Mrn (120) dominated, with incipient local 1M or $2M_1$ ordering; a few 1M crystals coexisted with disordered ones.
- At 502°C: crystals with more or less faulted 1M structure coexisted near others with severely faulted $2M_1$-dominant stacking.
- At 630°C: almost all crystals were $2M_1$ with incipient stacking faults often located in the middle of the sequence, i.e., at the locus of the nucleus.

Generally, for a number of crystals observed at each temperature, increasing temperature was found to promote the 1Mrn (120)→1M→$2M_1$ trend of BS conversions, in agreement with past x-ray diffraction studies (Yoder and Eugster, 1955; Velde, 1965).

For crystals examined along c^* from cores to the rims, increasing stacking disorder was found to accompany conditions of increasing supersaturation at constant temperature. Furthermore, packing of the first few layers (in the core) during the nucleation stage is commonly unrelated to the subsequent layer arrangement, in contrast to the hypothesis of Takeda and Ross (1975) in which once a packing style starts, it would be expected to continue during further growth. Although the suggestion of Takeda and Ross (1975) appears reasonable, Amouric and Baronnet (1983) suggested from the above experiments that system parameters such as growth temperature and supersaturation may play an important role.

These experiments decisively support the faulted-matrix model of spiral growth for micas (see §7.4.2). The existence of BS order affected by stacking faults within the initial platelets can be observed directly by HRTEM. Combined with many examples of spiral growth for synthetic micas as produced from replica techniques of CTEM, it is clear that electron microscopy has contributed significantly to define polytypic relations of clay-sized crystals of mica.

7.6.3.2 Polytypes, stacking faults and thin crystals

It is difficult to define a stacking sequence and stacking faults in crystallites a few layers thick. For example, Figure 7-24 illustrates a sequence of layers: - - - - + - corresponding to either of the two enantiomorphous Zvyagin sequences:

$$| A A A A B A | \text{ or } | \overline{B}\,\overline{B}\,\overline{B}\,\overline{B}\,\overline{A}\,\overline{B} | ,$$

with vertical bars representing the {001} faces. In terms of basic structures and stacking faults, this limited sequence can be interpreted in two ways: (1) a 1M sequence with a single layer in the wrong position (layer displacement fault), (2) the syntactic coalescence of one 1M and one $2M_1$ stacking sequences, each of three layers. This ambiguity concerning layer stacking interpretation becomes more cumbersome as the number of layers in the crystallite decreases. Obviously a significant periodicity is necessary to establish an ordered BS and likewise sufficient periodicity is needed to establish the location of a stacking fault. This problem complicates the interpretation for thin crystallites and thin slabs of consecutively intergrown polytypes, as in mixed-layer silicates with coarse slabs (Fig. 7-7a and b). Thus, at very early stages of growth, it is difficult to interpret the formation of a polytype given the initial two, three, or four layers of the platelet nucleus. Thomson (1981) argued from configuration entropy considerations that stacking disorder prevails in thin mica crystals at equilibrium, whereas larger crystallites are more ordered. The above HRTEM observations of synthetic micas appear to support this based on statistically considerations, the core of the crystals being regularly non-periodic with respect to a more ordered rim (Amouric and Baronnet, 1983). However, the nucleation stage corresponds to a large departure from equilibrium conditions.

7.6.3.3 Ostwald ripening of polytypes

For crystals a few microns in size or less, solubility increases significantly with decrease of size, similar to what the vapor pressure does around liquid droplets at equilibrium conditions (Jones, 1913). If a population of crystals with different grain sizes coexists within an aqueous solution at a given temperature and global supersaturation, then Ostwald ripening may occur. Crystals smaller than a "critical size" dissolve whereas those greater in size grow (Ostwald, 1900; Bigelow and Trimble, 1927). In closed systems, a mass transfer occurs from dissolved grains to those that grow, with a decreasing rate with time. This process of dissolution/crystallization was observed by Baronnet (1974, 1982) for synthetic phlogopite and muscovite crystal populations. Based on TEM examination of grain-size distributions and habits of mica crystals from hydrothermal runs of varying duration, the ripening process explains the following: decay of the number of crystals, decrease in the specific surface area, broadening of grain-size distributions toward coarser sizes, crystal habit evolution from very flat to more isotropic, and coexistence of growth and dissolution forms. Coalescence of crystals took place along with the ripening, especially in the muscovite-water system. The concomitant 1Mrn(120) → 1M conversion for phlogopite and 1Mrn(120) → 1M → $2M_1$ sequence for muscovite was implied to occur during the ripening process (Baronnet, 1981). The assumed mechanism of polytype transformation is illustrated in Figure 7-25. At step one, progressive nucleation results in 1Mrn (120) crystals with various sizes. At step two, the smallest (C+D) crystallites dissolve, and material is transferred to growing crystals (A+B) as 1M overgrowths on both {001} faces. Owing to a decrease in supersaturation due mainly to the massive disappearance of the smallest crystals, the critical size increases. Therefore, previously-growing particles which are now among the smallest in the grain-size distribution undergo dissolution. Mica with 1Mrn (120) + 1M structures dissolves and redeposition at step three takes place in the $2M_1$ form under slow growth-rate conditions. As a whole, the ripening

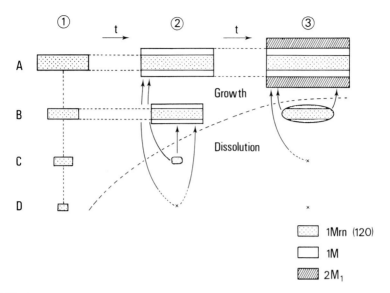

Figure 7-25. A speculative mechanism of solution-mediated polytype transformations for muscovite as based on the Ostwald ripening process. [From Baronnet (1981) Fig. 24, p. 114.]

process results in gradual polytype conversion, as detected by x-ray powder diffraction (Velde, 1965; Mukhamet-Galeyev et al., 1985). Dissolution/crystallization under ripening conditions is a process with small activation energy as compared to changes in stacking structure occurring through a solid-state mechanism. However, the above model lacks direct HRTEM confirmation of the microstructure of ripening crystallites. Surface micro-topographic (Tomura et al., 1979) and grain-size (Eberl and Srodon, 1988; Inoue et al., 1988) relations suggest that Ostwald ripening of natural dioctahedral materials like illite occurs in diagenetic or hydrothermal environments.

7.6.4 Local microstructures and large crystal size

As shown above, HRTEM study of mica polytypism in crystals of less than about a micron proved helpful in understanding early stages of crystal growth of layer-stacking sequences, because growth features and inferred mechanisms are unique on small basal faces. However, there is up to 11 orders of magnitude difference in surface area between a large crystal (e.g., a pegmatitic mica) and clay-sized micas. Accordingly, little information is available on the layer-stacking structures of medium-sized crystals (mm to cm). This section considers if current theories of polytypism are applicable to these crystals.

7.6.4.1 Homogeneity and description of microstructures

Many x-ray diffraction studies showed that single crystals of layer silicates exhibit intimate intergrowths of polytype structures (e.g., Ross et al., 1966; Rieder, 1970). HRTEM has been used to study various kinds of layer-stacking sequences (Table 7-5). Similar studies of a large number of species from a variety of geological occurrences has not been carried out. However, the following observations can be made:

• inhomogeneity in distribution of microstructures is ubiquitous in micas and chlorite,

Table 7-5. Microstructural features reported by HRTEM in some mica and chlorite "single crystals" Numbers between brackets correspond to references below.

| | Micas | | | | | Chlorite |
| | Trioctahedral | Dioctahedral | | | | |
	Biotite	Lepidolite	Muscovite	Phengite	Paragonite	Clinochlore
Ordered Basic Structures	1M (2,6,7), 2M$_1$(2,6), 3T (1,2,6,7)	3M$_2$ [1$\bar{2}$1] (11)	1M(1) 2M$_1$ (4)	2M$_1$ (10)	2M$_1$ (10), 3T(10)	II b-2 (8)
Syntactic coalescence	1M-2M$_1$ (6,7), 2M$_1$-3T (7), 1M-3T (6,7), 2M$_1$-1Mr (7)	3M$_2$-6M$_1$ (11)	no(1) no(4)	no(10)	2M$_1$-3T(10)	-
Semi-random disorder	1Mr n (120) (1,2,4)	yes (11)	no (1) no (4)	no (10)	1Mr n (120) (10)	yes (8,9)
Stacking faults	yes many, intrinsic, extrinsic types (1, 4, 6, 7)	yes (11)	yes(1) extrinsic yes(4) extrinsic	no (10)	yes, many	too disordered
Complex Polytypes	3Tc[02$\bar{2}$](2), 4M[2$\bar{2}\bar{2}$2](6), 5Tc[(0)$_3$22](2), 8Tc[0020$\bar{2}$02$\bar{2}$](2), 9Tc[0020$\bar{2}$0022$\bar{2}$](2), 13M(3),13Tc(3), 17Tc(3), 18Tc(3), 19M(3), 21Tc[($\bar{2}$2)$_4$(2$\bar{2}$)$_5$222] (2),etc...	6M$_1$[1$\bar{2}$1$\bar{1}$1$\bar{2}$1$\bar{1}$]	no (1) no (4)	no (10)	8M$_8$[(222)$_2$2$\bar{2}$] (10)	20L (12)
Quasi-periodic polytypes	yes (7)	?	?	?	?	?
Microtwinning	yes (2,5)	yes (11)	no(1) no (4)	no ?	no ?	yes (8)

(1) Iijima and Buseck, 1978.
(2) Bigi, 1990.
(3) Czank, 1986.
(4) Iijima and Zhu, 1982.
(5) Bell and Wilson, 1977, 1986.
(6) Baronnet and Kang, 1989.
(7) Baronnet and al 1992, in press.
(8) Spinnler et al, 1984.
(9) Bons and Schryvers, 1989.
(10) Ahn et al, 1985.
(11) Rule et al, 1987.
(12) Ferrow et al, 1990.

• stacking disorder is apparently more common than x-ray diffraction as shown (Fig. 7-26a),

• crystals that exhibit long-period or complex polytypes have both stacking disorder and stacking faults in ordered basic structures,

• complex polytypism in biotite is common (Bigi, 1990); complex polytypism in dioctahedral micas is uncommon although not unknown (Ahn et al., 1985); complex polytypism in chlorite (Ferrow et al., 1990), has been reported once.

• Where stacking faults are sufficiently separated from each other (Fig. 7-26b), their nature and relative frequency may be described as a function of the host basic structure (Baronnet et al., 1992). For example, the most frequent faults found in a Chinese biotite were:

Zv. ... C C C A C C C ...   (1) in the 1M basic structure
RTW... 0 0 2 $\bar{2}$ 0 0 ...
Zv. ... $\bar{A}$ $\bar{B}$ $\bar{A}$ $\bar{B}$ $\bar{C}$ $\bar{B}$ $\bar{A}$ $\bar{B}$ $\bar{A}$ (2) in the $2M_1$ basic structure
RTW... 2 $\bar{2}$ 2 2 $\bar{2}$ 2 $\bar{2}$ ...
Zv. ... A B C C B C A B C   (3) in the 3T basic structure.
RTW... 2 2 0 $\bar{2}$ 2 2 2 2 ...

The SFE of fault (2), $0\phi_1 + 2\phi_2$, is the smallest expected for the $2M_1$ basic structure. However, the SFE's of fault (1), $2\phi_1(2) + 3\phi_2$ and of fault (3), $1\phi_1(0) + 3\phi_2$ do not attain predicted minimums (Baronnet et al., 1981; Pandey et al., 1982). Such faults are all of layer-displacement type (see §7.3.4) and may be considered to be embrionic sequences of alternate basic structures in the host matrix [$2M_1$ in 1M for (1), 3T in $2M_1$ for (2) and 1M + $2M_1$ in 3T for (3)].

• For this Chinese biotite, *quasi-periodic polytypes* are observed (Fig. 7-26c) for the first time in a layer silicate (Baronnet et al., 1992). The biotite stacking involves regular alternation of thin slabs of two basic structures. The thickness of each slab is not strictly maintained along c*, and the diffraction pattern has incommensurate spots along $0k\ell$ rows. A striking feature is that each slab maintains its orientation throughout the stacking sequence (no twin orientation). Such quasi-periodic polytypes are usually located on the mesoscopic (001) boundaries between neighboring ordered basic structures (1M-$2M_1$, 1M-3T, rarely $2M_1$-3T) which form component slabs. The origin of this orientational order is puzzling because it is difficult to reconcile the origin of long-range interaction forces capable of producing this feature. Since the periodicity is only approximate, spiral growth cannot be invoked to explain such memory for orientation.

7.6.5   Chemical signatures of stacking microstructures

Layer stacking microstructures with respect to chemical variations are discussed below. Emphasis is on biotites and EDS chemical analyses. Since the technique of EDS analysis is described in Chapter 4, only details as to how it relates to polytypism are given here (see Baronnet et al., 1992 for more details).

7.6.5.1   *Dedicated analytical electron microscopy (AEM)*

Modern high resolution electron microscopes (100, 200, 300 keV) with a double-tilt, side-entry, HR specimen holder are easily modified to include an energy-dispersive x-ray spectrometer with a Si(Li) detector. If the take-off angle of x-ray s detected by the Si(Li) detector through the objective polepiece is high (~70°), shifting from the HR imaging and diffraction modes to the analytical mode is easy. This can be done without tilting the specimen between the two modes and then, preserving the contrast and the location of the microstructures to be analyzed.

To avoid compositional change due to beam damage, it is best to maintain electron beam diameters greater than 200 Å for, say, nA beam currents (Ahn et al., 1986), or use the STEM mode. For the same reason, a very short through-focus series of HR images is

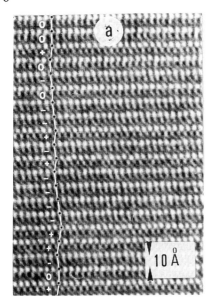

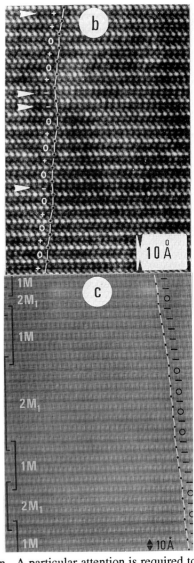

Figure 7-26. Structure images of biotite single crystals. (a) Semi-random, 1Mm (120) stacking sequence. (b) Isolated stacking faults (arrows) in a 2M₁ matrix (.. 0 + 0 + 0 ...). (c) Quasi-periodic polytype made by the regular alternation of 2M₁ and 1M slabs of varying thicknesses. [From Baronnet and Kang (1989) Fig. 4b, p. 486.]

recorded initially in principal zone-axis orientation. A particular attention is required to determine that the chemical analysis is not severely affected by electron channelling effects (see Chapter 5). The x-ray spectrum is then acquired (100 sec) and processed by a data reduction procedure based on the original Cliff and Lorimer method (Cliff and Lorimer, 1975). Experimental $k_{x,Si}$ factors (Cliff and Lorimer, 1985; Lorimer and Cliff, 1976; Mellini and Menichini, 1985) are pre-calibrated against layer-silicate standards which were checked for chemical homogeneity. After recording microchemical data, we shift back to the HR imaging mode to record a photograph of the area of impact of the electron beam on the microstructure.

7.6.5.2 Chemical differences among biotite basic structures

Comparing select biotite high-resolution images and relevant compositional data shows significant differences in composition between stacking-fault free 1M, stacking-fault free

Table 7-6. Microcompositions of the 1M, 2M₁ and 1Mrn (120) basic structures in a Chinese biotite single crystal. In cations percent analyzed. From Baronnet et al. (1989).

Elements	1M	2M$_1$	1Mrn (120)
Si	34.58 (.18)	35.58 (.29)	36.14 (1.57)
Al	19.14 (.22)	17.06 (.34)	16.95 (.72)
Mg	18.46 (.37)	20.43 (.46)	20.07 (.71)
Fe	13.66 (.40)	13.13 (.27)	13.00 (1.19)
K	13.05 (.88)	11.30 (.85)	11.79 (.90)
Ti	1.11 (.13)	2.47 (.11)	2.05 (.15)

$2M_1$ and the semi-random 1Mrn(120) basic structures (Baronnet et al., 1989, 1992). 1M differs noticeably from both $2M_1$ and 1Mrn(120), although the latter two are more alike. 1M is significantly enriched in total Al and poor in Ti relative to the two others (Table 7-6). Although it was not possible to distinguish between 1 Mrn (120) and quasi-periodic polytypes, it is thought that the latter maintain the respective compositions of the component slabs. Thus, they are able to accomodate intermediate chemical compositions between neighboring blocks of basic structures. If so, the modulated type ($1M-2M_1$) of structure-composition coupling resembles an unmixing phenomenon, whereas 1Mrn (120) is analogous to perfect chemical and structural mixing down to the layer scale. In micas, similar polytype-composition relationships were clearly identified with combined electron microprobe and light microscopy in $2M_1$ and 3T phengitic micas (Stöckhert, 1985) and chemical composition partially controls polytypism. Other results on biotite by TEM/AEM are difficult to interpret. Czank (1986) found from qualitative EDS data that ordered 1M has slightly more Fe and less Al than disordered sequences. These results contrast with our own. Furthermore, Bigi (1990) did not find any chemical differences between coexisting polytypes.

Since the differences in composition between the different types of microstructures are small to nonexistent, it is important to investigate biotite from different environments (magmatic, metamorphic, hydrothermal, etc.) with the same analytical procedure before generalizing further about microstructure-composition relationships. However, the role of departures from stoichiometry on polytypism has been emphasized, even for simple substances like SiC (Verma and Krishna, 1966).

7.6.6 Polytype parentage as an indicator of phase-transformation mechanisms

TEM studies of mineral reactions are discussed in Chapters 6, 9 and 13 of this volume. Therefore only a few mineralogical examples in which polytypism may be useful to support or discredit possible reaction mechanisms between different minerals are discussed here.

7.6.6.1 Solid-state transformation or dissolution/crystallization

In natural systems, there are two principal mechanisms for mineral reactions. If the parent and daughter phases have sufficient portions of their structures in common, then a topotactic solid-state transformation mechanism is possible (Buseck, 1983). The reaction may be either isovolumetric (without volume change) or heterovolumetric (with expansion or shrinkage), isochemical (without chemical composition change) or heterochemical (with chemical composition change, i.e., implying selective element diffusion to and from the reaction front). The reaction occurs within the parent crystal. For cases in which the parent requires structural reconstruction to transform into the daughter phase, a dissolution/crystallization mechanism is involved. The latter proceeds through dissolution of the parent (unstable) phase and nucleation and growth of the daughter (metastable or stable) phase, usually as separate or epitaxial grains. Most of the structural "memory" is usually lost during the process of dissolution (or melting) of the parent phase (Lebedev, 1975).

The solid-state transformation mechanism maintains the approxmate size and shape of the parent phase. On the other hand, the dissolution/crystallization mechanism may involve also the loss of these morphological characteristics.

TEM may efficiently assist in determining the mechanism involved or excluded for cases requiring structurally related reactants and products (examples of amphibole-layer silicate and layer silicate-layer silicate reactions are given below). In such a case, if a solid-state transformation mechanism is possible, the dissolution/crystallization mechanism may proceed as well. These two mechanisms may occur side by side, as observed by, for example, Baronnet and Onrubia (1988) and Amouric and Olives (1991).

To discriminate between the two mechanisms, the following is noted: if parent and daughter phases are are in close topotactic contact and their polytypes are identical or have at least rotational or translational operators of some subunits of their structure in common, then a solid-state reaction is likely. It is especially likely if the daughter polytype is not a usual structure of the product species. This criterion is not absolute in that a lateral overgrowth of the daughter phase on a seed of the parent phase may result in the same effect (Tairov and Tsvetkov, 1983). The laboratory vermiculitization of 1M or $2M_1$ phlogopite and biotite single crystals results in vermiculite with the corresponding stacking angles of TOT layers (de la Calle and Suquet, 1988; de la Calle et al., 1976), as shown by single-crystal x-ray diffraction. On the other hand, the lack of any polytype inheritance usually indicates a dissolution/crystallization process.

7.6.6.2   The biotite-to-chlorite reaction

Two mechanisms are possible to explain the biotite-to-chlorite reaction (Iijima and Zhu, 1982; Veblen and Ferry, 1983; Olives Banos and Amouric, 1984; Eggleton and Banfield, 1985). Mechanism (1) involves the replacement of the interlayer (I) of the mica layer by a brucite-like octahedral (O) sheet (Fig. 7-27a). Therefore, each TOT-I mica layer ($\simeq$ 10 Å thick) transforms to one TOT-O chlorite layer ($\simeq$ 14 Å thick). This mechanism maintains each TOT talc-like sheet of mica to form chlorite. Thus, the sequence of intralayer shifts of successive mica layers can be maintained in the form of the shift sequence of talc-like subunits in the resulting chlorite. Mechanism (2) consists of the replacement of two mica layers by one single chlorite layer (Fig. 7-27a). In this case, any alternate mica layer loses its two tetrahedral sheets to form a brucite-like sheet, whereas all interlayer cations of the reacting mica are lost. Accordingly, alternate intralayer shifts of successive mica layers are maintained in the sequence of talc-like components of the final chlorite.

Figure 7-27b is a 2D-structure image of a sharp (001) interface between well ordered $2M_1$ biotite and semi-random chlorite. The series of projected staggers of the talc-like subunits (T in this figure) of chlorite are disordered. They violate the perfectly periodic

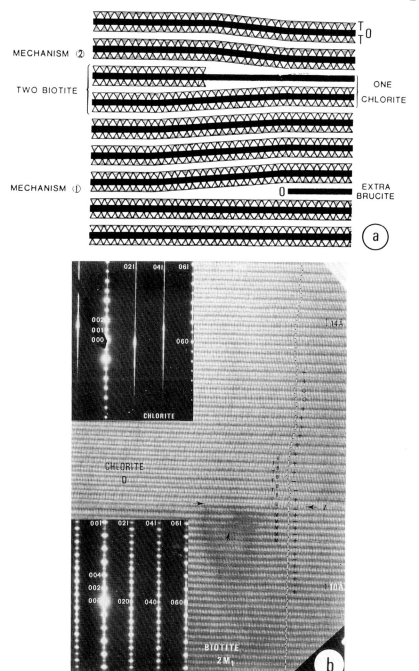

Figure 7-27. Parentage between biotite and chlorite. (a) Sketch of the two solid-state transformation mechanisms. [From Eggleton and Banfield (1985) Fig. 8, p. 907.] (b) (001) interface between 2M₁ biotite seen along [100] and semi-random chlorite. [From Baronnet and Kang (1989) Fig. 7, p. 489.]

zigzag pattern of biotite. Therefore, this chlorite probably formed by epitaxial overgrowth onto (001) of biotite rather than by a solid-state transformation mechanism.

7.6.6.3   The chlorite-to-lizardite reaction

Baronnet et al. (1988) observed a sample with one chlorite layer ($\simeq$ 14 Å) that had reacted laterally to two contiguous lizardite layers (2×7 Å) layers (Fig. 7-28a): -TOT-O-$\rightarrow$ -TO-TO-. This reaction may be explained by reversals perpendicular to the layers of every

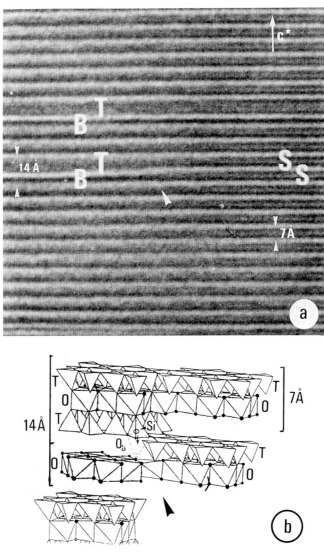

Figure 7-28. The nanoreaction of one chlorite layer into a pair of contiguous serpentine layers. (a) A one-dimensional HRTEM image of a lateral termination. (b) A diagrammatic sketch of the atomistic process of transformation. Tetrahedral (Si,Al)^{IV} should pass between the three O_Bs forming the base of the tetrahedra. A counter flux of protons is needed to balance the electrostatic charges and complete the serpentine structure. OH$^-$ is shown as solid circles.

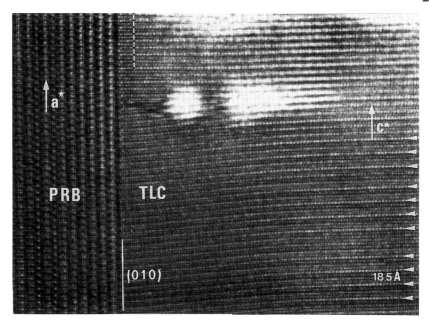

Figure 7-29. 2Or talc (TLC) in (010) contact with pyribole (PRB) from Chester, Vermont. Arrows point to contrasts marking the dominant two-layer periodicity of talc. [Photograph by Baronnet and Veblen, March 1991.]

alternate tetrahedral sheet of chlorite while maintaining the O_B network (Fig. 7-28b). These reversals are mainly due to the migration of Si,Al^{IV} into the region between the brucite-like and talc-like subunits of reacting chlorite. This process results in the retention of the non-zero talc unit stagger of chlorite as the interlayer shift vector of the resulting lizardite. On the other hand, 1T lizardite is the most common polytype of this serpentine (Bailey, 1988a). In this polytype (Mellini, 1982) tetrahedral rings of successive 7 Å layers are directly superimposed without any interlayer shift. From this, we can conclude that such a common polytype cannot result from a solid-state mechanism, involving limited Si-Al diffusion.

7.6.6.4   The orthoamphibole-to-talc reaction

The layer stacking sequences of talc are usually disordered as shown by HRTEM images (Akizuki and Zussman, 1978). However, Veblen and Buseck (1980) found two types of stacking sequences for talc displaying topotactic relationships with pyribole from Chester, Vermont. First, as shown in Figure 7-29, the stacking sequence of talc having (010) in common with pyribole may be quite ordered as indicated by white dots aligned on straight lines parallel to c^*. With the same reasoning as for micas (§7.5.5) this ordering is consistent with a 2Or or 1M stacking sequence observed along [100] or [$\overline{1}$00]. If the specimen is tilted from this exact orientation, then the true periodicity and sequence of talc appear as a moderately faulted 2Or sequence in images. In contrast to talc with layers parallel to the (210) planes of adjacent orthorhombic pyribole is disordered. In the former case, the 2Or polytype of talc is probably constrained by the alternate chain arrangement along a^* of the orthorhombic amphibole anthophyllite and/or pyriboles. This is more indicative of a solid-state reaction, and the persistance of such an unusual polytype would be indicative of the precursor phase from which it formed.

7.7 CONCLUSIONS

Transmission electron microscopy has made important contributions to the discovery of new polytypic substances, to the detailed nature of polytypic arrangements in crystals, and to constraints upon the theories that explain their origins. The layer-scale information that TEM provides for minerals, either by surface-replica techniques or by HRTEM imaging techniques, makes this method invaluable, and a complementary method to x-ray, neutron, and optical investigations of minerals. The electron diffraction techniques provide intermediate-scale information. Among the various types of micro-structures in minerals imaged by HRTEM, new details about polytype sequences, stacking faults and stacking disorder in single crystals are provided (e.g., see Mellini, 1989, and this volume). Not only are they potentially able to characterize the geological occurrences, but these observations may be used also to constrain the possible type of formation mechanism. Such microstructural data complement crystal-chemical data for polytypic minerals. The composition-microstructure relationship as reported for biotite basic structures is indicative of a polysomatic trend; these relationships emphasize the idealized character of the definition of polytypism. The large heterogeneity in the distribution of polytype micro-structures as found in most, if not all, large crystals of polytypic substances requires a suitable explanation. Is this distribution due to a response to fluctuating external conditions? If so, this information might be of use to the geologist, and more comprehensive theories of polytypism should be developed. Heterogeneous microstructural features may be also the result of some chaotic or periodic solutions of a deterministic model of growth (e.g., Vignoles, 1992) for nearby values of internal parameters. In this last case, the interest for the geologist would be largely decreased. Eventually, there is a great need of additional HRTEM data on the layer-stacking microstructures of polytypic minerals, either natural or synthetic. A better knowledge or control of the thermodynamic and kinetic conditions of their formation will hopefully provide clues for a better understanding of the subtle phenomenon of polytypism in minerals.

ACKNOWLEDGMENTS

The author is greatly indebted to Don Peacor and an anonymous reviewer for theri thorough and very constructive reviews of the manuscript. Peter Buseck gave precious advice and encouragement and he and Paul Ribbe kindly finalized the presentation of the chapter. Sincere thanks are due to Jean-Paul Gauthier for donating the TEM micrograph of opal and to David Veblen for permitting me to record the pyribole-talc micrograph from the famous Chester specimen. Marc Amouric contributed much to the experiments on synthetic micas. Most of the HRTEM and SAED micrographs reported herein were taken by the author with the JEOL-2000FX electron microscope at the electron microscopy and microanalysis facilities at CRMC[2]. Excellent maintenance of the machine and technical advice by Serge Nitsche are much appreciated. The author also gratefully acknowledges technical assistance on the manuscript by Dominique Destre, Bertrand Devouard, Simone Hanania, Marie-Claude Toselli, Isabelle Motz and Francis Quintric.

REFERENCES

Ahn, J.H., Peacor, D.R. and Essene, E.J. (1985) Coexisting paragonite-phengite in blueschist eclogite: a TEM study. Am. Mineral. 70, 1193-1204.

Ahn, J.H., Peacor, D.R. and Essene, E.J. (1986) Cation-diffusion induced characteristic beam damage in transmission electron microscope images of micas. Ultramicroscopy 19, 375-382.

Akizuki, M. and Zussman, J. (1978) The unit cell of talc. Mineral. Mag. 42, 107-110.

Amelinckx, S. (1952) La croissance hélicoïdale des cristaux de biotite. Compt. Rend. Acad. Sci. Paris 234, 971-973.

Amelinckx, S. and Dekeyser, W. (1953) Le polytypisme des minéraux micacés et argileux. I.: observations et interprétations. Compt. Rend. Congr. Geol. Int'l Algiers, 1952, 18, 9-22.

Amouric, M. (1981) Polytypisme et ordre-désordre dans les micas dioctaédriques. Apports de la microscopie électronique haute résolution au rôle des conditions de germination. Dr. es Sci. Thesis, University of Aix-Marseille III.

Amouric, M. and Baronnet, A. (1983) Effect of early nucleation conditions on synthetic muscovite polytypism as seen by high resolution transmission electron microscopy. Phys. Chem. Minerals 9, 146-159.

Amouric, M. and Baronnet, A. and Finck, C. (1978) Polytypisme et désordre dans les micas dioctaédriques synthétiques. Etude par imagerie de réseau. Mat. Res. Bull. 13, 627-634.

Amouric, M., Gianetto, I. and Proust, D. (1988) 7, 10 and 14 Å mixed-layer phyllosilicates studied structurally by TEM in pelitic rocks of the Piemontese zone (Venezuela). Bull. Mineral. 104, 298-313.

Amouric, M., Mercuriot, G. and Baronnet, A. (1981) On computed and observed HRTEM images of perfect mica polytypes. Bull. Mineral. 104, 298-313.

Amouric, M. and Olives, J. (1991) Illitization of smectite as seen by high resolution transmission electron microscopy. Eur. J. Mineral. 3, 831-835.

Amouric, M. and Parron, C. (1985) Structure and growth mechanism of glauconite as seen by high-resolution transmission electron microscopy. Clays and Clay Minerals 33, 473-482.

Angel, R.J. (1986) Polytypes and polytypism. Z. Krist. 176, 193-204.

Angel, R.J., Price, G.D. and Yeomans, J. (1985) The energetics of polytypic structures: further applications of the ANNNI model. Acta Cryst. B41, 310-319.

Bailey, S.W. (1963) Polymorphism of the kaolin minerals. Am. Mineral. 49, 1196-1209.

Bailey, S.W. (1984) Classification and structures of the micas. In: Micas, S.W. Bailey, ed. Rev. Mineral. 13, 1-12.

Bailey, S.W. (1988a). Polytypism of 1:1 layer silicates. In: Hydrous Phyllosilicates (exclusive of micas), S.W. Bailey, ed., Rev. Mineral. 19, 9-27.

Bailey, S.W. (1988b) Chlorites: structures and crystal chemistry. In: Hydrous Phyllosilicates. S.W. Bailey, ed. Rev. Mineral. 19, 347-403.

Bailey, S.W., Frank-Kamenetskii, V.A., Goldstaub, S., Kato, A., Pabst, A., Schultz, H., Taylor, H.F.W., Fleisher, M. and Wilson, A.J.C. (1977) Report Int'l Mineral. Assoc. (IMA)—Int'l Union of Crystallogr. (IUCr) Joint Comm. on Nomenclature. Acta Cryst. A 33, 681-684.

Bak, P. and von Boehm, J. (1980) Ising model with solitons, phasons and the devil's staircase. Rep. Prog. Phys. 45, 587-629.

Bandy, M.C. (1938) Mineralogy of three sulphate deposits of northern Chile. Am. Mineral. 23, 669-760.

Baronnet, A. (1972) Growth mechanisms and polytypism in synthetic hydroxyl-bearing phlogopite. Am. Mineral. 57, 1272-1293.

Baronnet, A. (1973) Sur les origines des dislocations vis et des spirales de croissance dans les micas. J. Crystal Growth 19, 193-198.

Baronnet, A. (1974) Etude en microscopie électronique des premiers stades de croissance d'un mica synthétique, la phlogopite hydroxylée. Phénomènes de coalescence et de mûrissement dans le système fermé conservatif: K_2O-6MgO- Al_2O_3-$6SiO_2$-excès H_2O. High Temp. High Press. 6, 675-685.

Baronnet, A. (1975a) Growth spirals and complex polytypism in micas. I. Polytypic structure generation. Acta Cryst. A 31, 345-355.

Baronnet, A. (1975b) L'aspect croissance du polymorphisme et du polytypisme dans les micas synthétiques d'intérêt pétrologique. Fortschr. Mineral. 52, 203-216.

Baronnet, A. (1976) Polytypisme et polymorphisme dans les micas. Contribution à l'étude du rôle de la croissance cristalline. Dr. Sci. Thesis, University of Aix-Marseille III.

Baronnet, A. (1978) Some aspects of polytypism in crystals. Prog. Crystal Growth Charact. 1, 151-211.

Baronnet, A. (1980) Polytypism in micas: a survey with emphasis on the crystal growth aspect. In: Current Topics in Materials Science, E. Kaldis, ed., Vol. 5, 447-548.

Baronnet, A. (1981) Les transformations polytypiques. In: Les Transformations de Phases dans les Solides Minéraux, V. Gabis and M. Lagache, ed., Soc. Franç. Min. Crist., vol. 1, chap. 2, 51-121.

Baronnet, A. (1982) Ostwald ripening in solution. The case of calcite and mica. Estudios Geologicos 38, 185-198.

Baronnet, A. (1989) Polytypism and crystal growth of inorganic crystals. In: Crystal Growth in Science and Technology, H. Arend and J. Hulliger, ed. Plenum Pub. Corp., 197-204.

Baronnet, A., Amouric, M. and Chabot, B. (1976) Mécanismes de croissance, polytypisme et polymorphisme de la muscovite hydroxylée synthétique. J. Crystal Growth 32, 37-59.

Baronnet, A. and Kang, Z.C.(1989) About the origin of mica polytypes. Phase Transitions, 16/17, 477-493.

Baronnet, A., Nitsche, S. and Kang, Z.C. (1992) Layer stacking microstructures in a biotite single crystal.

284

A combined HRTEM-AEM study. Phase Transitions, in press.

Baronnet, A. and Onrubia, Y. (1988) Combined powder XRD, HRTEM and AEM studies of mica-chlorite-serpentines-talc mixed-layering and related phase transformation processes. Z. Krist. 185, 115.

Baronnet, A., Pandey, D. and Krishna, P. (1981) Application of the faulted-matrix model to the growth of polytype structures in micas. J. Crystal Growth 52, 963-968.

Basset, G. (1958) A new technique for decoration of cleavage and slip steps on ionic crystal surfaces. Phil. Mag. 3, 1042-1045.

Baumhauer, H. (1912) Uber die Kristalle des Carborundums. Z. Krist. 50, 33-39.

Baumhauer, H. (1915) Uber die verschiedenen Modifikationen des Carborundums und die Erscheinung der Polytypie. Z. Krist. 55, 249-259.

Bell, I.A. and Wilson, C.J.L. (1977) Growth defects in metamorphic biotite. Phys. Chem. Minerals 2, 153-169.

Bernal, J.D. (1924) The structure of graphite. Proc. Roy. Soc. (London), Ser. A 106, 749-773.

Bigelow, S.L. and Trimble, H.M. (1927) The relation of vapour pressure to particle size. J. Phys. Chem. 31, 1798-1816.

Bigi, S. (1990) Microstrutture in biotiti naturali. Ph.D. Thesis, University of Modena, 114 p.

Bons, A.J. and Schryvers, D. (1989) High resolution electron microscopy of stacking irregularities in chlorites from the Central Pyrénées. Am. Mineral. 74, 1113-1123.

Bragg, W.L. and Claringbull G.F. (1965) Crystal Structures of Minerals. Bell, London.

Burton, W.K., Cabrera, N. and Frank, F.C. (1951) The growth of crystals and the equilibrium structure of their surfaces. Phil. Trans. Roy. Soc. 243, 299- 358.

Buseck, P.R. (1983) Electron microscopy of minerals. Am. Scientist 71, 175-185.

Buseck, P.R. and Cowley, J.M. (1983) Modulated and intergrowth structures in minerals and electron microscope methods for their study. Am. Mineral. 68, 18-40.

de la Calle, C. and Suquet, H. (1988) Vermiculite. In: Hydrous Phyllosilicates (exclusive of micas), S.W. Bailey, ed., Rev. Mineral. 19, 455-496.

de la Calle, C., Dubernat, J., Suquet, H., Pezerat, H., Gaultier, J.P. and Mamy, J. (1976) Crystal structure of two layer Mg-vermiculites and Na,Ca-vermiculites. In: Proc. Int. Clay Conf., Mexico, 1975, S.W. Bailey, ed., Applied Publishing Ldt, Illinois, 201-209.

Cliff, G. and Lorimer, G.W. (1975) The quantitative analysis of thin specimens. J. Microscopy 103, 203-207.

Debye, P. and Scherrer, P. (1916) Interferenzen an regellos orientierten Teilchen in Röntgenlicht. Z. Physik 17, 277-283.

Ducellier, F. (1991) Obtention et simulation d'images 2D des phyllosilicates en MET à haute résolution 200 KV. DEA report. University of Aix-Marseille III, p. 1-85.

Eberl, D.D. and Srodon, J. (1988) Ostwald ripening and interparticle-diffraction effects for illite crystals. Am. Mineral. 73, 1335-1345.

Eggleton, R.A. and Banfield, J.F. (1985) The alteration of granitic biotite to chlorite. Am. Mineral. 70, 902-910.

Ergun, S. and Alexander, L.E. (1962) Crystalline forms of carbon: A possible hexagonal polymorph of diamond. Nature 195, 765-767.

Evans, Jr., H.T. and Mc Knight, E.T. (1959) New wurtzite polytypes from Joplin, Missouri. Am. Mineral. 44, 1210-1218.

Farkas-Jahnke, M. (1983) Structure determination of polytypes. In: Polytype Structures. Prog. Crystal Growth Charact. 7, 163-210.

Ferrow, E.A., London, D., Goodman, K.S. and Veblen, D.R. (1990) Sheet silicates of the Lawler Peak granite, Arizona: chemistry, structural variations and exsolution. Contrib. Mineral. Petrol. 105, 491-501.

Ferrow, E.A. and Roots, M. (19889) A preparation technique for TEM specimens; application to synthetic Mg-chlorite. Eur. J. Mineral., 1, 815-819.

Flörke, O.W. (1955) Strukturanomalien bei Tridymit und Cristobalit. Bel. Dtsh. Keram. Ges. 32, 369-381.

Forty, A.J. (1952) The growth of cadmium iodide crystals. II. A study of the height of growth steps on cadmium iodide. Phil. Mag. 43, 377-392.

Frank, F.C. (1951a) The growth of carborundum: dislocations and polytypism. Phil. Mag. 42, 1014-1021.

Frank, F.C. (1951b) Crystal dislocations - Elementary concepts and definitions. Phil. Mag. 42, 809-819.

Frondel, C. and Marvin, U.B. (1967) Lonsdaleite a hexagonal polymorph of diamond. Nature 214, 587-589.

Frondel, C. and Palache, C. (1948) Three new polymorphs of zinc sulphide. Science 107, 602.

Frondel, C. and Palache, C. (1950) Three new polymorphs of zinc sulphide. Am. Mineral. 35, 29-42.

Frondel, C. and Wickman, F.E. (1970) Molybdenite polytypes in theory and occurrence. II. Some naturally-occurring polytypes of molybdenite. Am. Mineral. 55, 1857-1875.

Gauthier, J.P. (1980) Using RHEED to study polytypism. J. Microscopy 119, 189-197.

Gauthier, J.P. and Michel, P. (1977). Polytypisme du carbure de silicium: diffraction électronique et simulation optique. Acta Cryst. A 33, 676-677.

Guinier, A., Bokij, G.B., Boll-Dornberger, K., Cowley, J.M., Durovic, S., Jagodzinski, H., Krishna, P., de Wolff, P.M., Zvyagin, B.B., Cox, D.E., Goodman, P., Hahn, Th., Kuchitsu, K. and Abrahams, S.C. (1984) Nomenclature of polytype structures, report of the International Union of Crystallography Ad-Hoc Committee on the Nomenclature of Disordered, Modulated and Polytype Structures. Acta Cryst. A 40, 399- 404.

Guthrie, G.D. and Veblen, D.R. (1989) High-resolution transmission electron microscopy of mixed-layer illite/smectite: computer simulations. Clays and Clay Minerals 37, 1-11.

Hartman, P. and Perdok, W.G. (1955) Relations between structure and morphology of crystals. Acta Cryst. 8, 49-52.

Hassel, O. and Mark, H. (1924) Über die Kristallstruktur des Graphites. Z. Physik 25, 317-337.

Haussühl, S. and Müller, G. (1963) Neues ZnS-Polytypen (9R, 12R und 21R) in mesozoischen Sedimenten NW-Deutschlands. Mineral. Petrogr. Mitt. 9, 28- 39.

Hirth, J.P. and Lothe, J. (1968) Theory of Dislocations. McGraw-Hill, New York.

Iijima, S. and Buseck, P.R. (1978) Experimental study of disordered mica structures by high-resolution electron microscopy. Acta Cryst. A 34, 709-719.

Inoue, A., Velde, B., Meunier, A. and Touchard, G. (1988) Mechanism of illite formation during smectite-to-illite conversion in a hydrothermal system. Am. Mineral. 73, 1325-1334.

Jagodzinski, H. (1954) Fehlordnungs Erscheinungen und ihr Zusammenhang mit der Polytypie des SiC. Neues Jahrb. Mineral. Monatsh. 3, 49-65.

Jepps, N.W. and Page, T.F. (1983) Polytypic transformations in silicon carbide. In: Polytype Structures, P. Krishna, ed. Prog. Cryst. Growth Charact. 7, 259-308.

Jones, W.C. (1913) Uber die Beziehung zwischen geometrischer Form und Dampfdruck, Lösligkeit, und Formenstabilität. Annal. Phys. 41, 441- 448.

Kiflawi, I. (1972) Ph.D. Thesis, the Hebrew University of Jerusalem.

Kitagawa, R., Takeno, S. and Sunagawa, I. (1983) Surface microtopographies of sericite crystals formed in different environmental conditions. Mineral. J. 11, 282-296.

Krishna, P., ed. (1983) Crystal Growth and Characterization of Polytype Structures. Pergamon Press, Oxford.

Krishna, P. and Verma, A.R. (1965) On the deduction of silicon carbide polytypes from screw dislocations. Z. Krist. 121, 36-54.

Lebedev, V.I. (1975) Does the phenomenon of transformation (structure inheritance) during the process of conversion of layered silicates exist in nature and experimental synthesis ? Intern. Geology Rev. 18, 1129-1138.

Lippens, B.C. (1961) Structure and Texture of Aluminas. Delft.

Lipson, H. and Stokes, A.R. (1942) A new structure of carbon. Nature 149, 328.

Lorimer, G.W. and Cliff, G. (1976) Analytical electron microscopy of minerals. In: Electron Microscopy in Mineralogy, Wenk, H.-R., ed., Springer-Verlag, Berlin, p. 506-519.

Mardix, S., Kalman, Z.H. and Steinberger, I.T. (1968) Periodic slip processes and the formation of polytypes in zinc sulfide crystals. Acta Cryst. A 24, 464- 469.

McLarnan, T.J. (1981) The number of polytypes in close-packings and related structures. Z. Krist. 155, 269-291.

Mellini, M. (1982) The crystal structure of lizardite 1T: hydrogen bonds and polytypism. Am. Mineral. 67, 587-598.

Mellini, M. (1989) High resolution transmission electron microscopy and geology. Adv. Electronics Electr. Phys. 76, 281-326.

Mellini, M. and Menechini, R. (1985) Proportionality factors for thin film TEM/EDS microanalysis of silicate minerals. Rend. Soc. Ital. Mineral. Petrol. 40, 261-266.

Mellini, M., Trommsdorff, V. and Campagnoni, R. (1987) Antigorit polysomatism: behavior during progressive metamorphism. Contrib. Mineral. Petrol. 97, 2, 147-155.

Mitchell, R.S. (1957) A correlation between theoretical screw dislocations and the known polytypes of silicon carbide. Z. Krist. 109, 1-28.

Mogami, K., Nomura, K., Miyamoto, M., Takeda, H. and Sadanaga, R. (1978) On the number of distinct polytypes of mica and SiC with a prime layer-number. Can. Mineral. 16, 427-435.

Monroe, E.A., Sass, D.B. and Cole, S.H. (1969) Stacking faults and polytypism in opal, $SiO_2 \cdot nH_2O$. Acta Cryst. A 25, 578-580.

Morimoto, N. (1978) Incommensurate superstructures in transformation of minerals. Recent Prog. Natur. Sci. Japan 3, 183-206.

Mukhamet-Galeyev, A.P., Pokrovskiy, V.A., Zotov, A.V., Ivanov, I.P. and Samotoin, N.D. (1985) Kinetics and mechanism of hydrothermal crystallization of $2M_1$ muscovite: an experimental study. Intern. Geol. Rev. 27, 1352-1364.

Nadeau, P.H. (1985) The physical dimensions of fundamental clay particles. Clay Minerals 20, 499-514.

Newberry, R.J.J. (1979a) Polytypism in molybdenite (I): a non-equilibrium impurity- induced phenomenon. Am. Mineral. 64, 758-767.

Newberry, R.J.J. (1979b) Polytypism in molybdenite (II): relationships between polytypism, ore deposition/alteration stages and rhenium contents. Am. Mineral. 64, 768-775.

O'Keefe, M.A., Buseck, P.R. and Iijima, S. (1978) Computed crystal structure images for high resolution electron microscopy. Nature 274, 322-324.

Olives-Banõs, J. (1985) Biotites and chlorites as interlayered biotite-chlorite crystals. Bull. Mineral. 108, 635-641.

Olives-Banõs, J. and Amouric, M. (1984) Biotite chloritization by interlayer brucitization as seen by HRTEM. Am. Mineral. 69, 869-871.

Olives-Banõs, J., Amouric, M., de Fouquet, C. and Baronnet, A. (1983) Interlayering and interlayer slip in biotite as seen by HRTEM. Am. Mineral. 68, 754-758.

Pandey, D. (1981) X-ray diffraction study of solid-state transformations in close-packed structures. Proc. Indian Natn. Sci. Acad. 47 A, suppl. 1, 78-99.

Pandey, D., Baronnet, A. and Krishna, P. (1982) Influence of stacking faults on the spiral growth of polytype structures in micas. Phys. Chem. Minerals 8, 268- 278.

Pandey, D. and Krishna, P. (1975a) On the spiral growth of polytype structures in SiC from a faulted matrix. I. Polytypes based on the 6H structure. Mat. Sci. Eng. 20, 234-249.

Pandey, D. and Krishna, P. (1975b) Influence of stacking faults on the growth of polytype structures. I. Cadmium iodide polytypes. Phil. Mag. 31, 1113-1132.

Pandey, D. and Krishna, P. (1976) On the spiral growth of polytype structures in SiC from a faulted matrix. II: Polytypes based on the 4H and 15R structures. Mater. Sci. Eng. 26, 53-63.

Pandey, D. and Krishna, P. (1982) Polytypism in close-packed structures. Current Topics in Materials Science 9, 415-491.

Pandey, D. and Krishna, P. (1983) The origin of polytypic structures. In: Polytype Structures, Prog. Crystal Growth Charact. 7, 213-257.

Plançon, A. (1981) Diffraction by layer structures containing different kinds of layers and stacking faults. J. Appl. Cryst. 14, 300-304.

Price, G.D. and Yeomans, J. (1984) The application of the ANNNI model to polytypic behaviour. Acta Cryst. B 40, 448-454.

Ramesesha, S. (1984) An explanation of the phenomenon of polytypism. Promana 23, 745-750.

Ramsdell, L.S. and Kohn, J.A. (1951) Disagreement between crystal symmetry and x- ray diffraction data as shown by a new type of silicon carbide, 10H. Acta Cryst. 4, 111-113.

Rayner, J.H. and Brown, G. (1966) Triclinic form of talc. Nature 212, 1352-1353.

Reynolds, R.C., Jr. (1980) Interstratified clay minerals. Ch. 3 in: G.W. Brindley and G. Brown, eds, Crystal Structure of Clay Minerals and their X-Ray Identification, The Mineralogical Society, London.

Reynolds, R.C., Jr. (1988) Mixed layer chlorite minerals. In: Micas, S.W. Bailey, ed., Rev. Mineral. 19, 601-629.

Rieder, M. (1970) Lithium-iron micas from the Krushne Hory Mountains (Erzgebirge): twins, epitactic overgrowths and polytypes. Z. Krist. 132, 161-184.

Ross, M., Takeda, H. and Wones, D.R. (1966) Mica polytypes: description and identification. Science 151, 191-193.

Rule, A.C., Bailey, S.W., Livi, K.J.T. and Veblen, D.R. (1987) Complex stacking sequences in a lepidolite from Tørdal, Norway. Am. Mineral. 72, 1163- 1169.

Saalfeld, H. and Jarchow, O. (1968) Die Kristallstruktur von Norstrandites Al(OH)$_3$. N. Jb. Min. Abh. 109, 185-191.

Salje, E., Palosz, B. and Wruck, B. (1987) *In situ* observation of the polytypic phase-transition 2H-12R in PbI$_2$ - Investigations of the thermodynamic, structural and dielectric properties. J. Phys. C: Solid State Physics 20, 4077-4096.

Sarna, I. and Trigunayat, G.C. (1989) Phase Transitions 16/17, 543.

Schneer, C.J. (1955) Polymorphism in one dimension. Acta Cryst. 8, 279-285.

Schwamm, S., Mazel, A., Dorignac, D. and Sevely, J. (1991) HREM identification of "one-dimensionally-disordered" polytypes in the SiC (CVI) matrix of SiC/SiC composites. Microsc. Microanal. Microstruct. 2, 59-73.

Seaman, D.M. and Hamilton, H. (1950) Occurrence of polymorphous wurtzite in Western Pennsylvania and Eastern Ohio. Am. Mineral. 35, 43-50.

Sebastian, M.T. and Krishna, P. (1987) Single crystal diffraction studies of stacking faults in close-packed structures. Prog. Crystal Growth Charact. 14, 103- 183.

Selke, W. and Duxbury, P.M. (1984) The mean-field theory of the three-dimensional ANNNI model. Z. Physik B 57, 49-58.

Sella, C., Conjeaud, P. and Trillat, J. (1960) Nouvelle méthode d'étude par microscopie électronique de la structure superficielle des faces de clivages d'halogénures alcalins. IV Int'l Kongr. Elektronen-mikroskopie 1, 508-512.

Skattula, W. and Horn, L. (1960) Ein einfaches, hochauflösendes Abdruckverfahren für die Elektronen-mikroskopie. Experim. Tech. Phys. 8, 1-9.

Smith, J., Yeomans, J.M. and Heine, V. (1984) A new theory of polytypism. In: Proc. NATO A.S.I., Modulated Structure Materials, T. Tsakalakos, ed., 95-105.

Smith, J.V. and Yoder, H.S., Jr. (1956) Experimental and theoretical studies of the mica polymorphs. Mineral. Mag. 31, 209-235.

Spinnler, G.E., Self, P.G., Iijima, S. and Buseck, P.R. (1984) Stacking disorder in clinochlore chlorite. Am. Mineral. 69, 252-263.

Spinnler, G.E., Veblen, D.R. and Buseck, P.R. (1982) Complexities in antigorite modulations Geol. Soc. Am. Abstr. Prog. 14, 623.

Steinberger, I.T. (1983) Polytypism in zinc sulfide. In: Polytype Structures, P. Krishna, ed., Prog. Cryst. Growth Charact., Vol. 7, 7-53.

Stemple, I.S. and Brindley, G.W. (1960) A structural study of talc and talc-tremolite relations. J. Am. Ceram. Soc. 43, 34-42.

Stöckert, B. (1985) Compositional control on the polymorphism ($2M_1$-3T) of phengitic white mica from high pressure parageneses of the Sesia Zone (lower Aosta Valley, Western Alps; Italy). Contrib. Mineral. Petrol. 89, 52-58.

Sunagawa, I. (1977) Natural crystallization. J. Crystal Growth 42, 214-223.

Sunagawa, I. and Koshino, Y. (1975) Growth spirals on kaolin group minerals. Am. Mineral. 60, 407-412.

Tairov, Y.M. and Tsvetkov, V.F. (1983) Progress in controlling the growth of polytypic crystals. In: Polytype Structures, P. Krishna, ed., Prog. Cryst. Growth Charct. Vol.7, 111-161.

Takeda, H. and Ross, M. (1975) Mica polytypism: dissimilarities in the crystal structures of coexisting 1M and $2M_1$ biotite. Am. Mineral. 60, 1030- 1040.

Takéuchi, Y. and Haga, N. (1971) Structural transformations of trioctahedral sheet silicates. Slip mechanisms of octahedral sheets and polytypic changes of micas. J. Mineral. Soc. Japan, Spec. Paper 1, 74-87.

Thompson, J.B., Jr. (1978) Biopyriboles and polysomatic series. Am. Mineral. 63, 239-249.

Thompson, J.B., Jr. (1981) Polytypism in complex crystals: contrasts between mica and classical polytypes. In: Structure and Bonding in Crystals. II, M. O'Keefe and A. Navrotsky, eds, Academic Press, New York, p. 167-196.

Tiwari, R.S., Rai, A.K. and Srivastava, O.N. (1974) Calculation of stacking-fault energy of polytype structure. Phys. Rev. B 9, 5155-5161.

Tiwari, R.S., Rai, A.K. and Srivastava, O.N. (1975) Evaluation of the stacking fault energy of polytype structures. Its relevance in the stabilization of polytypes. Phys. Stat. Sol. (a) 31, 419-430.

Tomura, S. (1985) Growth mechanisms and morphological variations of phyllosilicate crystals under different growth environments and conditions. Ph.D. dissertation, Tohoku University, Sendai, 165 p.

Tomura, S., Kitamura, M. and Sunagawa, I. (1979) Surface microtopography of metamorphic white micas. Phys. Chem. Minerals 5, 65-81.

Trigunayat, G.C. (1991) A survey of the phenomenon of polytypism in crystals. Solid State Ionics 48, 3-70.

Trigunayat, G.C. and Chadha, G.K. (1971) Progress in the study of polytypism in crystals. Phys. Stat. Sol. (a) 4, 9-42.

Ungemach, H. (1935a) Sur certains minéraux sulfatés du Chili. Bull. Soc. Franç. Minéral. Crist. 58, 97-221.

Ungemach, H. (1935b) Sur la syntaxie et la polytypie. Z. Krist. 91, 1-22.

Vand, V. and Hanoka, J.I. (1967) Epitaxial theory of polytypism. Observations on the growth of PbI_2 crystals. Mat. Res. Bull. 2, 241-251.

Van Landuyt, J., Van Tendeloo, G. and Amelinckx, S. (1983) High resolution electron microscopy of polytypes. In: Polytype Structures, Prog. Crystal Growth Charact. 7, 343-377.

Veblen, D.R. (1983) Microstructures and mixed layering in intergrown wonesite, chlorite, talc, biotite and kaolinite. Am. Mineral. 68, 566-580.

Veblen, D.R. (1991) Polysomatism and polysomatic series: a review and applications. Am. Mineral. 76, 801-826.

Veblen, D.R. and Buseck, P.R. (1980) Microstructures and reaction mechanisms in biopyriboles. Am. Mineral. 65, 599-623.

Veblen, D.R. and Ferry, J.M. (1983) A TEM study of the biotite-chlorite reaction and comparison with petrologic observations. Am. Mineral. 68, 1160-1168.

Velde, B. (1965) Experimental determination of muscovite polymorph stabilities. Am. Mineral. 50, 436-449.

Verma, A.R. (1951) Observations on carborundum of growth spirals originating from screw dislocations.

288

Phil. Mag. 42, 1005-1013.

Verma, A.R. (1953) Crystal Growth and Dislocations. Butterworths, London.

Verma, A.R. and Krishna, P. (1966) Polymorphism and Polytypism in Crystals. Wiley, New York.

Verma, A.R. and Trigunayat, G.C. (1973) Solid State Chemistry, C.N.R. Rao, ed., Marcel Dekker.

Vignoles, G.L. (1992) Atomic relaxation and dynamical generation of ordered and disordered CVI SiC polytypes. J. Crystal Growth, in press.

Weltner, W. (1969) On polytypism and internal rotation. J. Chem. Phys. 51, 2469-2483.

Wenk, H.R., Muller, W.F., Liddell, N.A. and Phakey, P.P. (1976) Polytypism in wollastonite. In: Electron Microscopy in Mineralogy, H.-R. Wenk, ed., p.324-331.

Yoder, H.S., Jr. and Eugster, H.P. (1954) Phlogopite synthesis and stability range. Geochim. Cosmochim. Acta 6, 157-185.

Yoder, H.S., Jr. and Eugster, H.P. (1955) Synthetic and natural muscovites. Geochim. Cosmochim. Acta 8, 225- 280.

Yonebayashi, K. and Hattori, T. (1979) Growth spirals on etched kaolinite crystals. Clay Science 5, 177-188.

Zvyagin, B.B. (1967) Electron Diffraction Analysis of Clay Mineral Structures. Plenum, New York.

Zvyagin, B.B. (1988) Polytypism of crystal structures. Comput. Math. Applic. 16, 569-591.

Chapter 8.   MINERAL DEFINITION BY HRTEM: PROBLEMS AND OPPORTUNITIES

Fred M. Allen

Engelhard Corporation, Iselin, New Jersey 08830, U.S.A.

8.1   INTRODUCTION

This chapter reviews ways in which high-resolution transmission electron microscopy (HRTEM) has been used to study minerals and mineral-like phases that are difficult to characterize using conventional methods for structural and chemical analysis. Three types of applications are discussed. The first involves the definition of mineral solid solutions and intergrowths exhibiting chemical and/or structural disorder perceived at the macroscopic scale, the second concerns the identification of phases found to exist only at the microscopic scale, and the third has to do with the determination of structures of known minerals.

HRTEM studies of minerals have provided considerable insight into the complex microstructure and microchemistry of so-called minerals that are either too small, occur too infrequently, or are insufficiently ordered to characterize using conventional methods such as x-ray or neutron diffraction, electron probe microanalysis, or optical microscopy. These results have caused us to question seriously the meaning of such basic technical words as mineral, crystal, and phase that have been with us for so long and have macroscopic connotations built into them. How do we, if at all, name and classify "minerals" that are noncrystalline, exhibit long-range structural disorder in one, two, or three dimensions, or that are nonstoichiometric, or consist of finely intergrown phases? What do we call microscopic regions existing in a disordered host that have unique, homogeneous, although short-range ordered, structures? Are they minerals? Are they phases or merely defects? How do we assess their significance in terms of frequency of occurrence? How do we deal with the fact that they cannot be mechanically separated and isolated from their host because of their size or because they form coherent (continuous) boundaries with neighboring regions? All of these questions perhaps lead up to a more basic question: what really constitutes a "fundamental geologic unit"?

The purpose of this chapter is not to answer these questions, but to encourage the reader to think about them in the context of what HRTEM has revealed to us about minerals from a highly localized perspective. Here I provide a review of some of the work that has been done over the past 20 years in which some of the questions posed above have been addressed. Studies are included that discuss the problems and opportunities associated with attempting to define minerals at the microscopic scale. Consideration is given to the impact or significance of microscopic phenomena on the macroscopic properties and behavior of minerals.

8.2   MINERAL DEFINITION AT THE MACROSCOPIC SCALE

Minerals are generally regarded as naturally occurring crystalline phases defined on the basis of their macroscopic physical properties. As phases, or more specifically bulk phases in the "Gibbsian" or classical equilibrium thermodynamic sense, a mineral is said to be homogeneous with respect to its macroscopic physical properties and separable from other so-called phases (and external surroundings) by a physically distinct or discontinuous boundary. Minerals have definite chemical compositions when analyzed at the macroscopic scale, but as exemplified by solid solutions, the chemical composition may vary within fixed limits. As crystalline phases, whether chemical end members or solid solutions, minerals have internal atomic structures said to possess long-range order on average in three dimensions. The average structure is defined at the microscopic (near atomic) scale and represented by a unit cell that when repeated over an essentially infinite number of atomic distances generates microscopic crystallites, which in turn comprise macroscopic single crystals, polycrystalline aggregates, or powders.

The two most fundamental macroscopic properties of a mineral are its chemical composition and crystallographic symmetry. Together these serve as the basis for naming and classifying minerals. Other macroscopic properties (e.g., optical, electrical, magnetic, thermal, mechanical, etc.) can be considered manifestations of a mineral's average structure.

In practice, minerals are operationally defined using so-called conventional methods to analyze their chemical compositions and determine their crystallographic properties. Chemical composition refers to the relative concentrations of atoms of different elements making up a mineral from which an average stoichiometry can be derived and expressed as a formula unit. Depending on the method used, a compositional analysis may also include a determination of oxidation states and/or atomic coordination environments. Crystallogaphic properties include the point group and space group symmetry of a mineral, its unit cell parameters, number of formula units per cell, average structure, etc.

A variety of instrumental methods, in addition to classical wet chemical methods, are available for analyzing a mineral's chemical composition. Probably the methods in widest use today are x-ray fluorescence analysis and electron probe microanalysis. Crystallographic properties are generally determined with x-ray diffraction (XRD) or neutron diffraction, both powder and single crystal methods. It is common practice and generally a good idea to assess the "homogeneity" of a mineral specimen using some form of optical microscopy (e.g., reflected light or polarized transmitted light) before it is chemically analyzed or examined by powder or single crystal diffraction methods.

There are over 3,500 recognized mineral species and countless mineral varieties that have been operationally defined as crystalline phases using the conventional methods listed above. Guidelines for naming minerals have been published by the Commission on New Minerals and Mineral Names (CNMMN) of the International Mineralogical Association (IMA), in addition to other sources, and these are summarized in the Appendix (§8.9).

Presently, conventional microscopy, diffraction, and analytical methods can be used to study minerals as crystalline phases in the classical sense as long as a macroscopic specimen exists. Macroscopic refers to the quantity of material available, expressed in terms of either the size of a single crystal, mass (volume) of material as a polycrystalline aggregate or powder, number of crystallites, and so on. To put things into perspective, the unaided human eye can resolve objects as small as 10-100 μm, which serves as a practical lower limit for the macroscopic scale with respect to the size of a specimen. A petrographic microscope and electron microprobe can be used to examine small grains (single crystals or crystal fragments) or small regions in a larger crystal that are on the order of a few micrometers across, but generally a macroscopic quantity (statistically significant number) of grains or regions are sampled. For conventional powder XRD, using a flat-plate or capillary specimen, at least 1 mg or so of crystalline material is needed to obtain measurable diffracted intensity (1 μg using a synchrotron source). For single-crystal XRD, a suitable crystal must be on the order of 100 μm across (10 μm with a rotating anode or synchrotron source). Crystallites comprising a powder or single crystal specimen must consist of at least six or so unit cell translations in any one dimension to act as a coherent scattering domain (serving as an operational definition for long-range order; see chapter 12 in Klug and Alexander, 1974). Depending on the unit cell size and scattering power of the atoms in the cell, coherent domains as small as 2 to 20 nm across can be detected by XRD if present in sufficient quantities. Coherent domains are more often on the order of 100 to 1,000 nm (0.1-1 μm) across in minerals and represent literally millions to billions of unit cells.

8.3 DEFINING MINERALS USING HRTEM

HRTEM offers the unique ability to observe minerals (or any solid material) directly, in real space, at or close to the atomic scale, i.e., the scale at which they are ultimately defined. Whereas conventional methods provide information about the average macro-scopic structural and chemical properties of a mineral specimen, HRTEM can be used to describe the microstructural and microchemical properties. HRTEM combines various modes of imaging, electron diffraction, and chemical microanalysis in a manner that complements the conventional methods of optical microscopy, x-ray and neutron diffraction, and instrumental chemical analysis. With modern HRTEM instruments, lattice or structure images of very small crystals (crystallites) or very small regions in larger crystals can be obtained with 0.2-0.3 nm (2-3 Å) resolution (less than 0.2 nm with dedicated intermediate- or high-voltage HRTEM instruments). Diffraction patterns (SAED, CBED) and chemical analyses (EDS, EELS) can be taken from areas a few micrometers to a few nanometers across (and less than 1 nm on dedicated STEM instruments).

Basically two types of HRTEM studies involving minerals have been reported in the literature. One type may be considered "microcrystallographic" in nature, dealing more with mineral properties, and the other "micropetrographic," the focus being on mineral behavior. Each type of study addresses a different aspect of mineral definition, as outlined below.

8.3.1 Microcrystallographic studies

Microcrystallographic studies of minerals using HRTEM have revealed a variety of "imperfections" or defects (local disruptions of long-range order) in a many minerals otherwise thought of as ideally crystalline phases, e.g., twins (orientation and translation variants), point defects and extended defects (dislocations and stacking faults), microprecipitates and inclusions, distortions and partings, grain boundaries, and so on. These defects typically originate while a mineral is growing from a melt or solution, or once grown, transforming, deforming, or chemically reacting in the solid state. Some minerals that exist at high temperatures as chemically disordered solid solutions may exhibit substitutional, positional, or distortional ordering of atoms upon cooling. Other minerals when they cool may undergo exsolution (phase separation). The trade-off between thermodynamic and kinetic effects dictates whether a mineral forms as a stable or meta-stable phase under conditions of chemical and textural equilibrium or nonequilibrium. Metastable phases commonly exhibit complicated microstructures that are difficult or impossible to interpret using classical equilibrium thermodynamics (in the context of Gibbs phase rule).

Other microcrystallographic problems arise from the (metastable) growth or transformation of minerals that: (1) are structurally or chemically related to other minerals (e.g., polymorphs), (2) exhibit structural variants (e.g., polytypes), or (3) have structural units or modules (clusters, rods, chains, layers of atoms) in common with other minerals (e.g., polysomes). Polytypism and polysomatism are frequently encountered in layered minerals such as the phyllosilicates. The biopyriboles probably represent the best known example of polysomatism in minerals, forming a continuous series relating the inosilicates and phyllosilicates that includes single-chain pyroxene, double-chain amphibole, triple-chain jimthompsonite and chesterite, up through the "infinite chain" or sheet-silicate talc or mica structure.

Polytypic and polysomatic minerals often form coherent (or semicoherent) mixtures possessing long-range structural disorder in one, two, or three dimensions, but also containing short-range ordered configurations (domains, regions, stacks, packets, etc.) generally ranging up to tens to hundreds of Ångströms in size. Depending on their size and/or frequency of occurrence, these short-range ordered configurations are regarded as either defects (ordered regions a few units cells wide formed by random twinning, faulting, rotation, etc.) or else microphases (ordered regions usually more than a few unit cell translations wide that are statistically nonrandom in occurrence). Because of their small size and low abundance, they might not be detectable by x-ray or neutron diffraction. In some cases, short-range ordered configurations have been observed that do not correspond to known structures of existing minerals. The mixtures themselves are generally described as either single-phase solid solutions (possessing totally random structural disorder) or else multi-phase intergrowths (with nonrandom disorder).

One final example of microcrystallographic work has been to elucidate the structures of layer-misfit (corrugated or modulated) structures exhibiting superperiodicities of longer wavelength (tens to hundreds of Ångströms) than the basic unit cell repeat. Such structures are common among the layered sulfosalt minerals and sheet silicates (e.g., antigorite serpentine).

8.3.2 Micropetrographic studies

A number of HRTEM studies have been done, especially over the last decade, in which the focus has not been so much on defining the structure and chemistry of minerals, but rather on identifying them as phases and describing their relations with other minerals. The goal of these studies has been to decipher the formation history of a mineral or rock in the context of thermodynamics and kinetics. Very often the mechanism of mineral growth, transformation, deformation, or chemical reaction will be inferred from microstructural and microtextural evidence. [See Putnis and McConnell (1980) for an excellent review of mineral behavior in light of conventional TEM observations.]

The systems under consideration for these micropetrographic studies typically represent micrometer- to millimeter-sized portions of rocks that formed under non-equilibrium conditions or in which equilibrium is attained on only a fine scale. Most of these micropetrographic studies have focussed on rocks formed at relatively low temperatures (in which crystal growth is extremely slow or arrested, resulting in fine-grained, often structurally and chemically disordered metastable minerals). Included are studies of sedimentary rocks resulting from weathering or diagenesis, low-grade metamorphic rocks (shales, slates), and products of hydrothermal alteration or deposition.

Not surprisingly, there are few HRTEM studies of minerals formed in plutonic igneous rocks or medium- to high-grade metamorphic rocks because equilibrium or near-equilibrium is often attained on a relatively large scale, and the minerals that form are generally sufficiently ordered and homogeneous to define at the macroscopic scale using conventional characterization methods. HRTEM has been used, however, on numerous occasions to study the relations between "synthetic minerals" produced in the laboratory under varied conditions of pressure, temperature, and compostion. Because it is sometimes difficult or impossible to attain equilibrium, especially textural equilibrium, on laboratory time scales, these fine-grained synthetic materials frequently exhibit a wide and unpredictable variety of textures and phases with complex microstructures.

What follows is a rather broad survey of the literature designed to give the reader an overview of the microcrystallographic and micropetrographic problems that have been addressed using HRTEM in which the definition of a mineral or phase has been brought into question. I have divided the problems into two categories: problems of chemical order-disorder and problems of structural order-disorder. Other general references covering aspects of order-disorder in minerals investigated by HRTEM are given at the end of this chapter.

8.4 CHEMICAL DISORDER EXAMINED BY HRTEM

Minerals that have long-range ordered structures but exhibit chemical disorder resulting from atomic substitutions on one or more crystallographic sites can usually be defined quite satisfactorily using conventional methods. The biggest problem is specifying the contents of sites on which "mixing" occurs. Normally, these are given as weighted averages of two or more atom types (or atoms and vacancies).

In general, HRTEM imaging of chemical variations in minerals at the near-atomic level is difficult unless there is some concomitant structural change that gives rise to sufficient interference contrast. Structural changes basically include changes in the size or shape of a unit cell or the topological arrangement of atoms within a cell. Even slight shifts in atomic positions or lattice distortions accompanying a chemical substitution can produce rather significant interference or strain contrast effects.

A long-range disordered chemical solid solution is in effect a random ensemble of all thermodynamically accessible ordered unit cell configurations of different atoms occupying equivalent positions in a structure. Differently ordered unit cells comprising such a solid solution cannot be imaged as a rule unless domains representing a particular configuration exist and the atomic number difference between scatterers mixing on a site is sufficiently large. Symmetry changes incurred by varying the topochemical arrangement of atoms on positions that are otherwise symmetrically equivalent on average do not, in general, produce effects that can be easily distinguished by HRTEM.

HRTEM has proven effective for examining the structures of several minerals that form chemical solid solution series and exhibit partial to complete long-range disorder and short-range order in which the size of the unit cell is affected and/or the structural framework (topology) changes slightly but is more or less continuous on average. HRTEM has also helped to identify fine-scale microprecipitates and inclusions in some minerals reflecting the initial stages of phase separation (exsolution). The method has been of limited use, however, for investigating topochemical effects that give rise to symmetry variations at the domain scale but involve no change in unit cell size or topology. Table 8-1 gives examples of minerals described in the literature using HRTEM that qualify as chemical solid solutions but which exhibit fine-scale ordering or phase separation.

Table 8-1. HRTEM Studies of Chemically Disordered Minerals

ORDERING: Superstructures
Substitutional Ordering

Pyrrhotite	Pierce & Buseck 1974; Nakazawa et al. 1975; Nguyen et al. 1985; Pósfai & Dódony 1990; Dódony & Pósfai 1990
Bornite-Digenite	Pierce & Buseck 1978
Zinkenite	Smith 1986
Baumhauerite	Pring et al. 1990
Calcian Dolomite	Reeder 1981; Reeder & Nakajima 1982; Van Tendeloo et al. 1985; Meike et al. 1988
Laihunite	Kitamura et al. 1984; Kondoh et al. 1985; Banfield et al. 1990
Synthetic Mullite	Nakajima & Ribbe 1981a; Ylä-Jääski & Nissen 1983; Hamid Rahman & Weichert 1990; O'Keefe & Kilas 1991; Epicier 1991
Armenite	Armbruster & Czank 1992
Synthetic Amphiboles	Gribelyuk et al. 1988; Skogby & Ferrow 1989
Muscovite	Werner & Heydenreich 1988
Prehnite	Akizuki 1987
Calcic Plagioclase	Nakajima et al. 1977; Kumao et al. 1981, 1987

Substitutional and/or Positional Ordering

Synthetic (Au,Ag)-Tellurides	Van Tendeloo et al. 1983a, 1983b, 1984
Ilmenite	Wang et al. 1990
Perovskite	White et al. 1985
Loparite	Hu et al. 1992
Synthetic Hollandites	Bursill & Grzinic 1980; Zandbergen et al. 1987; Wu et al. 1991
Cancrinite-Vishnevite	Hassan & Grundy 1984; Hassan & Buseck 1992
Sodalite (Nosean & Huayne)	Hassan & Buseck 1989a, 1989b

Substitutional and/or Distortional Ordering

Al-Goethite	Mann et al. 1985
Synthetic Melilite	Czank & Seifert 1986; Seifert et al. 1987; Iishi et al. 1990; Van Heurck et al. 1992
Synthetic Tridymite Derivative Structures	Barbier & Fleet 1987, 1988; Barbier & Neuhausen 1990

ORDERING: Cell Size Unchanged

Andalusite	Hamid Rahman 1987
Vesuvianite	Veblen & Wiechmann 1991; Allen & Burnham 1992
K-feldspar	Eggleton & Buseck 1980b
Scapolite	Hassan & Buseck 1988

Null Results (No Evidence of Ordering)

Rhodizite	Pring et al. 1986
Grossular	Allen & Buseck 1988

EXSOLUTION
Microprecipitates and Inclusions

Nitrogen-Bearing Platelets and Voidites in Diamond	Bursill & Glaisher 1985; Bruley & Brown 1989; Barry 1991
Iron-Bearing Platelets in Rutile	Banfield & Veblen 1991
Magnetite/Coulsonite	Wang et al. 1989
Garnet in OPX	Reiche & Bautsch 1985
GP Zones in OPX	Nord 1980
Exsolved Pyroxenes	Champness & Lorimer 1974; Vander Sande & Kohlstedt 1974; Buseck et al. 1980; Livi & Veblen 1989; Feuer et al. 1989; Skrotzki et al. 1991
OPX Megacrysts	Veblen & Bish 1988
OPX in Sapphirine	Seifert et al. 1992
Amphibole Exsolution	Smelik & Veblen 1991; Smelik et al. 1991
Wonesite Exsolution	Veblen 1983a
Altered Biotite	Veblen & Ferry 1983
Cu-Rich Biotite	Ilton & Veblen 1988

Null Results (No Evidence of Phase Separation)

Au-Rich Arsenopyrite	Cabri et al. 1989
K-Rich Clinopyroxene	Harlow & Veblen 1991

8.4.1 Ordered superstructures

Some minerals, e.g., sulfides, feldspars and feldspathoids, form superstructures that result from ordering of different types of atoms, atoms and vacancies, or clusters of atoms that "mix" on or around one or more crystallographic sites in a structure that on average remains topologically continuous. The ordering may be substitutional, positional, or distortional in nature and may occur as part of a phase transformation on cooling or form directly during growth in a chemically fluctuating (nonstoichiometric) environment.

Superstructures in minerals may occur as short-range ordered twin domains (orientation or translation variants), tens to hundreds of Ångströms in size. Until these domains reach a critical size and abundance, they are difficult or impossible to detect using conventional diffraction methods. Sometimes a specimen will actually be a continuous intergrowth of two or more different superstructures each of which constitutes a distinct thermodynamic phase. Again, the superstructures comprising these intergrowths are likely to be indistinguishable using conventional methods so such a specimen may often be regarded a single phase on average when perceived at the macroscopic scale.

8.4.1.1 Sulfides

In an early HRTEM study of the nonstoichiometric iron sulfide mineral pyrrhotite, Pierce and Buseck (1974) used both bright-field and dark-field HRTEM techniques successfully to image superstructures linked to Fe-vacancy ordering. The superstructures observed are interpretable through an antiphase model. A 5C superstructure results from an ordered sequence of antiphase domains (1-, 2-, and 3-subcells thick with different arrangements of Fe-vacancy layers), while the higher temperature NC type results from a disordered sequence. Nakazawa et al. (1975) observed similarly ordered arrangements of metal vacancies in 4C pyrrhotite. Rotational twin boundaries and antiphase boundaries appear as narrow bands in HRTEM images. Ordering of vacancies presumably takes place on cooling, during which time a phase transition also occurs that involves a change in the average symmetry (hexagonal to monoclinic). Twinning serves the purpose of minimizing strain in the structure during the transformation. Nguyen et al. (1985) report finely intergrown 3C and 4C superstructures in pyrrhotite and Dódony and Pósfai (1990) describe stacking disorder in intergrown 4C and 5C superstructures in which different stacking sequences are separated from each other by low energy barriers.

Pierce and Buseck (1978), again using both bright-field and dark-field HRTEM imaging, discovered intergrown superstructures in bornite-digenite resulting from compositional (Cu,Fe-vacancy) ordering and positional displacements. They observed vacancy clustering and various ordered or semi-ordered arrangements of vacancies (separated into vacancy-rich and vacancy-poor clusters). Also apparent are distortions resulting from structural dissimilarities between clusters and structural discontinuities due to mismatch of lattice spacings. Pierce and Buseck interpreted various stages in superstructuring observed upon cooling in terms of unmixing. Initial superstructuring involves unmixing of vacancy-rich and vacancy-poor regions, the vacancies being still disordered on a sub-unit cell level. Stages that follow then represent successive short-range ordering schemes of the vacancies within the framework of the compositional modulation. Ordering of vacancies may itself contribute heavily to the arresting of the unmixing process by lowering the free energy of the crystal below that required to overcome the activation energy of continued exsolution.

Superstructures in bornite-digenite arise from a combination of compositional and structural modulations and short-range vacancy ordering. According to Pierce and Buseck, if this is due to arrested exsolution, bornite and digenite are not phases in the classical sense, but are metastable "mixtures" of structures; however, bornite and digenite are found to exhibit equilibrium-like relationships with other minerals as determined by many synthesis experiments.

8.4.1.2 Feldspars and feldspathoids

The tectosilicate minerals that have aluminosilicate framework structures with added cations and sometimes anions (single ions or clusters) to maintain electrostatic charge balance also typically form superstructures and show evidence of short-range ordered domains in antiphase relationships.

In a HRTEM study of plagioclase feldspar, Kumao et al. (1981, 1987) used a combination of imaging and optical diffraction techniques to show ordering of Ca and Na atoms in antiphase domains. In an example of both substitutional and positional ordering, Hassan and Grundy (1984) attributed superstructures in the feldspathoid series cancrinite-vishnevite to the orientation of SO_4^{2-} groups and also slight changes in chemistry on unit-cell scale. Ordering of cations (K^+, Na^+, Ca^{2+}) and/or anions (H_2O, SO_4^{2-}, CO_3^{2-}) is thought to take place in channels. In sodalite (nosean and huayne), Hassan and Buseck (1989a,b) detected cluster ordering and antiphase domains caused by modulations of framework oxygen atoms. In synthetic feldspathoid materials, (Na,K) and (Ca,Sr) substitutions are found to slightly distort the framework topology (Barbier and Fleet, 1987, 1988; Barbier and Neuhausen, 1990).

Another tectosilicate with an aluminosilicate framework structure is scapolite (marialite-meionite series). Hassan and Buseck (1988) found HRTEM evidence of anti-phase domains in intermediate scapolites related to substitutional ordering of Cl^- ions and CO_3^{2-} clusters (coupled with Na^+ and Ca^{2+}), although the framework structure remained continuous (no change in cell size).

8.4.2 Exsolved phases

A chemically disordered mineral, instead of ordering as a single phase, can sometimes separate into distinct phases upon cooling. These phases are described as products of exsolution. Under equilibrium conditions, phase separation generally occurs by the discontinuous process of nucleation and growth. Under nonequilibrium conditions, when sluggish kinetics are a factor, phase separation involves the continuous process of spinodal decomposition. Whereas a discontinuous process results in the formation of discrete phases, a continuous process often follows the Ostwald step rule and results in the formation of short-range ordered (transitional) microphases comprising a metastable intergrowth.

Exsolved microphases tend to form coherently (or semicoherently) intergrown platelets or lamellae that can be readily imaged with conventional TEM and HRTEM. Often these microscopic phases are regarded as defects. If they are extremely small in size or low in abundance, these microphases may escape detection by XRD or optical microscopy and be ignored. However, their presence can pose problems if a mineral specimen is chemically analyzed at the macroscopic scale. Such a mineral may appear non-stoichiometric which could be significant if chemical data are to be used for purposes of geothermometry or geobarometry.

8.4.2.1 Diamond

HRTEM has been used to study nitrogen-bearing platelets and voidites in diamond (Bursill and Glaisher, 1985; Bruley and Brown, 1989; Barry, 1991). Voidites are thought to be the dissociation product of precipitated platelet defects that arise from the discontinuous precipitation of nitrogen on dislocations, including dislocations bounding platelets. The density of nitrogen in the voidites places constraints on the depth of diamond formation.

8.4.2.2 Rutile

Banfield and Veblen (1991) used HRTEM to examine iron-bearing platelets in non-stoichiometric rutile. Retrograde precipitation of Fe results in hematite lamellae one or two unit cells wide. These may be viewed as extended defects of rhombohedral oxide structure in rutile. Variations in Fe content with metamorphic grade are accommodated by increased thickness and abundance of the lamellae. Other minerals with the rutile structure or its simple derivatives (e.g., stishovite, cassiterite, pyrolusite, marcasite), may also have lamellae with the rhombohedral oxide structure that accomodates nonstoichiometry.

8.4.2.3 Pyriboles

Exsolution is a common phenomenon in pyroxenes and amphiboles and has been studied extensively over the years using optical microscopy and conventional TEM. HRTEM has been employed more recently to study these important rock-forming minerals in an effort to better define lamellae-host interface structures and coherency and to interpret exsolution mechanisms.

Nord (1980) observed calcium-enriched regions, 5-6 unit cells thick, in lunar and terrestrial orthopyroxenes that correspond to Guinier-Preston (GP) zones. GP zones represent the first stage in a precipitation series and are followed by the formation of metastable transition phases. At the other size extreme, Veblen and Bish (1988) described the complex structure of orthopyroxene megacrysts. Multiphase mixtures of different types of lamellae are observed, comprising what can be considered pyroxene "rocks." The microstructure is consistent with subsolidus formation of lamellae in the orthopyroxene rather than formation by coprecipitation.

Smelik and Veblen (1991) and Smelik et al. (1991) report on exsolution in amphibole using HRTEM, providing descriptions of the interface structure and coherency of cummingtonite/glaucophane (host) and exsolved calcic amphibole. In the latter study, Smelik et al. consider the fundamental question of whether the complex range of observed actinolite-hornblende intergrowths represents equilibrium crystallization and therefore a miscibility gap. They determined that coexisting amphiboles can be the result of such a gap or else disequilibrium processes and kinetically hindered crystallization accompanying changing metamorphic conditions. But the observed exsolution microstructures clearly show there is in fact a miscibility gap.

8.5 STRUCTURAL DISORDER EXAMINED BY HRTEM

8.5.1 Polymorphic intergrowths

HRTEM has been used to examine polymorphs (phases with the same composition but different structures) intergrown at the unit cell level in only a few minerals. Fine-scale

Table 8-2. HRTEM Studies of Polymorphic Intergrowths

Rutile/alpha-PbO$_2$- Type TiO$_2$	Kusaba et al. 1988
TiO$_2$(B)/Anatase	Banfield et al. 1991
beta-/gamma-Spinel	Rubie & Brearley 1990; Brearley et al. 1992
beta-/gamma-Ca$_2$SiO$_4$	Eberhard et al. 1991
Synthetic (Mn,Mg)- Amphiboles	Maresch & Czank 1988

polymorphic intergrowths can result from incomplete phase transformations on cooling or by direct growth of metastable phases under nonequilibrium conditions. Table 8-2 lists minerals and some related synthetic materials exhibiting polymorphic intergrowths that have been studied by HRTEM. (An excellent non-mineralogical example of using HRTEM in conjunction with other methods to define intimately intergrown polymorphs is given by zeolite beta, a synthetic zeolite used in industry as a catalyst. It took over twenty years to solve the structure of this material, and HRTEM observations were crucial. See the article by Newsam et al., 1988, included in the reference list.)

8.5.1.1 TiO$_2$ polymorphs

There are seven known polymorphs of TiO$_2$, three of which (rutile, anatase, and brookite) are defined as minerals. Banfield et al. (1991) discovered the first natural occurrence of the TiO$_2$ polymorph known as TiO$_2$(B) occurring as thin, oriented lamellae, tens to hundreds of Ångströms wide, in crystals of anatase imaged by HRTEM. The new phase occupies only a small volume of the anatase and is barely detectable by x-ray diffraction. The TiO$_2$(B) structure can be generated from the anatase structure by regular shear to produce oriented intergrowths. TiO$_2$(B) lamellae may either have formed by simultaneous growth with anatase or have remained after the incomplete prograde replacement of TiO$_2$(B) by anatase. The irreversible TiO$_2$(B) " anatase and anatase " rutile reactions may be useful indicators of peak temperatures if it can be establisted that the three minerals are common in low-grade metamorphic rocks. Reactions involving precursor phases may be responsible for the formation of TiO$_2$ phases in nature.

8.5.1.2 Mg$_2$SiO$_4$ polymorphs

Rubie and Brearley (1990) and Brearley et al. (1992) used HRTEM to examine phase relations among high-pressure/high-temperature run-pruducts of olivine and Mg$_2$SiO$_4$ polymorphs (gamma- to beta-spinel transition). Beta-phase and metastable spinel (gamma) form from Mg$_2$SiO$_4$ olivine (alpha) by an incoherent nucleation and growth mechanism involving nucleation on olivine grain boundaries. With time, the spinel progressively transforms to the stable beta-phase. Different types of intergrowths of beta-phase and spinel are observed: (1) spinel grains with lamellae of beta-phase, (2) grains of beta-phase with lamellae of spinel, and (3) grains in which beta-phase and spinel are separated by a sharp but irregular interphase boundary. Structural disorder in spinel is interpreted as coherently intergrown lamellae of beta-phase (1-6 unit cells thick), and faulted regions in beta-phase are interpreted as oriented lamellae of spinel (1-3 unit cells thick). The quasi-cubic-close-packed oxygen sublattices are continuous between both phases. Observed micro-structures, similar to those in shocked meteorites, show that gamma-to-beta transformation can occur either by diffusion-controlled growth or by a martensitic (shearing) mechanism, depending on how far P-T conditions deviate from their values at phase equilibrium.

8.5.1.3 *Prehnite*

Akizuki (1987) used HRTEM to study optically heterogeneous "crystals" of prehnite that exhibit complex structural growth sectors. Different polymorphs occur as coherently intergrown twin lamellae at the unit cell and domain level. The polymorphs are metastably produced by a charge-balance effect related to Al-Si ordering on the crystal surface during two-dimensional growth. Crystals examined by x-ray diffraction appear more or less homogeneous on average. Whereas relatively thick crystals exhibit considerable strain contrast (as seen by optical microscopy), extremely thin ion-milled specimens show much less contrast because the strain is presumably released.

8.5.2 Polytypic intergrowths

There are many examples of minerals with layered structures that form different polytypes, i.e., phases with the same composition but different stacking sequences. Different polytypes are regarded as structural modifications, each of which is built up by stacking layers of (nearly) identical structure and composition.

HRTEM studies show that different polytypes can sometimes occur as short-range ordered stacking sequences only tens to hundreds of Ångströms thick, coherently inter-grown in an otherwise long-range disordered host. At the macroscopic scale, when examined with x-ray diffraction, structurally disordered polytypes are looked upon as statistical ensembles, representing a random distribution of all possible microscopically ordered configurations. Typically, these short-range ordered sequences are too small to act as coherent diffracting domains. Their frequency of occurrence may also be insufficient for detection by x-ray diffraction.

The role or possible significance of these short-range ordered configurations is unclear at this time. Perhaps they represent statistical anomalies that are bound to occur in a long-range disordered specimen from time to time. Maybe they are phases that are locally stabilized by slight chemical differences, or else metastable transition phases representing kinetically favorable configurations that incrementally lower the free energy of a transforming or reacting phase.

Fine-scale polytypic intergrowths are commonly observed among the sheet silicates (including clay minerals), and a large number of HRTEM studies have been done describing their microstructures. Sometimes intergrowths occur as very small discrete particles; other times they are very small regions in larger grains. Long-range disordered intergrowths exhibiting regions of short-range order have structures said to be nonrandomly disordered. Examples of minerals forming intergrown poltypes that have been examined by HRTEM are listed in Table 8-3.

8.5.2.1 *Inosilicates*

Coherently intergrown polytypes have been observed by HRTEM in several chain silicate minerals that contain stacking faults and exhibit unit cell twinning. Iijima and Buseck (1975) and Buseck and Iijima (1975) describe intergrown twin-related polytypes in enstatite. They inferred aspects of the geological history of enstatites containing both orthoenstatite (OREN) and clinoenstatite (CLEN) from the abundance and character of the

Table 8-3. HRTEM Studies of Polytypic Intergrowths

Sulfides
Sphalerite and Wurtzite	Akizuki 1981, 1983; Fleet 1983

Oxides
Zirconolite, Zirkelite, and Related Phases	White et al. 1984; White 1984

Orthosilicates
Clinozoisite/Zoisite	Ray et al. 1986
Sursassite/Pumpellyite	Mellini et al. 1984
Chloritoid	Jefferson & Thomas 1978; Banfield et al. 1989
Orientite	Mellini et al. 1986
Ardennite	Pasero & Reinecke 1991

Inosilicates
Enstatite	Buseck & Iijima 1975; Buseck et al. 1980
Pyroxene in Interplanetary Dust Particles	Bradley et al. 1983
Sapphirine	Christy & Putnis 1988; Christy 1989
Rhönite	Bonaccorsi et al. 1990
Synthetic $Mg_4Ga_8Ge_2O_{20}$	Barbier 1990

Planar Phyllosilicates
Kaolinite	Chen & Lu 1990
Muscovite and Biotite	Iijima & Buseck 1978; Amouric et al. 1981
Synthetic Muscovite	Amouric & Barronet 1983
Illite	Lonker & Fitz Gerald 1990
Paragonite	Ahn et al. 1985
Lepidolite	Rule et al. 1987
Margarite	Yau et al. 1984
Chlorite	Veblen & Ferry 1983; Spinnler et al. 1984; Amouric 1987; Bons & Schryvers 1989
Wonesite, Chlorite, Biotite	Veblen 1983b

Non-Planar Phyllosilicates
Pyrosmalite Group	Iijima 1982b; Takéuchi et al. 1983
Stilpnomelane	Crawford et al. 1977

twinning. Measuring the width of CLEN regions within OREN and noting the degree of CLEN twinning is used to distinguish between CLEN that formed by solid state transformation (a) directly from OREN in the absence of shearing, (b) from OREN by shearing, and (c) by temperature quench from the protoenstatite stability field. Bradley et al. (1983) observed pyroxene platelets in interplanetary dust particles consisting of pervasively intergrown OREN and CLEN, together with material possessing extreme stacking disorder. Based on the occurrence of such intergrowths, it is likely the platelets formed by inversion from protoenstatite during cooling.

Christy and Putnis (1988) and Christy (1989) describe intergrowths of polytypes separated by stacking faults in sapphirine. Based on a statistical analysis of polytype occurrence and calculation of stacking fault energies using the ANNNI model (Price and Yeomans, 1984; Angel et al., 1985), they addressed the question of whether individual polytypes represent defects or phases. Bonaccorsi et al. (1990) describe disordered intergrowths in rhönite, a mineral structurally similar to sapphirine, and report a new ordered polytype. Barbier (1990) found disordered intergrowths of twin-related polytypes in synthetic $Mg_4Ga_8Ge_2O_{20}$ which also has a structure related to sapphirine.

8.5.2.2 Phyllosilicates

Random and nonrandom disordered polytypic intergrowths are commonly encountered in planar sheet silicates, especially micas and chlorites. Several workers have attempted to use HRTEM to determine polytypes of these minerals. Iijima and Buseck (1978) discuss the use of lattice images compared to structure images for determining polytypes of muscovite and biotite. Amouric et al. (1981) used image simulations to show that contrast details are very sensitive to experimental details like defocus and specimen thickness in muscovite and phlogopite. Slight departures from optimum imaging conditions can lead to erroneous apparent stacking sequences as well as apparent thicknesses of component single layers. Spinnler et al. (1984) use HRTEM to distinguish one layer polytypes in clinochlore chlorite.

Polytypic intergrowths have also been observed for some nonplanar sheet silicates. Takéuchi et al. (1983) describe intergrowths of different polytypes seen with HRTEM in the relatively rare pyrosmalite group minerals. Pyrosmalite is a manganese-bearing, modulated sheet silicate with a 1:1 serpentine-like structure. Stacking disorder hindered the identification and structure refinement of these minerals by x-ray diffraction. Polytypes are built from pyrosmalite and mcGillite unit layers. The frequency of stacking mistakes may be related to the Mn content, which controls the lateral fit between tetrahedral and octahedral sheets. The substitution of Mn for Fe^{2+} causes local strain in the crystal structure (caused by a significant difference in cation size). Stacking mistakes or the occurrence of various stacking sequences ("incipient twinning") are thought to be instrumental in releasing such strain. In an earlier study, Iijima (1982b) had identified intergrowths of multi-layer polytypes in mcGillite (manganpyrosmalite), but the origin of polytypism in this case did not appear to be controlled by chemical composition.

Crawford et al. (1977) found ordered and disordered multi-layer intergrowths of polytypes in stilpnomelane, a modulated sheet silicate with a 2:1 mica-like structure. Stilpnomelane structures probably arise from some form of ordering scheme in what is usually a very severely disordered mineral, where usually randomly arranged displacement vectors are forced into a more regular disposition by an outside influence. Chemical substitution may ultimately give rise to these ordered polytypes.

8.5.3 Polysomatic intergrowths

Ordered and disordered polysomatic intergrowths have been observed using HRTEM in a wide variety of minerals. Unless otherwise noted, the word intergrowth is used here in a generic sense to describe a structurally disordered system (or mixture) that represents either a single-phase, defect-ridden, solid solution or multi-phase "assemblage" of microscopic phases (a "mineral rock"). It is not known a priori whether the short-range ordered configurations comprising an intergrowth should be considered defects or phases. In clay minerals or layered materials, these intergrowths exhibit structural disorder in one dimension, normal to the layers, and are more commonly referred to as interstratifications or mixed-layers. Minerals that exhibit chemical twinning may also be considered polysomatic intergrowths.

The short-range ordered regions comprising an intergrowth are typically tens to hundreds of Ångströms in size, too small and usually not repeated enough times to detect by x-ray diffraction. They represent different configurations of modules or building blocks contained in "end member" unit cells: polyhedra, rods, chains, layers, etc. The intergrowth arises from "modular mixing," giving rise to what can be described as a structural solid solution (analogous to atomic mixing and chemical solid solution). These ordered regions are usually separated by what appear to be coherent boundaries (stacking faults), sometimes making it impossible to conceptually separate one region from another. Totally random disordered intergrowths therefore comprise a continuous series in which every short-range ordered configuration has an equal probability of occurring. In nonrandom disordered intergrowths, certain ordered configurations have higher probabilities of occurring than others, in which case they are considered phases.

Polysomatic intergrowths are commonly found in natural or synthetic systems formed under nonequilibrium conditions. These short-range ordered configurations therefore represent metastable microphases (also sometimes called precursor, transitional, or intermediate phases). They are favored by kinetics and represent the continuous nature of transformations or reactions under nonequilibrium conditions according to the Ostwald step rule (Morse and Casey, 1988; Eberl and Srodon, 1988; Merriman et al., 1990; Jiang et al., 1990a,b; Baxter Grubb et al., 1991; Livi and Veblen, 1992). Such phases may also be stabilized by local chemical effects.

When examined at the macroscopic scale using conventional characterization methods, these intergrowths pose several structural and chemical problems that could cause one to misidentify a mineral or imprecisely interpret its crystal chemistry or phase relations. Long-range structural disorder in one or more dimensions may complicate the determination of symmetry and unit cell parameters by x-ray diffraction. Electron microprobe analyses often indicate nonstoichiometry resulting from structural and chemical disorder, thereby making it hard to define a formula unit. Microscopic examination in transmitted or reflected light often reveals anomalous optical properties.

Structural disorder is manifested in minerals in different ways. Certain manganese oxides exhibit intergrowths of octahedral chains of different widths affecting the size and shape of tunnels. In pyroxenes and amphiboles, disorder results from variations of the width of tetrahedral chains, and in pyroxenoids, the periodicity of chains. In planar phyllosilicates, the stacking of layers is affected, and in non-planar phyllosilicates the width of layers or strips may be disordered. In our quest to define minerals more precisely, it will probably be useful and necessary to incorporate some new words and terms describing the type and degree of structural disorder into our system of mineral nomenclature.

Table 8-4 lists minerals forming polysomatic intergrowths that have been described using HRTEM. Some examples are given below for selected minerals to give an overview of the problems encountered. For more details, the reader should consult the individual papers, or else see Buseck and Veblen (1988) for an in-depth review of many of the minerals listed.

8.5.3.1 Sulfosalts

In a HRTEM study of the sartorite ($PbAs_2S_4$) group minerals, Pring (1990) found disordered liveingite, compositionally intermediate between dufrenoysite ($Pb_2As_2S_5$) and liveingite ($Pb_9As_{13}S_{28}$), to be a disordered intergrowth of sartoritelike units (baumhauerite) and dufrenoysitelike units (liveingite). The disordered intergrowth represents a transitional stage in the replacement of dufrenoysite by the more As-rich mineral liveingite. Disordered intergrowths may be quite widespread in sulfide minerals, particularly those subjected to moderate- or lower-temperature metamorphism or hydrothermal alteration during the late stages of their history. The occurrence of this type of nonequilibrium behavior in sulfides provides direct evidence of replacement in the paragenesis of sulfide deposits.

8.5.3.2 Iron-oxyhydroxides

Amouric et al. (1986) used HRTEM to examine hard iron-oxide crusts (ferricretes) formed as a result of lateritic weathering. Pisolites near the top of the weathering profile consist of a core rich in Al-hematite rimmed by Al-goethite. Poorly crystalline kaolinite presumably formed from a well-crystallized precursor. An unidentified phase (layered, Fe-rich, mica-like mineral with a 2M superstructure) was likely in equilibrium with iron-hydroxides, unlike kaolinite. This other phase was probably stabilized because it was able to incorporate more Fe^{3+} than kaolinite. Macroscopic mineralogical trends across the weathering profile are about the same as those revealed by TEM studies inside the pisolites of the iron-crust zone as successive mineral "neoformations." These neo-formations are merely responses at any time to an evolving chemical system; therefore, such well indurated iron-crust zones are not "end-products" of weathering, but rather intermediate products of a still evolving system.

Boudeulle and Muller (1988) also studied the details of the laterization process, involving several stages of goethite formation, with HRTEM. The close associations of hematite and goethite either in the same particle (intermediate phase) or as intergrowths may mean the stability fields of the two minerals are joined or overlap. However, hematite seems to be a precursor for goethite growth in hematite-goethite transformation involving a dissolution-crystallization mechanism. Slow kinetic processes resulting in the metastable persistence of hematite would explain the failure of classical thermodynamic approaches to describe iron-oxyhydroxide behavior in lateritization.

8.5.3.3 Orthosilicates

Banfield et al. (1990) used HRTEM to describe ordered and disordered intergrowths of olivine and its weathering product laihunite. Different superstructures of laihunite are observed related to ordering of Fe^{3+}, Fe^{2+}, and vacancies. A sizable portion of an optically homogeneous olivine crystal may be composed of laihunite without radically changing its composition. Excess of silicon and low-octahedral cation totals in microprobe and TEM analyses may indicate the presence of ordered or disordered olivine/laihunite intergrowths. Laihunite will significantly change the thermodynamic, magnetic and transport properties of olivine. Its presence is interpreted to indicate an episode of subsolidus oxidation of the lava.

Banfield et al. (1989) examined chloritoid, which is an orthosilicate with a layered structure, intergrown with chlorite, stilpnomelane, paragonite, and berthierine. There is a dimensional similarity between the unpolymerized tetrahedral layer of chloritoid and the tetrahedral sheets of the other minerals. It is possible that chloritoid re-equilibrated to more Fe-rich compositions stable at lower temperatures accompanied by the formation of intergrown chlorite lamellae that are relatively rich in Mg. However, detailed chemical budgets between chloritoid, the various sheet silicates, and metamorphic fluids could not be calculated with any certainty.

8.5.3.4 Inosilicates

Biopyriboles exhibiting chain-width order-disorder were among the first minerals to be examined in detail with HRTEM. Much has been learned and inferred over the years as a result of numerous HRTEM studies about the structures and relations of pyroxenes, amphiboles, and micas from both a crystallographic and thermodynamic perspective. The pyroxenoids, which exhibit chain-perioidicity order-disorder, have also been scrutinized with HRTEM.

Table 8-4. HRTEM Studies of Polysomatic Intergrowths

Sulfosalts
    Cylindrite/Franckeite   Williams & Hyde 1988
    Aikinite/Krupkaite   Pring 1989
    Sartorite Group   Pring 1990
    Schafarzikite Group   Mellini et al. 1981

Oxides
    Anatase/$TiO_2(B)$ Banfield & Veblen 1992
    Goethite/Hematite   Amouric et al. 1986; Boudeulle & Muller 1988

Manganese Oxides (tunnel-size disorder)
    Hollandite/Romanechite   Turner & Buseck 1979
    Todorokite   Turner & Buseck 1981; Turner et al. 1982
    Nsutite   Turner & Buseck 1983
    Pyrolusite/Mn-Oxides   Rask & Buseck 1986; Amouric et al. 1991
    Lithiophorite   Manceau et al. 1990
    Bixbyite-Braunite   de Villiers & Buseck 1989
    Series

Carbonates
    Bastnaesite-Synchisite   Van Landuyt & Amelinckx 1975
    Series

Oxyborates (chemical twinning)
    Synthetic and Natural   Bovin & O'Keefe 1981; Bovin et al. 1981a;
    Pinakiolite-Related   Bovin & Norrestam 1990
    Oxyborates
    Sb-rich Pinakiolite/   Hansen et al. 1988
    Pinakiolite

Orthosilicates
    Olivine/Laihunite   Banfield et al. 1990
    Humites   Müller & Wenk 1978; White & Hyde 1982a, 1982b,
                                  1983; Yau & Peacor 1986
    Chloritoid/Sheet   Banfield et al. 1989
    Silicates
    Orientite   Mellini et al. 1986

Inosilicates (chain-width disorder; chain-periodicity disorder)
    Mixed-Chain Pyriboles   Veblen et al. 1977; Veblen & Buseck 1979a; Buseck
    (general)   et al. 1980; Veblen 1981
    Altered Pyroxenes   Nakajima & Ribbe 1980, 1981b; Veblen & Buseck
                                  1981; Akai 1982; Eggleton & Boland 1982
    Enstatite/Lizardite   Le Gleuher et al. 1990
    Double- and Triple-   Schumacher & Czank 1987; Welch et al. 1992
    Chain Pyriboles
    Fibrous Amphiboles   Hutchison et al. 1975; Veblen 1980; Cressey et al.
                                  1982
    Nephrite (Actinolite)   Jefferson et al. 1978; Mallinson et al. 1980
    Altered Anthophyllite   Veblen & Buseck 1980
    Synthetic (Mn,Mg)-   Maresch & Czank 1988
    Amphiboles
    Synthetic Tremolitic   Ahn et al. 1991
    Amphiboles
    Crocidolite   Ahn & Buseck 1991
    Heated Talc   Konishi & Akai 1991
    Synthetic Na-(Co,Ni)   Gribelyuk et al. 1988
    Chain Silicates
    Pyroxenoids   Jefferson et al. 1980; Ried & Korekawa 1980; Czank &
                                  Liebau 1980; Smith et al. 1981; Jefferson & Pugh
                                  1981; Czank & Simons 1983; Aikawa 1984; Veblen 1985
    Pyroxenes/Pyroxenoids   Ried 1984; Veblen 1985; Angel 1986; Livi & Veblen 1992
    Babingtonite   Czank 1981
    Sapphirine/Surinamite   Christy & Putnis 1988

305

Planar Phyllosilicates (layer-stacking disorder)

Mixed-Layer Phyllo-silicates (general)	Page 1980; Knipe 1981; Barber 1985; Vali & Köster 1986; Klimentidis & Mackinnon 1986; Amouric 1987; Guthrie & Veblen 1989b
Paragonite/Phengite	Ahn et al. 1985
Muscovite/Biotite	Iijima & Zhu 1982
Goethite/Kaolinite	Jefferson et al. 1975; Boudeulle & Muller 1988
Muscovite/Kaolinite	Lee et al. 1975a; Jiang & Peacor 1991; Robertson & Eggleton, 1991
Fuchsite/Kaolinite	Singh & Gilkes 1991
Fe-Mica/Kaolinite	Amouric et al. 1986
Biotite/Kaolinite	Ahn & Peacor 1987
Kaolinite/Halloysite	Singh & Gilkes 1992
Talc/Chlorite	Bons & Schryvers 1989
Biotite/Chlorite	Iijima & Zhu 1982; Olives Baños et al. 1984; Veblen & Ferry 1983; Olives Baños & Amouric 1984; Yau et al. 1984; Maresch et al. 1985; Olives Baños 1985; Eggleton & Banfield 1985; Amouric et al. 1988; Mellini et al. 1991
Chlorite/Muscovite	Franceschelli et al. 1986; Bons et al. 1989
Chlorite/Wonesite	Veblen, 1983b
Chlorite/Illite	Lee & Peacor 1983; Lee et al. 1984; Lee & Peacor 1985
Donbassite/Other Sheet Silicates	Ahn et al. 1988; Ahn & Buseck 1988
Goethite/Smectite (Iddingsite)	Eggleton 1984; Smith et al. 1987
Illite/Smectite	Yoshida 1973; Page & Wenk 1979; Eggleton & Buseck 1980a; Bell 1986; Klimentidis & Mackinnon 1986; Ahn & Peacor 1986a, 1989; Vali & Köster 1986; Yau et al. 1987a, 1987b; Huff et al. 1988; Guthrie & Veblen 1989a, 1990; Ahn & Buseck 1990; Veblen et al. 1990; Jiang et al. 1990b; Środoń et al. 1990; Amouric & Olives 1991; Buatier et al. 1992 *Related Studies of Illite/Smectite:* Zen 1962; Aagaard & Helgeson 1983; Garrels 1984; Nadeau et al. 1984a, 1984b, 1984c, 1985a, 1985b, 1987; Nadeau 1985; Rosenberg et al. 1985; Sawhney & Reynolds 1985; Corbató & Tettenhorst 1987; Decarreau et al. 1987; Eberl et al. 1987; Mackinnon 1987; Sass et al. 1987; Altaner & Bethke 1988; Inoue et al. 1988; Eberl & Środoń 1988; Altaner & Vergo 1988; Eberl et al. 1988; Meunier & Velde 1989; Jiang et al. 1990a; Tettenhorst et al. 1990; Rosenberg et al. 1990; Güven 1991; Lindgreen et al. 1991; Kitagawa & Matsuda 1992
Kaolinite/Illite/ Smectite	Lee et al. 1975b; Ahn & Peacor 1985b; Banfield & Eggleton 1990
Chlorite/Illite/ Smectite	Ahn & Peacor 1985a
Chlorite/Corrensite	Shau et al. 1990
Chlorite/Smectite	Bettison-Varga et al. 1991
Fe-Smectite/Glauconite	Buatier et al. 1989
Biotite/Vermiculite	Banfield & Eggleton 1988
Chlorite/Vermiculite	Wada & Kakuto 1989
Phlogopite/Serpentine	Sharp et al. 1990
Chlorite/Lizardite	Lee & Peacor 1983; Wicks 1986
Chlorite/Serpentine/ Mica	Amouric et al. 1988
Serpentine/Amphiboles/ Phyllosilicates	Veblen & Buseck 1979b
Carbonaceous Chondrite Matrix Phyllosilicates	Mackinnon & Buseck 1979; Mackinnon 1982; Barber et al. 1983; Tomeoka & Buseck 1983, 1985; Mackinnon & Zolensky 1984; Zolensky & Mackinnon 1986

Non-Planar Phyllosilicates (layer-width and strip-width disorder)

Antigorite	Mellini et al. 1987
Minnesotaite	Ahn & Buseck 1989

Veblen et al. (1977) were among the first to describe disorder in the sequence of double-and triple-chains in altered amphibole specimens, and found isolated chains wider than triple. This structural disorder helps to explain why asbestiform amphiboles are fibrous. "Single-crystals" examined with x-ray diffraction actually turned out to be multi-phase mixtures of amphibole and new pyriboles (e.g., jimthompsonite and chesterite). Diffuse streaking in diffraction photographs indicates disorder. Presumably, these crystals possess many mistakes in the sequence of double and triple chains, regions of no apparent order in the chain sequence, or areas with chains even wider than triple. Veblen et al. noted that it is essential that the extent of chain-width disorder be evaluated since pyroxenes and amphiboles are used extensively by geologists to assess rock histories and formation temperatures and pressures.

Veblen and Buseck (1979a) conducted a statistical analysis of chain-width order-disorder in altered anthophyllite using a runs-probability test and found that certain chain-width sequences were the result of nonrandom growth processes and could be considered new microscopic phases, whereas others were random and constituted defects. If the minerals chesterite and jimthompsonite are stable, their stability ranges are probably between those of anthophyllite and talc. Ordered minerals may be produced metastably during reaction as a result of constraints imposed by the reaction mechanisms. The occurrence of chesterite appears to be controlled by the chemical potentials of Fe, Ca and Mg; newly observed long-range ordered structures are probably the result of some nonrandom reaction or growth mechanism, rather than being thermodynamically stabilized.

Maresch and Czank (1988) used HRTEM to examine synthetic (Mn,Mg)-amphiboles. They describe chain-multiplicity faults (CMFs) and chain-arrangement faults (CAFs) and the occurrence of ordered and disordered intergrowths. They also apply a statistical factor used to quantify defect structure. Changes in symmetry and faulting were examined as a function of run time and composition. CAFs affect the apparent symmetry and CMFs lead to significant inhomogeneity. Both effects must be considered when interpreting crystal chemistry and phase relations.

Ahn et al. (1991) evaluated chain-width disorder in synthetic tremolitic amphiboles and performed a statistical analysis of defect structures. Changes in the number of chain-width defects occur as a function of P, T, X. The abundance of triple chains is inversely proportional to that of double-chain structures, suggesting that the triple-chain silicate may be an important precursor to calcic amphibole synthesis. Temperature appears to be the most critical experimental variable affecting the number of chain-width defects. Structural defects in minerals are important in controlling diffusion properties and chemical reactions. Structural disorder caused by random intergrowths of various chain-silicate structures will increase the Gibbs free energy of synthetic amphiboles relative to defect-free natural ones, at least within the amphibole stability field. Besides the compositional aspect, the structural disorder caused by the intergrowth of various chain widths can play a critical role in modifying thermodynamic properties of synthetic amphiboles. The extent to which the variation in chain-width defects influences the thermodynamic properties of synthetic tremolite cannot yet be quantified.

Jefferson and Pugh (1981) examined synthetic (Fe,Mn)-pyroxenoids using HRTEM. They found intergrowths of wollastonite and pyroxmangite repeats in the rhodonite structure. Pyroxenoid was originally produced in a glassy state and annealed for different times. The frequency and variety of stacking faults are found to decrease as annealing time increases. Crystallization of pyroxenoid structures is, in effect, a continual process where the structure type adopted gradually changes as the temperature is lowered, via a series of disordered intergrowth structures, to an equilibrium, or near-equilibrium arrangement. The material appears to adapt continually in the crystallization process, by means of metastable intermediate states, until the appropriate structural type is attained, with no evidence of the more normal nucleation and growth from the initial glassy product.

Veblen (1985) used HRTEM to describe chain-periodicity order-disorder in natural Mn-pyroxenoids containing metastable intermediate phases formed by incomplete reactions. Statistical analysis confirmed the occurrence of an ordered intermediate phase. Reaction of pyroxene to pyroxenoid (johannsenite to rhodonite) forms oriented lamellar intergrowths during the replacement of intermediate phases. Reactions can take place by lamellar or bulk reactions. Pyroxmangite (7-repeat) may be a metastable intermediate phase, because the reaction involving calcium loss would be expected to favor formation of rhodonite (5-repeat).

8.5.3.5 Phyllosilicates

Layer-stacking order-disorder (interstratification, mixed-layering, etc.) in the phyllosilicates has been the subject of numerous x-ray diffraction and TEM investigations. Common layer types defined in terms of the (001) lattice spacing include: 5-Å (brucite/gibbsite), 7-Å (serpentine/kaolinite), 9-Å (talc/pyrophyllite), 10-Å (mica), 12-Å (smectite), and 14-Å (chlorite). The mixing of layers typically occurs as a result of incomplete (metastable) reactions that accompany weathering or diagenetic processes at relatively low temperatures, or else low- to medium-grade metamorphic reactions. Mixing can occur both between layers and along layers.

Generally, mixed-layer intergrowths are randomly disordered, but short-range ordered (nonrandom) sequences or stacks are occasionally observed, tens to hundreds of Ångströms thick, that are interpreted as metastable intermediate phases. These microscopic phases are usually too small and/or occur too infrequently in a single specimen to detect by x-ray diffraction. However, there are several examples of regular interstratifications found to exist at the macroscopic scale that can be characterized using conventional methods and named as minerals (e.g., rectorite, corrensite, kulkeite).

Disordered interlayering in phyllosilicates also generally gives rise to nonstoichiometry, making it difficult to obtain accurate single-phase chemical compositions. Failure to recognize fine-scale compositional heterogeneity and the presence of microscopic phases can interfere with thermodynamic calculations in which a single-phase solid solution model is assumed.

8.5.3.5.1 Kaolinitized mica.
Jiang and Peacor (1991) and Robertson and Eggleton (1991) describe kaolinitized muscovite based on HRTEM observations. Kaolinite occurs as packets of layers which alternate with packets of muscovite or smectite-like layers. The alternation with the micaceous phase(s) is somewhat regular, i.e., micaceous layers are separated by approximately equal numbers of kaolinite layers. The smectitelike phase may be ordered illite/smectite that appears to be a direct "along-layer" alteration product of muscovite. Smectite may have formed as an intermediate phase during the hydrothermal kaolinitization of muscovite. Hydrothermal topotactic alteration of muscovite to kaolinite with smectite as an intermediate phase proceeds as: muscovite " illite " smectite " kaolinite (Ostwald step rule).

Ahn and Peacor (1987) examined kaolinitized biotite and found kaolinite occurring in two distinct modes: layers interstratified with biotite and 2-layer units irregularly interlayered with biotite. Some 2-layer kaolinite units (14 Å) terminate at single biotite layers (10 Å), implying a reaction of one biotite layer to two kaolinite layers. The alteration mechanism consists of complete dissolution of biotite and crystallization of kaolinite at linear boundaries. Ordered mixed layering of phyllosilicates may involve either primary growth or replacement.

8.5.3.5.2 *Biotite/chlorite.* A number of workers have used HRTEM to characterize intergrowths of biotite/chlorite. Veblen and Ferry (1983) found that chloritization of biotite forms disordered mixed-layer intergrowths. They deduced chemical consequences of reaction mechanisms from TEM observations compared with electron microprobe and petrographic observations. Examination of partially choritized biotite shows that alteration to chlorite proceeded by a process in which brucitelike layers replaced mica layers, with a consequential volume loss of 30%. No evidence was observed of the volume increasing mechanism in which extra brucitelike layers are inserted into the biotite interlayer.

Olives Baños and Amouric (1984) and Amouric et al. (1988) found that biotite chloritization involves essentially the replacement of the plane of potassium ions by a brucitelike sheet. The process is a chemical transformation that occurs in the interlayer levels of biotite ("interlayer brucitization"). It is a different mechanism of chloritization from that ("talc brucitization") observed by Veblen and Ferry (1983). Interlayer brucitization produces chlorite layers interstratified in biotite, and all intermediate states between biotite and chlorite can be observed. Ordered structures, such as the 1:1 biotite-chlorite structure, may be locally formed.

Olives Baños (1985) also examined interstratified biotite and chlorite. He showed that the energy of mixing of the layers is nearly equal to zero, confirming that ordered 1:1 "biochlorites" may be stable crystals. Layer sequences are generally disordered as seen in HRTEM images, but the ordered 1:1 biochlorite is frequently observed.

Maresch et al. (1985) used HRTEM to examine highly altered chlorite that actually consisted of ordered and disordered intergrowths of chlorite and biotite. Biotitization of chlorite involves transformation of chlorite into biotite. The mechanism involves the replacement of brucite-like octahedral sheets in chlorite by K^+ interlayer sheets (related to hydrothermal activity in a retrograde path of the metamorphic cycle).

Mellini et al. (1991) examined pleochroic "brown chlorite" resembling biotite previously described as oxychlorite. It forms as a retrograde alteration product of chlorite/biotite. Ambiguous results were obtained when observed with optical microscopy, x-ray diffraction, and electron microprobe analysis. Grains actually consist of chlorite matrix interleaved with minor, very thin biotite lamellae, and 16-Å hydrated chlorite interleaved with lamellae of hematite or of a hematite precursor. The 16-Å phase exhibits wavy lattice fringes indicative of locally variable swelling (interpreted as chlorite layers interstratified with water layers).

8.5.3.5.3 *Illite/chlorite.* Ordered and disordered intergrowths of illite and chlorite have been examined with HRTEM. These are particularly interesting because they represent mixed layering of dioctahedral and trioctahedral phyllosilicates. Lee and Peacor (1983), Lee et al. (1984), and Lee and Peacor (1985) describe illite/chlorite intergrowths and find that the mixed layering is random at an individual layer level and separates into discrete packets of illite and chlorite layers as burial metamorphism (diagenesis) proceeds. The entire sequence of textural, chemical and structural changes involving illites and chlorites is consistent with a general trend from heterogeneous metastable structures toward homogeneous, defect-free grains approaching an equilibrium state with their local environment.

8.5.3.5.4 Altered feldspar. HRTEM has been used to identify and define fine- scale phyllosilicate alteration products in feldspar. Page and Wenk (1979) found that plagioclase alters to smectite, which in turn changes through a series of intermediate phyllosilicate phases to sericite. Intermediate phyllosilicates are highly disordered, only a few unit cells wide, with curved lattice planes and abundant interlayering. Before they coarsen they transform into new phases until stable sericite is grown.

Eggleton and Buseck (1980a) proposed a mechanism for microcline alteration based on HRTEM observations. A 10-Å layer silicate phase shows an irregular stacking sequence, including 10-, 20- and 30-Å sequences. This may represent a mixture of layer silicates with different morphologies: montmorillonite (thin crinkled crystal), illite (modulated crystal) and muscovite (flat, uniform crystal). At this level of crystallization it is also possible that these distinctions have little meaning; the modulations of one crystal may reflect variations in chemistry across it, giving rise to a 10-Å layer that is "illite" in one area and "montmorillonite" in another. Such modulated structures may be the precursors to interstratified illite/montmorillonite of soils, as the K and Al proportions in a given layer dictate whether it will ultimately become expandable. At the early stage of alteration, the only crystalline phases are feldspar and 10-Å layer silicate, and it is assumed that these are in equilibrium. The beginning of feldspar weathering is an equilibrium process leading to the formation of an illite/montmorillonite mixed-layer mineral.

Banfield and Eggleton (1990) describe the sequential formation of weathering products in plagioclase and K-feldspar. An alteration layer <1 μm thick on feldspar surfaces is termed "protocrystalline." The alteration sequence involves transitional metastable phases, and there is strong chemical control on which phases occur.

8.5.3.5.5 Illite/smectite. Mixed-layer illite/smectite represents one of the most, if not the most, controversial problems to have been addressed in light of conventional mineralogical theory and practice. It has caused quite a stir relating to how one goes about defining the word phase (see Altaner and Vergo, 1988, and Eberl et al., 1988). The problem has been extensively investigated from both crystallographic and thermodynamic standpoints. Questions about the structural state of mixed-layer illite/smectite, whether it represents a single-phase solid solution or multi-phase intergrowth, whether it is stable or metastable, and so on, have been asked and answered and the answers debated in the literature over the past ten years. In general, there is considerable interest in understanding the mechanism by which smectite converts to illite during diagenesis. There are several proposed mecahnisms: Ostwald ripening of fundamental illite particles, layer-by-layer transformation, and dissolution/recrystallization.

X-ray diffraction and TEM have been used to define the structural state (degree of order-disorder) of mixed-layer illite/smectite samples from which the above formation mechanisms have been inferred and on which thermodynamic calculations have been based. There are still several basic questions remaining to be fully answered, however, that pertain to x-ray and TEM methodology: sample preparation artifacts, reconciling definitions of particle boundary coherency, relevance of short-range order, etc. The controversy is still in progress. I will not elaborate further on the specific details of mixed-layer illite/smectite, but encourage the interested reader to consult the references given in Table 8-4.

8.5.3.5.6 Chlorite/smectite. Mixed-layer chlorite/smectite specimens generally exhibit abnormal optical properties and chemical compositions. Shau et al. (1990) used HRTEM to define the structural state of corrensite, which is an ordered 1:1 mixed-layer chlorite/smectite. "Expandable chlorite" is a mixed-layer chlorite/corrensite. Assemblages of trioctahedral phyllosilicates tend to occur as intergrown discrete phases, such as chlorite-corrensite, corrensite-smectite, or chlorite-corrensite-smectite. A model for the corrensite crystal structure suggests that corrensite should be treated as a unique phase rather than as an ordered intergrowth of chlorite and smectite. The proportion of corrensite (or smectite-like) layers relative to chlorite layers in low-grade rocks is inferred to be controlled principally by Fe/Mg ratio in the fluid or the bulk rock and by temperature. Compositional variations of "chlorites" in low-grade rocks, which appear to correlate with temperature or metamorphic grade, may reflect variable proportions of mixed-layered components.

Bettison-Varga et al. (1991) examined mixed-layer chlorite/smectite. Smectite and chlorite layers are present in abundances similar to those predicted with XRD techniques. Regular alternation of chlorite and smectite occurs at the unit-cell scale; sequences of 1:1 chlorite and smectite layers (24 Å spacing) tend to cluster in what may otherwise appear to be disordered mixtures, suggesting the existence of a corrensite end-member having thermodynamic significance.

8.5.3.5.7 Serpentine/amphiboles/phyllosilicates. Veblen and Buseck (1979b) observed the serpentine minerals chrysotile, lizardite, and antigorite intimately intergrown with each other and with talc, chlorite, and amphibole in incompletely reacted chain silicates. They described several new variations in serpentine planar and roll structures, as well as regions of mixed-layer serpentine and talc. Serpentine-talc-chlorite-amphibole intergrowths demonstrate that retrograde hydration reactions of chain silicates can result in complex disequilibrium mineral mixtures.

8.5.3.5.8 Carbonaceous chondrite matrix phyllosilicates. Ordered and disordered intergrowths of 7-Å serpentine (S) and 5-Å brucite (B) layers have been observed by several workers in altered carbonaceous chondrite meteorites using HRTEM (Mackinnon and Buseck, 1979; Mackinnon, 1982; Barber et al. 1983; Tomeoka and Buseck, 1983, 1985; Mackinnon and Zolensky, 1984; Zolensky and Mackinnon, 1986). The 10-Å poorly characterized phase (PCP) is Fe-rich tochilinite (also referred to as FESON phase), which is interlayered mackinawite and brucite. The 17-Å SBB phase is interlayered tochilinite and serpentine (FESON + cronstedite).

8.6 SHORT-RANGE ORDERED REGIONS
IN DISORDERED INTERGROWTHS

As discussed above, structurally disordered "minerals" may be described either as single-phase solid solutions or else as multi-phase intergrowths. An HRTEM image of a structurally disordered specimen, whether a lattice image or structure image, can be quite complicated and inhomogeneous in appearance. One should not assume *a priori* that each homogeneous region in an image consitutes a distinct thermodynamic phase. It may be more appropriate to start by calling such ordered regions defects. Statistical significance tests like those described by Veblen and Buseck (1979a) and Maresch and Czank (1988) can be used to determine whether a particular ordered configuration seen in HRTEM images is likely random or nonrandom in occurrence. If it can be shown statistically that a configuration is the product of a nonrandom growth process, then perhaps it is acceptable to call it a phase. The discontinuity (lack of coherency) across a boundary between two differently ordered regions may also serve as a basis for defining a phase.

Phases that are microscopic in size and/or quantity generally cannot be treated as bulk phases that are defined on the basis of their macroscopic properties. There is no way, for example, to mechanically separate such phases and determine their thermodynamic properties. Certain aspects of phase behavior may be assessed inside the electron microscope as a result of electron beam heating or, better yet, under controlled conditions with a high-temperature holder. However, establishing the stability region for such phases may be difficult or impossible because some seem to form as metastable intermediate phases or are stabilized by local chemistry.

How should we refer to these structurally disordered intergrowths and short- range ordered polymorphic, polytypic, or polysomatic configurations? Are the intergrowths varieties of mineral species? What if they represent a polysomatic combination of two species? Are they mineraloids because they are not long-range ordered in three dimensions?

Is it useful, let alone appropriate, to isolate short-range ordered microphases and assign them names by determining their chemical composition and crystallographic properties using HRTEM and associated diffraction and analytical methods? Are such microphases minerals in their own right, i.e., fundamental geologic units, or is the disordered intergrowth of which they are a part the mineral? Maybe they should be called "micro-minerals" that never or only rarely form crystalline regions large or abundant enough to detect using conventional methods [Imagine a display of micro-minerals in your local museum!]. How many identical unit cells of some structure does it take before a substance is considered a mineral? In other words, when does short-range order really become long-range order? All of these questions stem from having the ability to make HRTEM observations of structurally disordered mineral intergrowths.

Regardless of what the mineralogical community ultimately decides to do regarding the naming and classification of structurally disordered intergrowths and short-range ordered microphases, there is no denying that their existence is significant and should be noted. Long-range structural disorder and short-range order do often have a pronounced effect on macroscopic crystallochemical and thermodynamic properties of a solid, in which case HRTEM observations do become relevant.

Although microphases observed by HRTEM may be volumetrically insignificant, it is generally recognized that they can play important roles in the mechanisms of mineral formation. They often exist as metastable precursor, intermediate, or transitional phases that form under nonequilibrium conditions. As such, when some short-range ordered configuration observed in HRTEM images is found to be statistically significant in occurrence, then that phase is presumably there for a reason and its structure should be reported, regardless of whether or not it can be detected by conventional diffraction methods. Is it reasonable to require a macroscopic quantity of so-called crystalline material before a phase is defined as a mineral? Triple-chain slabs are commonly observed in HRTEM images of altered pyroxenes and amphiboles occurring as defects or microphases. However, triple-chain and mixed triple- and double-chain sequences did not receive mineral names until found in nature in sufficient amounts to be defined by x-ray diffraction, optical, and electron microprobe methods. Veblen and Buseck (1979a) found several other long-period mixed triple-and double-chain sequences to be statistically significant in occurrence in an altered amphibole constituting distinct (metastable) phases (see Chapter 6). These phases, to date, have not warranted mineral names, yet they may play an important role as geological units. How should these units be defined and classified?

Also, how do we refer to a microphase that is amorphous but seems to be pervasive in a specimen, e.g., forming layers on crystal surfaces or occupying regions between grain boundaries? Whereas short-range ordered configurations may be considered "micro-minerals," perhaps these amorphous phases may be called "micro-mineraloids" and deserve to be acknowledged.

Table 8-5 lists short-range ordered polymorphs, polytypes, and polysomes observed by HRTEM and reported in the literature that do not exist as bulk phases corresponding to known minerals. Also listed are reports of microphases exhibiting limited short-range order or which appear amorphous, but are described as playing key roles as metastable precursor, transitional or intermediate phases in mineral transformation and reaction mechanisms.

8.6.1 New polymorphs

Banfield et al. (1991) describe the structure of the first reported natural occurrence of the TiO_2 polymorph called $TiO_2(B)$. The phase occurs in microscopic quantities intergrown with anatase and could not be adequately defined using x-ray diffraction. They determined the structure using an approach that combined analytical electron microscopy, electron diffraction, high-resolution imaging, image simulations, and distance-least-squares refinement. The mineral is presently unnamed, but it will be interesting to see if this combined microscopy/modelling approach is accepted as a viable way of defining a new mineral species.

8.6.2 New polytypes

Polytypes are regarded as varieties of mineral species. Distinct polytypes identified at the macroscopic scale, e.g., by XRD, are generally considered phases that may or may not have well-defined stability fields. A number of minerals exhibit fine-scale intergrown polytypes, some of which are only found to exist at the microscopic scale. It is not usually known if these microscopic polytypes act as defects or phases.

Iijima and Buseck (1975) and Buseck and Iijima (1975) described theoretical and observed polytypes of orthoenstatite and clinoenstatite resulting from unit-cell twinning. They identified new polytypes on the basis of at least three unit cell repeats. A corrected procedure for defining enstatite polytypes is given by Buseck et al. (1980). Long-period stacking sequences appear to be a result of random growth processes based on the runs-probability test of Veblen and Buseck (1979a). Christy and Putnis (1988) applied the same statistical test to show that two unique polytypes identified in HRTEM images of sapphirine are most likely distinct (metastable) phases.

8.6.3 New polysomes

Quite a few minerals forming polysomatic intergrowths contain unique short-range ordered configurations of modules that are only observed at the microscopic scale. These are generally interpreted as metastable intermediate phases resulting from sluggish kinetic effects or else as phases stabilized by local chemistry.

Using HRTEM Turner and Buseck (1979) observed isolated structures with quadruple and septuple chain-widths as coherent intergrowths with manganese oxide minerals hollandite and romanechite. In todorokite, Turner and Buseck (1981) observed structures with triple-chains in one direction and chain-widths ranging from triple to septuple in the other direction (todorokite is nominally 3×3 Mn-octahedra).

Table 8-5. Unique Short-Range Ordered Configurations Observed by HRTEM

New Polymorphs
TiO$_2$(B)	Banfield et al. 1991

New Polytypes
Zirconolite	White 1984
Chloritoid	Jefferson & Thomas 1978
Orientite	Mellini et al. 1986
Ardennite	Pasero & Reinecke 1991
Enstatite	Iijima & Buseck 1975; Buseck & Iijima 1975; Buseck et al. 1980
Sapphirine	Christy & Putnis 1988
Rhönite	Bonaccorsi et al. 1990
Paragonite	Ahn et al. 1985
Stilpnomelane	Crawford et al. 1977
McGillite (Manganpyrosmalite)	Iijima 1982b; Takéuchi et al. 1983

New Polysomes
Hollandite/Romanechite	Turner & Buseck 1979
Todorokite	Turner & Buseck 1981
Bixbyite-Braunite	de Villiers & Buseck 1989
Bastnaesite-Synchisite	Van Landuyt & Amelinckx 1975
Synthetic Pinakiolite Related Oxyborates	Bovin & Norrestam 1990
Mg-Humite	White & Hyde 1982a
Jerrygibbsite-Leucophoenicite	Yau & Peacor 1986
Pyriboles	Veblen & Buseck 1979a
Nephrite	Mallinson et al. 1980
Heated Talc	Konishi & Akai 1991
Pyroxenoids	Veblen 1985
Synthetic Fe-Rhodonite	Ried & Korekawa 1980
Ferrosilite III	Czank & Simons 1983
Biotite/Kaolinite	Ahn & Peacor 1987
Altered Feldspar	Eggleton & Buseck 1980a
Ba-Treated Muscovite	Brown & Rich 1968
Phlogopite/Chlorite	Yau et al. 1984
Biotite/Chlorite	Olives Baños & Amouric 1984; Maresch et al. 1985
Chlorite/Illite	Lee & Peacor 1985
Poorly Characterized Phase (PCP)	Mackinnon & Buseck 1979; Mackinnon 1982; Barber et al. 1983; Tomeoka & Buseck 1983, 1985; Mackinnon & Zolensky 1984
Minnesotaite	Ahn & Buseck 1989

Other Microscopic Phases
Ordered Regions in Carbon	Buseck et al. 1988; Buseck et al. 1992
Hematite/Goethite	Boudeulle & Muller 1988
Mn-Oxides	Rask & Buseck 1986
Olivine	Eggleton 1984
Altered Feldspar and Mica	Banfield & Eggleton 1990; Romero et al. 1992a
Clay Precursor	Tazaki 1986; Tazaki & Fyfe 1987; Tazaki et al. 1987; Tazaki et al. 1992; Romero et al. 1992b; Zhou et al. 1992
"Fe-Mica"	Jefferson et al. 1975; Amouric et al. 1986
Chlorite	Bons et al. 1990
Na-Nontronite	Stucki & Tessier 1991
Glauconite	Amouric & Parron 1985

In the La-carbonate bastnaesite-synchisite series, Van Landuyt and Amelinckx (1975) identified three new mixed-layer compounds. According to these authors, a stacking sequence corresponds to a different compound if it produces a well-defined diffraction pattern and a lattice image exhibiting regions containing at least ten identical repeat sequences.

As mentioned earlier, in an examination of altered anthophyllite, Veblen and Buseck (1979a) reported unique short-range ordered mixed-chain intergrowths with statistically nonrandom occurrences. These sequences include: (2233), (233), (232233), (222333), (2332323), (2333), (433323), (2234), and (43332343332423), where (2) represents double-chain amphibole, (3) is triple-chain jimthompsonite, and (32) is alternating triple- and double-chain chesterite. In heated talc, Konishi and Akai (1991) found local structures with 4-6 repeated (21) sequences (alternating double- and single-chain) observed in disordered pyribole. A statistical test suggests that the (21) structure is unstable and formed by chance as a result of the reaction mechanism.

In kaolinite/biotite Ahn and Peacor (1987) noted the occurrence of four repeats of 2:1 ordered sequences of kaolinite and biotite within a mixed-layer crystal. These were found to be insignificant according to a statistical test. Ordered interstratifications are also observed by HRTEM in several other mixed-layer phyllosilicates:

1:1 illite-chlorite (Lee and Peacor, 1985),

2:1 phlogopite-chlorite (Yau et al., 1984),

1:1 and 2:1 biotite-chlorite (Olives Baños and Amouric, 1984; Maresch et al., 1986),

1:2 serpentine-brucite (SBB phase in PCPs; FESON + cronstedite) (Mackinnon and Buseck, 1979; Mackinnon, 1982; Tomeoka and Buseck, 1983, 1985; Mackinnon and Zolensky, 1984).

8.6.4 Other microscopic phases

Buseck et al. (1988) described ordered regions in structurally disordered carbon that show little or no evidence of crystallinity by powder x-ray diffraction measurements. The evident structural heterogeneity of the carbonaceous material within Pre-cambrian kerogen samples suggests that bulk measurements, be they chemical or structural, provide only averaged values and thus potentially hide important information residing in these samples. Unraveling the elemental and isotopic chemistry of the mixtures that are reflected in these bulk values can provide much information regarding the origin and subsequent history of the kerogen fractions.

Eggleton (1984) described a "phase M" formed as an alteration product of olivine. The metastable hexagonal phase probably consists of close-packed, metal-oxygen octa-hedra, similar to brucite. It is a precursor to goethite formation. (Banfield et al., 1990, find no such metastable phase in weathered olivine and show that "phase M" may be just an imaging artifact.)

In altered feldspar, Banfield and Eggleton (1990) describe a short-range ordered "protocrystalline" phase forming an alteration layer <1 μm thick on feldspar surfaces. It is depleted in Ca, Na, K and Si and enriched in Fe and occurs alone on K-feldspar and as an intergrowth or mixture with clay (only a few unit cells wide) on plagioclase. This meta-stable transition phase is considered a precursor of clay. Although not volumetrically abundant, it represents an essential step in the conversion of the feldspar structure to smectite.

Tazaki (1986) and Tazaki and Fyfe (1987) also describe a clay precursor phase in altered feldspar. It is a poorly ordered hydrated Fe-rich silicate phase with ultrathin 150-200 Å circular form and 14-20 Å lattice spacings, or long, curled fiber forms with varied spacings. The phase appears to represent the first step before the development of normal crystalline clays. It occurs on feldspar surfaces, possibly forming by stripping off K-ions from the feldspar framework and replacing with protons. Units from the feldspar framework are rearranged into primitive polymerlike, clay precursors which can accommodate considerable iron.

Amouric et al. (1986) identified an Fe-rich, alkali-poor, micalike phase (10 Å spacing) with 2M superstructure that accompanies goethite. Bons et al. (1990) reported a 70-100 Å thick crystalline phase occupying incoherent boundaries between grains of chlorite. Finally, Amouric and Parron (1985) define a "phase X" that exists as precursor phase to glauconite.

8.7   STRUCTURE DETERMINATIONS
OF KNOWN MINERALS BY HRTEM

HRTEM and associated TEM methods have been used to confirm and refine existing structure models of known minerals, propose new structure models, help define new minerals, and even discredit minerals. Combined HRTEM and x-ray diffraction studies have proven to be most effective for elucidating structures of long-range disordered minerals. Examples illustrating some of these special HRTEM applications are given in Table 8-6.

HRTEM and image simulations have been used to refine the models of modulated (corrugated) structures of cylindrite/franckeite (Williams, 1986; Williams and Hyde, 1988; Wang and Kuo, 1991), antigorite (Spinnler and Otten, 1988), and minnesotaite (Guggenheim and Eggleton, 1986; Ahn and Buseck, 1989). The average structure model of leucophoenicite determined by x-ray diffraction has been confirmed (White and Hyde, 1983), as have the structures of ordered interstratified phyllosilicates rectorite (1:1 illite/smectite) (Klimentidis and Mackinnon, 1986; Ahn and Peacor, 1986b) and corrensite (1:1 chlorite/smectite) (Klimentidis and Mackinnon, 1986; Shau et al., 1990).

Wang et al. (1990) used high-angle annular dark-field (HAADF) imaging to examine compositional layering in franckeite. (HAADF imaging is a powerful high-resolution STEM technique that is sensitive to atomic-number (Z) contrast.) Downing et al. (1990) used image reconstruction techniques to determine oxygen positions in the staurolite structure.

New structure models have been proposed for several sulfosalt minerals (lengenbachite, valleriite, potosiite) and synthetic Sn-Sb-Se phases for which suitable single crystals could not be obtained. Similar studies have been done, combining single-crystal x-ray diffraction and HRTEM, for chemically-twinned oxyborates (ludwigite, takéuchiite) and synthetic Mg-Mn-B-O phases.

Mellini et al. (1985) used HRTEM to define the structure of carlosturanite, a chain-silicate forming a polysomatic series with serpentine. Mellini et al. (1986) determined the structure of orientite using single-crystal x-ray diffraction and HRTEM. Iijima (1982a) and Ozawa

Table 8-6. Structure Determinations of Known Minerals By HRTEM

Confirming/Refining an Existing Structure Model

Molybdenite	Shiojiri et al. 1991
Lead Oxides	Wang et al. 1992
Cylindrite/Franckeite	Williams 1986; Williams & Hyde 1988; Wang & Kuo 1991; Wang et al. 1990
Leucophoenicite	White & Hyde 1983
Staurolite	Downing et al. 1990
Sursassite	Mellini et al. 1984
Cordierite	Epicier et al. 1991
Antigorite	Spinnler & Otten 1988
Greenalite	Guggenheim et al. 1982
Minnesotaite	Guggenheim & Eggleton 1986; Ahn & Buseck 1989
Rectorite	Klimentidis & Mackinnon 1986; Ahn & Peacor, 1986b
Corrensite	Klimentidis & Mackinnon 1986; Shau et al. 1990

Proposing a New Structure Model

Sulfosalts	
Lengenbachite	Williams & Pring 1988
Valleriite	Wang & Buseck 1990
Potosiite	Kissin & Owens 1986
Synthetic Sn-Sb-Se Phases	Smith & Parise 1985
Oxyborates	
Ludwigite	Norrestam et al. 1989
Takéuchiite	Bovin et al., 1981b; Norrestam & Bovin 1987
Synthetic Mg-Mn-B-O Phases	Cooper & Tilley 1986
Orientite	Mellini et al. 1986
Carlosturanite	Mellini et al. 1985
McGillite	Iijima 1982a; Ozawa et al. 1983
10-Å PCP	Mackinnon & Zolensky 1984

Radiation Damaged (Metamict) Minerals

Titanite	Hawthorne et al. 1991
Thorite-Group	Lumpkin & Chakoumakos 1988
Zircon	Yada et al. 1981, 1987; Chan & Buseck 1982; Chakoumakos et al. 1987; Murakami et al. 1991
Pyrochlore-Group	Lumpkin & Ewing 1988
Fluorapatite	Paul & Fitzgerald 1992

Defining New Minerals with XRD and HRTEM

Sb-rich Pinakiolite	Hansen et al. 1988
Kulkeite	Schreyer et al. 1982

Discrediting Minerals

Al-rich Gedrite	Ferrow & Ripa 1990
Eastonite	Livi & Veblen 1987
Mountain Wood	Subbanna et al. 1986
Baumite	Guggenheim & Bailey 1989, 1990

---

et al. (1983) used a combination of methods to more precisely define the structure of mcGillite (manganpyrosmalite). X-ray diffraction revealed a unusually long c-axis, so the structure was examined using selected area electron diffraction and HRTEM. The presence of diffuse rods parallel to **c*** suggests extensive stacking disorder caused by rotation twinning. Crystals of mcGillite (and the related mineral friedelite) do not grow to sufficient size to give sharp x-ray diffraction maxima because of the phenomenon of "incipient twinning." Repeated twinning is universal, and twin domains are commonly only a few unit layers thick. Friedelite may be regarded as a disordered equivalent of mcGillite (no regular structure can be identified by XRD).

Mackinnon and Zolensky (1984) used HRTEM to define the structure of the 10-Å, poorly characterized phase (PCP) found in carbonaceous chondrites. The structure corresponds to Fe-rich tochilinite (interlayered mackinawite and brucite).

HRTEM has also proved useful for defining the structures of metamict minerals (titanite, thorite-group, zircon, pyrochlore-group, fluorapatite) that have been damaged to varying degrees by radiation (naturally or artificially induced). These minerals, or more properly, mineraloids, may contain fission tracks and typically exhibit aperiodic domain structures (domains typically 50-100 Å across).

HRTEM has been used in conjunction with x-ray diffraction to define the structures of new minerals such as Sb-rich pinakiolite (Hansen et al., 1988), a Sb-end-member (not formally named) of the oxyborate mineral pinakiolite, as well as the phyllosilicate mineral kulkeite (Schreyer et al., 1982) which is an ordered 1:1 chlorite/talc mixed-layer structure.

Several minerals have been discredited or renamed based on HRTEM observations. Ferrow and Ripa (1990) describe an Al-rich gedrite that is an intergrowth of chlorite and lizardite. Livi and Veblen (1987) found that the type locality specimen of eastonite is actually a lamellar intergrowth of phlogopite and lizardite. It represents retrograde replacement of phlogopite by serpentine along (001) layers. "Mountain wood" is considered a partly fibrous morphological variant of anthophyllite (more brittle than asbestos). A sample examined by Subbanna et al. (1986) consists of intimate brucite/anthophyllite intergrowths (coexisting asbestiform anthophyllite does not show brucite intergrowths). Brucite lamellae vary from 20-200 Å wide. Mountain wood is thought to be a precursor to the asbestiform anthophyllite. Guggenheim and Bailey (1989, 1990) had the Zn- and Mn-rich serpentine mineral baumite discredited by showing that it is actually a coherent intergrowth of 7-Å lizardite (at least two polytypes) and modulated 1:1 layer silicate similar to those in the greenalite-caryopilite series with 14-Å chlorite structure.

8.8 CONCLUSIONS

This chapter has attempted to show the range of problems and opportunities associated with using HRTEM to define minerals, in particular, those minerals that cannot be fully characterized using conventional characterization methods. Minerals that are fine-grained, exist in small quantities, and are insufficiently ordered are ideally suited for study by HRTEM.

HRTEM has been used with limited success to address problems involving chemical disorder. Conventional TEM seems to provide as much, if not more, information in these cases. HRTEM has proved quite effective, however, for dealing with problems of structural disorder. Much has been learned about minerals with so-called modular structures (e.g., chain and sheet silicates, oxides, sulfosalts), that form polymorphic, polytypic, or polysomatic intergrowths.

Minerals that exhibit structural and chemical variations on a unit-cell and domain sized scale typically represent metastable phases that have grown, transformed, deformed, or reacted to form new minerals under nonequilibrium conditions (kinetically rather thermodynamically controlled). Mineral formation processes that take place under nonequilibrium conditions tend to be continuous rather than discontinuous in nature so structural and chemical variations may occur on a fine scale. Often microscopic metastable phases form that represent intermediate steps in a kinetically controlled mineral formation process. Such processes are often encountered during weathering, diagenesis and low-grade metamorphism, hydrothermal alteration or deposition; basically, any situation in which crystallization occurs at relatively low temperatures.

It is important to understand that the phenomena of chemical and structural disorder are not confined to naturally formed minerals, but are frequently encountered in a wide range of solid materials synthesized in the laboratory and produced in industry, including metals, ceramics, catalysts, semiconductors, superconductors, and so on. Disorder is often found to have a significant effect on the behavior and performance of so-called crystalline materials, sometimes in a positive way, others times in an adverse way. With synthetic materials, the goal is often to be able to tailor the degree of disorder to suit a particular application. Much of what is learned about synthetic materials regarding disorder and the specific problems of noncrystallinity (no unit cell), nonstoichiometry (no formula unit), metastable formation (no stable thermodynamic control), can be readily applied to minerals. The modern mineralogist benefits greatly by keeping an eye on what is happening in other related fields of materials science.

It will be interesting to see where the science of mineralogy goes as we close out the 20th century. Just over 200 years ago, René Haüy proposed his "integral molecule" concept to explain the internal structure of crystals. The concept survives almost in its original form today as the unit cell theory of modern crystallography. Think about how far we have progressed in our ability to analyze the chemical composition and determine the crystallographic properties of minerals. Mineralogists have gone from using reflecting goniometry, wet chemical analysis, and optical microscopy, to relying on such methods as x-ray diffraction, electron probe microanalysis, and transmission electron microscopy. At the beginning of the 19th century, there were only 200 or so known mineral species; as we begin the 21st century, there probably will be close to 4,000.

Over the years, mineralogists, materials scientists, crystallographers, crystal chemists, and the like, have been defining structural states for solids that lie between the two familiar extremes, i.e., crystalline and noncrystalline (amorphous). Various words appear in the literature describing different crystal/crystallite sizes or degrees of long-range order (periodicity). Most of the words are operationally defined, i.e., based on observations or measurements made with a specific method. Examples of such words are: microcrystal-line, nanocrystalline, cryptocrystalline, protocrystalline, paracrystalline, quasicrystalline, aperiodic, nonperiodic, and so on.

Until now, the science of mineralogy has only been concerned with the structural extremes. There are two categories for naturally-occurring materials, i.e., crystalline minerals and noncrystalline mineraloids. One may ask if there is a need to define some new words that more accurately describe polysomatic intergrowths exhibiting structural and chemical disorder. Should there be more explicit rules for naming intermediate members of a polysomatic series as mineral species? Several authors have addressed this topic (see Guggenheim and Eggleton, 1988, p. 713; Buseck and Veblen, 1988, pp. 356-361; de Villiers and Buseck, 1989, p. 1335). Perhaps the time has arrived for the mineralogical community to review this matter collectively, for example, by posing the question to the IMA/CNMMN for consideration.

The mineralogist of the future will no doubt be well versed in several areas of materials science, and will need to combine various methods of structural and chemical characterization to help solve problems. Fortunately, HRTEM does not have quite the same reputation it did 10 or 20 years ago as being very difficult to do and the theory being too complicated. Modern HRTEM instruments are largely computer controlled, and the basic theory behind most imaging, diffraction, and analytical methods has been worked out. HRTEM instruments are now becoming more readily accessible to researchers through inter-laboratory use and national facilities. Although HRTEM is still far from being considered a routine characterization method, it should be employed whenever possible to aid in the definition of specimens that are extremely fine-grained, exist in limited supply, or exhibit considerable structural disorder.

HRTEM information becomes even more useful when combined with data from other sources. The standard way of solving new synthetic zeolite structures, for example, is to use x-ray diffraction and nuclear magnetic resonance spectroscopy along with HRTEM (Thomas and Vaughan, 1989). The trend of conducting complementary studies will no doubt continue as a way of defining microscopic or structurally disordered materials. Expect to see more HRTEM studies combined with Rietveld powder x-ray diffraction studies and synchrotron x-ray diffraction studies involving small volumes of sample or small single crystals (1-10 μm across). Also HRTEM and scanning tunneling/atomic force microscopy may prove to be a powerful pair for imaging the internal and external structure of minerals.

We should look forward to seeing new HRTEM techniques emerge that will find applications in mineralogy. Convergent-beam electron diffraction (see Chapter 2) is just beginning to find wider use as a way of determining the symmetry of minerals that exist as small crystals or that contain small ordered regions in larger disordered crystals. Other techniques to watch for in the near future that will be applied more to minerals are Z-contrast (atomic-number-contrast) imaging, high-temperature HRTEM, and electron holography (image reconstruction).

As methods for doing "electron crystallography" continue to improve, mineral definition will take on new dimensions. What will be important here is that a consistent set of jargon words and terms be developed that mean the same things when using different methods. Because most technical words are operationally defined, communication problems sometimes arise when too many methods employ the same jargon. Consider for example, terms like defect and phase; long-range order and short-range order; coherent, semicoherent, incoherent boundary. Microscopists, spectroscopists, and diffractionists all use these terms, but the meanings may not be quite the same for each application. A concerted effort should be made to ensure the consistent use of these basic words and terms so that data from different sources can be compared meaningfully.

How we ultimately choose to name and classify structurally disordered minerals or minerals only observed at the microscopic scale is still a topic for debate. The matter should continue to be discussed by researchers. In the meantime, until some consensus is reached, observations about such "non-ideal" minerals need to be recorded. Who knows, but maybe someday you'll be able to show off your valuable collection of "micro-minerals" that can only be seen with a high- esolution transmission electron microscope.

8.9 APPENDIX

PUBLISHED GUIDELINES FOR NAMING MINERALS

Minerals are generally regarded as crystalline phases formed as the result of geological processes. As a (bulk) crystalline phase in the classical sense, a mineral must satisfy the conditions of long-range structural order in three dimensions, and homogeneity with respect to its macroscopic physical and chemical properties.

The essential parts in the definition of a mineral are its chemical composition and crystallographic properties. Together these serve as the basis for naming and classifying minerals. As elements or compounds, minerals generally have chemical compositions that are definite (stoichiometric), but which may vary within fixed limits (as exemplified by solid solutions). Observing or measuring other physical properties, such as optical character, or specifying mode of occurrence contributes to the description of a mineral, and may subsequently aid in its identification, but the information is not usually adequate for establishing uniqueness.

A substance formed as the result of a geological process warrants a mineral species name if certain criteria are satisfied that demonstrate uniqueness with respect to composition or crystal structure. The Commission on New Minerals and Mineral Names (CNMMN) of the International Mineralogical Association (IMA) is the professional organization that presently controls mineral nomenclature. Nickel and Mandarino (1987) summarize the procedures involved in naming a new mineral species and discuss criteria used to establish uniqueness. The CNMMN imposes no rigid rules to define whether or not a compositional or crystallographic difference is sufficiently large to require a new mineral name. They consider each new mineral proposal on its own merits. The following general guidelines are offered for naming minerals:

"... a general guideline for compositional criteria is that at least one major structural site should be occupied by a different chemical component than that which occurs in the equivalent site in an existing mineral. But if the presence of an element occurring in a relatively minor amount stabilizes the structure, or if its presence in an occupied site effects a structural change or size difference, then consideration may be given to create a new name for such a mineral."

" ... a crystallographic difference sufficiently large to jusify the creation of a new mineral name is one in which the structure of the mineral is topologically different from that of an existing one."

Polymorphs are regarded as distinct mineral species but polytypes and superstructures are not. Regular interstratifications of clay minerals must satisfy certain other operational criteria to be considered as distinct mineral species (see Bailey, 1981 or 1982). Other CNMMN/IMA guidelines relating to mineral nomenclature may be found in Nickel (1992a or 1992b), Dunn (1990), Bayliss and Levinson (1988), and Mandarino (1987). For more information on polytypes see Guinier et al. (1984) and for clay minerals also see Martin et al. (1991) and Bailey (1980a or 1980b, 1982).

The CNMMN does not specify the experimental methods that should be used when defining a new mineral species. These days, electron probe microanalysis and X-ray diffraction are the most widely used methods for identifying and defining minerals as phases. Together they provide a sound basis for operationally defining minerals in terms of their average chemistry and structure.

ACKNOWLEDGMENTS

I am grateful to many people for sharing their opinions, insights, and suggestions on mineral and phase definition. People at Engelhard include T. Dombrowski, T. Gegan, M. Hamil, M. Hobson, S. Hollis, J. St. Amand, K. Thrush, R. Truitt, and F. von Trentini. I am most grateful to my Engelhard colleague J. Lampert for his ongoing support and advice, as well as professional expertise. Others to thank for useful discussions include R. Cohen (Geophysical Laboratory), G. Guthrie (Los Alamos National Laboratory), P. Heaney (Princeton University), R. Reynolds (Dartmouth College) and E. Salje (Cambridge University). J. Post (Smithsonian Institution) and D. Veblen (The Johns Hopkins University) provided helpful reviews of the chapter on short notice. My appreciation goes to Engelhard management for giving me the time and space to work on this project, as well as MAX group members for their interest and technical help, and to M. Fedors for providing me with a comprehensive literature search. I am indebted to Peter Buseck for asking me to participate in this MSA short course and for his patience and encouragement. I also thank my wife, Barbara, my daughters, Eve and Emma, and other family members for putting up with my obsessive behavior over the past year to see this project through.

GENERAL REFERENCES

Barbier, J., Hiraga, K., Otero-Diaz, L.C., White, T.J., Williams, T.B. and Hyde, B.G. (1985) Electron microscope studies of some inorganic and mineral oxide and sulphide systems. Ultramicroscopy 18, 211-234.

Buseck, P.R. (1983) Electron microscopy of minerals. Am. Scientist 71, 175-185.

Buseck, P.R. (1984) Imaging of minerals with the TEM. Bull. Electron Microscopy Soc. Am. 14, 47-53.

Buseck, P.R. and Cowley, J.M. (1983) Modulated and intergrowth structures in minerals and electron microscope methods for their study. Am. Mineral. 68, 18-40.

Buseck, P.R. and Iijima, S. (1974) High resolution electron microscopy of silicates. Am. Mineral. 59, 1-21.

Buseck, P.R. and Veblen, D.R. (1978) Trace elements, crystal defects and high resolution electron microscopy. Geochim. Cosmochim. Acta 42, 669-678.

Buseck, P.R. and Veblen, D.R. (1981) Defects in minerals as observed with high-resolution transmission electron microscopy. Bull. Minéral. 104, 249-260.

Buseck, P.R. and Veblen, D.R. (1988) Mineralogy. In High-Resolution Transmission Electron Microscopy, P.R. Buseck, J.M. Cowley, L. Eyring, eds., Oxford University Press, New York, 308-377.

Drits, V.A. (1987) Electron Diffraction and High-Resolution Electron Microscopy of Mineral Structures. Springer-Verlag, New York, 304 pp.

Thomas, J.M., Jefferson, D.A., Mallinson, L.G., Smith, D.J. and Crawford, E.S. (1978-79) The elucidation of the ultrastructure of silicate minerals by high resolution electron microscopy and x-ray emission microanalysis. Chemica Scripta 14, 167-179.

Veblen, D.R. (1985) Extended defects and vacancy non-stoichiometry in rock-forming minerals. In Point Defects in Minerals, Schock, R.N., ed., American Geophysical Union, Washington D.C., 122-131.

Veblen, D.R. (1985) Direct TEM imaging of complex structures and defects in silicates. Ann. Rev. Earth Planet. Sci. 13, 119-146.

Zussman, J. (1987) Minerals and the electron microscope. Mineral. Mag. 51, 129-138.

SPECIFIC REFERENCES

Aagaard, P. and Helgeson, H.C. (1983) Activity/composition relations among silicates and aqueous solutions: II. Chemical and thermodynamic consequences of ideal mixing of atoms on homological sites in montmorillonites, illites, and mixed-layer clays. Clays Clay Minerals 31, 207-217.

Ahn, J.H. and Buseck, P.R. (1988) Al-chlorite as a hydration reaction product of andalusite: a new occurrence. Mineral. Mag. 52, 396-399.

Ahn, J.H. and Buseck, P.R. (1989) Microstructures and tetrahedral strip-width order and disorder in Fe-rich minnesotaites. Am. Mineral. 74, 384-393.

Ahn, J.H. and Buseck, P.R. (1990) Layer-stacking sequences and structural disorder in mixed-layer illite/smectite: Image simulations and HRTEM imaging. Am. Mineral. 75, 267-275.

Ahn, J.H. and Buseck, P.R. (1991) Microstructures and fiber-formation mechanisms of crocidolite asbestos. Am. Mineral. 76, 1467-1478.

Ahn, J.H. and Peacor, D.R. (1985a) Transmission electron microscopic study of diagenetic chlorite in Gulf Coast argillaceous sediments. Clays Clay Minerals 33, 228-236.

Ahn, J.H. and Peacor, D.R. (1985b) Transmission electron microscopic study of the diagenesis of kaolinite in Gulf Coast argillaceous sediments. In Proc. Int'l Clay Conf., Denver, L.G. Schultz, H. van Olphen, and F.A. Mumpton, eds., The Clay Minerals Society, Bloomington, IN, 151-157.

Ahn, J.H. and Peacor, D.R. (1986a) Transmission and analytical electron microscopy of the smectite-to-illite transition. Clays Clay Minerals 34, 165-179.

322

Ahn, J.H. and Peacor, D.R. (1986b) Transmission electron microscope data for rectorite: implications for the origin and structure of "fundamental particles." Clays Clay Minerals 34, 180-186.

Ahn, J.H. and Peacor, D.R. (1987) Kaolinitization of biotite: TEM data and implications for an alteration mechanism. Am. Mineral. 72, 353-356.

Ahn, J.H. and Peacor, D.R. (1989) Illite/smectite from Gulf Coast shales: a reappraisal of transmission electron microscope images. Clays Clay Minerals 37, 542-546.

Ahn, J.H., Burt, D.M. and Buseck, P.R. (1988) Alteration of andalusite to sheet silicates in a pegmatite. Am. Mineral. 73, 559-567.

Ahn, J.H., Peacor, D.R. and Essene, E.J. (1985) Coexisting paragonite-phengite in blueschist eclogite: a TEM study. Am. Mineral. 70, 1193-1204.

Ahn, J.H., Cho, M., Jenkins, D.M. and Buseck, P.R. (1991) Structural defects in synthetic tremolitic amphiboles. Am. Mineral. 76, 1811-1823.

Aikawa, N. (1984) Lamellar structure of rhodonite and pyroxmangite intergrowths. Am. Mineral. 69, 270-276.

Akai, J. (1982) Polymerization process of biopyribole in metasomatism at the Akatani ore deposit, Japan. Contrib. Mineral. Petrol. 80, 117-131.

Akizuki, M. (1981) Investigation of phase transition of natural ZnS minerals by high resolution electron microscopy. Am. Mineral. 66, 1006-1012.

Akizuki, M. (1983) Investigation of phase transition of natural ZnS minerals by high resolution electron microscopy: reply. Am. Mineral. 68, 847-848.

Akizuki, M. (1987) Al,Si order and the internal texture of prehnite. Can. Mineral. 25, 707-716.

Allen, F.M. and Burnham, C.W. (1992) A comprehensive structure-model for vesuvianite: symmetry variations and crystal growth. Can. Mineral. 30, 1-18.

Allen, F.M. and Buseck, P.R. (1988) XRD, FTIR, and TEM studies of optically anisotropic grossular garnets. Am. Mineral. 73, 568-584.

Altaner, S.P. and Bethke, C.M. (1988) Interlayer order in illite/smectite. Am. Mineral. 73, 766-774.

Altaner, S.P. and Vergo, N. (1988) Sericite from the Silverton caldera, Colorado: discussion. Am. Mineral. 73, 1472-1474.

Amouric, M. (1987) Growth and deformation defects in phyllosilicates as seen by HRTEM. Acta Crystallogr. B43, 57-63.

Amouric, M. and Baronnet, A. (1983) Effect of early nucleation conditions on synthetic muscovite polytypism as seen by high resolution transmission electron microscopy. Phys. Chem. Minerals 9, 146-159.

Amouric, M. and Parron, C. (1985) Structure and growth mechanism of glauconite as seen by high-resolution transmission electron microscopy. Clays Clay Minerals 33, 473-482.

Amouric, M. and Olives, J. (1991) Illitization of smectite as seen by high- resolution transmission electron microscopy. Eur. J. Mineral. 3, 831-835.

Amouric, M., Gianetto, I. and Proust, D. (1988) 7, 10 and 14 Å mixed-layer phyllosilicates studied structurally by TEM in pelitic rocks of the Piemontese zone (Venezuela). Bull. Minéral. 111, 29-37.

Amouric, M., Mercuriot, G., Baronnet, A. (1981) On computed and observed HRTEM images of perfect mica polytypes. Bull. Minéral. 104, 298-313.

Amouric, M., Parc, S. and Nahon, D. (1991) High-resolution transmission electron microscopy study of Mn-oxyhydroxide transformations and accompanying phases in a lateritic profile of Moanda, Gabon. Clays Clay Minerals 39, 254-263.

Amouric, M., Baronnet, A., Nahon, D. and Didier, P. (1986) Electron microscopic investigation of iron oxyhydroxides and accompanying phases in lateritic iron- crust pisolites. Clays Clay Minerals 34, 45-52.

Angel, R.J. (1986) Transformation mechanisms between single-chain silicates. Am. Mineral. 71, 1441-1454.

Angel, R.J., Price, G.D. and Yeomans, J. (1985) The energetics of polytypic structures: further applications of the ANNNI model. Acta Crystallogr. B41, 310-319.

Armbruster, T. and Czank, M. (1992) H_2O ordering and superstructures in armenite, $BaCa_2Al_6Si_9O_{30} \cdot 2H_2O$: a single-crystal x-ray and TEM study. Am. Mineral. 77, 422-430.

Bailey, S.W. (1980a) Summary of recommendations of AIPEA nomenclature committee. Clay Minerals 15, 85-93.

Bailey, S.W. (1980b) Summary of recommendations of AIPEA nomenclature committee on clay minerals. Am. Mineral. 65, 1-7.

Bailey, S.W. (1981) A system of nomenclature for regular interstratifications. Can. Mineral. 19, 651-655.

Bailey, S.W. (1982) Nomenclature for regular interstratifications. Am. Mineral. 67, 394-398.

Bailey, S.W., Brindley, G.W., Kodama, H. and Martin, R.T. (1982) Report of the Clay Minerals Society Nomenclature Committee for 1980-1981. Nomenclature for regular interstratifications. Clays Clay Minerals 30, 76-78.

Banfield, J.F. and Eggleton, R.A. (1988) Transmission electron microscope study of biotite weathering. Clays Clay Minerals 36, 47-60.

Banfield, J.F. and Eggleton, R.A. (1990) Analytical transmission electron microscope studies of plagioclase, muscovite, and K-feldspar weathering. Clays Clay Minerals 38, 77-89.

Banfield, J.F. and Veblen, D.R. (1991) The structure and origin of Fe-bearing platelets in metamorphic rutile. Am. Mineral. 76, 113-127.

Banfield, J.F. and Veblen, D.R. (1992) Conversion of perovskite to anatase and $TiO_2(B)$: a TEM study and the use of fundamental building blocks for understanding relationships among the TiO_2 minerals. Am. Mineral. 77, 545- 557.

Banfield, J.F., Karabinos, P. and Veblen, D.R. (1989) Transmission electron microscopy of chloritoid: intergrowth with sheet silicates and reactions in metapelites. Am. Mineral. 74, 549-564.

Banfield, J.F., Veblen, D.R. and Jones, B.F. (1990) Transmission electron microscopy of subsolidus oxidation and weathering of olivine. Contrib. Mineral. Petrol. 106, 110-123.

Banfield, J.F., Veblen, D.R. and Smith, D.J. (1991) The identification of naturally occurring $TiO_2(B)$ by structure determination using high-resolution electron microscopy, image simulation, and distance-least-squares refinement. Am. Mineral. 76, 343-353.

Barber, D.J. (1985) Phyllosilicates and other layer-structured materials in stony meteorites. Clay Minerals 20, 415-454.

Barber, D.J., Bourdillon, A. and Freeman, L.A. (1983) Fe-Ni-S-O layer phase in C2M carbonaceous chondrites - a hydrous sulfide? Nature 305, 295-297.

Barbier, J. (1990) $Mg_4Ga_8Ge_2O_{20}$: A new synthetic analog of the mineral sapphirine. Phys. Chem. Minerals 17, 246-252.

Barbier, J. and Fleet, M.E. (1987) Investigation of structural states in the series $MGaSiO_4$, $MAlGeO_4$, $MGaGeO_4$ (M = Na,K). J. Solid State Chem. 71, 361-370.

Barbier, J. and Fleet, M.E. (1988) Investigation of phase relations in the $(Na,K)AlGeO_4$ system. Phys. Chem. Minerals 16, 276-285.

Barbier, J. and Neuhausen, J. (1990) Phase formation in the system $(Ca,Sr)Al_2O_4$. Eur. J. Mineral. 2, 273-282.

Barry, J.C. (1991) HRTEM of {100} platelets in natural type 1aA diamond at 1.7 Å resolution: a defect structure refinement. Phil. Mag. A 64, 111-135.

Bayliss, P. and Levinson, A.A. (1988) A system of nomenclature for rare-earth mineral species: revision and extension. Am. Mineral. 73, 422-423.

Baxter Grubb, S.M., Peacor, D.R. and W.-T., Jiang (1991) Transmission electron microscope observations of illite polytypism. Clays Clay Minerals 39, 540-550.

Bell, T.E. (1986) Microstructure in mixed-layer illite/smectite and its relationship to the reaction of smectite to illite. Clays Clay Minerals 34, 146-154.

Bettison-Varga, L., Mackinnon, I.D.R. and Schiffman P. (1991) Integrated TEM, XRD and electron microprobe investigation of mixed-layer chlorite-smectite from the Point Sal ophiolite, California. J. Metamorphic Geol. 9, 697-710.

Bonaccorsi, E., Merlino, S. and Pasero, M. (1990) Rhönite: structural and microstructural features, crystal chemistry and polysomatic relationships. Eur. J. Mineral. 2, 203-218.

Bons, A.J. and Schryvers, D. (1989) High-resolution electron microscopy of stacking irregularities in chlorites from the central Pyrenees. Am. Mineral. 74, 1113-1123.

Bons, A.J., Drury, M.R., Schryvers, D. and Zwart, H.J. (1990) The nature of grain boundaries in slates. Implications for mass transport processes during low temperature metamorphism. Phys. Chem. Minerals 17, 402-408.

Boudeulle, M. and Muller, J.-P. (1988) Structural characteristics of hematite and goethite and their relationships with kaolinite in a laterite from Cameroon. A TEM study. Bull Minéral. 111, 149-166.

Bovin, J.-O. and Norrestam, R. (1990) A crystal growth scenario from HREM studies of pinakiolite related oxyborates. Microsc. Microanal. Microstruct. 1, 365-372.

Bovin, J.-O. and O'Keeffe, M. (1981) Electron microscopy of oxyborates. II. Intergrowth and structural defects in synthetic crystals. Acta Crystallogr. A37, 35-42.

Bovin, J.-O., O'Keeffe, M. and O'Keefe, M. (1981a) Electron microscopy of oxyborates. I. Defect structures in the minerals pinakiolite, ludwigite, orthopinakiolite and takéuchiite. Acta Crystallogr. A37, 28-35.

Bovin, J.-O., O'Keeffe, M. and O'Keefe, M. (1981b) Electron microscopy of oxyborates. III. On the structure of takéuchiite. Acta Crystallogr. A37, 42-46.

Bradley, J.P., Brownlee, D.E. and Veblen, D.R. (1983) Pyroxene whiskers and platelets in interplanetary dust: evidence of vapour phase growth. Nature 301, 473-477.

Brearley, A.J., Rubie, D.C. and Ito, E. (1992) Mechanisms of the transformation between the α, ß and γ polymorphs of Mg_2SiO_4 at 15 GPa. Phys. Chem. Minerals 18, 343-358.

324

Brown, J.L. and Rich, C.I. (1968) High-resolution electron microscopy of muscovite. Science 161, 1135-1137.

Bruley, J. and Brown, L.M. (1989) Quantitative electron energy-loss spectroscopy microanalysis of platelet and voidite defects in natural diamond. Phil. Mag. A59, 247-261.

Buatier, M., Honnorez, J. and Ehret, G. (1989) Fe-smectite-glauconite transition in hydrothermal green clays from the Galapagos spreading center. Clays Clay Minerals 37, 532-541.

Buatier, M.D., Peacor, D.R. and O'Neil, J.R. (1992) Smectite-illite transition in Barbados accretionary wedge sediments: TEM and AEM evidence for dissolution/crystallization at low temperature. Clays Clay Minerals 40, 65-80.

Bursill, L.A. and Glaisher, R.W. (1985) Aggregation and dissolution of small and extended defect structures in Type Ia diamond. Am. Mineral. 70, 608-618.

Bursill, L.A. and Grzinic, G. (1980) Incommensurate superlattice ordering in the hollandites $Ba_xTi_{8-x}Mg_xO_{16}$ and $Ba_xTi_{8-2x}Ga_{2x}O_{16}$. Acta Crystallogr. B36, 2902-2913.

Buseck, P.R. and Iijima, S. (1975) High resolution electron microscopy of enstatite. II: Geological application. Am. Mineral. 60, 771-784.

Buseck, P.R. and Veblen, D.R. (1988) Mineralogy. In High-Resolution Transmission Electron Microscopy, P.R. Buseck, J.M. Cowley, L. Eyring, eds., Oxford University Press, New York, 308-377.

Buseck, P.R., Huang, B.-J. and Miner, B. (1988) Structural order and disorder in Precambrian kerogens. Organic Geochem. 12, 221-234.

Buseck, P.R., Nord, G.L., Jr. and Veblen, D.R. (1980) Subsolidus phenomena in pyroxenes. In Pyroxenes, C.T. Prewitt, ed., Rev. Mineral. 7, 117-211.

Buseck, P.R., Tsipursky, S.J. and Hettich, R. (1992) Fullerenes from the geological environment. Science 257, 215-217.

Cabri, L.J., Chryssoulis, S.L., de Villiers, J.P.R., Gilles LaFlamme, J.H. and Buseck, P.R. (1989) The nature of "invisible" gold in arsenopyrite. Can. Mineral. 27, 353-362.

Chakoumakos, B.C., Murakami, T., Lumpkin, G.R. and Ewing, R.C. (1987) Alpha-decay-induced fracturing in zircon: the transition from the crystalline to the metamict state. Science 236, 1556-1559.

Champness, P.E. and Lorimer, G.W. (1974) A direct lattice-resolution study of precipitation (exsolution) in orthopyroxene. Philos. Mag. 30, 357-366.

Chan, I. and Buseck, P.R. (1982) Metamictization in zircon. In 40th Ann. Proc. Elec. Microsc. Soc. Amer., Washington, D.C., G.W. Bailey, ed., Claitor's Publishing Div., Baton Rouge, LA, 618-619.

Chen, Q.Q. and Lu, X. (1990) The use of high resolution electron microscopy in studying clay minerals. In Proc. XIIth Int'l Cong. Elec. Microsc., Seattle, WA, L.D. Peachey and D.B. Williams, eds., San Francisco Press, San Francisco, CA, 490-491.

Chisholm, J.E. (1973) Planar defects in fibrous amphiboles. J. Mater. Sci. 8, 475-483.

Christy, A.G. (1989) A short-range interaction model for polytypism and planar defect placement in sapphirine. Phys. Chem. Minerals 16, 343-351.

Christy, A.G. and Putnis, A. (1988) Planar and line defects in the sapphirine polytypes. Phys. Chem. Minerals 15, 548-558.

Cooper, J.J. and Tilley, R.J.D. (1986) New oxyborates in the Mg-Mn-B-O system. J. Solid State Chem. 63, 129-138.

Corbató, C.E. and Tettenhorst, R.T. (1987) Analysis of illite-smectite interstratification. Clay Minerals 22, 269-285.

Crawford, E.S., Jefferson, D.A. and Thomas, J.M. (1977) Electron-microscope and diffraction studies of polytypism in stilpnomelane. Acta Crystallogr. A33, 548-553.

Cressey, B.A., Whittaker, E.J.W., Hutchison, J.L. (1982) Morphology and alteration of asbestiform grunerite and anthophyllite. Mineral. Mag. 46, 77-87.

Czank, M. (1981) Chain periodicity faults in babingtonite, $Ca_2Fe^{2+}Fe^{3+}H[Si_5O_{15}]$. Acta Crystallogr. A37, 617-620.

Czank, M. and Liebau, F. (1980) Periodicity faults in chain silicates: a new type of planar lattice fault observed with high resolution electron microscopy. Phys. Chem. Minerals 6, 85-93.

Czank, M. and Seifert, F. (1986) Commensurate-incommensurate phase transition in åkermanites, $Ca_2(Mg,Fe)[Si_2O_7]$. In Proc. XIth Int'l Cong. Elec. Microsc., Kyoto, Japan, 1693-1694.

Czank, M. and Simons, B. (1983) High resolution electron microscopic studies on ferrosilite III. Phys. Chem. Minerals 9, 229-234.

Decarreau, A., Colin, F., Herbillon, A., Manceau, A., Nahon, D., Paquet, H., Trauth-Badaud, D. and Trescases, J.J. (1987) Domain segregation in Ni-Fe-Mg-smectites. Clays Clay Minerals 35, 1-10.

de Villiers, J.P. and Buseck, P.R. (1989) Stacking variations and nonstoichiometry in the bixbyite-braunite polysomatic mineral group. Am. Mineral. 74, 1325-1336.

Dódony, I. and Pósfai, M. (1990) Pyrrhotite superstructures. Part II: A TEM study of 4C and 5C structures. Eur. J. Mineral. 2, 529-535.

Downing, K.H., Meisheng, H., Wenk, H.-R. and O'Keefe, M.A. (1990) Resolution of oxygen atoms in staurolite by three-dimensional transmission electron microscopy. Nature 348, 525-528.

Dunn, P.J. (1990) The discreditation of mineral species. Am. Mineral. 75, 928-930.

Eberhard, E., Fröhlich, A. and Rahman, S.H. (1991) The crystallographic orientation-relationship between ß- and γ-Ca$_2$SiO$_4$ determined by HRTEM. N. Jahrb. Mineral. Abhand. 163, 87-92.

Eberl, D.D. and Srodon, J. (1988) Ostwald ripening and interparticle-diffraction effects for illite crystals. Am. Mineral. 73, 1335-1345.

Eberl, D.D., Srodon, J., Lee, M. and Nadeau, P.H. (1988) Sericite from the Silverton caldera, Colorado: reply. Am. Mineral, 73, 1475-1477.

Eberl, D.D., Srodon, J., Lee, M., Nadeau, P.H. and Northrop, H.R. (1987) Sericite from the Silverton caldera, Colorado: correlation among structure, composition, origin, and particle thickness. Am. Mineral, 72, 914-934.

Eggleton, R.A. (1984) Formation of iddingsite rims on olivine: a transmission electron microscope study. Clays Clay Minerals 32, 1-11.

Eggleton, R.A. and Banfield, J.F. (1985) The alteration of granitic biotite to chlorite. Am. Mineral. 70, 902-910.

Eggleton, R.A. and Boland, J.N. (1982) Weathering of enstatite to talc through a sequence of transitional phases. Clays Clay Minerals 30, 11-20.

Eggleton, R.A. and Buseck, P.R. (1980a) High resolution electron microscopy of feldspar weathering. Clays Clay Minerals 28, 173-178.

Eggleton, R.A. and Buseck, P.R. (1980b) The orthoclase-microcline inversion: a high-resolution transmission electron microscope study and strain analysis. Contrib. Mineral. Petrol. 74, 123-133.

Epicier, T. (1991) Benefits of high-resolution electron microscopy for the structural characterization of mullites. J. Am. Ceram. Soc. 74, 2359-2366.

Epicier, T., Esnouf, C., Guille, J. and Werckmann, J. (1991) High-resolution electron microscopy of α-cordierite (indialite). Materials Lett. 11, 389-395.

Ferrow, E.A. and Ripa, M. (1990) Al-poor and Al-rich orthoamphiboles: a Mössbauer spectroscopy and TEM study. Mineral. Mag. 54, 547-552.

Feuer, H., Schröpfer, L., Fuess, H. and Jefferson, D.A. (1989) High resolution transmission electron microscope study of exsolution in synthetic pigeonite. Eur. J. Mineral. 1, 507-516.

Fleet, M.E. (1983) Investigation of phase transition of natural ZnS minerals by high resolution electron microscopy: discussion. Am. Mineral. 68, 845-846.

Franceschelli, M., Mellini, M., Memmi, I. and Ricci, C.A. (1986) Fine-scale chlorite-muscovite association in low-grade metapelites from Nurra (NW Sardinia), and the possible misidentification of metamorphic vermiculite. Contrib. Mineral. Petrol. 93, 137-143.

Garrels, R.M. (1984) Montmorillonite/illite stability diagrams. Clays Clay Minerals 32, 161-166.

Gribelyuk, M.A., Zakharov, N.D., Hillebrand, R. and Makarova, T.A. (1988) Microstructure of chain silicates investigated by HRTEM. Acta Crystallogr. B44, 135-142.

Guggenheim, S. and Bailey, S.W. (1989) An occurrence of a modulated serpentine related to the greenalite-caryopilite series. Am. Mineral. 74, 637-641.

Guggenheim, S. and Bailey, S.W. (1990) Baumite discredited. Am. Mineral. 75, 705.

Guggenheim, S. and Eggleton, R.A. (1986) Structural modulations in iron-rich and and magnesium-rich minnesotaite. Can. Mineral. 24, 479-497.

Guggenheim, S. and Eggleton, R.A. (1988) Crystal chemistry, classification, and identification of modulated layer silicates. In Hydrous Phyllosilicates Exclusive of the Micas, S.W. Bailey, ed., Rev. Mineral. 19, 675-725.

Guggenheim, S., Bailey, S.W., Eggleton, R.A. and Wilkes, P. (1982) Structural aspects of greenalite and related minerals. Can. Mineral. 20, 1-18.

Guinier, A., Bokij, G.B., Boll-Dornberger, K., Cowley, J.M., Durovic, S., Jagodzinski, H., Krishna, P., de Wolff, P.M., Zvyagin, B.B., Cox, D.E., Goodman, P., Hahn, Th., Kuchitsu, K. and Abrahams, S.C. (1984) Nomenclature of polytype structures. Report of Int'l Union of Crystallography *Ad Hoc* Committee on the nomenclature of disordered, modulated and polytype structures. Acta Crystallogr. A40, 399-404.

Guthrie, G.D., Jr. and Veblen, D.R. (1989a) High-resolution transmission electron microscopy of mixed-layer illite/smectite: computer simulations. Clays Clay Minerals 37, 1-11.

Guthrie, G.D., Jr. and Veblen, D.R. (1989b) High-resolution transmission electron microscopy applied to clay minerals. In Spectroscopic Characterization of Minerals and Their Surfaces, L.M. Coyne, S.W.S. McKeever and D.F. Blake, eds., Sym. Ser. 415, Am. Chem. Soc., Washington, D.C., 75-93.

Guthrie, G.D., Jr. and Veblen, D.R. (1990) Interpreting one-dimensional high- resolution transmission electron micrographs of sheet silicates by computer simulation. Am. Mineral. 75, 276-288.

Güven, N. (1991) On a definition of illite/smectite mixed-layer. Clays Clay Minerals 39, 661-662.

Hamid Rahman, S. (1987) HRTEM observation of disorder in andalusite, Al_2SiO_5. Z. Kristallogr. 181, 127-133.

Hamid Rahman, S. and Weichert, H.-T. (1990) Interpretation of HREM images of mullite. Acta Crystallogr. B46, 139-149.

Hansen, S., Hålenius, U. and Lindqvist, B. (1988) Antimony-rich pinakiolite from Långban, Sweden: a new structural variety. N. Jahrb. Mineral. Monatsh. 5, 231-239.

Harlow, G.E. and Veblen, D.R. (1991) Potassium in clinopyroxene inclusions from diamonds. Science 251, 652-655.

Hassan, I. and Buseck, P.R. (1988) HRTEM characterization of scapolite solid solutions. Am. Mineral. 73, 119-134.

Hassan, I. and Buseck, P.R. (1989a) Incommensurate-modulated structure of nosean, a sodalite-group mineral. Am. Mineral. 74, 394-410.

Hassan, I. and Buseck, P.R. (1989b) Cluster ordering and antiphase domain boundaries in hauyne. Can. Mineral. 27, 173-180.

Hassan, I. and Buseck, P.R. (1992) The origin of the superstructure and modulations in cancrinite. Can. Mineral. 30, 49-59.

Hassan, I. and Grundy, H.D. (1984) The character of the cancrinite-vishnevite solid-solution series. Can. Mineral. 22, 333-340.

Hawthorne, F.C., Groat, L.A., Raudsepp, M., Ball, N.A., Kimata, M., Spike, F.D., Gaba, R., Halden, N.M., Lumpkin, G.R., Ewing, R.C., Greegor, R.B., Lytle, F.W., Ercit, T.S., Rossman, G.R., Wicks, F.J., Ramik, R.A., Sherriff, B.L., Fleet, M.E. and McCammon, C. (1991) Alpha-decay damage in titanite. Am. Mineral. 76, 370-396.

Hu, M., Wenk, H.-R. and Sinitsyna, D. (1992) Microstructures in natural perovskites. Am. Mineral. 77, 359-373.

Huff, W.D., Whiteman, J.A. and Curtis, C.D. (1988) Investigation of a K-bentonite by x-ray powder diffraction and analytical transmission electron microscopy. Clays Clay Minerals 36, 83-93.

Hutchison, J.L., Irusteta, M.C. and Whittaker, E.J.W. (1975) High-resolution electron microscopy and diffraction studies of fibrous amphiboles. Acta Crystallogr. A31, 794-801.

Iijima, S. (1982a) High-resolution electron microscopy of mcGillite. I. One-layer monoclinic structure. Acta Crystallogr. A38, 685-694.

Iijima, S. (1982b) High-resolution electron microscopy of mcGillite. II. Polytypism and disorder. Acta Crystallogr. A38, 695-702.

Iijima, S. and Buseck, P.R. (1975) High resolution electron microscopy of enstatite. I: Twinning, polymorphism, and polytypism. Am. Mineral. 60, 758- 770.

Iijima, S. and Buseck, P.R. (1978) Experimental study of disordered mica structures by high-resolution electron microscopy. Acta Crystallogr. A34, 709-719.

Iijima, S. and Zhu, J. (1982) Electron microscopy of a muscovite-biotite interface. Am. Mineral. 67, 1195-1205.

Iishi, K., Fujino, K. and Furukawa, Y. (1990) Electron microscopy studies of åkermanites $(Ca_{1-x}Sr_x)_2CoSi_2O_7$ with modulated structure. Phys. Chem. Minerals 17, 467-471.

Ilton, E.S. and Veblen, D.R. (1988) Copper inclusions in sheet silicates from porphyry Cu deposits. Nature 334, 516-518.

Inoue, A., Velde, B., Meunier, A. and Touchard, G. (1988) Mechanism of illite formation during smectite-to-illite conversion in a hydrothermal system. Am. Mineral. 73, 1325-1334.

Jefferson, D.A. and Pugh, N.J. (1981) The ultrastructure of pyroxenoid chain silicates. III. Intersecting defects in a synthetic iron-manganese pyroxenoid. Acta Crystallogr. A37, 281-286.

Jefferson, D.A. and Thomas, J.M. (1978) High resolution electron microscopic and x-ray studies of non-random disorder in an unusual layered silicate (chloritoid). Proc. Royal Soc. Lond. A 361, 399-411.

Jefferson, D.A., Tricker, M.J. and Winterbottom, A.P. (1975) Electron-microscopic and Mössbauer spectroscopic studies of iron-stained kaolinite minerals. Clays Clay Minerals 23, 355-360.

Jefferson, D.A., Mallinson, L.G., Hutchison, J.L. and Thomas, J.M. (1978) Multiple-chain and other unusual faults in amphiboles. Contrib. Mineral. Petrol. 66, 1-4.

Jefferson, D.A., Pugh, N.J., Alario-Franco, M., Mallinson, L.G., Millward, G.R. and Thomas, J.M. (1980) The ultrastructure of pyroxenoid chain silicates. I. Variation of the chain configuration in rhodonite. Acta Crystallogr. A36, 1058-1065.

Jiang, W.-T. and Peacor, D.R. (1991) Transmission electron microscopic study of the kaolinitization of muscovite. Clays Clay Minerals 39, 1-13.

Jiang, W.-T., Essene, E.J. and Peacor, D.R. (1990a) Transmission electron microscopic study of coexisting pyrophyllite and muscovite: direct evidence for the metastability of illite. Clays Clay Minerals 38, 225-240.

Jiang, W.-T., Peacor, D.R., Merriman, R.J. and Roberts, B. (1990b) Transmission and analytical electron microscopic study of mixed-layer illite/smectite formed as an apparent replacement product of diagenetic illite. Clays Clay Minerals 38, 449-468.

Kissin, S.A. and Owens, D.R. (1986) The properties and modulated structure of potosiite from the Cassiar District, British Columbia. Can. Mineral. 24, 45-50.

Kitagawa, R. and Matsuda, T. (1992) Microtopography of regularly-interstratified mica and smectite. Clays Clay Minerals 40, 114-121.

Kitamura, M., Shen, B., Banno, S. and Morimoto, N. (1984) Fine textures of laihunite, a nonstoichiometric distorted olivine-type mineral. Am. Mineral. 69, 154-160.

Klimentidis, R.E. and Mackinnon, I.D.R. (1986) High-resolution imaging of ordered mixed-layer clays. Clays Clay Minerals 34, 155-164.

Klug, H.P. and Alexander, L.E. (1974) X-ray Diffraction Procedures for Polycrystalline and Amorphous Materials, John Wiley and Sons, New York, 966 pp.

Knipe, R.J. (1981) The interaction of deformation and metamorphism in slates. Tectonophysics 78, 249-272.

Kondoh, S., Kitamura, M. and Morimoto, N. (1985) Synthetic laihunite ($\square_x Fe^{2+}_{2-3x} Fe^{3+}_{2x} SiO_4$), an oxidation product of olivine. Am. Mineral. 70, 737-746.

Konishi, H. and Akai, J. (1991) Depolymerized pyribole structures derived from talc by heating. Phys. Chem. Minerals 17, 569-582.

Kumao, A., Nissen, H.-U. and Wessicken, R. (1987) Structure images and superstructure model of calcic plagioclase. Acta Crystallogr. B43, 326-333.

Kumao, A., Hashimoto, H., Nissen, H.-U. and Endoh, H. (1981) Ca and Na positions in labradorite feldspar as derived from high-resolution electron microscopy and optical diffraction. Acta Crystallogr. A37, 229-238.

Kusaba, K., Kikuchi, M., Fukuoka, K. and Syono, Y. (1988) Anisotropic phase transition of rutile under shock compression. Phys. Chem. Minerals 15, 238-245.

Lee, J.H. and Peacor, D.R. (1983) Intralayer transitions in phyllosilicates of Martinsburg shale. Nature 303, 608-609.

Lee, J.H. and Peacor, D.R. (1985) Ordered 1:1 interstratification of illite and chlorite: a transmission and analytical electron microscopy study. Clays Clay Minerals 33, 463-467.

Lee, J.H., Peacor, D.R., Lewis, D.D. and Wintsch, R.P. (1984) Chlorite- illite/muscovite interlayered and interstratified crystals: a TEM/STEM study. Contrib. Mineral. Petrol. 88, 372-385.

Lee, S.Y., Jackson, M.L. and Brown, J.L. (1975a) Micaceous vermiculite, glauconite, and mixed-layered kaolinite-montmorillonite. Examination by ultramicrotomy and high resolution electron microscopy. Soil Sci. Soc. Amer. Proc. 39, 793-800.

Lee, S.Y., Jackson, M.L. and Brown, J.L. (1975b) Micaceous occlusions in kaolinite observed by ultramicrotomy and high resolution electron microscopy. Clays Clay Minerals 23, 125-129.

Le Gleuher, M., Livi, K.J.T., Veblen, D.R., Noack, Y. and Amouric. M. (1990) Serpentinization of enstatite from Pernes, France: reaction microstructures and the role of system openness. Am. Mineral. 75, 813-824.

Lindgreen, H., Garnæs, J., Hansen, P.L., Besenbacher, F., Lægsgaard, E., Stensgaard, I., Gould, S.A.C. and Hansma, P.K. (1991) Ultrafine particles of the North Sea illite/smectite clay minerals investigated by STM and AFM. Am. Mineral. 76, 1218-1222.

Livi, K.J.T. and Veblen, D.R. (1987) "Eastonite" from Easton, Pennsylvania: a mixture of phlogopite and a new form of serpentine. Am. Mineral. 72, 113-125.

Livi, K.J.T. and Veblen, D.R. (1989) Transmission electron microscopy of interfaces and defects in intergrown pyroxenes. Am. Mineral. 74, 1070-1083.

Livi, K.J.T. and Veblen, D.R. (1992) An analytical electron microscopy study of pyroxene-to-pyroxenoid reactions. Am. Mineral. 77, 380-390.

Lonker, S.W. and Fitz Gerald, J.D. (1990) Formation of coexisting 1M and 2M polytypes in illite from an active hydrothermal system. Am. Mineral. 75, 1282-1289.

Lumpkin, G.R. and Chakoumakos, B.C. (1988) Chemistry and radiation effects of thorite-group minerals from the Harding pegmatite, Taos County, New Mexico. Am. Mineral. 73, 1405-1419.

Lumpkin, G.R. and Ewing, R.C. (1988) Alpha-decay damage in minerals of the pyrochlore group. Phys. Chem. Minerals 16, 2-20.

Mackinnon, I.D.R. (1982) Ordered mixed-layer structures in the Mighei carbonaceous chondrite matrix. Geochim. Cosmochim. Acta 46, 479-489.

Mackinnon, I.D.R. (1987) The fundamental nature of illite/smectite mixed-layer clay particles: a comment on papers by P.H. Nadeau and coworkers. Clays Clay Minerals 35, 74-76.

Mackinnon, I.D.R. and Buseck, P.R. (1979) New phyllosilicate types in a carbonaceous chondrite matrix. Nature 280, 219-220.

328

Mackinnon, I.D.R. and Zolensky, M.E. (1984) Proposed structures for poorly characterized phases in C2M carbonaceous chondrite meteorites. Nature 309, 240-242.

Mallinson, L.G., Jefferson, D.A., Thomas, J.M. and Hutchison, J.L. (1980) The internal structure of nephrite: experimental and computational evidence for the coexistence of multiple-chain silicates within an amphibole host. Phil. Trans. Royal Soc. Lond. A 295, 537-552.

Manceau, A., Buseck, P.R., Miser, D., Rask, J. and Nahon, D. (1990) Characterization of Cu in lithiophorite from a banded Mn ore. Am. Mineral. 75, 490-494.

Mandarino, J.A. (1987) The check-list for submission of proposals for new minerals to the Commission on New Minerals and Mineral Names, Int'l Mineralogical Assoc. Can. Mineral. 25, 775-783.

Mann, S., Cornell, R.M. and Schwertmann, U. (1985) The influence of aluminum on iron oxides: XII. High-resolution transmission electron microscopic (HRTEM) study of aluminous goethites. Clay Minerals 20, 255-262.

Maresch, W.V. and Czank, M. (1988) Crystal chemistry, growth kinetics and phase relationships of structurally disordered (Mn^{2+},Mg)-amphiboles. Fortschr. Mineral. 66, 69-121.

Maresch, W.V., Massonne, H.-J. and Czank, M. (1985) Ordered and disordered chlorite/biotite interstratifications as alteration products of chlorite. N. Jahrb. Mineral. Abhand. 152, 79-100.

Martin, R.T., Bailey, S.W., Eberl, D.D., Fanning, D.S., Guggenheim, S., Kodama, H., Pevear, D.R., Srodon, J. and Wicks, F.J. (1991) Report of the Clay Minerals Society nomenclature committee: revised classification of clay minerals. Clay and Clay Minerals 39, 333-335.

Meike, A., Wenk, H.-R., O'Keefe, M.A. and Gronsky, R. (1988) Atomic resolution microscopy of carbonates. Interpretation of contrast. Phys. Chem. Minerals 15, 427-437.

Mellini, M., Ferraris, G. and Compagnoni, R. (1985) Carlosturanite: HRTEM evidence of a polysomatic series including serpentine. Am. Mineral. 70, 773-781.

Mellini, M., Merlino, S. and Pasero, M. (1984) X-ray and HRTEM study of sursassite: crystal structure, stacking disorder, and sursassite-pumpellyite intergrowth. Phys. Chem. Minerals 10, 99-105.

Mellini, M., Merlino, S. and Pasero, M. (1986) X-ray and HRTEM structure analysis of orientite. Am. Mineral. 71, 176-187.

Mellini, M., Trommsdorff, V. and Compagnoni, R. (1987) Antigorite polysomatism: behavior during progressive metamorphism. Contrib. Mineral. Petrol. 97, 147-155.

Mellini, M., Amouric, M., Baronnet, A. and Mercuriot, G. (1981) Microstructures and nonstoichiometry in schafarzikite-like minerals. Am. Mineral. 66, 1073-1079.

Mellini, M., Nieto, F., Alvarez, F. and Gomez-Pugnaire, M.T. (1991) Mica-chlorite intermixing and altered chlorite from the Nevado-Filabride micaschists, Southern Spain. Eur. J. Mineral. 3, 27-38.

Merriman, R.J., Roberts, B. and Peacor, D.R. (1990) A transmission electron microscope study of white mica crystallite size distribution in a mudstone to slate transitional sequence, North Wales, UK. Contrib. Mineral. Petrol. 106, 27-40.

Meunier, A. and Velde, B. (1989) Solid solutions in I/S mixed-layer minerals and illite. Am. Mineral. 74, 1106-1112.

Morse, J.W. and Casey, W.H. (1988) Ostwald processes and mineral paragenesis in sediments. Am. J. Sci. 288, 537-560.

Müller, W.F. and Wenk, H.-R. (1978) Mixed-layer characteristics in real humite structures. Acta Crystallogr. A34, 607-609.

Murakami, T., Chakoumakos, B.C., Ewing, R.C., Lumpkin, G.R. and Weber, W.J. (1991) Alpha-decay event damage in zircon. Am. Mineral. 76, 1510-1532.

Nadeau, P.H. (1985) The physical dimensions of fundamental clay particles. Clay Minerals 20, 499-514.

Nadeau, P.H., Tait, J.M., McHardy, W.J. and Wilson, M.J. (1984a) Interstratified XRD characteristics of physical mixtures of elementary clay particles. Clay Minerals 19, 67-76.

Nadeau, P.H., Wilson, M.J., McHardy, W.J. and Tait, J.M. (1984b) Interstratified clays as fundamental particles. Science 225, 923-925.

Nadeau, P.H., Wilson, M.J., McHardy, W.J. and Tait, J.M. (1984c) Interparticle diffraction: a new concept for interstratified clays. Clay Minerals 19, 757- 769.

Nadeau, P.H., Wilson, M.J., McHardy, W.J. and Tait, J.M. (1985a) The conversion of smectite to illite during diagenesis: evidence from some illitic clays from bentonites and sandstones. Mineral. Mag. 49, 393-400.

Nadeau, P.H., Wilson, M.J., McHardy, W.J. and Tait, J.M. (1985b) Interstratified clays as fundamental particles: a reply. Clays Clay Minerals 33, 560.

Nadeau, P.H., Wilson, M.J., McHardy, W.J. and Tait, J.M. (1987) The fundamental nature of interstratified illite/smectite clay particles: a reply. Clays Clay Minerals 35, 77-79.

Nakajima, Y. and Ribbe, P.H. (1980) Alteration of pyroxenes from Hokkaido, Japan to amphibole, clays, and other biopyriboles. N. Jahrb. Mineral. Monatsh. 6, 258-268.

Nakajima, Y. and Ribbe, P.H. (1981a) Twinning and superstructure of Al-rich mullite. Am. Mineral. 66, 142-147.

Nakajima, Y. and Ribbe, P.H. (1981b) Texture and structural interpretation of the alteration of pyroxene to other biopyriboles. Contrib. Mineral. Petrol. 78, 230-239.

Nakajima, Y., Morimoto, N. and Kitamura, M. (1977) The superstructure of plagioclase feldspars. Electron microscopic study of anorthite and labradorite. Phys. Chem. Minerals 1, 213-225.

Nakazawa, H., Morimoto, N. and Watanabe, E. (1975) Direct observation of metal vacancies by high-resolution electron microscopy. Part I: 4C type pyrrhotite (Fe_7S_8). Am. Mineral. 60, 359-366.

Newsam, J.M., Treacy, M.M.J., Koetsier, W.T. and de Gruyter, C.B. (1988) Structural characterization of zeolite beta. Proc. Royal Soc. Lond. A, 420, 375-405.

Nguyen, T.A., Hobbs, L.W. and Buseck, P.R. (1985) High-resolution observation of twinning in $Fe_{1-x}S$ crystals. In 43rd Ann. Proc. Elec. Microsc. Soc. Amer., Louisville, KY, G.W. Bailey, ed., San Francisco Press, San Francisco, CA, 224-225.

Nickel, E.H. (1992a) Solid solutions in mineral nomenclature. Mineral. Mag. 56, 127-130.

Nickel, E.H. (1992b) Nomenclature for mineral solid solutions. Am. Mineral. 77, 660-662.

Nickel, E.H. and Mandarino, J.A. (1987) Procedures involving the IMA Commission on New Minerals and Mineral Names, and guidelines on mineral nomenclature. Can. Mineral. 25, 353-377.

Nord, G.L., Jr. (1980) The composition, structure, and stability of Guinier-Preston zones in lunar and terrestrial orthopyroxene. Phys. Chem. Minerals 6, 109-128.

Norrestam, R. and Bovin, J.-O. (1987) The crystal structure of takéuchiite, $Mg_{1.71}Mn_{1.29}BO_5$. A combined single crystal x-ray and HRTEM study. Z. Kristallogr. 181, 135-149.

Norrestam, R., Dahl, S. and Bovin, J.-O. (1989) The crystal structure of magnesium-aluminum ludwigite, $Mg_{2.11}Al_{0.31}Fe_{0.53}Ti_{0.05}Sb_{0.01}BO_5$, a combined single crystal x-ray and HREM study. Z. Kristallogr. 187, 201-211.

O'Keefe, M.A. and Kilas, R. (1991) Comments on "HRTEM-Bildkontrastsimulation von Strukturen mit Punktdefekten in speziellen Lagen". Z. Kristallogr. 194, 125-128.

Olives Baños, J. (1985) Biotites and chlorites as interlayered biotite-chlorite crystals. Bull. Minéral. 108, 635-641.

Olives Baños, J. and Amouric, M. (1984) Biotite chloritization by interlayer brucitization as seen by HRTEM. Am. Mineral. 69, 869-871.

Olives Baños, J., Amouric, M., De Fouquet, C. and Baronnet, A. (1984) Interlayering and interlayer slip in biotite as seen by HRTEM. Am. Mineral. 68, 754-758.

Ozawa, T., Takéuchi, Y., Takahata, T., Donnay, G. and Donnay, J.D.H. (1983) The pyrosmalite group of minerals. II. The layer structure of mcGillite and friedelite. Can. Mineral. 21, 7-17.

Page, R.H. (1980) Partial interlayers in phyllosilicates studied by transmission electron microscopy. Contrib. Mineral. Petrol. 75, 309-314.

Page, R. and Wenk, H.-R. (1979) Phyllosilicate alteration of plagioclase studied by transmission electron microscopy. Geology, 7, 393-397.

Pasero, M. and Reinecke, T. (1991) Crystal chemistry, HRTEM analysis and polytype behavior of ardennite. Eur. J. Mineral. 3, 819-830.

Paul, T.A. and Fitzgerald, P.G. (1992) Transmission electron microscopic investigation of fission tracks in fluorapatite. Am. Mineral. 77, 336-344.

Pierce, L. and Buseck, P.R. (1974) Electron imaging of pyrrhotite superstructures. Science 186, 1209-1212.

Pierce, L. and Buseck, P.R. (1978) Superstructuring in the bornite-digenite series: a high-resolution electron microscopy study. Am. Mineral. 63, 1-16.

Pósfai, M. and Dódony, I. (1990) Pyrrhotite superstructures. Part I: Fundamental structures of the NC (N = 2,3,4 and 5) type. Eur. J. Mineral. 2, 525-528.

Price, G.D. and Yeomans, J.M. (1984) The application of the ANNNI model to polytypic behaviour. Acta Crystallogr. B40, 448-454.

Pring, A. (1989) Structural disorder in aikinite and krupkaite. Am. Mineral. 74, 250-255.

Pring, A. (1990) Disordered intergrowths in lead-arsenic sulfide minerals and the paragenesis of the sartorite-group minerals. Am. Mineral. 75, 289-294.

Pring, A., Din, V.K., Jefferson, D.A. and Thomas, J.M. (1986) The crystal chemistry of rhodozite: a re-examination. Mineral. Mag. 50, 163-172.

Pring, A., Birch, W.D., Sewell, D., Graeser, S., Edenharter, A. and Criddle, A. (1990) Baumhauerite-2a: a silver-bearing mineral with a baumhauerite-like supercell from Lengenbach, Switzerland. Am. Mineral. 75, 915-922.

Putnis, A. and McConnell, J.D.C. (1980) Principles of Mineral Behaviour. Elsevier, New York, 257 pp.

Rask, J.H. and Buseck, P.R. (1986) Topotactic relations among pyrolusite, manganite, and Mn_5O_8: A high-resolution transmission electron microscopy investigation. Am. Mineral. 71, 805-814.

Ray, N.J., Putnis, A. and Gillet, P. (1986) Polytypic relationship between clinozoisite and zoisite. Bull. Minéral. 109, 667-685.

330

Reeder, R.J. (1981) Electron optical investigation of sedimentary dolomites. Contrib. Mineral. Petrol. 76, 148-157.
Reeder, R.J. and Nakajima, Y. (1982) The nature of ordering and ordering defects in dolomite. Phys. Chem. Minerals 8, 29-35.
Reiche, M. and Bautsch, H.-J. (1985) Electron microscopical study of garnet exsolution in orthopyroxene. Phys. Chem. Minerals 12, 29-33.
Ried, H. (1984) Intergrowth of pyroxene and pyroxenoid; chain periodicity faults in pyroxene. Phys. Chem. Minerals 10, 230-235.
Ried, H. and Korekawa, M. (1980) Transmission electron microscopy of synthetic and natural fünferketten and siebenerketten pyroxenoids. Phys. Chem. Minerals 5, 351-365.
Robertson, I.D.M. and Eggleton, R.A. (1991) Weathering of granitic muscovite to kaolinite and halloysite and of plagioclase-derived kaolinite to halloysite. Clays Clay Minerals 39, 113-126.
Romero, R., Robert, M., Elsass, F. and Garcia, C. (1992a) Evidence by transmission electron microscopy of weathering microsystems in soils developed from crystalline rocks. Clay Minerals 27, 21-33.
Romero, R., Robert, M., Elsass, F. and Garcia, C. (1992b) Abundance of halloysite neoformation in soils developed from crystalline rocks. Contribution of transmission electron microscopy. Clay Minerals 27, 35-46.
Rosenberg, P.E., Kittrick, J.A. and Aja, S.U. (1990) Mixed-layer illite/smectite: a multiphase model. Am. Mineral. 75, 1182-1185.
Rosenberg, P.E., Kittrick, J.A. and Sass, B.M. (1985) Implications of illite/smectite stability diagrams: a discussion. Clays Clay Minerals 33, 561-562.
Rubie, D.C. and Brearley, A.J. (1990) Mechanism of the $\gamma-\beta$ phase transformation of Mg_2SiO_4 at high temperature and pressure. Nature 348, 628-631.
Rule, A.C., Bailey, S.W., Livi, K.J.T. and Veblen, D.R. (1987) Complex stacking sequences in a lepidolite from Tørdal, Norway. Am. Mineral. 72, 1163-1169.
Sass, B.M., Rosenberg, P.E. and Kittrick, J.A. (1987) The stability of illite/smectite during diagenesis: an experimental study. Geochim. Cosmochim. Acta 51, 2103-2115.
Sawhney, B.L. and Reynolds, R.C., Jr. (1985) Interstratified clays as fundamental particles: a discussion. Clays Clay Minerals 33, 559.
Schreyer, W., Medenbach, O., Abraham, K., Gebert, W. and Müller, W.F. (1982) Kulkeite, a new metamorphic phyllosilicate mineral: ordered 1:1 chlorite/talc mixed-layer. Contrib. Mineral. Petrol. 80, 103-109.
Schumacher, J.C. and Czank, M. (1987) Mineralogy of triple-and double-chain pyriboles from Orijärvi, southwest Finland. Am. Mineral. 72, 345-352.
Seifert, F., Ackermand, D. and Czank, M. (1992) A coherent orthopyroxene exsolution from sapphirine. Z. Kristallogr. 199, 99-111.
Seifert, F., Czank, M., Simons, B. and Schmahl, W. (1987) A commensurate-incommensurate phase transition in iron-bearing åkermanites. Phys. Chem. Minerals 14, 26-35.
Sharp, T.G., Otten, M.T. and Buseck, P.R. (1990) Serpentinization of phlogopite phenocrysts from a micaceous kimberlite. Contrib. Mineral. Petrol. 104, 530- 539.
Shau, Y.-H., Peacor, D.R. and Essene, E.J. (1990) Corrensite and mixed-layer chlorite/corrensite in metabasalt from northern Taiwan: TEM/AEM, EPMA, XRD, and optical studies. Contrib. Mineral. Petrol. 105, 123-142.
Shiojiri, M., Isshiki, T., Enomoto, S., Kobayashi, E. and Takahashi, N. (1991) High-resolution electron microscopy observations of the layer structures and stacking faults in molybdenite crystals. Phil. Mag. A, 64, 971-980.
Singh, B. and Gilkes, R.J. (1991) Weathering of a chromian muscovite to kaolinite. Clays Clay Minerals 39, 571-579.
Singh, B. and Gilkes, R.J. (1992) An electron optical investigation of the alteration of kaolinite to halloysite. Clays Clay Minerals 40, 212-229.
Skogby, H. and Ferrow, E. (1989) Iron distribution and structural order in synthetic calcic amphiboles studied by Mössbauer spectroscopy and HRTEM. Am. Mineral. 74, 360-366.
Skrotzki, W., Müller, W.F. and Weber, K. (1991) exsolution phenomena in pyroxenes from the Balmuccia Massif, NW-Italy. Eur. J. Mineral. 3, 39-61.
Smelik, E.A. and Veblen, D.R. (1991) Exsolution of cummingtonite from glaucophane: a new orientation for exsolution lamellae in clinoamphiboles. Am. Mineral. 76, 971-984.
Smelik, E.A., Nyman, M.W. and Veblen, D.R. (1991) Pervasive exsolution with the calcic amphibole series: TEM evidence for a miscibility gap between actinolite and hornblende in natural samples. Am. Mineral. 76, 1184-1204.
Smith, D.J., Jefferson, D.A. and Mallinson, L.G. (1981) The ultrastructure of pyroxenoid chain silicates. II. Direct structure imaging of the minerals rhodonite and wollastonite. Acta Crystallogr. A37, 273-280.

Smith, K.L., Milnes, A.R. and Eggleton, R.A. (1987) Weathering of basalt: formation of iddingsite. Clays Clay Minerals 35, 418-428.

Smith, P.P.K. (1986) Direct imaging of tunnel cations in zinkenite by high-resolution electron microscopy. Am. Mineral. 71, 194-201.

Smith, P.P.K. and Parise, J.B. (1985) Structure determination of $SnSb_2S_4$ and $SnSb_2Se_4$ by high-resolution electron microscopy. Acta Crystallogr. B41, 84-87.

Spinnler, G.E. and Otten, M.T. (1988) HRTEM of antigorite: the structure as a polysomatic series. In 46th Ann. Proc. Elec. Microsc. Soc. Amer., Milwaukee, WI, G.W. Bailey, ed., San Francisco Press, San Francisco, CA, 566-567.

Spinnler, G.E., Self, P.E., Iijima, S. and Buseck, P.R. (1984) Stacking disorder in clinochlore chlorite. Am. Mineral. 69, 252-263.

Srodoh, J., Andreoli, C., Elsass, F. and Robert, M. (1990) Direct high-resolution transmission electron microscopic measurement of expandability of mixed-layer illite/smectite in bentonite rock. Clays Clay Minerals 38, 373-379.

Stucki, J.W. and Tessier, D. (1991) Effects of iron oxidation state on the texture and structural order of Nanontronite gels. Clays Clay Minerals 39, 137-143.

Subbanna, G.N., Kutty, T.R.N. and Anantha Iyer, G.V. (1986) Structural intergrowth of brucite in anthophyllite. Am. Mineral. 71, 1198-1200.

Takéuchi, Y., Ozawa, T. and Takahata, T. (1983) The pyrosmalite group of minerals. III. Derivation of polytypes. Can. Mineral. 21, 19-27.

Tazaki, K. (1986) Observation of primitive clay precursors during microcline weathering. Contrib. Mineral. Petrol. 92, 86-88.

Tazaki, K. and Fyfe, W.S. (1987) Primitive clay precursors formed on feldspar. Can. J. Earth Sci. 24, 506-527.

Tazaki, K., Fyfe, W.S. and van der Gaast, S.J. (1987) Growth of clay minerals in natural and synthetic glasses. Clays Clay Minerals 37, 348-354.

Tazaki, K., Tiba, T., Aratani, M. and Miyachi, M. (1992) Structural water in volcanic glass. Clays Clay Minerals 40, 122-127.

Tettenhorst, R.T., Corbató, C.E. and Haller, R.I. (1990) The I-S contact in 10-17 Å interstratified clay minerals. Clay Minerals 25, 437-445.

Thomas, J.M. and Vaughan, D.E.W. (1989) Methodologies to establish the structure and composition of new zeolitic molecular sieves. J. Phys. Chem. Solids, 50, 449-467.

Tomeoka, K. and Buseck, P.R. (1983) A new layered mineral from the Mighei carbonaceous chondrite. Nature 306, 354-356.

Tomeoka, K. and Buseck, P.R. (1985) Indicators of aqueous alteration in CM carbonaceous chondrites: microtextures of a layered mineral containing Fe, S, O and Ni. Geochim. Cosmochim. Acta 49, 2149-2163.

Turner, S. and Buseck, P.R. (1979) Manganese oxide tunnel structures and their intergrowths. Science 203, 456-458.

Turner, S. and Buseck, P.R. (1981) Todorokites: a new family of naturally occurring manganese oxides. Science 212, 1024-1027.

Turner, S. and Buseck, P.R. (1983) Defects in nsutite (γ-MnO_2) and dry-cell battery efficiency. Nature 304, 143-146.

Turner, S., Siegel, M.D. and Buseck, P.R. (1982) Structural features of todorokite intergrowths in manganese nodules. Nature 296, 841-842.

Vali, H. and Köster, H.M. (1986) Expanding behavior, structural disorder, regular and random irregular interstratification of 2:1 layer-silicates studied by high-resolution images of transmission electron microscopy. Clay Minerals 21, 827-859.

Vander Sande, J.B. and Kohlstedt, D.L. (1974) A high-resolution electron microscopy study of exsolution lamellae in enstatite. Phil. Mag. 29, 1041-1049.

Van Heurck, C., Van Tendeloo, G. and Amelinckx, S. (1992) The modulated structure of melilite $Ca_2ZnGe_2O_7$. Phys. Chem. Minerals 18, 441-452.

Van Landuyt, J. and Amelinckx, S. (1975) Multiple beam direct lattice imaging of new mixed-layer compounds of the bastnaesite-synchisite series. Am. Mineral. 60, 351-358.

Van Tendeloo, G., Gregoriades, P. and Amelinckx, S. (1983a) Electron microscopy studies of modulated structures in $(Au,Ag)Te_2$: Part I. Calaverite $AuTe_2$. J. Solid State Chem. 50, 321-334.

Van Tendeloo, G., Gregoriades, P. and Amelinckx, S. (1983b) Electron microscopic studies of modulated structures in $(Au,Ag)Te_2$: Part II. Sylvanite $AgAuTe_4$. J. Solid State Chem. 50, 335-361.

Van Tendeloo, G., Amelinckx, S. and Gregoriades, P. (1984) Electron microscopic studies of modulated structures in $(Au,Ag)Te_2$. III. Krennerite. J. Solid State Chem. 53, 281-289.

Van Tendeloo, G., Wenk, H.-R. and Gronsky, R. (1985) Modulated structures in calcian dolomite: a study by electron microscopy. Phys. Chem. Minerals 12, 333-341.

Veblen, D.R. (1980) Anthophyllite asbestos: microstructures, intergrown sheet silicates, and mechanisms of fiber formation. Am. Mineral. 65, 1075-1086.

Veblen, D.R. (1981) Non-classical pyriboles and polysomatic reactions in biopyriboles. In Amphiboles and Other Hydrous Pyriboles - Mineralogy, D.R. Veblen, ed., Rev. Mineral. 9A, 189-236.

Veblen, D.R. (1983a) Exsolution and crystal chemistry of sodium mica wonesite. Am. Mineral. 68, 554-565.

Veblen, D.R. (1983b) Microstructures and mixed layering in intergrown wonesite, chlorite, talc, biotite, and kaolinite. Am. Mineral. 68, 566-580.

Veblen, D.R. (1985) TEM study of a pyroxene-to-pyroxenoid reaction. Am. Mineral. 70, 885-901.

Veblen, D.R. and Bish, D.L. (1988) TEM and x-ray study of orthopyroxene megacrysts: microstructures and crystal chemistry. Am. Mineral. 73, 677-691.

Veblen, D.R. and Buseck, P.R. (1979a) Chain-width order and disorder in biopyriboles. Am. Mineral. 64, 687-700.

Veblen, D.R. and Buseck, P.R. (1979b) Serpentine minerals: intergrowths and new combination structures. Science 206, 1398-1400.

Veblen, D.R. and Buseck, P.R. (1980) Microstructures and reaction mechanisms in biopyriboles. Am. Mineral. 65, 599-623.

Veblen, D.R. and Buseck, P.R. (1981) Hydrous pyriboles and sheet silicates in pyroxenes and uralites: intergrowth microstructures and reaction mechanisms. Am. Mineral. 66, 1107-1134.

Veblen, D.R. and Ferry, J.M. (1983) A TEM study of the biotite-chlorite reaction and comparison with petrologic observations. Am. Mineral. 68, 1160-1168.

Veblen, D.R. and Wiechmann, M.J. (1991) Domain structure of low-symmetry vesuvianite from Crestmore, California. Am. Mineral. 76, 397-404.

Veblen, D.R., Buseck, P.R. and Burnham, C.W. (1977) Asbestiform chain silicates: new minerals and structural groups. Science 198, 359-365.

Veblen, D.R., Guthrie, G.D., Jr., Livi, K.J.T. Livi, and Reynolds, R.C., Jr. (1990) High-resolution transmission electron microscopy and electron diffraction of mixed-layer illite/smectite: experimental results. Clays Clay Minerals 38, 1-13.

Wada, K. and Kakuto, Y. (1989) "Chloritized" vermiculite in a Korean ultisol studied by ultramicrotomy and transmission electron microscopy. Clays Clay Minerals 37, 263-268.

Wang, S. and Buseck, P.R. (1990) HRTEM study of valleriite, a hydroxide-bearing copper sulfide. In Proc. XIIth Int'l Cong. Elec. Microsc., Seattle, WA, L.D. Peachey and D.B. Williams, San Francisco Press, San Francisco, CA, 462-463.

Wang, S. and Kuo, K.H. (1991) Crystal lattices and crystal chemistry of cylindrite and franckeite. Acta Crystallogr. A47, 381-392.

Wang, S., Liu, J., Buseck, P.R. and Cowley, J.M. (1990) Compositional observations of franckeite using high-angle annular dark-field microscopy. In Proc. XIIth Int'l Cong. Elec. Microsc., Seattle, WA, L.D. Peachey and D.B. Williams, ed., San Francisco Press, San Francisco, CA, 398-399.

Wang, Y.G., Ye, H.Q., Kuo, K.H. and Guo, J.G. (1992) The defects and intergrowth of lead oxides revealed by HRTEM. J. Appl. Crystallogr. 25, 199-204.

Wang, Y.G., Ye, H.Q., Ximen, L.L. and Kuo, K.H. (1989) A HREM study of the intergrowth of magnetite and coulsonite. Acta Crystallogr. A45, 264-268.

Wang, Y.G., Ye, H.Q., Ximen, L.L. and Kuo, K.H. (1990) A HREM study of nonbasal twinning and superlattices in ilmenite. J. Appl. Crystallogr. 23, 82-87.

Welch, M.D., Rocha, J. and Klinowski, J. (1992) Characterization of polysomatism in biopyriboles: double-chain/triple-chain lameller intergrowths. Phys. Chem. Minerals 18, 460-468.

Werner, P. and Heydenreich, J. (1988) Experimental study of structure modification of $2M_1$-muscovite by high-resolution electron microscopy. Phys. Stat. Sol. (a), 107, 807-816.

White, T.J. (1984) The microstructure and microchemistry of synthetic zirconolite, zirkelite and related phases. Am. Mineral. 69, 1156-1172.

White, T.J. and Hyde, B.G. (1982a) Electron microscope study of the humite minerals: I. Mg-rich specimens. Phys. Chem. Minerals 8, 55-63.

White, T.J. and Hyde, B.G. (1982b) Electron microscope study of the humite minerals: I. Mn-rich specimens. Phys. Chem. Minerals 8, 167-174.

White, T.J. and Hyde, B.G. (1983) An electron microscope study of leucophoenicite. Am. Mineral. 68, 1009-1021.

White, T.J., Segall, R.L., Barry, J.C. and Hutchison, J.L. (1985) Twin boundaries in perovskite. Acta Crystallogr. B41, 93-98.

White, T.J., Segall, R.L., Hutchison, J.L. and Barry, J.C. (1984) Polytypic behavior of zirconolite. Proc. Royal Soc. Lond. A 392, 343-358.

Wicks, F.J. (1986) Lizardite and parent enstatite: a study by x-ray diffraction and transmission electron microscopy. Can. Mineral. 24, 775-788.

Williams, T.B. (1986) Structural examination of sulpho-salt minerals by HRTEM. In Proc. XIth Int'l Cong. Elec. Microsc., Kyoto, Japan, 1705-1706.

Williams, T.B. and Hyde, B.J. (1988) Electron microscopy of cylindrite and franckeite. Phys. Chem. Minerals 15, 521-544.

Williams, T.B. and Pring, A. (1988) Structure of lengenbachite: a high-resolution transmission electron microscope study. Am. Mineral. 73, 1426-1433.

Wu, X.-J., Fujiki, Y., Ishigame, M. and Horiuchi. S. (1991) Modulation mechanism and disorder structure in hollandite-type crystals. Acta Crystallogr. A47, 405-413.

Yada, K., Tanji, T. and Sunagawa, I. (1981) Application of lattice imagery to radiation damage investigation in natural zircon. Phys. Chem. Minerals 7, 47-52.

Yada, K., Tanji, T. and Sunagawa, I. (1987) Radiation induced lattice defects in natural zircon ($ZrSiO_4$) observed at atomic resolution. Phys. Chem. Minerals 14, 197-204.

Yamada, N., Ohmasa, M. and Horiuchi, S. (1986) Textures in natural pyrolusite, ß-MnO_2, examined by 1 MV HRTEM. Acta Crystallogr. B42, 58-61.

Yau, Y.-C. and Peacor, D.R. (1986) Jerrygibbsite-leucophoenicite mixed layering and general relations between the humite and leucophoenicite families. Am. Mineral. 71, 985-988.

Yau, Y.-C., Peacor, D.R. and McDowell, S.D. (1987a) Smectite-to-illite reactions in Salton Sea shales: a transmission and analytical electron microscopy study. J. Sed. Petrol. 57, 335-342.

Yau, Y.-C., Anovitz, L., Essene, E. and Peacor, D.R. (1984) Phlogopite-chlorite reaction mechanisms and physical conditions during retrograde reactions in the Marble Formation, Franklin, New Jersey. Contrib. Mineral. Petrol. 88, 299-306.

Yau, Y.-C., Peacor, D.R., Essene, E.J., Lee, J.H., Kuo, L.-C. and Cosca, M.A. (1987b) Hydrothermal treatment of smectite, illite, and basalt to 460°C: comparison of natural with hydrothermally formed clay minerals. Clays Clay Minerals 35, 241-250.

Ylä-Jääski, J. and Nissen, H.-U. (1983) Investigation of superstructures in mullite by high resolution electron microscopy and electron diffraction. Phys. Chem. Minerals 10, 47-54.

Yoshida, T. (1973) Elementary layers in the interstratified clay minerals as revealed by electron microscopy. Clays Clay Minerals 21, 413-420.

Zandbergen, H.W., Everstijn, P.L.A., Mijlhoff, F.C., Renes, G.H. and Ijdo, D.J.W. (1987) Composition, constitution and stability of the synthetic hollandites $A_xM_{4-2x}N_{2x}O_8$, M = Ti,Ge,Ru,Zr,Sn and N = Al,Sc,Cr,Ga,Ru,In and the system $(A,Ba)_xTi_yAl_zO_8$ with A = Rb,Cs,Sr. Mater. Res. Bull. 22, 431-438.

Zen, E-An (1962) Problem of the thermodynamic status of the mixed layer minerals. Geochim. Cosmochim. Acta 26, 1055-1067.

Zhou, Z., Fyfe, W.S., Tazaki, K. and van der Gaast, S.J. (1992) The structural characteristics of palogonite from DSDP site 335. Can. Mineral. 30, 75-81.

Zolensky, M.E. and Mackinnon, I.D.R. (1986) Microstructures of cylindrical tochilinites. Am. Mineral. 71, 1201-1209.

Chapter 9. DIAGENESIS AND LOW-GRADE METAMORPHISM OF SHALES AND SLATES

Donald R. Peacor Department of Geological Sciences, University of Michigan, Ann Arbor, Michigan 48109 U.S.A.

9.1 INTRODUCTION

The most significant minerals in low-grade pelitic rocks—mudstones, shales, and slates—are clay minerals. This important group of rocks was terra incognita until the advent of modern TEMs because clays and associated minerals such as feldspars, quartz, and carbonates are generally so fine-grained that individual grains cannot be resolved by conventional means. They have therefore been characterized traditionally by techniques that measure bulk properties averaged over many grains. For example, x-ray diffraction gives average results for all grains of a given species, even though there may be more than one origin for different populations of the same mineral; electron microprobe analyses commonly are subject to averaging of both heterogeneous single-phase materials and different homogeneous phases. It has therefore not generally been possible to characterize and to correlate structure and composition directly, except in those cases where unusually homogeneous clay minerals occur. However, it is exactly because clay minerals form at low temperatures that they are likely to be highly disordered and heterogeneous; direct correlation of composition and structure for individual grains is therefore especially important for clay minerals.

Recent advances in TEM technology permit direct characterization of clay minerals in three ways: (1) grain texture—size, shape, relation to other minerals—can be resolved directly in ion-milled samples, (2) crystal structure relations, including layer stacking sequences, of individual grains can be determined directly using SAED, lattice-fringe imaging, and structure imaging, and (3) AEM analyses with accuracies approaching those of EMPA can be obtained and directly correlated with specific structures. TEM/STEMs are therefore capable of providing the kinds of characterization of textures, compositions, and crystal structures of clays and other minerals in shales that have been conventionally obtained only for optically-resolvable minerals.

Where shales undergo sequential changes during diagenesis through low-grade metamorphism, key samples defined by field relations and conventional analysis techniques such as XRD can be chosen that represent all stages of a given sequence. Study of such samples allows, through comparison of relations for sequential features, definition of the mechanisms and processes by which changes occur. For example, Li et al. (in prep.) have recently been able to characterize five different types of white mica in Welsh shales by their unique compositions and textures and to correlate each type with diagenetic and tectonic events. Such analyses can now be carried out routinely.

9.2 CLAY MINERALS – SPECIAL TECHNIQUES AND PROBLEMS

9.2.1 Specimen preparation

As is the case for x-ray diffraction, clay mineral structure types are most easily identified through observation of $00l$ sequences of reflections in SAED patterns. Such reflections can be used to form lattice fringe images whose characteristic spacings and, in

336

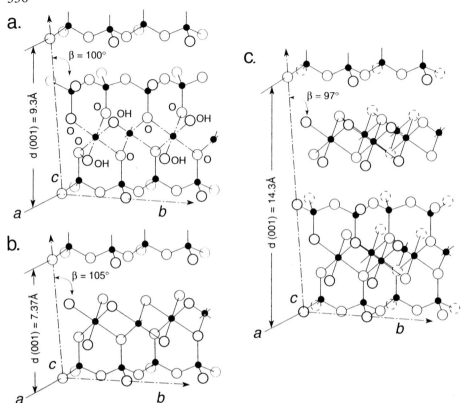

Figure 9-1. Diagrams of crystal structures of (a) pyrophyllite (2:1 structure; two tetrahedral sheets and one octahedral sheet) (b) kaolinite (1:1 structure; one tetrahedral sheet and one octahedral sheet) (c) chlorite (2:1 layer + octahedral sheet). Tetrahedrally coordinated atoms (e.g., Si, Al) and octahedrally coordinated atoms (e.g., Al, Mg, Fe) are shown as black spheres, and O and OH as open spheres. The 1:1, 2:1, and chlorite structures are characterized by $d(001)$ values of approximately 0.7, 0.9-1.0, and 1.4 nm, respectively. Modified from Berry, Mason, and Dietrich (1983).

some cases, contrast differences, lead to direct identification of basic layer types. For example 1:1, 2:1, and mixed-layer sequences can be directly identified. Figure 9-1 shows the correspondence between structure type and interlayer spacing for three of the more significant kinds of clay mineral structures, and Table 9-1 contains a listing of values of $d_{(001)}$ for one-layer polytypes of some of the more important phyllosilicates occurring in pelitic rocks. In order to obtain SAED patterns and lattice fringe images that correspond to the layer patterns, it is necessary to have grains oriented with c^* normal to the electron beam, i.e., with {001} layers approximately normal to the specimen surface and parallel to the beam. Unfortunately, natural fine-grained clay or crushed separates invariably consist of grains that are planar or tabular parallel to {001}. Such grains lie with {001} parallel to the surface of formvar- or holey C-coated grids, and therefore with {001} perpendicular rather than parallel to the electron beam. Although some grains have the correct orientations, e.g., when they lie on holes in the coating or if their edges are curled as in smectite, they can generally be located only with difficulty.

Coherent samples can be prepared as petrographic thin sections oriented normal to predictable directions of preferred basal plane orientations; e.g., oriented normal to bedding and/or slaty cleavage. Such thin sections can first be studied using optical, SEM (BSE

Table 9-1. Interplanar spacings, $d(001)$, for one-layer polytypes of some of the more significant layered structures associated with pelites

Structure Unit		Example[1]	$d(001)$ (nm)
Oct Sheet		⌈Brucite	0.477
		⌊Gibbsite	0.485
1 Tet : 1 Oct		⌈Serpentine	0.730
		⌊Kaolinite	0.715
2 Tet : 1 Oct		⌈Talc	0.935
		⌊Pyrophyllite	0.920
	Mica	⌈Biotite	1.000
		⌊Muscovite	0.995
		Glauconite	1.01
		Illite	1.00
2 Tet : 1 Oct	Smectite	⌈Saponite	≈ 1.0[2]
+ cation/H_2O sheet		⌊Montmorillonite	≈ 1.0[2]
		Vermiculite	1.42
2 Tet : 1 Oct	Chlorite	Clinochlore	1.41
+ Oct Sheet			
Mixed Layer			
1 Trioct smectite		⌈Corrensite	≈ 2.4[2]
+ 1 trioct chlorite			
1 Dioct smectite		Rectorite	≈ 2.0[2]
+ 1 dioct mica		⌊(R1 I/S)	

[1]Bracketed minerals are isostructural except that the first is trioctahedral and the second dioctahedral.
[2]d-values for structures that become dehydrated and collapse in the high-vacuum TEM environment.

---

imaging combined with qualitative EDS analyses), and other techniques, in order to obtain an overview of specimen characteristics and to guarantee that the areas that are eventually chosen for TEM study are representative of the features for which the study is designed. If the thin section is attached with "sticky-wax" or some other low-melting temperature or soluble material, appropriate areas can be removed following attachment of an Al washer, with subsequent thinning by ion milling. The sample can be dimpled preceding attachment of the washer in order to minimize time of ion milling. Dimpling consists of grinding a small dimple in the thin section at the site to be surrounded by the Al washer, and is accomplished with commercial devices made for that purpose. It results in a significant decrease in the time necessary for ion-milling, which minimizes the relief caused by the differential thinning rates of different minerals. The Al washer is attached with epoxy that is not affected by the heat or solvent used to separate the TEM sample from the thin section. Friable samples can first be impregnated using, for example, high pressure impregnation or passive solvent-diffusion techniques (see discussion by McHardy and Bernie, 1987).

Powdered samples can be prepared as "sandwiches." The powder is allowed to settle through a fluid such as methanol onto a flat surface of already-hardened epoxy, followed by addition of another layer of epoxy after the sample is allowed to air-dry thoroughly.

Spurr low-viscosity embedding resin is the most commonly used medium. Ultra-thin sections can then be prepared in one of two ways: (1) An ordinary thin section can be prepared with the flat clay layer oriented normal to the section. A TEM sample is then prepared by separation of a 3-mm diameter area which is bisected by the clay layer, followed by ion milling. Because of differential thinning rates of resin and clay, such sections are commonly prepared only with difficulty. (2) A microtome can be directly used to produce slices that are electron transparent. Such slices can be supported on the surface of ordinary commercial Cu TEM grids. Grids of low atomic number, ideally Be, should be used if samples are to be used as sources of AEM data.

One of the more vexing problems in TEM analysis of clay minerals concerns the ambiguity in differentiation of illite and smectite lattice fringes. Unfortunately, smectite dehydrates either in the ion-mill or TEM environments. The contraction in structure that usually occurs causes the smectite 2:1 layer to have a spacing $d_{(001)} = 1.0$ nm. It therefore cannot ordinarily be differentiated from an illite layer (see below for section on image calculations). Because illite and smectite form mixed-layer structures, this is a particularly serious problem. Contraction does not occur in all cases (see review of Ahn and Peacor, 1986b), but even when some of the smectite layers collapse, 1.0 nm layers cannot be identified either as illite-like or smectite-like with certainty.

Much work has focused on finding a "magic bullet" that will cause smectite layers to remain expanded in the TEM environment. Many organic molecules will intercalate with smectite, giving rise to expansion, but such expansion may not be permanent in the TEM vacuum. However, treatment by n-alkylammonium ions has been shown to cause expansion that is retained in a vacuum. The n-alkylammonium ions can be prepared with various lengths of carbon atoms ($n = 7$-18), each giving rise to a different degree of expansion. Vali and Koster (1986) showed how clay separates could be treated with n-alkylammonium ions, embedded in Spurr resin, and thinned using an ultramicrotome. Vali et al. (1991) used such methods to show that differential expansion of layers can be related to interlayer charge, leading to the identification of different kinds of layers. Klimentidis and Mackinnon (1986) utilized both dodecylamine hydrochloride ($n = 12$ alkylammonium ion + HCl) and Spurr resin in separate mounts to obtain TEM images showing expansion of smectite layers. Vali and Hesse (1990) have improved on that method; they utilized alkylammonium ion treatment of separates embedded in Spurr resin, obtaining electron-transparent slices with an ultramicrotome. The slice that is mounted on a TEM grid is treated with n-alkylammonium ions. In order to compare textures before and after treatment, either parallel slices can be separately examined, or the same crystallites can be observed before and after expansion.

There is a serious drawback to most methods that utilize expansion, because the increase in volume causes decrepitation and spalling. Most methods therefore can only be used on separates; the textures of the original rocks or sediments are destroyed.

Šrodoń et al. (1990) have successfully used a unique method that allows preservation of the expanded smectite layers as well as original sample textures. Their treatment involves gentle rehydration of air-dried chips followed by embedding in, and impregnation with, L.R. White resin, which is of lower viscosity than Spurr resin. The treated chips are cut with a diamond microtome to produce electron-transparent slices. Srodon et al. successfully applied the method to bentonite samples, observing lattice fringes with spacings for smectite and illite layers in relative proportions that were in good agreement with predictions based on XRD data.

9.2.2   Identification of minerals through SAED and 00*l* lattice fringes

9.2.2.1   General relations

Most of the common clay minerals are pseudohexagonal (with some polytypes being hexagonal). They are consequently indexable on an orthohexagonal unit cell with translations **a** and **b** parallel to the layers and with $b = \sqrt{3}a$, whereas the translation **c** defines the periodicity of the layers, as shown in Figure 9-1. Clay minerals with {001} oriented parallel to the electron beam therefore give SAED patterns that include **c***. The diffraction pattern also includes a vector from the *hk*0 plane. The second vector in the SAED pattern may be **a*** or **b*** ($a^* = \sqrt{3}b^*$), or one of the pseudosymmetrically-related translations obtained by 60° rotations around **c***. If one of the vectors **a*** or **b*** (or pseudosymmetrical equivalents) is in the plane of the diffraction pattern, the diffraction pattern resembles those shown in Figures 9-2a and 9-2b, respectively. Thus, in general, rotation about **c*** (as accomplished with **c*** parallel to the tilt axis of the specimen holder)

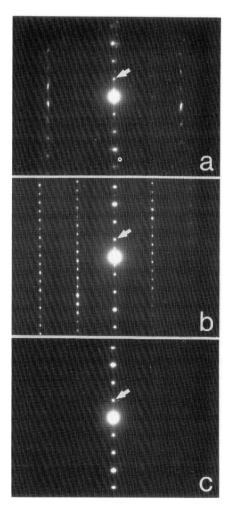

Figure 9-2. SAED patterns of micas. The 001 reflection with $d(001) = 1.0$ nm is indicated by an arrow in each case. (a) **a***-**c*** section (ortho-hexagonal indexing); 00*l* and 20*l* rows for which $k = 3N$. Both rows therefore show 1.0-nm periodicity even though the polytype in question is a 2-layer (2.0-nm) polytype. (b) [110]*-**c*** section; 00*l*, 11*l*, and 22*l* rows; 11*l* and 22*l* rows ($k \ne 3N$) show 2.0-nm periodicity. The 00*l* row has weak reflections showing 2.0-nm periodicity due to dynamical diffraction. This section is related pseudosym-metrically by 60° rotation around **c*** to the **b***-**c*** section that contains 00*l*, 02*l*, and 04*l* rows. (c) 00*l* reflections only, showing a 1.0-nm periodicity. Courtesy of W.-T. Jiang.

yields patterns resembling Figure 9-2a every 60°; the details of the patterns are determined by the true symmetry relations, with only one of the three patterns that are related by approximate rotations of 60° having the true vector $\mathbf{a}^*$. Likewise, patterns resembling Figure 9-2b occur every 60°, but at 30° intervals to those illustrated in Figure 9-2a. However, if the plane of the SAED pattern does not include a rational $hk0$ vector, then only the $00l$ sequence of reflections will occur (Fig. 9-2c).

The diffraction pattern in Figure 9-2c will directly yield an image with 001 lattice fringes, as will use of an objective aperture that includes only the $00l$ reflections (say, $l < 3$) of Figures 9-2a and 9-2b, but care must be used in the latter case as noted in the next section on polytypism.

The value of $d_{(001)}$ of any given pair of 001 fringes can be directly measured from a lattice fringe image. For example, it is a simple matter to differentiate between 1:1, 2:1, and chlorite-family minerals, as $d_{(001)}$ is approximately 0.7, 1.0, and 1.4 nm for each of those structure types, respectively, as illustrated in the lattice fringe image of Figure 9-1a. Unfortunately, smectite usually yields fringes that are approximately 1.0 nm in spacing as a result of dehydration of interlayer water. Smectite fringes therefore resemble those of nonexpandable 2:1 structures such as muscovite, leading to ambiguity in identification (Fig. 9-9, below).

Because of errors in direct measurement of $d_{(001)}$ from both diffraction patterns and lattice-fringe images, it is not generally possible to utilize the small differences in $d_{(001)}$ that exist for two different minerals of the same structure type; e.g., two 2:1 micas such as biotite and muscovite. However, where two such minerals occur in the same image, relative differences in d-values can be accurately determined. Lattice-fringe images of specific clay minerals may have other kinds of characteristic features; e.g., fringes of smectite that occurs in mudstones typically are wavy, subparallel, and discontinuous, whereas illite fringes are straight and parallel (e.g., Ahn and Peacor, 1986b); differences in beam damage effects lead to "mottled" textures that differ for different micas (Ahn and Peacor, 1986c), as further described below.

Mixed-layer clay minerals yield sequences of fringes that can often be used to identify sequences of layers. For example, Figure 9-3b shows fringes for corrensite with periodicity of 2.4 nm that consist of subfringes with separate 1.0 nm (dehydrated smectite) and 1.4 nm (chlorite) spacings. Mixed layering directly contributes to the $00l$ sequence of reflections (in contrast to polytypism; see below), as demonstrated in the inset diffraction pattern of Figure 9-3b. Ordered mixed layering produces a characteristic superperiodicity, whereas randomly intercalated layers give rise to diffuseness parallel to $\mathbf{c}^*$.

Mixed layering is well known to give rise to irrational sequences of $00l$ reflections in XRD and SAED patterns of mixed-layer illite/smectite, for example. Figure 9-17 (below) shows an SAED pattern of mixed-layer chlorite/corrensite that illustrates such an irrational reflection sequence. Such observations are common in trioctahedral mixed-layer minerals, as they consist of interlayered 1.4-nm chlorite-like layers and dehydrated and collapsed 1.0-nm smectite-like or vermiculite-like layers; i.e., the mixed layers have different spacings. However, dehydration of smectite layers generally causes the smectite-like layer in illite/smectite to have a thickness approximately equal to that of the illite layer (1.0 nm). Mixed-layer I/S therefore generally gives SAED patterns that have 1.0 nm periodicity. Because of imperfections in structure, only a single, diffuse $00l$ reflection may occur, non-$00l$ reflections being ill-defined, and very weak to unobservable, with discontinuous

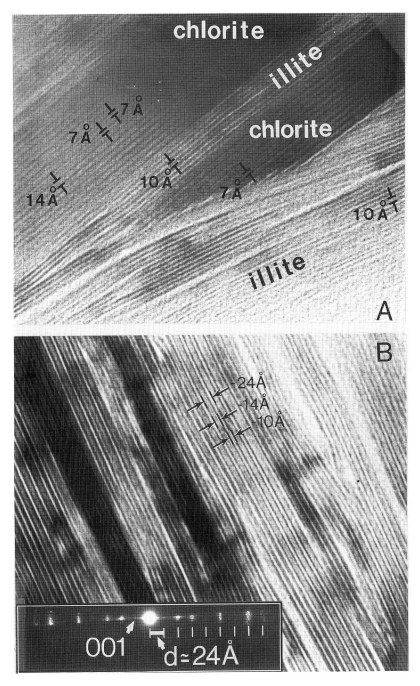

Figure 9-3. Typical 001 lattice fringe images of some phyllosilicates occurring commonly in pelites. (a) Packets of illite (1.0 nm) and chlorite (1.4 nm). The packets of chlorite have interlayers of 0.7-nm berthierine (depth of 2450 m, Case Western Reserve University Gulf Coast 6 well). (b) Corrensite (2.4 nm) with a structure consisting of ordered 1.4-nm (chlorite) and 1.0-nm (dehydrated smectite) layers. Sample from anchizone, Gaspé Peninsula. Courtesy of W.-T. Jiang.

streaking parallel to $\mathbf{c}^*$. Nevertheless, Veblen et al. (1990) demonstrated that $00l$ reflections can show the doubling and quadrupling of the $\mathbf{c}$-axis for R1 (..ISIS..) and R3 (..IISIIS..) I/S layer sequences, respectively, wherever there is a relatively high degree of order.

9.2.2.2 Calculated images

Most interpretations of one-dimensional lattice-fringe images formed using $00l$ reflections have been carried out simply on an intuitive basis. Correlations have commonly been made only between fringe spacings and known values of $d_{(001)}$ for different structure types, with little consideration of those experimental factors that might cause variations in such patterns. For example, Ahn and Peacor (1986b) observed small differences in fringe spacings in smectite, but the degree to which such differences could be correlated with differences in structure was not established. Although intuitive interpretations of one-dimensional fringe images are appropriate as long as they are not "over-interpreted," they may be in error where heterogeneity occurs, as in mixed layering. That is especially true in the case of mixed-layer illite/smectite, where collapse of the smectite layer causes illite and smectite layers to give similar fringe images.

In a recent series of papers, Guthrie and Veblen (1989a,b; 1990) have calculated images for a variety of typical phyllosilicates, including kaolinite, lizardite, chlorite, vermiculite, muscovite, and phlogopite, and mixed-layer illite/smectite, phlogopite/chlorite, lizardite/chlorite, and brucite/chlorite. Their calculations corresponded to the lens configuration of a Philips 420 TEM and therefore can be directly utilized only for instruments with similar optics. However, they also serve as a general guide for interpretation of images obtained with other instruments.

Guthrie and Veblen showed how the sequences of dark and light fringes vary according to a variety of conditions including specimen thickness, orientation, and focus. They noted that the appearance of fringes may change dramatically as imaging conditions change. Most importantly, they showed that optimum information regarding structure and composition are obtained at entirely different imaging conditions. *Structure* information is best obtained for the Scherzer underfocus condition. In general, bright fringes correspond to slabs of low charge density, whereas dark fringes correspond to slabs of high charge density, although there are exceptions to that rule.

Fringes may change dramatically as focus is changed, as shown in Figure 9-4, giving rise to possible misinterpretations. For example, underfocus images of chlorite may have 0.7-nm periodicity resembling images of 1:1 phyllosilicates; that is especially bothersome in that chlorite may be interlayered with 1:1 phyllosilicates such as berthierine (e.g., Amouric et al., 1988; Ahn and Peacor, 1985a). Guthrie and Veblen (1990) emphasize three different relations: (1) Where layers have different thicknesses, Scherzer focus conditions generally give fringes of correct thickness, but images obtained in other conditions may show constant thickness. (2) The number of layers in a packet may appear to be different than the actual number at a non-Scherzer focus condition. (3) Even though all layers in a packet are identical, Scherzer-focus images may appear to have fringes with different thicknesses when the structure at the packet boundaries is different, as for 1:1 layers. Great caution should therefore be used when interpreting fringes, noting that their appearance may vary dramatically, especially as a function of focus, but also of orientation and specimen thickness.

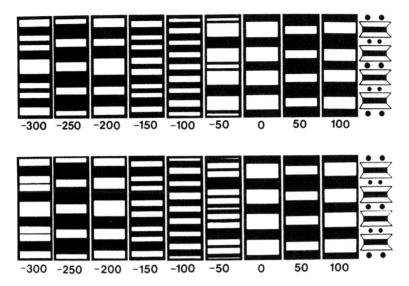

Figure 9-4. Calculated lattice fringe images for ordered, mixed-layered illite/smectite for various focus conditions (in nanometers). The images vary significantly with focus and display the true 2.0-nm periodicity only for certain focus conditions. Upper and lower sequences correspond to K- and Na-rich material, respectively. The correspondence with units of structure is shown on the right with H_2O (smectite-like) interlayers designated by large circles, and K- or Na-containing (mica-like) interlayers designated by smaller circles. Courtesy of G.D. Guthrie.

Guthrie and Veblen also showed that information regarding *composition* is best obtained for unusual overfocus conditions, corresponding to overfocus of approximately 100 nm, as compared to the underfocus of the Scherzer condition. Veblen et al. (1990) studied samples of I/S that had been shown by XRD to consist of well-ordered R1 and R3 material. Images obtained with the Scherzer condition and at other focus conditions showed 1.0-nm periodicity corresponding to the periodicity of the 2:1 structure. However, images obtained at appropriate overfocus conditions displayed contrast differences that correctly reflected the differences in composition between illite-like and smectite-like layers.

Ahn and Buseck (1990) studied samples of R1 and R3 I/S utilizing high resolution images obtained from a JEOL 4000 EX instrument, as compared with calculated two-dimensional images. They showed that for [110] and [100] directions, and the five pseudohexagonal-related equivalents of each, the calculated images of illite and smectite interlayers display distinct differences, corresponding to underfocus conditions (-20 nm) slightly different from the Scherzer condition (-49 nm). However, those differences could not be detected in the experimental images.

9.2.3 Polytypism and mixed-layering

Polytypism is discussed in detail by Baronnet in Chapter 7. Only a brief discussion that relates to some practical experimental procedures is given here.

The 00*l* sequence of reflections is identical for all polytypes of a given clay mineral, assuming that there is no contribution from non-00*l* reflections owing to dynamical diffraction, as described below. Thus all polytypes of mica yield the 00*l* sequence of reflections as shown in Figure 9-2c. This is so because such reflections correspond to the Fourier transform of the structure as projected onto c^*; i.e., they are a function only of the z coordinates and scattering factors of the atoms. Such a projection is independent of the shifts in the **a-b** plane that define polytypes.

Bailey (1988) has reviewed the relations between diffraction patterns and stacking sequences of important groups of clay minerals. Assuming that indexing is based on an orthohexagonal cell ($a = 0.53$, $b = 0.92$ nm), non-$00l$ reflections with $k = 3N$ give information regarding the periodicity primarily of octahedral sheets, whereas reflections for which $k \neq 3N$ have a periodicity that primarily reflects the stacking of tetrahedral sheets, i.e., the polytype. Thus the diffraction pattern in Figure 9-2a ($00l$ and $20l$ rows of reflections) contains no information regarding polytypism, but that in Figure 9-2b ($00l$, $11l$, and $22l$ rows) does, illustrating 2-layer (2.0 nm) periodicity. Polytypic layer periodicities can therefore be identified only when rotation about c^* gives rise to reflections of the appropriate indices.

Because the common phyllosilicates are all pseudohexagonal, the periodicities of the SAED patterns are repeated through rotations of 60° around c^*. However, although the periodicities are repeated, the intensities and relative positions of reflections differ as a function of the true symmetry. They are also affected by glide plane-generated extinctions in some cases. Bailey (1988) discusses such issues in detail.

If a two-dimensional SAED pattern includes reflections of type $k \neq 3N$, dynamical diffraction causes reflections with indices characteristic of the polytype periodicity to occur within the $00l$ sequence. Thus the $00l$ sequence shown in Figure 9-2b contains reflections that reflect 2.0-nm periodicity. A lattice fringe image of such a pattern may show alternating fringes with darker and lighter contrast, with the periodicity of fringe contrast being 2.0 nm, even though only $00l$ reflections are included in the objective aperture.

If the tilt axis of the specimen holder is parallel to c^* of a crystal that gives a diffraction pattern with non-$00l$ reflections, the crystal may be rotated about c^* so that non-$00l$ reflections are eliminated and only $00l$ reflections remain. That diffraction pattern will display only 1.0-nm periodicity, as there can be no contribution due to dynamical diffraction involving non-$00l$ reflections. Contrast differences between fringes in the corresponding lattice fringe image can therefore not be caused by polytypism, but must reflect differences in composition or structure.

As noted above, $00l$ reflections contain information about mixed layering. Diffuseness parallel to c^* of $00l$ reflections can be due to mixed layering, but diffuseness can also be caused by disorder in the polytypic layer sequence, if multiple diffraction occurs for those crystal orientations where non-$00l$ reflections define polytypic sequences. Ordered mixed layering of 1.0-nm layers with different compositions or structures yields 2.0-nm periodicity resembling that of a 2.0-nm, two-layer polytype affected by multiple diffraction. Thus, it is essential to obtain orientations of crystals that give rise only to $00l$ reflections and their corresponding lattice fringe images, if the effects of mixed layering are to be unambiguously separated from those of polytypism.

Characteristics of x-ray diffraction patterns of polytypes of two-layer and three-layer phyllosilicates have been well-described by Bailey (1988). For many such polytypes, the basic periodicity as defined by $k \neq 3N$ reflections is sufficient to characterize the stacking sequence, e.g., for the $3T$ polytype of mica. Ambiguity exists in other cases, e.g., for the three possible two-layer polytypes ($2O$, $2M_1$, and $2M_2$) of mica. In such cases, those two dimensional diffraction patterns (e.g., $\mathbf{a}^*$-$\mathbf{c}^*$ patterns) that are related by 60° rotations around $\mathbf{c}^*$ are different. Care must therefore be used in correlating the periodicity, symmetry, and intensities of specific two-dimensional SAED patterns that contain $\mathbf{c}^*$ with those of known polytypes before conclusions regarding polytypism can be reached.

Where layers are stacked such that they are randomly related by rotations of multiples of 60° as in the $1M_d$ polytype of mica, reflections are diffuse parallel to $\mathbf{c}^*$. Likewise, turbostratic stacking, wherein layers are entirely randomly related by rotation around $\mathbf{c}^*$ gives rise to diffuseness parallel to $\mathbf{c}^*$, but with each reflection spread into a "tube" coaxial with $\mathbf{c}^*$. Figure 9-5 shows the diffraction pattern of illite (or illite-rich R1 I/S) in a shale from the zone of diagenesis. Reflections are not continuously diffuse parallel to $\mathbf{c}^*$ as required of the ideal $1M_d$ polytype, but reflections are poorly defined, somewhat diffuse, and nonperiodic. Contributions from both $h0l$ and $0kl$ rows further show that layers are, at least to a limited degree, related by completely random rotations about $\mathbf{c}^*$.

Two-dimensional HRTEM images which are formed utilizing hkl as well as $00l$ reflections are capable of providing contrast variation that is sensitive to the stacking sequence for specific focus and orientation conditions, as verified by computed simulations. Thus, stacking sequences in muscovite (Iijima and Buseck, 1978), muscovite and phlogopite (Amouric et al., 1981), chlorite (Spinnler et al., 1984), and R1 and R3 I/S (Ahn and Buseck, 1990) have been directly interpreted through the sequences of spots corresponding to specific structure units.

9.2.4 Clay separates: {001} sections

Single crystals of clay minerals oriented with $\mathbf{c}^*$ parallel to the beam give $hk0$ diffraction patterns that have the geometry of a hexanet (e.g., Fig. 9-6). Because common clay minerals have similar values of a and b, the hexanets are similar for all such minerals and cannot be used in identification of specific species. Much information can nevertheless be obtained from clay minerals in such orientations.

The finely powdered material from ground-up coherent rocks (e.g., ordinary clay mineral separates prepared by gravity settling), disaggregated friable sediments, or original fine powdered material as in soils can be prepared using an ordinary formvar-coated copper grid. A prime advantage of such samples is that they can be prepared in a very few minutes, as compared with the lengthy preparation required for ion-milled or microtomed samples. The data provided by such samples are largely of value in providing a quick, qualitative overview. Where clay separates of coherent rocks are prepared, there is a complete loss of textural information. However, such samples may also provide unique observations of perfection of stacking sequence, crystal morphology, and particle thickness.

Coherent single crystals have layers that are related by rotations of 60° about $\mathbf{c}^*$, thus giving rise to $hk0$ SAED patterns that are hexanets. Turbostratic stacking, in which successive layers are related by random rotations, yields hexanets with random relative rotation angles, causing each single crystal reflection to be spread into a circle (Fig. 9-7). Guthrie and Veblen (1990) have pointed out that overlapping, separate grains may give similar patterns and that care must be used in differentiating the two different kinds of relations between layers or packets of layers. Nadeau et al. (1984c) obtained diffraction patterns of clay mineral separates as part of their studies that led to the definition of so-called elementary particles; indeed, they defined an "elementary particle" as one that gives a single-crystal electron diffraction pattern.

346

Figure 9-5. Lattice fringe and corresponding SAED pattern of illite or illite-rich I/S from the zone of diagenesis of a prograde sequence of pelites, Gaspé Peninsula, Quebec. The diffraction pattern shows the well-defined 1.0-nm sequence of $00l$ reflections, but diffuse, ill-defined, and non-periodic $0kl$ and $h0l$ rows of reflections that are typical of so-called $1M_d$ illite. Courtesy of W-T Jiang.

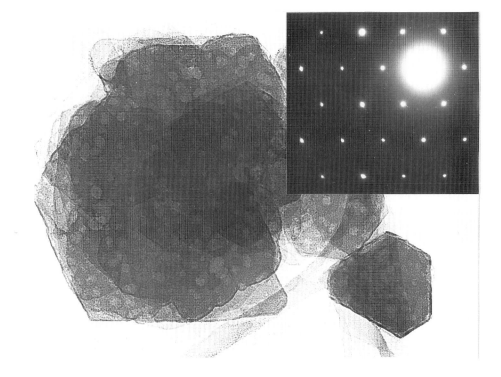

Figure 9-6. Hexagonal crystals of illite-rich I/S from a depth of 8000 ft, DeWitt Co., Texas. Crystals spread on a holey-C film. The diffraction pattern is a pseudo-hexagonal $hk0$ net, corresponding primarily to a single crystal. Courtesy of R.L. Freed.

Crystal morphology is clearly defined in TEM images of separates. For example, smectite usually displays a veil-like morphology (Fig. 9-7), whereas illite or I/S crystallites are flat (Fig. 9-6). Inoue et al. (1987, 1988, 1990) described the variation in shapes of crystals of euhedral pseudohexagonal illite and I/S crystallites produced during hydrothermal alteration of smectite. Yau et al. (1987b) showed that the illite precipitated in void space from convecting hydrothermal fluids occurs as euhedral crystals, whereas ion-milled samples showed only tabular cross-sections through crystals and thus failed to display the pseudohexagonal outlines.

Nadeau and coworkers (Nadeau et al., 1984a,b,c, 1985; Nadeau, 1985) have used Pt shadowing techniques to measure the thickness of individual particles of clay minerals, obtaining results that led to the concept of "fundamental particles" and "interparticle diffraction" (see discussion below). This technique utilizes clay mineral separates that are sedimented onto a mica cleavage surface. The sample is shadowed with platinum at an angle of 10°, and then carbon-coated. The carbon layer, with attached clay mineral particles is then floated on water and picked up on a TEM grid. Measurements of the length of the Pt shadow lead to a determination of particle thickness. Mackinnon (1987) has discussed the errors inherent in such measurements, noting that errors are significant relative to the conclusions reached regarding fundamental particles, and that assumptions about the nature of the particles were unwarranted. Nadeau et al.(1987) have replied, noting that the standard error of measurement is relatively small (±0.4 nm) and adequate for determining the range of thicknesses of fundamental particles. Inoue et al. (1987, 1990) used a somewhat simpler technique, in that the clay mineral separates were directly spread over a Ni TEM grid and shadowed with Pt-Pd.

Vali et al. (1991) have reviewed the general techniques for preparing separates with {001} parallel to sample surfaces, emphasizing the processes of cryofixation and freeze-etching. They compared such results with impregnated ultramicrotomed separates, noting that the impregnated samples consist in part of reaggregated, separated packets.

9.2.5   Specimen damage

Clay minerals are subject to various kinds of damage, including dehydration in the vacuum of both the ion mill and TEM column, structural damage due to direct specimen-ion or specimen-electron beam interaction, and heating of the specimen in either the ion mill or electron beam.

Damage is perhaps most clearly shown by AEM analyses of alkali-containing clay minerals. As detailed in Chapter 4, K, especially, is lost through diffusion during chemical analysis. However, other alkali elements are severely subject to the problem, and even elements such as Al are affected. The problem is exacerbated at thin edges and for small area analyses (van der Pluijm et al., 1988, Li and Peacor, in prep.). Methods of optimizing analytical data are described below and in Chapter 4.

Structural damage, in part associated with chemical diffusion, is revealed in several ways. In micas and some other phases a mottled contrast can be seen to gradually become more pronounced with exposure to the beam. As noted by Ahn and Peacor (1986c), mottling is due to strain caused by heterogeneity induced by diffusion of cations and/or dehydration that is associated with growing, lens-like separations of layers. Exceptionally large separations are shown in Figure 9-8. Such layer separations occur commonly along some {001} layers, possibly nucleating at original sites of structural defects or hydrated interlayers. Figure 9-6 shows pock-marked, beam-damaged illite crystals, with the

348

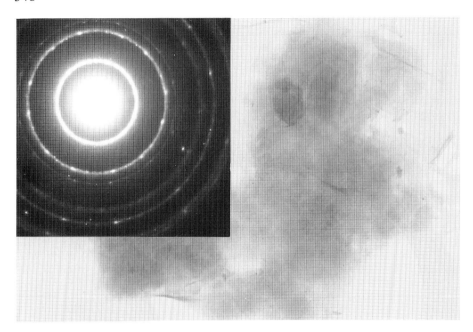

Figure 9-7. Smectite-rich, disordered I/S from a depth of 2000 ft, DeWitt Co., Texas. Typical smectite lacks specific form. The SAED pattern is powder-like, corresponding to random rotations of the SAED pattern of Figure 9-6 about **c***. Courtesy of R.L. Freed.

Figure 9-8. Illite (muscovite) 001 lattice fringe image showing lenticular layer separations caused by cation diffusion, and diagnostic, mottled contrast that results from strain caused by incipient layer separation. Illite is authigenic, oriented parallel to bedding of a pelite from the anchizone, central Wales. Courtesy of G. Li.

damaged areas occurring as small rounded areas with light contrast; such areas increase in size with continuous exposure to the beam.

Damage is also evident in the very rapid fading of diffraction patterns that may, in some clay minerals, occur in a few seconds. The loss of diffraction pattern is of course directly associated with loss of lattice fringes. Kaolinite-family minerals are especially notorious for becoming "diffraction-amorphous" very rapidly (e.g., Jiang and Peacor, 1991), although some minerals such as muscovite may continue to give satisfactory diffraction patterns after even hours of observation. Lee et al. (1986) noted that they could discriminate between authigenic illite and metamorphic illite-like mica by the rapid deterioration in diffraction patterns of the former, showing that the beam damage was directly linked to structural defects or local heterogeneity in composition or structure.

Samples that are rapidly beam-damaged should be initially viewed with the weakest possible beam that still allows images and/or diffraction patterns to be seen or recorded on film. The sample can then be quickly translated to an as-yet unexposed region that is structurally contiguous with the area to be imaged. When lattice-fringe images are desired, and especially where variations in layer orientation occur over the area to be studied, the high-resolution direct image (rather than the diffraction pattern) can be viewed as the sample is moved; when an appropriate area is encountered, focusing can be rapidly achieved and a photograph quickly obtained. Experienced operators who work quickly and decisively are able to obtain high resolution lattice fringe images even of kaolinite.

9.2.6   Chemical analyses of clay minerals

9.2.6.1   General relations

Techniques of analytical electron microscopy (AEM) are discussed in detail in Chapter 4. Only some relations that are especially important to clay mineral analyses will therefore be discussed here.

Van der Pluijm et al. (1988) showed, using ion-milled samples, that diffusion of alkalis from the area of dioctahedral micas directly affected by the beam is a function of beam current, time of analysis, specimen orientation, and thickness of sample. As specimen thickness increases, diffusion becomes much less significant, but atomic number (Z), absorption (A), and fluorescence (F) (or 'ZAF') effects become measureable at thicknesses of a few tens of nanometers. If ion-milled samples are used, analyses should generally be obtained away from thin edges, but avoiding electron-opaque areas where absorption is significant. Where clay separates are used, the individual grains may be as thin as a single 1.0-nm layer, and element diffusion is especially serious. In such cases, individual crystallites that are small enough to be included within the area of the defocused beam in TEM mode should be utilized, as diffusion does not cause element loss from the grain (Livi, pers. comm.).

The most serious effects occur in STEM mode with a point beam. As the area of a raster increases, the effects of alkali diffusion diminish rapidly. Therefore, analyses should be obtained using as large an area as possible. Analyses should, of course, only be obtained from areas for which TEM data indicate structural homogeneity.

The effects of diffusion have also been shown to be significant for non-alkali elements (Mackinnon, 1990; Li and Peacor, unpublished data). Even the Si/Al ratio has been shown to change slightly under adverse analytical conditions. Small differences in diffusion of alkalis, and therefore in the alkali/Si ratio) have been shown to occur for mica {001} orientations that are parallel or perpendicular to the beam. Lastly, experienced analysts are very aware of the relation between diffusion and defect density. For example, illite or I/S that forms early in diagenesis generally occur as small packets of layers that are rich in layer terminations, interlayer site vacancies, and other defects relative to mature mica; the effects of alkali diffusion in such phases are more severe, although there are no data showing a regular quantitative relation.

Mackinnon (1990) has shown that use of a liquid N_2 stage causes the effects of diffusion to be significantly reduced. Van der Pluijm et al. (1988) suggested a procedure designed to minimize the effects of diffusion. Analysis of an unknown should be obtained under exactly the same conditions as were used for standards, including utilization of the same area of beam-specimen interaction and counting time of analysis. A specific area should then be found which gives the same count rate as was obtained for the standard, the thicknesses of the unknown and standard thus being approximately the same. Although diffusion occurs in standard and unknown, it should occur approximately equally in both. The ratio of intensities should therefore be constant for identical concentrations. Unfortunately, the rate of diffusion is in part a function of defect density, structure, and composition, so the preceding method still gives rise to alkali loss when data for illite (defect-rich) are compared with standard data from muscovite (defect-poor).

9.2.6.2 *Interpretation of analytical data*

The concentration ratios obtained from AEM must be converted to a structural formula by normalizing to some number of atoms. It has been conventional to normalize analyses to the number of O atoms whose total charge includes components due both to O and OH (e.g., 7 O atoms for kaolinite). This method can give rise to serious errors, some of which are: (1) Despite care used during analysis, diffusion commonly gives rise to low values for some elements, especially alkalis; (2) the oxidation state of Fe (or of Mn) is generally unknown; (3) some elements (e.g., N in illite, Li in amphibole) may be undetected.

On the the other hand, some minerals have formulae for which some number of cations is known to be an integer. Normalization to cations is preferable in such cases, assuming that the concentrations of all cations have been accounted for accurately. Because Al can occur in both octahedral and tetrahedral sites of most common phyllosilicates, normalization can be carried out to the sum of octahedral plus tetrahedral sites. In the case of dioctahedral 2:1 minerals such as illite and micas such as muscovite and paragonite, analyses should be normalized to a sum of 6 (or 12) octahedral and tetrahedral cations, whereas trioctahedral 2:1 minerals should be normalized to 7 (or 14) cations. In the case of trioctahedral chlorite or serpentine-group minerals, normalization should be based on 10 (or 20) cations.

Normalization to cations in the case of 2:1 clay minerals containing volatile alkali cations (e.g., smectite, illite, micas) has a distinct advantage in that the resulting low relative concentrations of K and Na do not bias the results. Our experience with common chlorites and 2:1 phyllosilicates generally confirms that, where AEM analyses have been obtained for minerals whose compositions have also been determined by EMPA, normalization to cations is generally preferred.

The principal problem with normalization to cations is in the solid solution between dioctahedral and trioctahedral components. The data of many studies imply that micas and chlorite commonly are not ideally trioctahedral or dioctahedral, and some chlorites are di-trioctahedral or tri-dioctahedral. Nevertheless, crystal structure refinements of common rock-forming phyllosilicates have shown that site occupancies are those of ideal dioctahedral and trioctahedral minerals, although it can be argued that such refinements are carried out for phases that occur as relatively large crystals that form at temperatures well above those of the metastable, disordered clay minerals that form in weathering or diagenetic environments; the refined structures therefore may not reflect those of some clay minerals.

Normalization of AEM data for clay minerals to either cations or anions is therefore subject to error. AEM data should be normalized as consistent with the special factors affecting a particular mineral, and not on the basis of a classical, familiar procedure that is derived from other analytical techniques. Our bias is toward normalization to cations, as at least for common phyllosilicates in pelites, the results have invariably been reasonable with respect to crystal chemical requirements. However, results should be interpreted cautiously, with due regard for problems such as di-trioctahedral or tri-dioctahedral components in unusual chlorites. Resultant formulae should be carefully analyzed for bias, and where problems are evident, the quality of the analytical data or the normalization procedure should be questioned.

Some AEM-derived formulae of 2:1 sheet silicates and chlorite are shown in Table 9-2. Such formulae are typical of those that are obtained on a regular basis in our laboratory. As demonstrated by the comparison of AEM and electron microprobe data in Table 4-1 of Chapter 4, such formulae approach those obtained by electron microprobe analysis under ideal circumstances.

Table 9-2. Formulae of some common phyllosilicates as derived from AEM analyses.

	Paragonite	Smectite	I/S	Illite	Chlorite	Talc	Corrensite
Si	5.96	7.96	6.68	6.61	6.01	7.99	5.94
Al^{IV}	2.04	0.04	1.32	1.39	1.99	0.01	2.06
Al^{VI}	3.82	2.76	3.88	2.94	1.78	0.07	1.33
Fe	0.04	0.72	0.10	0.34	3.60	0.26	2.04
Mn	--	--	--	--	0.03	--	0.03
Mg	0.14	0.52	0.08	0.72	6.58	5.65	5.60
Ca	--	0.24	0.04	0.04	0.10	--	0.21
Na	1.68	0.14	0.24	0.15	--	--	--
K	0.10	0.30	0.98	1.56	--	--	0.27

9.3   DIOCTAHEDRAL PHYLLOSILICATES: ILLITE/SMECTITE

9.3.1   Identification of illite, smectite, and mixed-layer illite/smectite

9.3.1.1   General sequence of changes with depth

The sequence of changes that occur with increasing depth of sediment burial were brought into focus in a now classic paper by Hower et al. (1976). The study utilized samples from a core from Harris County, Texas, that was drilled in Gulf Coast sediments, a general sequence that is recognized as a "type" sequence. Based in large part on XRD data from separates, Hower et al. and subsequent workers showed that the sequence of dioctahedral 2:1 clay minerals is as follows: (1) The dominant clay mineral to depths of approximately 6000-9000 ft is randomly interstratified illite/smectite (I/S) with approximately 30% illite (commonly incorrectly referred to as "smectite"). (2) At slightly greater depths, a transition occurs over a narrow depth interval of approximately 1,000 ft, with the proportion of illite increasing to ~80%. The randomly interstratified material transforms to ordered R1 I/S (alternate illite-like and smectite-like layers; ...ISISIS...), with the proportion of illite layers increasing with depth. R3 I/S (...IIISIIIS...) has been identified in other localities. (3) The proportion of illite remains constant (~80% illite layers; commonly incorrectly referred to as "illite") for the deepest samples studied. The sequence has commonly been viewed as one in which smectite layers are individually replaced by illite. Thus, if R1 I/S is an early clay in the sequence, R2 I/S is not a possible later product. In past years, much of the debate regarding the conversion of smectite to illite has centered on the degree to which there is a solid-state transition component (as in layer-by-layer transition), in contrast to dissolution/crystallization, relations which can only be directly studied using high-resolution TEM techniques.

9.3.1.2   TEM observations of smectite/illite sequences

The first TEM studies of I/S in Gulf Coast sediments were those of Bell (1986) and Ahn and Peacor (1986b). Ahn and Peacor utilized ion-milled samples from the same suite studied by Hower et al. (1976) so as to reproduce original textures. Figure 9-9 shows a typical TEM image of shallow material, for which XRD indicates the presence of smectite with ~30% illite. It is typical of smectite-rich clay that occurs in shales. The lattice fringes are anastomozing and wavy, gradually changing orientation by several degrees over a few tens of nanometers. Layer terminations occur frequently within a packet of layers.

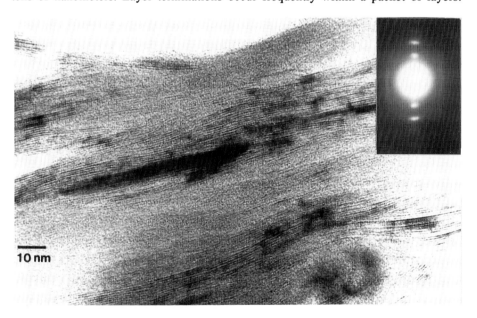

10 nm

Figure 9-9. Smectite-dominant, random mixed-layer I/S from a depth of 7110 ft, Brazoria Co., Texas. Wavy, anastomozing smectite fringes have variable contrast along fringes and variable spacings. Although layer terminations are common, and layers vary in orientation, the layers form a more or less continuous array. Lack of fringe contrast in some areas is due to variation in orientation of layers. Inset SAED pattern shows only diffuse 001 and 002 reflections, and powder-like non-00l reflections. Adapted from Freed and Peacor (1989a).

Subparallel packets intersect at small angles, with the layers of one packet terminating against another in one region; however, because of the change in orientation of the wavy layers, the layers of such packets may be parallel in other regions. It is possible to continuously move across parallel fringes or along individual fringes from one area of the image to another as in a single crystal, but the units at the beginning and end of such a traverse may have very different orientations. Ahn and Peacor thus referred to such a continuum as a "megacrystal." The contrast in fringes varies considerably from area to area, or even along a given packet of fringes, because of variations in orientation of layers relative to the plane of the image. The fringes have spacings of approximately 1.0 nm, having collapsed upon dehydration, but the spacings vary along layers. Such variation may reflect heterogeneity in composition or it may be an artifact of changing orientation of layers relative to the beam (Guthrie and Veblen, 1990). The electron diffraction pattern of such material is that of typical smectite, showing little more than a broad diffuse 001 reflection.

Figure 9-7 is an image of smectite-rich clay separates. The thin individual grains have surfaces that are not planar; the surface is irregular, having been described as "cornflake"- or "veil"-like in appearance. The edges of such grains are commonly curved upward, so that the layers are parallel to the beam, permitting diffraction patterns containing the $00l$ row and corresponding lattice fringe images to be obtained. The $hk0$ SAED patterns consist of complete circles, equivalent to rotation of an $hk0$ single-crystal pattern about c^*. Some patterns are spotty, with each spot corresponding to a single 1.0-nm layer or to only a small number of coherently-related layers, but such diffraction patterns indicate that most layers are randomly positioned relative to layers above and below; i.e., the stacking is largely turbostratic. The precise fraction of coherently-related layers in such sequences has yet to be determined in any specific case.

AEM data show that the composition of such material corresponds to that of smectite, having a typical net negative charge of ~0.3 (Freed and Peacor, 1992). Rather than having Ca or Na as the principal interlayer cation, as typical of much smectite, K is the dominant cation; such K may be utilized in the conversion of smectite to illite occurring at greater depths, and therefore need not be derived from feldspar.

The combination of composition, SAED patterns with $00l$ and $hk0$ reflections, texture, and correspondence to XRD patterns collectively demonstrates that the material is predominantly smectite, with layers related largely by turbostratic stacking.

Figure 9-10 is a lattice-fringe image of a specimen from an intermediate depth corresponding to the midway point in the smectite-to-illite transformation. The packet of layers for which the fringe width is indicated has fringes of relatively constant spacing and contrast that are characteristic of illite or ordered I/S. The surrounding fringes have the appearance of smectite. SAED patterns of such apparent illite-rich I/S show typical well-defined $00l$ reflections but there is also a diffuse, broad 1.0-nm reflection characteristic of smectite; the latter apparently is present because the area within the aperture included some of the surrounding smectite.

The $00l$ diffraction row of such crystallites shows the well-defined series of mica-like reflections for which $d_{(001)} = 1.0$ nm. When such crystallites are properly oriented to show reflections in rows parallel to c^*, such reflections are invariably of irregular shape, broad, weak, and diffuse parallel to c^*. Although such reflections may be related by 1.0-nm periodicity, they generally are irregularly spaced parallel to c^*. The crystallites have therefore been characterized as being the $1M_d$ polytype (Grubb et al., 1991).

354

Figure 9-10. Packet of authigenic illite or ordered I/S (inferred to correspond in shape to crystals as shown in Fig. 9-6) within a matrix of I/S. The 001 lattice fringes are straight and of constant spacing. Depth of 10,700 ft, Brazoria Co., Texas. Adapted from Freed and Peacor, 1989a.

Figure 9-6 is an image of a clay separate from a depth which yields material of the same sort as observed in the lattice-fringe image of Figure 9-10. Well-defined pseudo-hexagonal crystallites with diameters of ~50 nm are abundant. The $hk0$ nets in SAED patterns, if obtained on the edge of such crystallites, are usually hexanets, i.e., those of a single crystal in which the layers are related by rotations of 60°. Where the thin crystallites overlap, a spotty ring pattern is obtained, as consistent with separate crystallites related by random rotations around c^*. Such SAED patterns are at least in part caused by random stacking of separate grains on the TEM grid.

The compositions of such grains are invariably those of typical illite or ordered I/S. The net negative charge on the 2:1 layer, although variable, has an average value of ~0.75 (for an anion composition of $O_{10}(OH)_2$); i.e., it is considerably larger than that of the smectite that occurs at shallower depths. The increased charge is largely due to the higher concentration of ^{IV}Al which has an average value of ~0.6 per $O_{10}(OH)_2$.

XRD data, TEM images obtained using both separates and ion-milled samples, and AEM data are therefore consistent in showing that in samples that are intermediate in the smectite-to-illite transformation, pseudohexagonal crystallites of R1 I/S occur within a matrix of smectite-rich R0 I/S.

Figure 9-11 shows a lattice-fringe image from material occurring at a depth that exceeds that of the I/S transformation, for which XRD indicates the presence of I/S with ~80% illite-like layers, with no detectable smectite (or R0, smectite-rich I/S). In contrast to the smectite lattice fringes, these are relatively straight, with a nearly constant spacing of ~1.0 nm. Contrast is relatively uniform along layers, implying that all layers have approximately the same orientation relative to the image plane. The layers occur in relatively

Figure 9-11. Illite-dominant I/S, depth of 11,750 ft, Brazoria Co., Texas. Packets of the kind shown in Figure 9-10 have grown and entirely replaced the matrix of smectite-rich I/S shown in Figure 9-9. Courtesy of R.L. Freed.

well-defined packets of parallel layers. Layer terminations are relatively rare within packets, but the packets intersect in small-angle grain boundaries where the layers of one packet terminate against another packet.

Inoue et al. (1990) studied clay separates of bentonites for which the proportion of illite-like layers varied from 50 to 70%. They observed grains with two morphologies, flaky and elongated lathlike. Both kinds of grains had thicknesses of 3.0 to 10.0 nm, with no correlation of thickness with proportion of illite-like layers. However, Inoue et al. (1987, 1988) detected three different morphologies in hydrothermally altered samples from the Shinzan area of Japan. They observed a progression from flaky grains, to elongated laths, to hexagonal plates, that corresponded to a progression from 45 to 100% illite, increase in K (see below), and change in polytype from $1M_d$ to $1M$ to $2M_1$. Thicknesses of lath-shaped grains varied from approximately 30 to 50 nm. Both widths (0.3 to 0.6 μm) and lengths (0.2 to 0.9 μm) of all kinds of grains varied regularly with increase in proportion of illite-like layers, as a reflection of change to more equidimensional basal sections.

Glasmann et al. (1989) studied clay separates from wells in the North Sea. The clays varied regularly from random I/S with ~80% smectite to R3 I/S with ~5% smectite. Smectite-rich samples were characterized by extremely thin, irregular laths and plates less than 0.1 μm in diameter. Such material is replaced by elongated laths, the proportion of which increases as the proportions of R1 and R3 I/S increase.

Yau et al. (1987b, 1988) observed both separates and ion-milled samples from hydrothermally altered sediments from the Salton Sea area. Shallow samples were dominated by detrital smectite occurring as continuous, wavy, anastomozing layers, as in shallow Gulf Coast samples. Ion milled TEM specimens of hydrothermally altered samples contained apparent elongated crystals of $1M_d$ illite having c^* normal to the length, ~25 nm thick, and occupying pore space (Fig. 9-12a). However, dispersed samples are dominated by thin,

356

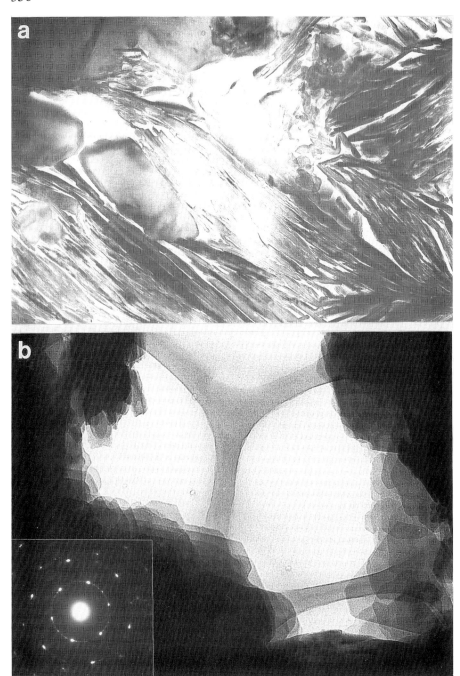

Figure 9-12. Authigenic illite from a depth of approximately 400 m, Salton Sea geothermal field, California. (a) Crystals of illite oriented "on-edge" with {001} parallel to the beam. Authigenic crystals fill pore space between detrital grains of quartz. (b) Low resolution TEM image of illite separates lying on holey-C. Pseudohexagonal crystals have {001} parallel to the image plane. SAED pattern of overlapping crystals is dominantly that of a single crystal. Courtesy of Y.-C. Yau.

pseudohexagonal plates and some laths (Fig. 9-12b), showing that the elongated grains observed in ion-milled samples are cross-sections of crystals with pseudohexagonal morphology. Similar morphologies of illite were observed in illite produced by hydrothermal synthesis with smectite as a starting material (Yau et al., 1987a).

9.3.1.3 Detection of I/S ordering in TEM images

The TEM images of Bell (1986) and Ahn and Peacor (1986b) failed to show the presence of ordered I/S in those samples for which XRD indicated the presence of R1 ordering. This occurred because the illite and collapsed smectite layers both have spacings of ~1.0 nm; only in samples dominated by layers of either illite or smectite could identification be made using average criteria such as composition or texture. Nevertheless, Ahn and Peacor (1986a) showed that rectorite, which consists of well-ordered smectite-like and illite-like layers of hydrothermal origin, gives lattice-fringe images showing alternating contrast in fringes with 1.0-nm spacing, giving rise to 2.0-nm superperiodicity. Those observations were obtained under the same experimental conditions and by the same operator for which the TEM images of authigenic Gulf Coast samples failed to indicate ordering.

The work of Guthrie and Veblen (1989a,b; 1990) and Veblen et al. (1990) was a key factor in defining the overfocus conditions with which the 2.0-nm periodicity could be observed. Ahn and Peacor (1989) subsequently reexamined their TEM images and were able to find a few rare examples of 2.0-nm periodicity out of thousands of images that had been obtained with sequences of through-focus images, including the correct overfocus condition with a JEOL 100CX instrument. Freed and Peacor (1989b) also only rarely observed such periodicity using both JEOL 100CX and Philips CM12 instruments. The ease of observation of ordering in some R1 I/S (e.g., rectorite) and difficulty in others is still an enigma, but may be related to the higher state of disorder in authigenic I/S (Ahn and Peacor, 1989). Nieto (unpublished data) also was only rarely able to observe 2.0-nm periodicity in R1 I/S in shales from the Vasco-Cantabrian Basin. Hansen and Lindgreen (1989) and Lindgreen and Hansen (1991), on the other hand, obtained images showing 2.0-nm periodicity that was inferred to be due to R1 I/S in ion-milled samples from North Sea sediments. Alt and Jiang (1991) made similar observations of I/S from a marine hydrothermal deposit. Likewise, Jiang et al. (1990) obtained images of R1, R2, and R3 I/S showing contrast variations in layers with 1.0-nm spacings for Welsh shales in which authigenic illite had been partially altered to I/S (Fig. 9-13).

The work of Srodon et al. (1990), in which gentle rehydration was followed by embedding in very low-viscosity resin, constitutes the only direct observations of illite-like and expanded smectite-like layers for which original texture was retained, and for which identification of illite-like and smectite-like layers was carried out without apparent ambiguity. Those results verified that XRD determinations of I/S ratios are valid when interparticle diffraction effects (Nadeau et al., 1984c) were considered.

TEM data have thus confirmed that powder XRD determinations of I/S ratios and the ordering state of the layers are accurate, apparently in large part because layers in original samples separate along smectite interlayers and are rearticulated in XRD samples along the same planes as expandable interlayers (Klimentidis and Mackinnon, 1986; Ahn and Peacor, 1989; Veblen et al., 1990; Srodon et al., 1990).

9.3.1.4 AEM analyses of smectite, illite, and I/S

There is a general paucity of AEM analyses of smectite, illite, and I/S, despite the

358

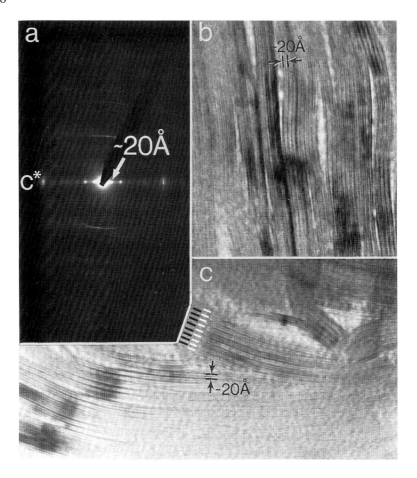

Figure 9-13. Lattice fringe images and diffraction pattern of R1 I/S that is inferred to have formed by hydrothermal alteration of illite in a Welsh slate. (a) SAED pattern with typically diffuse reflections indicating 2.0-nm periodicity. (b) Lattice fringe image showing superperiodicity of 2.0 nm superimposed on a 1.0-nm substructure. (c) Relatively well-ordered R1 I/S with 2.0-nm periodicity. Courtesy of W.-T. Jiang.

Table 9-3. Structural formulae of primary illite and secondary mixed-layer illite/smectite in pelite sample BRM202 from Wales.[1,2] (After Jiang et al., 1990)

	1	2	3	4	5	6	7	8	9	10
Si	6.52(13)[3]	6.40(11)	6.29(15)	6.23(14)	6.12(14)	6.12(14)	6.32(12)	6.24(12)	6.27(16)	6.25(12)
Al^{IV}	1.48(03)	1.60(03)	1.71(05)	1.77(05)	1.88(05)	1.88(05)	1.68(04)	1.76(04)	1.73(05)	1.75(04)
Al^{VI}	3.42(08)	3.33(07)	3.49(10)	3.33(09)	3.25(08)	3.33(09)	3.74(08)	3.63(08)	3.81(11)	3.69(08)
Fe^{2+}	0.22(02)	0.22(01)	0.31(03)	0.33(03)	0.42(03)	0.34(02)	0.08(01)	0.14(02)	0.04(01)	0.08(01)
Mg	0.36(04)	0.45(03)	0.20(04)	0.34(07)	0.33(04)	0.33(04)	0.18(02)	0.23(03)	0.15(03)	0.23(03)
Na	0.07(02)	0.10(02)	0.17(04)	0	0.20(03)	0.16(03)	0.15(03)	0.19(04)	0.47(06)	0.38(04)
K	1.31(05)	1.09(04)	1.20(06)	1.27(06)	1.17(06)	1.21(06)	0.70(03)	0.66(03)	0.33(02)	0.59(04)

[1] Each formula is normalized to a total of 12 cations in tetrahedral and octahedral sites and is an average of at least three scanning areas within the same grain.

[2] Formulae 1-6, 7-8, and 9-10 were derived from the analyses for areas characteristic of illite, R>1 illite/smectite, and ordered R1 illite/smectite, respectively.

[3] Numbers in parentheses correspond to two standard deviations based on counting statistics.

many studies that included HRTEM imaging. The literature is replete with bulk sample analyses and with electron microprobe analyses, but because clay mineral samples, and especially shales, are heterogeneous, such analyses are subject to contamination by other minerals, and/or averaging of compositions of heterogeneous clay minerals (e.g., illite in shales may include detrital and authigenic grains of different composition). Therefore, only AEM data are discussed here. However, I note first that EMPA and other bulk analytical data have shown that smectite generally has a small net negative charge (nnc) on the 2:1 layers of 0.25–0.4 (for an anion composition of $O_{10}(OH)_2$), illite, a range from ~0.55 to 0.95, and muscovite, a range from ~0.95 to 1 (see Srodon and Eberl, 1984, for a review based on non-AEM data). The nnc is provided both through substitutions of divalent cations (Mg and Fe) in octahedral sites and Al in tetrahedral sites, although the tetrahedral substitution is generally more significant.

The first AEM analyses of illite and glauconite from low-grade pelites were obtained by Ireland et al. (1983). They showed that normalized formulae corresponded to nearly ideal dioctahedral structures and that there is a significant range in solid solution of octahedral Fe^{3+} and Al. Their analyses appeared to verify the existence of a very large range in nnc of illite.

Jiang et al. (1990) obtained AEM analyses of I/S that was a hydrothermal alteration product of authigenic illite occurring in Welsh shales (Table 9-3). Those analyses appeared

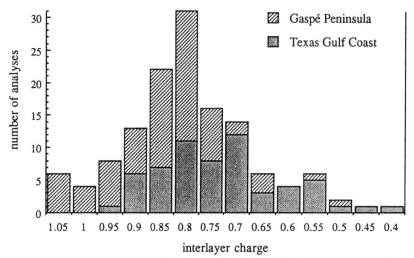

Figure 9-14. Plot of compositions of illite and ordered I/S as a function of interlayer charge for samples from the Gulf Coast and Gaspé Peninsula. Ideal muscovite has an interlayer charge of 1. The plot shows that illite has a broad range of compositions but has an average interlayer charge of approximately 0.8. Courtesy of W.-T. Jiang.

at first sight to be in error because the compositions were not weighted averages of the generally-accepted-compositions of smectite and illite; rather, the nnc was ~0.8 when due regard was taken for apparent interlayer Mg, and perhaps even larger if interlayer Mg is not present, values that are larger than an average of those of smectite and illite. Subsequent consideration of available rectorite analyses showed that rectorite, which consists of equal numbers of alternating smectite-like and illite-like layers, has an average composition that is very simliar to the Welsh I/S; i.e., the nnc is ~0.75. Huff et al. (1988) obtained similar results for K-bentonite consisting of R2 I/S. This presents an apparent enigma in that R1

I/S is generally inferred to be a mineral intermediate both in structure and formation sequence to smectite and illite. Assuming that illite has nnc as small as 0.55, the compositions of rectorite and Welsh I/S are apparently not intermediate to those of illite and smectite.

Nieto (unpublished data) subsequently obtained AEM data for a prograde sequence of smectite, I/S, and illite from the Vasco-Cantabrian Basin in Spain. Those data indicated that R1 I/S, illite, and I/S with intermediate ratios of smectite-like and illite-like layers have approximately the same composition. Jiang (pers. comm.) has subsequently collated AEM data for illite and mixed-layer I/S that have been obtained at the University of Michigan. For example, data from shales from the Gulf Coast and Gaspé Peninsula, Quebec, are plotted in Figure 9-14. Those data show that, although there is a wide range of compositions, in part because of varying proportions of substitutions of divalent cations (Mg and Fe) in octahedral sites and Al in tetrahedral sites, the analyses of illite and ordered I/S have average nnc of approximately 0.8. All such material consisted of the $1M_d$ polytype, and, based on texture and geological relations, was in its original state of formation.

On the other hand, similar plots of AEM analyses of smectite show a maximum with nnc of ~0.4. By contrast, 2-layer polytypes of so-called illite (muscovite or paragonite) found in higher grade anchizonal and epizonal shales and slates invariably have mica-like compositions, with nnc ~0.95. Such mica (commonly referred to as illite because of the small grain size) occurs ubiquitously as detrital grains in sediments of all grades of diagenesis and as grains preferentially aligned parallel to slaty cleavage in anchizonal and epizonal shales and slates. It invariably consists of a two-layer or other well-ordered polytype. The components of such mica (illite) are inferred to be derived through dissolution of authigenic illite or I/S having nnc = 0.7 to 0.8.

Inoue and colleagues have obtained very different results. Bentonites from Sweden, consisting of I/S with 50 to 70% illite-like layers, exhibited a regular increase in K (0.27 to 0.56 pfu) and Al (1.71 to 2.07 pfu) with increasing proportion of illite-like layers (Inoue et al., 1990). However, when Ca and Na (wet chemical analysis) were included in the total of interlayer cations, the total interlayer charge varied from 0.62 to 0.84, with no regular relation to proportion of illitelike layers. Clay separates were of two morphologies, flaky and lathlike, but no difference in composition was detected. Inoue et al. (1987, 1988) obtained analyses of I/S from the Shinzan hydrothermal area of Japan. They showed that there was a regular progression in K content from 0.29 to 0.80 pfu that parallels an increase in proportion of illite-like layers from 45 to 100%. Unfortunately, no analytical data were given for Na, whereas the Ca content was approximately constant (0.04 to 0.08 pfu). Tetrahedral Al contents varied regularly from 0.16 to 0.92 pfu. Inoue et al. observed three different morphologies, flaky, lathlike, and hexagonal, with K content smallest in the flaky material and greatest in the hexagonal material. Those results suggest that there is a regular change in composition of I/S that is directly related to the proportion of illite-like layers.

9.3.2 Particle theory *vs.* McEwan crystallites

There is considerable controversy concerning the relation between layers in smectite, I/S, and illite in original geologic samples, separates of such samples, and samples that are reconstituted for XRD, as summarized, for example, by Altaner et al. (1988). In one view, sequences of illite-like and smectite-like layers are visualized as continuous three-dimensional units which may be referred to as MacEwan crystallites that may cleave into separate units during grinding or dispersion in a liquid. Such sequences may have different degrees of coherency between layers, depending upon origin. The fundamental particle theory, on the other hand, is based on observations of separates, but implies that sequences of layers both in original rock samples and in XRD samples consist of aggregates of so-called "fundamental particles" of smectite (1.0 nm thick) and illite (≥2.0 nm thick) having only K interlayers.

Nadeau et al. (1984a,b,c; 1985) and Nadeau (1985) utilized Pt-shadowing of separates to measure the thicknesses of grains of a variety of clay minerals, including smectite, randomly and irregularly interstratified I/S, and illite. Although the thicknesses covered a range, the majority of grains had measured thicknesses of: (1) 1.0 nm for smectite, (2) 2.0 nm for R1 I/S with approximately equal numbers of illite-like and smectite-like layers, (3) 1.0–5.0 nm for interstratified I/S with >50% illite-like layers, and (4) 2.0–16.0 nm (mean value 7.0 nm) for authigenic illite having no expandable component. As defined by Nadeau et al. (1984b), a fundamental particle is an "individual or free" particle that gives a single-crystal $hk0$ diffraction pattern. They defined an elementary smectite particle as one being 1.0 nm thick, and an elementary illite particle as being 2.0 nm thick, with two 2:1 layers centered on a K interlayer.

Randomly interstratified I/S was inferred to consist of a collection of elementary illite and smectite particles, and rectorite to be an array of 2.0-nm thick illite particles. This led directly to the concept of interparticle diffraction (Nadeau et al., 1984c), wherein individual particles were inferred to aggregate into stacks in XRD mounts. The interface between any two illite particles was inferred to be hydrated and thus to behave like a smectite interlayer; a stack of illite particles was therefore inferred to give rise to alternating expandable hydrated and 1.0-nm K interlayers that gave a powder diffraction pattern analogous to that of rectorite.

The fundamental particle concept was seemingly at variance with TEM observations that utilized ion-milled samples of shales (e.g., Ahn and Peacor, 1986b). Such data indicated that smectite, illite, and I/S consist of packets with sequences of a few to many tens of layers (e.g., Figs. 9-9 through 9-11). However, although such images do show that layers do not occur as free or separate particles, they are not sufficient to prove that contiguous layers are coherently related; i.e., they may instead be related by random rotations about c^* (turbostraticly stacked). The fundamental particle theory implies that layers that are incoherently-related should be viewed as separate particles. It is of course debatable if that should be the case, but that cannot be discussed here; however, that layers of smectite occurring in shale are incoherently related was proven by the ring-like $hk0$ diffraction patterns obtained by Freed and Peacor (1992). That is sensible, in that such layers are assumed to have been transported by rivers and deposited as a flocculent; i.e., they were transported and sedimented as "elementary particles." However, similar observations were made by Masuda et al. (unpublished data) for smectite that was derived by direct alteration of volcanic ash in marine sediments; that is, regardless of origin, the layers of smectite appear to largely be turbostraticly-related. On the other hand, Banfield et al. (1991b) observed cross-fringes in (001) lattice fringe images of smectite, demonstrating at least limited coherency. More observations are necessary in order to determine the degree of coherency in stacking sequences and how it varies with geologic origin.

Evidence for a coherent relation between contiguous layers of I/S was directly provided by Veblen et al. (1990), who observed 0.45-nm fringes in addition to normal 1.0-nm fringes (Fig. 9-15) in images of R1 I/S. Such fringes are continuous across several 1.0-nm fringes, verifying that there is coherency between several sequential layers.

362

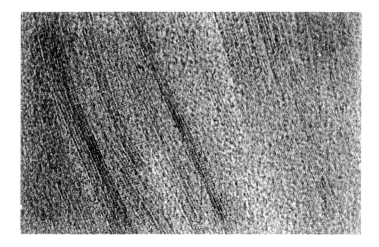

Figure 9-15. High resolution TEM image of R1 I/S from hydrothermally altered rock, Shinzan Area, Akita Prefecture, Japan. Cross-fringes with $d = 0.45$ nm transect the basic 1.0-nm fringes, implying that layers with mutual cross-fringes are coherently related. Courtesy of G.D. Guthrie.

The latter observations were obtained on I/S of hydrothermal origin and may not be valid for authigenic I/S. However, Ahn and Buseck (1990) obtained high-resolution images of Gulf Coast I/S and were able to demonstrate that sequences of at least several layers were coherently stacked. Lindgreen and Hansen (1991) obtained images from North Sea I/S with cross-fringes that demonstrated coherency over several layers. Freed has observed cross-fringes in I/S packets from Gulf Coast samples, using a JEOL 4000 HRTEM. The single crystal $hk0$ diffraction patterns of I/S obtained by Freed and Peacor (1992) for grains from Gulf Coast samples that were several layers thick also provide evidence for a high degree of coherency; i.e., regularly interstratified I/S (R = 1,2,3) has been observed to occur as coherent sequences of at least several layers forming "crystals" in natural samples wherever such observations have been made. Much work remains to be done to determine if that is the case in all authigenic I/S and to determine the relation between crystal thickness and conditions of diagenesis. It is now clear, however, that smectite and I/S are not simple assemblages of fundamental particles, but have varying degrees of coherency that may have important implications for conditions of formation.

As noted by Ahn and Peacor (1989) and Veblen et al. (1990), the natural sequences of layers may become disaggregated during preparation of XRD samples. Ahn and Peacor (1986a) verified that rectorite easily cleaves along smectite interlayers, giving rise to 2.0-nm thick units. When such separate so-called "particles" aggregate in XRD samples, they must stack with original smectite-like interlayers in contact, giving rise to a sequence of layers similar to that in the natural sample.

9.3.3 I/S and illite during late diagenesis

Until recently, research on pelites and their major clay minerals has focused largely on diagenesis during early burial metamorphism, as in Gulf Coast or North Sea sediments on the one hand, or tectonically-stressed sediments that give rise to shale/slate/schist transitions that normally have been the domain of structural geology. However, there have been few studies of sequences in which I/S has been observed to transform to mature micas, in rocks transitional between burial and tectonic effects.

Recently, Jiang and Peacor (in prep.) have observed the following prograde sequence in pelitic sediments of the Gaspé Peninsula: (1) Zone of diagenesis – I/S and illite, with nnc = 0.7 and $1M_d$ polytypism in crystals approximately 20 nm thick in bedding-parallel orientation. (2) Anchizone – muscovite (phengite) with nnc = 0.95, $2M_1$ polytypism in crystals approximately 60 nm thick in bedding-parallel orientation. Intermediate samples show mixtures of both. (3) Epizone – mature, large crystals of muscovite (phengite) in non-bedding orientations. The relations imply that immature I/S that forms by reaction of detrital smectite transforms during a second reaction to mature white mica during late burial metamorphism (latter stages of diagenesis). Li et al. (in prep.) and Nieto (unpublished data) have made similar observations of pelites from Wales and Spain, respectively. This is compatible with observations in Gulf Coast sediments (Hower et al., 1976) for which I/S with approximately 80% illite was found to undergo no further change subsequent to the smectite-to-illite transformation, to depths of at least 18,000 ft. This may have been the result of insufficient burial to promote the second stage of reaction, or, as Hower et al. proposed, to exhaustion of K-feldspar which may have served as a source of K for the smectite-to-illite reaction.

9.3.4 "Back-reaction" of illite to smectite

It is generally assumed that I/S and illite that occur in pelites form as prograde products of reactions involving smectite or other clays. However, Jiang et al. (1990) have shown that I/S with a variety of illite/smectite ratios and ordering sequences formed as a direct, although localized, hydrothermal replacement of illite in Welsh pelites. Similarly, Nieto (in prep.) has shown that I/S that is the principal clay mineral in some shales from Spain is a hydrothermal alteration product of reactant illite on a regional basis. The TEM/AEM data of both Jiang et al. and Nieto demonstrate how such processes can be characterized and differentiated from prograde relations, in part on the basis of textural relations, whereas bulk sample observations as obtained by XRD cannot provide such information. Appropriate TEM characterization is therefore essential in many cases, if sequences of clay minerals are to be properly interpreted.

9.3.5 Summary of prograde changes

It is important to differentiate the changes that occur in permeable sandstones and siltstones where filamentous or "hairy" illite occurs as a precipitate in pore space on the one hand, and on the other, diagenesis of compact, impermeable shales and bentonite in which the fine-grained minerals form a continuous matrix for which little or no pore space is observed. Almost all of the TEM observations described above and in this summary relate to the latter.

TEM observations show that smectite (both pure and with up to 50% randomly interstratified illite layers – R0 I/S) occurs largely as a continuous matrix of wavy, anastomozing layers with frequent layer terminations. Layers are generally turbostratically related, whether formed directly from volcanic ash or as a rearticulated sequence of detrital layers of terrestrially weathered material, although there are exceptions as demonstrated by Banfield et al. (1991b). Observations of separates of several different diagenetic sequences have shown that thin lathlike or pseudohexagonal euhedral to subhedral plates mark the appearance of the first detectable ordered (R1) I/S with approximately 50% illite-like layers (Inoue et al. 1987, 1988, 1990; Freed and Peacor, 1992; Lindgreen and Hansen, 1991; Glasmann et al. 1989; Yau et al., 1987b), coexisting with anhedral smectite. As the proportion of illite increases, the thickness and proportion of plates increases (e.g., Nadeau et al., 1985; Inoue et al., 1987). TEM observations of samples in which original layer relations are preserved verify the existence of ordered mixed-layered I/S with the proportion of illite-like layers approximately equal to that observed by XRD (Veblen et al., 1990; Srodon et al., 1990). HRTEM has demonstrated that such I/S occurs as sequences of coherent layers whenever such observations have been attempted (Ahn and Buseck, 1990; Veblen et al., 1990; Lindgreen and Hansen, 1991).

Observations of ion-milled samples of clay-rich mudstones show a continuous array of illite and smectite having a complete range of proportions of illite-like and smectite-like layers, depending on grade of diagenesis. However, Ahn and Peacor (1986b) and Freed and Peacor (1989a, 1992) showed that apparent packets of I/S occur within the matrix of subparallel smectite (R0 I/S) layers, the packets being larger with increase in proportion of illite-like layers detected by XRD. Freed and Peacor interpreted images of such packets as being cross sections through the hexagonal plates seen in separates, as demonstrated by Yau et al. (1987b, 1988) for shales and siltstones of the Salton Sea. As the transformation proceeds, the proportion of illite-like layers increases and the I/S packets increase in size, with the formation of an array of interlocking packets of illite or ordered I/S that are subparallel to bedding and with thicknesses of a few up to tens of nanometers (Lee et al., 1985; Freed and Peacor, 1989a).

Buatier et al. (1992) have studied sediments from a drill core extending through recently-accumulated sediments of the Barbados accretionary complex. Smectite is the dominant clay mineral at all depths to approximately 700 m. However, starting at 600 m (T = 20 to 30°C) packets of illite-like material were occasionally observed. Such packets were up to 10 nm thick, occurred within a matrix of wavy smectite layers, gave typical $1M_d$ illite (or I/S) diffraction patterns, and had an illite-like composition as shown by AEM analyses. Such packets were interpreted to represent the onset of the transition of smectite to illite.

Whitney and Northrop (1988) have reviewed the literature relating to mechanisms by which smectite transforms to illite, noting the apparently competing notions of "replacement" and dissolution/crystallization. The mechanism for replacement presumably would have a significant solid-state component in that both reactant smectite and product illite are composed of topologically similar 2:1 layers. Based in part on O isotope data, Whitney and Northrop outlined a three-step process involving a "transformation" to random I/S in initial stages, but with dissolution/crystallization giving rise to final I/S. Amouric and Parron (1985) emphasized that formation of glauconite (Fe-rich equivalent of illite) was by direct crystallization from fluids subsequent to dissolution of precursor minerals. Ahn and Peacor (1986b) emphasized that only dissolution of smectite and crystallization of I/S or illite is a viable mechanism, and Nadeau et al. (1985) made similar conclusions but related them to their fundamental particle theory. Buatier et al. (1989) showed that Fe-rich flaky smectite transformed to lathlike glauconite by dissolution and crystallization in sediments less than 30 m deep, presumably at near-surface temperatures. The recent recognition that I/S and illite occur commonly as subhedral to euhedral thin crystals or packets of layers, with compositions of 2:1 layers different from those of anhedral, turbostraticly-layered smectite, rather conclusively implies a dissolution/crystallization mechanism for ordered I/S.

Clauer et al. (1990) have utilized O isotope data in conjunction with TEM observations of crystal morphology to show that in shallow marine sediments detrital smectite with a flaky shape undergoes dissolution, with subsequent crystallization of lathlike smectite. Further studies that combine stable O and H isotopic data with TEM observations are essential to an understanding of reaction mechanisms. Such data should be obtainable with the new laser techniques, which permit isotope data to be obtained from selected small volumes.

Yau et al. (1988) emphasized that complete dissolution of smectite could result in transport of components and crystallization of illite at a distant site in permeable systems, as in siltstones and sandstones. In relatively impermeable shales, however, Ahn and Peacor (1986b) inferred that dissolution of smectite and crystallization of illite could occur continuously across the boundary of a growing I/S or illite packet until the resultant I/S or illite completely occupied the volume previously occupied by smectite, as observed in images of ion-milled samples. There are many examples of layer-by-layer replacement of one phyllosilicate by another (e.g., Banfield and Eggleton, 1988; Yau et al., 1984; Jiang and Peacor, 1991; Olives Baños, 1985; Veblen and Ferry, 1983), but Ahn and Peacor (1987) inferred that such a process is one of dissolution and crystallization across the linear discontinuity between product and reactant layer(s), with flux of reactants and products along the discontinuity. Thus, the scale of dissolution and crystallization through fluids may vary from that of massive dissolution, transport, and crystallization in new sites (neomorphism), to progressive local dissolution of individual smectite layers with crystallization of parallel layers of illite. In the latter case, the crystal volumes and orientations of crystal structure elements may be closely related; indeed significant portions of the reactant structure may be retained in the product. Such reactions have therefore been referred to as solid state reactions even though water plays a central role.

9.4   ILLITE AND CHLORITE CRYSTALLINITY: LATE DIAGENESIS THROUGH LOW-GRADE METAMORPHISM

9.4.1   General relations

"Illite crystallinity" is a term that is synonymous with the Kubler index, the half-height width of the 1.0-nm peak on diffractometer traces of illite (Kubler, 1964, 1967, 1968), whereas "chlorite crystallinity" is a measure of the half-height width of the 1.4-nm chlorite peak (e.g., Frey, 1987; Arkai, 1991). The Kubler index varies regularly with grade of diagenesis and low grade metamorphism of pelitic rocks and is one of the prime factors in estimating grade in the sequence of diagenesis through anchizonal to epizonal grades (see reviews by Frey, 1987; Kisch 1989).

Although the Kubler index is generally inferred to vary primarily as a function of the size and perfection of illite crystals, there is a paucity of data showing the significance of each of the various contributing factors. Little is known also about causes of change in chlorite crystallinity. Pelites may contain several different 2:1 phyllosilicates with spacings of approximately 1.0 nm, including illite (muscovite), paragonite, and pyrophyllite, as well as detrital phyllosilicates inherited from older rocks. Care must therefore be used to avoid contributions to peak breadth from such minerals, as cautioned by Merriman et al. (1990).

There are three principal factors that may contribute to peak broadening of a single mineral species. These are: (1) Small crystallite size, as approximately given by the Scherrer equation:

$$\beta = K \lambda / [N d_{(hkl)} \cos \theta] ,$$

where β is the peak broadening in radians 2θ, K a constant, $d_{(hkl)}$ the interplanar spacing (approximately 1.0 nm for mica), and N is the number of unit cells in the direction in which thickness is measured in the diffracting crystallites. Only the thickness parallel to c^* affects the breadth of 00l reflections. (2) Crystal strain, commonly referred to as lattice strain, includes the effects of any defects such as point defects, dislocations, and planar defects. Such defects can be produced either during growth of a crystal or as a result of stress, such stress generally having a tectonic component in geological materials. Polytypism in phyllosilicates does not contribute to peak broadening of 00l x-ray reflections, as it causes no change in the z coordinates of atoms. (3) Heterogeneity of composition or structure between layers, giving rise to disorder parallel to c^*. Such heterogeneity may be caused by small differences in composition among layers of a single mineral grain, or by mixed layering involving a second type of layer. Each of those factors is considered in turn in the following sections.

9.4.2 Crystal size

Crystallinity has been qualitatively correlated with measurements of illite crystal thickness by Kreutzberger and Peacor (1988). They observed a decrease in Kubler index from 0.47 to 0.37°2θ for illite in a limestone and the clay-rich shale resulting from pressure-induced solution of the limestone. Crystallites increased in average thickness from 45 to 1,000 nm, giving rise to a difference in width based on the Scherrer equation of 0.17° 2θ. This value is larger than the measured difference, implying that crystal thickness must be the principal contributing factor in this case.

Merriman et al. (1990) have carried out a quantitative study of the relation between crystal thickness and the Kubler index. They used samples from the zone of diagenesis, anchizone, and epizone of a prograde sequence of pelites in Wales. They utilized ion-milled samples, measuring crystallite size as the thickness of individual packets of layers within which there were no discontinuities. They determined particle size distributions, for which the modes were 8, 37.5, and 125 layers for the diagenetic zone, anchizone, and epizone, respectively (Fig. 9-16). This compares with values of 7, 29, and 63 layers, respectively, as predicted from the measured Kubler indices, demonstrating that increasing crystallite thickness is the principal cause of decrease in the Kubler index for these minerals. Merriman et al. (1990) further note that only crystals <100 layers thick contribute to peak broadening under normal diffractometer conditions, so that the Kubler index changes very little through the epizone, the largest changes occurring with crystal growth in the anchizone, concomitant with cleavage development.

Several TEM studies of chlorite and/or illite have demonstrated a qualitative decrease in thickness of packets with decreasing grade in ion-milled samples that have preserved original crystals, although such changes have not been correlated with crystallinity indices. For example, Ahn et al. (1988) showed that chlorite crystals derived by alteration of volcanic glass in New Zealand are only 10 to 30 nm thick, whereas the mica packets were 5 to 30 nm thick. Similar packet thicknesses have been observed by Yau et al. (1987b) for Salton Sea area sediments and Ahn and Peacor (1986b) for phyllosilicates derived through the formation of I/S from smectite; indeed such packet sizes are characterisitic of crystals that have formed during early diagenesis. Lee et al. (1984) summarized such relations for chlorite and muscovite, showing that early diagenesis is marked by small packets that may be only a few layers thick and mixed-layered, with crystal size increasing to the micron range in slates. Eberl et al. (1990) have determined particle thickness as a function of grade, suggesting that particle size distributions are a function of Ostwald ripening with increasing metamorphism in pelites and bentonites in the anchizone and higher grades; no attempt was made to correlate those results with illite crystallinity, however. Much work

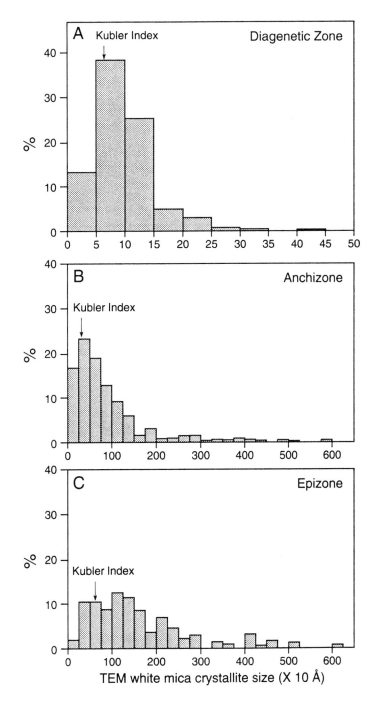

Figure 9-16. Distribution of packet thicknesses of illite in pelites in a mudstone to slate prograde sequence, North Wales, as determined using TEM images of ion-milled samples, and demonstrating the progressive increase in packet thickness with grade. The value corresponding to the Kubler index is the mean packet thickness calculated from XRD 001 peak width via the Scherrer equation. After Merriman et al. (1990).

remains to be done in quantifying the relations between crystal sizes, metamorphic grade, and crystallinity indices.

9.4.3 Crystal strain

Defects can be produced during growth or in response to stress that is generated either during burial or tectonic activity, but the term crystal strain is generally used to describe the defect state regardless of origin. It has been generally observed that clay minerals forming at low temperatures are rich in defects and that the proportion of defects generally decreases with increasing grade (e.g., Lee et al., 1985). Layer terminations have been described by many investigators, and their significance in fluid and solute transport was emphasized by Ahn and Peacor (1986a, 1987). However, there are no systematic or quantitative studies of defects in clay minerals as a function of metamorphic grade or other geologic parameters, with the exception of the work of Bons and coworkers (Bons, 1988; Bons et al, 1990; Bons and Schryvers, 1989). On the other hand, stress-induced defects have been studied using TEM by many investigators (e.g., Amouric, 1987; Goodwin and Wenk, 1990). Because such defects are discussed in Chapter 12, they are not further described here. Suffice it to say that the kinds and origins of defects in clay minerals formed during diagenesis and low grade metamorphism represent a rich but untapped source of information.

9.4.4 Crystal heterogeneity

Mica and chlorite may be heterogeneous in both composition and structure. Structural heterogeneity consists principally of mixed layering ranging from the scale of one or a small number of layers of one type randomly interlayered within another dominant type, through random or regular mixed layering of approximately equal numbers of two kinds of layers. Such interlayering can occur as a result of alteration of a preexisting, homogeneous mineral. However, examples of primary intergrowths formed upon growth in apparent prograde sequences (e.g., Fig. 9-3a) are especially common [e.g., berthierine, mica, and chlorite (Amouric et al., 1988; Ahn et al., 1988; Lee and Peacor, 1985); micas, kaolinite, talc, and chlorite (Veblen, 1983); chlorite and mica (Lee et al., 1984); paragonite and phengite (Ahn et al., 1985; Shau et al., 1990a); pyrophyllite and muscovite (Jiang et al., 1990); kaolinite and I/S (Ahn and Peacor, 1985b)]. Such interlayering is common only in relatively low grade rocks. Because it gives rise to diffuseness parallel to c^*, it may contribute to breadths of 00l peaks in XRD patterns.

Smectite interlayers in illite are especially common in diagenetic illite but are absent in higher-grade rocks. However, the interfaces between illite grains may behave like smectite and give rise to interparticle diffraction (Nadeau et al., 1984c). Eberl et al. (1987) noted that the effect is serious where grains are especially thin, again as in low grade rocks, and they accounted for such effects in diffraction patterns of "sericite." Shau et al. (1990b) showed that smectite or vermiculite are commonly interlayered with chlorite in low grade rocks, causing significant variations in diffraction patterns that are interpreted to be only of chlorite. Such mixed-layering may one of the most significant, if not the primary, cause of variation in chlorite crystallinity indices.

Ahn and Peacor (1986b) noted that smectite in shallow Gulf Coast shales was heterogeneous in composition relative to illite in deeper rocks. They observed variations in thickness of smectite-like layers that were ascribed to variations in composition of smectite, although such conclusions are problematical (Guthrie and Veblen, 1990). Similarly, Vali and Hesse (1990) have noted that differential expansion of smectite layers implies heterogeneity. In general, minerals that form in low-grade environments may be metastable and heterogeneous, with homogeneity in composition between and within grains increasing with higher grade. Differences in composition between layers result in diffuseness parallel to c^*, whereas the resultant slight variations in d-values contribute directly to increased peak breadth. Such qualitiative observations remain to be thoroughly investigated and should be a focus of future TEM research.

9.4.5 Summary

Although there are many TEM studies that have included observations of crystal defects, size, and heterogeneity, the study of Merriman et al. (1990) of the effect of crystal size is the only detailed study that has been directed at an understanding of changes in illite crystallinity. Although that study verified that crystal size is the dominant cause of variation in illite crystallinity indices, qualitative observations imply that crystal defects and interlayering and chemical heterogeneity also can contribute to peak broadening. However, the significance of all three factors decreases with increasing grade, consistent with increasing order, so that the relative proportion of each may be similar at all grades. The relative and absolute effects of each factor as a function of grade or other variables remain to be quantified.

9.5 PROGRADE CHANGES IN TRIOCTAHEDRAL PHYLLOSILICATES

TEM studies of chlorite and its low-grade precursors have focused on two different kinds of occurrence: (1) Chlorite in pelitic sediments, normally in direct association with illite. (2) Chlorite as a hydrothermal alteration product of glass, olivine, pyroxenes, and other phases of basalts.

9.5.1 Trioctahedral phyllosilicates in basalt

Shau et al. (1990b) showed that trioctahedral phyllosilicates in a basalt from Taiwan that had been initially characterized as chlorite actually consist of packets as small as 20 nm thick of discrete chlorite, corrensite, and mixed-layered chlorite/corrensite (Fig. 9-17). The AEM-determined compositions (Table 9-4) show that excellent charge balance is obtained when analyses of pure chlorite are normalized to a trioctahedral chlorite formula. However, mixed-layer material invariably shows minor Ca, Na, or K and has an excess of octahedral Al relative to tetrahedral Al, i.e., it is not charge balanced. Such materials are typical of low-grade trioctahedral "chloritic" materials, and they have been referred to as "swelling chlorite." Indeed, Shau et al. (1990b) noted that chlorite in low-grade rocks (zone of diagenesis or anchizone) may contain some expandable smectite and vermiculite layers more commonly than not and that such interlayers are a principal cause of apparent changes in composition or crystallinity index with grade.

Shau et al. (1990b) inferred that corrensite is a structurally and thermodynamically unique phase, akin to rectorite among dioctahedral phyllosilicates. This implies that the normal prograde sequence should consist of saponite (smectite)—mixed-layered saponite/corrensite—corrensite—mixed-layered corrensite/chlorite—chlorite; i.e., mixed-layered structures intermediate to chlorite and smectite were inferred not to be (in general) simple mixtures of chlorite-like and smectite-like layers. However, Shau et al. concluded that disordered mixed-layered chlorite/smectite could form metastably in some cases.

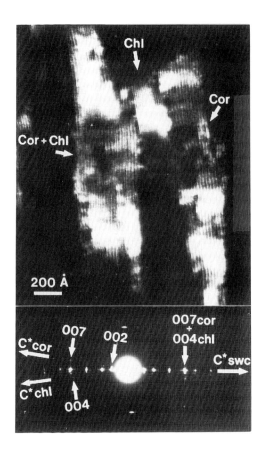

Figure 9-17. Lattice fringe image of coexisting packets of discrete chlorite (Chl), discrete corrensite (Cor), and mixed-layer chlorite/corrensite occurring in a metabasalt, Taiwan. The SAED pattern has rational chlorite and corrensite reflections and irrational mixed-layer chlorite/corrensite reflections (swc). Courtesy of Y.-H. Shau.

Table 9-4. Formulae derived from AEM analyses of 20 to 50-nm thick packets consisting of chlorite, mixed-layer corrensite/chlorite, and corrensite, from the Junghua metabasalt, Taiwan.[1,2] (After Shau et al., 1990b)

	Chlorite	Mixed-layer Corrensite/Chlorite	Corrensite	
Si	3.03(3)	3.44(3)	3.62(3)	6.46(5)
Al^{IV}	0.97(1)	0.56(1)	0.38(1)	1.54(3)
Al^{VI}	0.99(1)	1.05(1)	1.14(2)	1.17(2)
Fe^{2+}	2.17(2)	1.75(2)	1.54(2)	2.74(4)
Mg	2.76(4)	2.88(4)	2.84(3)	5.07(5)
Mn	0.016(4)	0.014(3)	0.011(3)	0.020(5)
Ca	0.049(4)	0.070(4)	0.101(5)	0.180(9)
K	n.d.	n.d.	n.d.	n.d.

[1]Numbers in parentheses correspond to one standard deviation.
[2]Formulae normalized to 14 O (assuming a chlorite formula), except for the right-hand corrensite formula, which is normalized to 25 O (ideal corrensite).

Shau and Peacor (1992) studied the sequence of trioctahedral materials in DSDP Drillhole 504, for which the sequence of hydrothermal alteration products is, starting from the top of the basalt sequence: (1) saponite to a depth of 624 m, (2) mixed-layered corrensite/chlorite, chlorite, and corrensite from 624 to 965 m, and (3) chlorite, talc, and mixed-layered talc/chlorite from 725 to 1076 m. Figure 9-18 shows some typical lattice-fringe images. The sequence is not a regular, continuous function of depth (temperature), occurrences being related to rock permeability in addition to depth. In addition, a given sample has alteration features that occurred at different times and temperatures during the time of spreading from the ridge, and occurring in several different modes (e.g., alteration of glass, olivine, or pyroxene, and in veins). A discontinuity occurs at 571 m between common occurrences of saponite and of corrensite. The latter corresponds to mixed-layer materials intermediate to saponite and corrensite, although both mixed-layer saponite/chlorite (rare) and saponite/corrensite (Fig. 9-18d) were found.

AEM analyses showed a regular increase in Fe relative to Mg, and decrease in Si relative to tetrahedral Al, in the sequence saponite—mixed-layered chlorite/smectite—

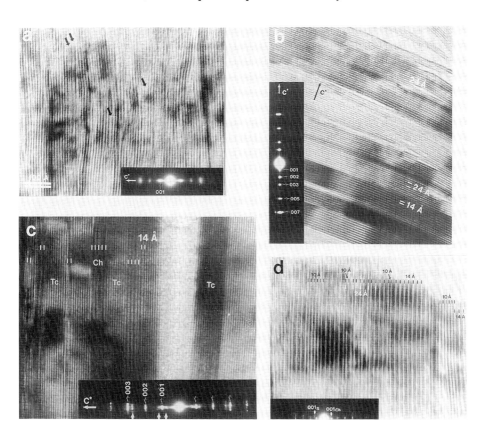

Figure 9-18. Lattice fringe images of trioctahedral clay minerals occurring as alteration products in basalts of DSDP core 504B, Costa Rica Rift, Pacific Ocean.   (a) Smectite (saponite), with typical wavy layers and common layer terminations; Arrows point to 1.4-nm chlorite fringes (600 m depth). (b) Discrete packets of chlorite (1.4-nm) and corrensite (2.4 nm); SAED pattern corresponds to corrensite (640 m depth). (c) Dominant talc (1.0 nm) with interlayered chlorite (1.4 nm). SAED patterns show rational talc reflections (black arrows) and irrational mixed-layered talc/chlorite (white arrows). (1030 m depth) (d) Interlayered chlorite (1.4 nm), saponite (1.0 nm), and corrensite (2.4 nm). (600 m depth). Courtesy of Y.-H. Shau.

corrensite—mixed-layered chlorite/corrensite—chlorite, which was interpreted by Shau et al. (1990b) to be related to the misfit between tetrahedral and octahedral sheets in 2:1 layers. In addition, K was found to be an important interlayer cation in saponite and corrensite, implying that expandable trioctahedral clays may be important sinks for K in ocean basalts.

Bettison-Varga et al. (1992) have observed similar relations in a sequence from the Point Sal ophiolite, California. However, on the basis of counting of sequences of layers (1.0 and 1.4 nm), they interpreted mixed-layer materials as mixed-layered saponite/chlorite, rather than as mixed-layered chlorite/corrensite. However, Shau and Peacor (1992) concluded that "The stacking sequence is therefore expressed as either a random mixed-layer chlorite/smectite in "crystals" of >100 nm thick, or an assemblage of chlorite, smectite, and corrensite with "crystal" size generally <15 nm in thickness," based on analysis of sequences of 1.0- and 1.4-nm layers. Both Shau et al. and Bettison-Varga et al. interpreted XRD patterns of mixed-layer clays in terms of the observed layer sequences.

9.5.2   Trioctahedral phyllosilicates in pelites

Ahn et al. (1985a) showed that thin packets of chlorite (frequently containing interlayered 0.7-nm berthierine) occur within dominant I/S that results from replacement of smectite. In part because the reactant smectite was shown to be more Fe- and Mg-rich than the product I/S, such chlorite was inferred to form contemporaneously with I/S. Ahn et al. (1988) showed that packets of illite, chlorite (with some berthierine interlayers), and mixed-layered illite-chlorite form simultaneously by alteration of volcanic ash. Similarly, the shale occurring at Lehigh Gap, Pennsylvania, was shown to contain complex intergrowths of illite and smectite ranging from thin packets of each to an ordered 1:1 mixed-layered material (Lee et al. 1984; Lee and Peacor, 1985). Such material was assumed to result from earlier diagenesis involving detrital smectite as a reactant. Subsequent tectonic stress was inferred to promote thick packets of illite and chlorite during formation of slaty cleavage by dissolution of authigenic illite and smectite (Lee et al., 1985, 1986). Many other pelitic shales and slates have been studied by TEM, with chlorite observed as subsidiary to white micas, and with an origin presumably related to detrital smectite as a parent.

In most of the studied sequences, white micas are the dominant minerals relative to trioctahedral phases. However, Jiang et al. (in prep.) and Merriman et al. (personal comm.) have observed another kind of occurrence in shale to slate sequences in the Gaspé Peninsula and the Southern Highlands of Scotland, respectively. Trioctahedral phyllo-silicates are major minerals, but coexist with separate dioctahedral smectite, I/S, illite or muscovite. In sediments in the zone of diagenesis, relict biotite, presumably of vulcanogenic origin, is shown to have been directly replaced principally by corrensite (Fig. 9-19a). A similar origin for chlorite in mica-chlorite stacks was documented by Li et al. (unpublished data) for Welsh pelites and for the copper-bearing Nonesuch Formation at the White Pine copper mine, Michigan. However, in the latter cases it is not known if corrensite had first replaced biotite as a precursor to chlorite. The alteration mechanisms apparently are of the progressive layer replacement type, as documented, for example, by Veblen and Ferry (1983). Jiang et al. (in prep.) have shown that, as white micas evolve with decreasing crystallinity index, corrensite derived at least in large part by replacement of biotite evolves through mixed-layered corrensite/chlorite and then to chlorite (Fig. 9-19b), in the same sequence as observed in altered basalts. The epizonal assemblage is

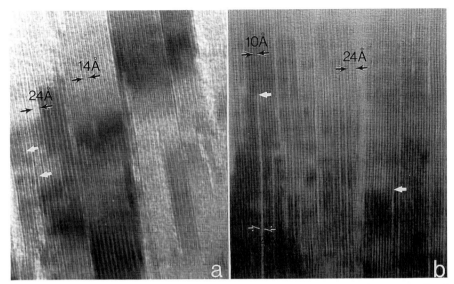

Figure 9-19. Lattice fringe images of trioctahedral phyllosilicates from a prograde pelite sequence, Gaspé Peninsula, Quebec. (a) Detrital biotite (1.0-nm fringes) partially replaced by corrensite (2.4 nm) from the zone of diagenesis, Gaspé Peninsula, Quebec. Arrows point to terminations of 1.0-nm biotite fringes at advancing 1.4-nm chlorite-like fringes. (b) Higher grade (anchizone) sample from the Gaspé Peninsula showing coexisting chlorite and corrensite. Courtesy of W.-T. Jiang.

characterized by muscovite and chlorite, each occurring in separate large packets oriented parallel to cleavage.

9.6 GENERAL TRENDS IN DIAGENESIS THROUGH LOW-GRADE METAMORPHISM OF PELITES

9.6.1 Prograde sequences: Heterogeneity to homogeneity

A requirement of stable chemical equilibrium is that phases in a given system be chemically and structurally homogeneous above some critical size, normally considered to be on the order of a few hundred nanometers, and that the phases obey the Gibbs phase rule. TEM/AEM techniques have shown that the textures in shales undergoing diagenesis exemplify lack of such stable equilibrium in the following ways: (1) Individual phyllosilicate crystals (packets of layers) are generally from one or a few layers to a few tens of layers in thickness. (2) Packets of many minerals may coexist, in violation of the phase rule; e.g., I/S, smectite, kaolinite, chlorite, and berthierine in Gulf Coast sediments, or chlorite, corrensite, and mixed-layer corrensite/chlorite in basalts. (3) More than one generation of a given mineral may occur; e.g., common detrital muscovite plus authigenic illite are normal associates. Li et al. (in prep.) have noted the occurrence of five kinds of illite/muscovite in Welsh shales, each with distinctive compositions, textures, and structures. (4) Individual packets are chemically and structurally heterogeneous, with mixed layering common and varying in extent both within and among packets. (5) Stacking sequences tend to be disordered, with the $1M_d$ polytype dominant in illite or I/S. (6) Extended defects such as layer terminations are common. (7) Individual minerals may be metastable. Lippmann (1981, 1982) emphasized that the increasing range of solid solution among phases such as clay minerals, with decrease in temperature of formation below that of the greenschist facies, demonstrated their inherent metastability. Jiang et al.

374

(1990) inferred that coexistence of pyrophyllite and muscovite (Fig. 9-20) under equilibrium conditions provided specific proof of that assertion in the case of illite, as the composition of illite is approximately intermediate to those of pyrophyllite and muscovite; that implies that smectite and I/S are likewise metastable phases.

By contrast, pelites that formed under epizonal conditions display relations that are not unlike those in greenschist facies rocks: (1) Individual phyllosilicate crystals commonly are several hundred nanometers to up to several microns in thickness, and much larger in width. (2) The number of minerals is relatively small, as consistent with the Gibbs phase rule. Chlorite and one or more micas known to be related by solvi (e.g., muscovite-paragonite-biotite) may coexist. (3) Grains of a given mineral species tend to be of one (metamorphic) origin, with grains having an origin in some earlier event (e.g., detrital or authigenic grains), having been eliminated. All grains of the same species tend to have approximately the same composition. (4) Individual packets of phyllosilicates are homogeneous, and mixed layering is rare; indeed, TEM operators may be frustrated with the relatively uninteresting, simple textural relations. (5) Stacking sequences are ordered, with the $2M_1$ polytype being dominant in white micas. (6) Except in grains subject to tectonic stress, defects such as layer terminations are uncommon. (7) The number and compositions of minerals approach those that are generally considered to represent stable equilibrium under greenschist facies conditions; i.e., muscovite (phengite) and/or paragonite, chlorite, albite, and quartz in the simplest pelites.

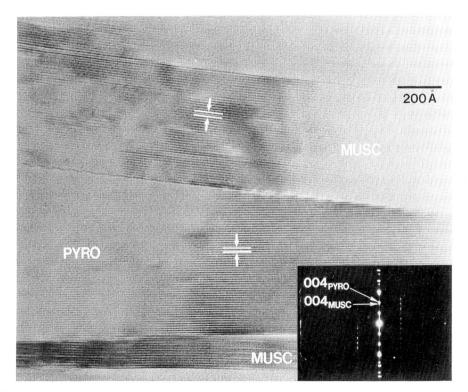

Figure 9-20. Lattice fringe image of separate packets of pyrophyllite (pyro) and muscovite (musc), Witwaters-rand, Republic of South Africa. The slightly larger value of $d(001)$ of muscovite leads to splitting of 001 reflections in the SAED pattern of both minerals. Courtesy of W.-T. Jiang.

9.6.2 Sequences of changes

In some respects, the entire sequence can be interpreted as one in which weathering, transportation, and deposition, or alteration of sedimented volcanic ash, initially produce a heterogeneous assemblage dominated by smectite-family minerals, which may themselves be heterogeneous in composition, especially with respect to Al and/or ferromagnesium components. Such materials transform through a series of changes toward a small number of homogeneous, ordered, phases as conditions approach those of the greenschist facies, the sequence being:

(1) Deposition of detrital phases, including glass and smectite.

(2) Smectite or R0 I/S forms a continuous matrix of primarily turbostraticly-stacked layers, or "megacrystals." Smectite is usually K-rich due to interaction with pore fluids and/or dissolution/crystallization.

(3) Growth of subhedral to euhedral $1M_d$ lathlike or pseudohexagonal crystals of ordered I/S or illite as packets of layers within a smectite-rich matrix, with eventual formation of an interlocking array of small $1M_d$ ordered I/S or illite crystals having a composition with nnc ~ 0.8.

(4) Recrystallization of bedding-parallel illite or ordered I/S to form larger interlocking crystals of $2M_1$ white mica (nnc ~ 0.95) consisting of muscovite (phengite) and/or paragonite and chlorite oriented preferentially parallel to bedding. This is the final event that has been identified in diagenetic settings that are free of tectonic stress.

(5) Progressive dissolution of bedding-parallel mica and chlorite with crystallization of mica and chlorite in a cleavage-parallel orientation. Crystal size of mature $2M_1$ micas and chlorite increases with increasing intensity of metamorphism in transition to phyllites, and crystals tend to be more defect-free and chemically and structurally homogeneous.

9.6.3   General conclusions, conjecture, and some suggestions for future work

In conjunction with other methods, TEM observations have demonstrated that there is a progressive sequence of changes from heterogeneous, metastable phases toward a system with phases in stable equilibrium during diagenesis and low-grade metamorphism. Such a sequence is consistent with the Ostwald Step Rule, in which a system with metastable phases progresses through a series of steps toward the state of stable equilibrium. TEM data demonstrate that virtually all reactions are mediated through a fluid phase, although the scale of such processes can vary from that of linear defects to large rock volumes. Where a reaction occurs by progressive replacement, as in replacement of biotite by corrensite, the closely-related reactant and product structures have orientations in common. However, most specific transformation processes have yet to be described, little is known about the reaction interfaces, and the nature of structural units that are common to both reactants and products and that are retained without modification are undefined.

The conclusion that the clay minerals in pelitic rocks are in states of metastable equilibrium, based in large part on TEM data, implies that the reactions through which they progress with increasing grade are controlled not by equilibrium considerations, but by kinetic factors (e.g., Jiang et al., 1990; Ahn and Peacor, 1986b). In part because there are so few relations that can be used to determine temperatures of reaction in sediments and

376

low-grade metamorphic rocks, clay mineral reactions continue to commonly be used as geothermometers. However, geothermometers as extensively used in metamorphic rock systems, for example, require systems to be in stable equilibrium, with all other relevant variables controlled. One such factor is indeed temperature, but water/rock ratio may be at least as significant, and the effects of solution chemical factors may be critical. It is therefore essential that the specific processes involving changes in clay minerals be detailed, in large part through TEM observations. Only then can the relevant variables be defined so that progressive changes in clay minerals can be utilized as accurate indicators of variables such as temperature of formation.

The degree to which the sequences of reactions described above are "step-like," rather than continuous functions of temperature, has yet to be determined. Eberl et al. (1990) have reviewed the concept of "Ostwald ripening," which is consistent with continuous increase in grain size with increasing time, but that is at least in part inconsistent with the "punctuated equilibrium" concept in which reactions are seen as activated by periodic tectonic events (e.g., Ohr et al., 1991; see review by Oliver, 1986). Much TEM work remains to be done in order to define such relations, in part involving the changing sizes, shapes and orientations of crystals in relation to their evolving compositions, and in conjunction with radioisotope data that may restrict the timing of such reactions.

ACKNOWLEDGMENTS

This paper has greatly benefitted from the constructive reviews of D. R. Veblen, R. C. Reynolds, and R. L. Freed. It has been made possible only through the efforts of the many students and colleagues who have allowed me to share in the joys of their discoveries. To Jung Ho Ahn, Lancy Yau, Jung Lee, Bob Freed, Wei-Teh Jiang, Eric Essene, Dick Merriman, Brin Roberts, Fernando Nieto, Martine Buatier, Gejing Li, and many others, I extend my deep appreciation for their contributions to this effort. This project was supported by NSF Project EAR-9104565.

REFERENCES

Ahn, J.H. and Buseck, P.R. (1990) Layer-stacking sequences and structural disorder in mixed-layer illite/smectite: Image simulations and HRTEM imaging. Am. Mineral. 75, 267-275.
Ahn, J.H. and Peacor, D.R. (1985a) Transmission electron microscopic study of diagenetic chlorite in Gulf Coast argillaceous sediments. Clays & Clay Minerals 33, 228-236.
Ahn, J.H. and Peacor, D.R. (1985b) Transmission electron microscopic study of the diagenesis of kaolinite in Gulf Coast argillaceous sediments. In: Proceedings of the International Clay Conference, Denver, L.G. Schultz, H. van Olphen, and F.A. Mumpton, eds., Clay Minerals Society, Bloomington, Indiana, 151-157.
Ahn, J.H. and Peacor, D.R. (1986a) Transmission electron microscope data for rectorite: Implications for the origin and structure of "fundamental particles." Clays & Clay Minerals 34, 180-186.
Ahn, J.H. and Peacor, D.R. (1986b) Transmission and analytical electron microscopy of the smectite-to-illite transition. Clays & Clay Minerals 34, 165-179.
Ahn, J.H. and Peacor, D.R. (1986c) Cation-diffusion-induced characteristic beam damage in transmission electron microscope images of micas. Ultramicrosc. 19, 375-382.
Ahn, J.H. and Peacor, D.R. (1987) Kaolinitization of biotite: TEM data and implications for an alteration mechanism. Am. Mineral. 72, 353-356.
Ahn, J.H. and Peacor, D.R. (1989) Illite/smectite from Gulf Coast shales: A reappraisal of transmission electron microscope images. Clays & Clay Minerals 37, 542-546.
Ahn, J.H., Peacor, D.R. and Coombs, D.S. (1988) Formation mechanisms of illite, chlorite and mixed-layer illite-chlorite in Triassic volcanogenic sediments from the Southland Syncline, New Zealand. Contrib. Mineral. Petrol. 99, 82-89.

Ahn, J.H., Peacor, D.R. and Essene, E.J. (1985) Coexisting paragonite-phengite in blueschist eclogite: A TEM study. Am. Mineral. 70, 1193-1204.

Alt, J.C. and Jiang, W.-T. (1991) Hydrothermally precipitated mixed-layer illite-smectite in recent massive sulfide deposits from the sea floor. Geology 19, 570-573.

Altaner, S.P., Weiss, C.A., Jr. and Kirkpatrick, R.J. (1988) Evidence from ^{29}Si NMR for the structure of mixed-layer illite/smectite clay minerals. Nature 331, 699-702.

Amouric, M., (1987) Growth and deformation defects in phyllosilicates as seen by HRTEM. Acta Crystallogr. B43, 57-63.

Amouric, M., Mercuriot, G. and Baronnet, A. (1981) On computed and observed HRTEM images of perfect mica polytypes. Bull. Mineral. 104, 298-313.

Amouric, M., Gianetto, I. and Proust, D. (1988) 7, 10 and 14 Å mixed-layer phyllosilicates studied structurally by TEM in pelitic rocks of the Piemontese zone (Venezuela). Bull. Mineral. 111, 29-37.

Amouric, M. and Parron, C. (1985) Structure and growth mechanism of glauconite as seen by high-resolution transmission electron microscopy. Clays & Clay Minerals 33, 473-482.

Arkai, P. (1991) Chlorite crystallinity: an empirical approach and correlation with illite crystallinity, coal rank and mineral facies as exemplified by Palaeozoic and Mesozoic rocks of northeast Hungary. J. Meta. Geol. 9, 723-734.

Bailey, S.W. (1988) X-ray diffraction identification of the polytypes of mica, serpentine, and chlorite. Clays & Clay Minerals 36, 193-213.

Banfield, J.F. and Eggleton, R.A. (1988) Transmission electron microscope study of biotite weathering. Clays & Clay Minerals 36, 47-60.

Banfield, J.F., Jones, B.F. and Veblen, D.R. (1991a) An AEM-TEM study of weathering and diagenesis, Albert Lake, Oregon: I. Weathering reactions in the volcanics. Geochim. Cosmochim. Acta 55, 2781-2793.

Banfield, J.F., Jones, B.F. and Veblen, D.R. (1991b) An AEM-TEM study of weathering and diagenesis, Albert Lake, Oregon: II. Diagenetic modification of the sedimentary assemblage. Geochim. Cosmochim. Acta 55, 2795-2810.

Bell, T.E. (1986) Microstructure in mixed-layer illite/smectite and its relationship to the reaction of smectite to illite. Clays & Clay Minerals 34, 146-154.

Bettison-Varga, L., Mackinnon, I.D.R. and Schiffman, P. (1992) Integrated TEM, XRD, and electron microprobe investigation of mixed-layer chlorite/smectite from the Point Sal ophiolite, California. J. Metamorphic Geol. (in press)

Bons, A.J. (1988) Deformation of chlorite in naturally deformed low-grade rocks. Tectonophysics 154, 149-165.

Bons, A. ., Drury, M.R., Schryvers, D. and Zwart, H.J. (1990) The nature of grain boundaries in slates. Phys. Chem. 17, 402-408.

Bons, A.J. and Schryvers, D. (1989) High-resolution electron microscopy of stacking irregularities in chlorites from the central Pyrenees. Am. Mineral. 74, 1113-1123.

Buatier, M., Honnorez, J. and Ehret, G. (1989) Fe-smectite-glauconite transition in hydrothermal green clays from the Galapagos spreading center. Clays & Clay Minerals 37, 532-541.

Buatier, M.D., Peacor, D.R. and O'Neil, J.R. (1992) Smectite-illite transition in Barbados accretionary wedge sediments: TEM and AEM evidence for dissolution/crystallization at low temperature. Clays & Clay Minerals 40, 65-80.

Clauer, N., O'Neil, J.R., Bonnot-Courtois, C. and Holtzapffel, T. (1990) Morphological, chemical, and isotopic evidence for an early diagenetic evolution of detrital smectite in marine sediments. Clays & Clay Minerals 38, 33-46.

Curtis, C.D., Hughes, C.R., Whiteman, J.A. and Whittle, C.K. (1985) Compositional variation within some sedimentary chlorites and some comments on their origin. Mineral. Mag. 49, 375-386.

Eberl, D.D., Srodon, J., Lee, M., Nadeau, P.H. and Northrop, H.R. (1987) Sericite from the Silverton caldera, Colorado: Correlation among structure, composition, origin, and particle thickness. Am. Mineral. 72, 914-934.

Eberl, D.D., Srodon, J., Kralik, M., Taylor, B.E., Peterman, Z.E. (1990) Ostwald ripening of clays and metamorphic minerals. Science 248, 474-477.

Freed, R.L. and Peacor, D.R. (1989a) Geopressured shale and sealing effect of smectite to illite transition. Am. Assoc. Petrol. Geol. Bull. 73, 1223-1232.

Freed, R.L. and Peacor, D.R. (1989b) TEM lattice fringe images with R1 ordering of illite/smectite in Gulf Coast pelitic rocks (abstract) G. S. A. Abs. with Prog. 21, A16.

Freed, R.L. and Peacor, D.R. (1992) Diagenesis and the formation of authigenic illite-rich I/S crystals in Gulf Coast shales: TEM study of clay separates. J. Sed. Pet., 62

Frey, M. (ed.) (1987) Low temperature metamorphism. Blackie and Son Ltd., Glasgow.

Glasmann, J.R., Larter, S., Briedis, N.A. and Lundegard, P.D. (1989) Shale diagenesis in the Bergen High area, North Sea. Clays & Clay Minerals 37, 97-112.

378

Goodwin, L.B. and Wenk, H.-R. (1990) Intracrystalline folding and cataclasis in biotite of the Santa Rosa mylonite zone: HVEM and TEM observations. Tectonophysics 172, 201-214.

Grubb, S.M.B., Peacor, D.R. and Jiang, W.-T. (1991) Transmission electron microscope observations of illite polytypism. Clays & Clay Minerals 39, 540-550.

Guthrie, G.D., Jr. and Veblen, D.R. (1989a) High-resolution transmission electron microscopy of mixed-layer illite/smectite: Computer simulations. Clays & Clay Minerals 37, 1-11.

Guthrie, G.D., Jr. and Veblen, D.R. (1989b) High resolution transmission electron microscopy applied to clay minerals. In L.M. Coyne, S.W.S. McKeever, and D.F. Blake, eds., Spectroscopic characterization of minerals and their surfaces. Sym. Ser. 415, Am. Chem. Soc., Washington, D. C.

Guthrie, G.D., Jr. and Veblen, D.R. (1990) Interpreting one-dimensional high-resolution transmission electron micrographs of sheet silicates by computer simulation. Am. Mineral. 75, 276-288.

Hansen, P.L. and Lindgreen, H. (1989) Mixed-layer illite/smectite diagenesis in Upper Jurassic claystones from the North Sea and onshore Denmark. Clay Minerals 24, 197-213.

Huff, W.D., Whiteman, J.A. and Curtis, C.D. (1988) Investigation of a K-bentonite by X-ray powder diffraction and analytical transmission electron microscopy. Clays & Clay Minerals 36, 83-93.

Hower, J., Eslinger, E.V., Hower, M.E. and Perry, E.A. (1976) Mechanism of burial metamorphism of argillaceous sediments: I. Mineralogical and chemical evidence. G.S. A. Bull. 87, 725-737.

Iijima, S. and Buseck, P.R. (1978) Experimental study of disordered mica structures by high-resolution electron microscopy. Acta Crystallogr. A34, 709.

Inoue, A., Kohyama, H., Kitagawa, R. and Watanabe, T. (1987) Chemical and morphological evidence for the conversion of smectite to illite. Clays & Clay Minerals 35, 111-120.

Inoue, A., Velde, B., Meunier, A. and Touchard, G. (1988) Mechanism of illite formation during smectite-to-illite conversion in a hydrothermal system. Am. Mineral. 73, 1325-1334.

Inoue, A., Watanabe, T., Kohyama, N. and Brusewitz, A.M. (1990) Characterization of illitization of smectite in bentonite beds at Kinnekulle, Sweden. Clays & Clay Minerals 38, 241-249.

Ireland, B.J., Curtis, C.D. and Whiteman, J.A. (1983) Compositional variation within some glauconites and illites and implications for their stability and origins. Sedimentology 30, 769-786.

Jiang, W.-T., Essene, E.J. and Peacor, D.R. (1990) Transmission electron microscopic study of coexisting pyrophyllite and muscovite: Direct evidence for the metastability of illite. Clays & Clay Minerals 38, 225-240.

Jiang, W.-T. and Peacor, D.R. (1991) Transmission electron microscopic study of the kaolinitization of muscovite. Clays & Clay Minerals 39, 1-13.

Jiang, W.-T. and Peacor, D.R., Merriman, R.J. and Roberts, B. (1990) Transmission and analytical electron microscopic study of mixed-layer illite-smectite formed as an apparent replacement product of diagenetic illite. Clays & Clay Minerals 38, 449-468.

Kisch, H.J., (1989) Calibration of the anchizone: a critical comparison of illite "crystallinity" scales used for definition. J. Metamorphic Geol. 8, 31-46.

Klimentidis, R.E. and Mackinnon, I.D.R. (1986) High-resolution imaging of ordered mixed-layer clays. Clays & Clay Minerals 34, 155-164.

Kreutzberger, M.E. and Peacor, D.R. (1988) Behavior of illite and chlorite during pressure solution of shaly limestone of the Kalkberg Formation, Catskill, New York. J. Struct. Geol. 10, 803-811.

Kubler, B. (1964) Le argiles, indicateurs de metamorphisme. Rev. Inst. Fr. Pet. 19, 1093-1112.

Kubler, B. (1967) Le cristallinite de l'illite et les zones tout a fait superieures du metamorphisme. In Etages tectoniques. Colloque de Neuchatel, 105-121.

Kubler, B. (1968) Evaluation quantitative du metamorphisme par la cristallinite de l'illite. Bull. Cent. Rech. Pau. SNPA 2, 385-397.

Lee, J.H., Ahn, J.H. and Peacor, D.R. (1985) Textures in layered silicates: Progressive changes through diagenesis and low-temperature metamorphism. J. Sediment. Petrol. 55, 0532-0540.

Lee, J.H. and Peacor, D.R. (1985) Ordered 1:1 interstratification of illite and chlorite: A transmission and analytical electron microscopy study. Clays & Clay Minerals 33, 463-467.

Lee, J.H., Peacor, D.R., Lewis, D.D. and Wintsch, R.P. (1984) Chlorite-illite/muscovite interlayered and interstratified crystals: A TEM/STEM study. Contrib. Mineral. Petrol. 88, 372-385.

Lee, J.H., Peacor, D.R., Lewis, D.D. and Wintsch, R.P. (1986) Evidence for syntectonic crystallization for the mudstone to slate transition at Lehigh Gap, Pennsylvania, U.S.A. J. Struct. Geol. 8, 767-780.

Lindgreen, H. and Hansen, P.L. (1991) Ordering of illite-smectite in upper Jurassic claystones from the North Sea. Clay Minerals 26, 105-125.

Lippmann, F. (1981) Stability diagrams involving clay minerals: in 8th Conference on Clay Mineralogy and Petrology, Teplic 1979, J. Konata, ed., Univerzita Karlova, Praha, Czechoslovakia, 153-171.

Lippmann, F. (1982) The thermodynamic status of clay minerals: in Proc. Int. Clay Conf. Bologna, Pavia, 1981, H. van Olphen and F. Veniale, eds., Elsevier, New York, 475-485.

Mackinnon, I.D.R. (1987) The fundamental nature of illite/smectite mixed-layer clay particles: A comment on papers by P.H. Nadeau and coworkers. Clays & Clay Minerals 35, 74-76.

Mackinnon, I.D.R. (1990) Low-temperature analyses in the analytical electron microscope. In I.D.R. Mackinnon and F.A. Mumpton, eds., CMS Workshop Lectures, vol. 2, Electron-Optical Methods in Clay Science. Clay Minerals Society, Evergreen, Colorado, 90-106.

McHardy, W. . (1987) Scanning electron microscopy. In M.J. Wilson, ed., A Handbook of Determinative Methods in Clay Mineralogy, Chapman and Hall, New York.

Merriman, R.J., Roberts, B. and Peacor, D.R. (1990) A transmission electron microscope study of white mica crystallite size distribution in a mudstone to slate transitional sequence, North Wales, U.K. Contrib. Mineral. Petrol. 106, 27-40.

Nadeau, P.H. (1985) The Physical dimensions of fundamental clay particles. Clay Minerals 20, 499-514.

Nadeau, P.H., Tait, J.M., McHardy, W.J. and Wilson, M.J. (1984a) Interstratified XRD characteristics of physical mixtures of elementary clay particles. Clay Minerals 19, 114-122.

Nadeau, P.H., Wilson, M.J., McHardy, W.J. and Tait, J.M. (1984b) Interstratified clays as fundamental particles. Science 225, 923-925.

Nadeau, P.H., Wilson, M.J., McHardy, W.J. and Tait, J.M. (1984c) Interparticle diffraction: A new concept for interstratified clays. Clay Minerals 19, 757-769.

Nadeau, P.H., Wilson, M.J., McHardy, W.J. and Tait, J.M. (1985) The conversion of smectite to illite during diagenesis: evidence from some illitic clays from bentonites and sandstones. Mineral. Mag. 49, 393-400.

Nadeau, P.H., Wilson, M.J., McHardy, W.J. and Tait, J.M. (1987) The fundamental nature of inter-stratified illite/smectite clay particles: A reply. Clays & Clay Minerals 35, 77-79.

Ohr, M., Halliday, A.N. and Peacor, D.R. (1991) Sr and Nd isotopic evidence for punctuated clay diagenesis, Texas Gulf Coast. Earth & Plan. Sci. Letters 105, 110-126.

Oliver, J. (1986) Fluids expelled tectonically from orogenic belts: their role in hydrocarbon migration and other geological phenomena, Geology 14, 99-102.

Olives Baños, J. (1985) Biotites and chlorites as interlayered biotite-chlorite crystals. Bull. Minéral. 108, 635-641.

Shau, Y.-H., Feather, M.E., Essene, E.J. and Peacor, D.R. (1990a) Genesis and solvus relations of submicroscopically intergrown paragonite and phengite in a blueschist from northern California. Contrib. Mineral. Petrol. 105, 7-72.

Shau, Y.-H., Peacor, D.R. and Essene, E.J. (1990b) Corrensite and mixed-layer chlorite/corrensite in metabasalt from northern Taiwan: TEM/AEM, EMPA, XRD, and optical studies. Contrib. Mineral. Petrol. 105, 123-142.

Shau, Y.-H. and Peacor, D.R. (1992) Phyllosilicates in hydrothermally altered basalts from DSDP Hole 504B, Leg 83Ωa TEM and AEM study. Contrib. Mineral. Petrol., submitted.

Spinnler, G.E., Self, P.G., Iijima, S. and Buseck, P.R. (1984) Stacking disorder in clinochlore chlorite. Am. Mineral. 69, 252-263.

Srodon, J., Andreoli, C., Elsass, F. and Robert, M. (1990) Direct high-resolution transmission electron microscopic measurement of expandability of mixed-layer illite/smectite in bentonite rock. Clays & Clay Minerals 38, 373-379.

Srodon, J. and Eberl, D.D. (1984) Illite. in Reviews in Mineralogy 13, Micas. S. W. Bailey ed., 495-544.

Vali, H. and Koster, H.M. (1986) Expanding behaviour, structural disorder, regular and random irregular interstratification of 2:1 layer-silicates studied by high-resolution images of transmission electron microscopy. Clay Minerals 21, 827-859.

Vali, H. and Hesse, R. (1990) Alkylammonium ion treatment of clay minerals in ultrathin section: A new method for HRTEM examination of expandable layers. Am. Mineral. 75, 1443-1446.

Vali, H., Hesse, R. and Kohler, E.E. (1991) Combined freeze-etch replicas and HRTEM images as tools to study fundamental particles and the multiphase nature of 2:1 layer silicates. Am. Mineral. 76, 1973-1984.

van der Pluijm, B.A., Lee, J.H. and Peacor, D.R. (1988) Analytical electron microscopy and the problem of potassium diffusion. Clays & Clay Minerals 36, 498-504.

Veblen, D.R. (1983) Microstructures and mixed layering in intergrown wonesite, chlorite, talc, biotite, and kaolinite. Am. Mineral. 68, 566-580.

Veblen, D.R. and Ferry, J.M. (1983) A TEM study of the biotite-chlorite reaction and comparison with petrologic observations. Am. Mineral. 68, 1160-1168.

Veblen, D.R., Guthrie, G.D., Jr., Livi, K.J.T. and Reynolds, R.C., Jr. (1990) High-resolution trans-mission electron microscopy and electron diffraction of mixed-layer illite/smectite: Experimental results. Clays & Clay Minerals 38, 1-13.

Whitney, G. and Northrop, H.R. (1988) Experimental investigation of the smectite to illite reaction: Dual reaction mechanisms and oxygen-isotope systematics. Am. Mineral. 73, 77-90.

Yau, Y.-C., Anovitz, L.M., Essene, E.J. and Peacor, D.R. (1984) Phlogopite-chlorite reaction mechanisms and physical conditions during retrograde reactions in the Marble Formation, Franklin, New Jersey. Contrib. Mineral. Petrol. 88, 299-306.

Yau, Y.-C., Peacor, D.R., Essene, E.J., Lee, J.H., Kuo, L.-C. and Cosca, M.A. (1987a) Hydrothermal treatment of smectite, illite, and basalt to 460°C: Comparison of natural with hydrothermally formed clay minerals. Clays & Clay Minerals 35, 241-250.

Yau, Y.-C., Peacor, D.R. and McDowell, S.D. (1987b) Smectite-to-illite reactions in Salton Sea shales: A transmission and analytical electron microscopy study. J. Sediment. Petrol. 57, 335-342.

Yau, Y.-C., Peacor, D.R., Beane, R.E., Essene, E.J. and McDowell, S.D. (1988) Microstructures, formation mechanisms, and depth-zoning of phyllosilicates in geothermally altered shales, Salton Sea, California. Clays & Clay Minerals 36, 1-10.

Chapter 10.   CARBONATES:
GROWTH AND ALTERATION MICROSTRUCTURES

Richard J. Reeder   Department of Earth and Space Sciences,
State University of New York at Stony Brook, Stony Brook, New York 11794, U.S.A.

10.1   INTRODUCTION

As with so many other mineral groups, the application of transmission electron microscopy and electron diffraction to carbonate minerals has made possible a new level of understanding of their *real* structures and, increasingly, is yielding new insight to the processes by which they crystallize and react. The widespread and varied occurrences of carbonates have fostered interest from many areas, not only within earth sciences, but in other physical sciences and in biological sciences as well. The greatest activity within earth sciences comes from researchers dealing with low-temperature environments characteristic of surface to moderate burial conditions. Wide diversity is found here, ranging from vast quantities of marine carbonate sediments to thick sequences of limestone and dolomite throughout the rock record.

Some carbonate minerals are formed directly as the result of biologic processes. Indeed, carbonates are the principal common mineral group that records extensive biologic activity throughout a significant portion of geologic time. Marine carbonate sediments, for example, are composed largely of calcite and aragonite that comprise the hard skeletal parts of certain invertebrates.

In view of their importance, the majority of the TEM applications reviewed in this chapter relate to formation and alteration of "low-temperature" carbonates. And quite predictably, virtually all attention has been accorded to the common carbonates—calcite, dolomite, and, to a lesser extent, aragonite. This emphasis reflects the overwhelming geologic importance of the marine Ca-Mg-CO$_3$ system. In comparison, fewer TEM studies have been done on metamorphic and igneous carbonates. Most of the work on metamorphic carbonates has been done in the context of deformation microstructures, which have been reviewed by Wenk et al. (1983) and by Wenk (1985), and consequently are not covered here. Only one major study (Barber and Wenk, 1984) has addressed microstructures in carbonatite carbonates. It is also surprising that only a few studies have examined synthetic or experimentally-modified samples, in spite of the enormous amount of synthesis work done, particularly with the Ca-Mg carbonates.

The ideal structures of the common carbonate minerals (e.g., calcite, dolomite, and aragonite) are rather simple, and, along with the other isostructural carbonates, have been extensively characterized (cf. Reeder, 1983). Despite their relative simplicity, early x-ray diffraction and geochemical studies of natural, low-temperature Ca-Mg carbonates revealed considerable complexity (e.g., Goldsmith and Graf, 1958). Experimental studies of phase relations (e.g., Graf and Goldsmith, 1955; Goldsmith and Heard, 1961) suggested that much of the observed compositional variation and structural complexity represented metastability. With the techniques available at that time, only a limited understanding of the real structures of such carbonates was possible. TEM has revealed the true structural details of natural carbonates, and has further provided insight to their reactivity and stability.

The formation and persistence of metastable phases is now regarded as common behavior in many low-temperature systems. Explanations usually involve sluggish kinetics or the relative ease of forming simple, sometimes disordered, structures rather than more complex, ordered ones. However, we will see from TEM observations that at least some common metastable Ca-Mg carbonates exhibit structures with more complex ordering than their stable counterparts.

If the low temperatures and sluggish kinetics are responsible for the formation of these complex microstructures, then these same factors must be credited, at least partly, with preserving them. Once formed, during growth for example, there is no evidence that any significant reorganization occurs in the solid state (excepting certain types of plastic deformation such as twinning and dislocation glide). This low-temperature behavior contrasts with that expected under metamorphic conditions, where long-range diffusion allows for exsolution, or short-range movements allow solid-state transformations such as between calcite and aragonite. Of course, both stable and metastable carbonates may react with aqueous solutions and undergo changes. But the reactions quite clearly involve dissolution and precipitation at low-temperatures, even though they may occur on a very fine scale, as in the case of replacement. Such diagenetic processes, and indeed the stabilization of predominately metastable marine carbonate sediments into *more stable* carbonate rocks, are a principal concern of carbonate geochemists.

The goals of this chapter are to demonstrate how application of transmission electron microscopy has provided information, not previously attainable, regarding (1) the *real* structures of common carbonate minerals and (2) their mode of formation. Microstructures in carbonates were previously reviewed by Wenk et al. (1983). Prior to that review the majority of TEM applications to carbonates had focused on deformation mechanisms (see e.g., Barber and Wenk, 1976; Wenk et al., 1983). Only since the early 1980s has interest shifted increasingly to non-deformational microstructures. In the period since the 1983 review, the TEM literature on carbonates has nearly tripled, with most of the new work relating to growth microstructures in low-temperature samples.

Because many of the microstructures documented in low-temperature carbonates originate during the growth process, this chapter also serves to illustrate the important role that TEM can play in characterizing crystal growth mechanisms. We will see that in the case of low-temperature solution growth, considerable variety exists.

10.2 SAMPLE-RELATED CONSIDERATIONS

10.2.1 Sample preparation

Foils of anhydrous carbonates are generally prepared and treated in the same manner as most anhydrous silicates. Ion milling at 4-6 kV and with a specimen current density up to 20-30 $\mu A/mm^2$ produces quite acceptable electron-transparent specimens, having a thin amorphous surface layer. The amorphous layer may affect high-resolution imaging, and it is customary to remove much of this layer by final thinning at 10-12° incidence and at low voltages. There is generally little evidence for introduction of artifacts during ion milling.

In sedimentary carbonates, fluid inclusion voids are among the most common features observed by TEM. There is some evidence that specimen heating during ion milling, which possibly may be as high as 120°C (Bahnck and Hull, 1990), induces stretching of inclusions. This is discussed in more detail later (see §10.5.8.2), but it is worth noting here that the principal result of stretching seems to be the production of dislocations surrounding the inclusion.

The traditional method of crushing grains for dispersal on a holey carbon film is not particularly effective with rhombohedral carbonates like dolomite and calcite. Owing to their perfect cleavage on three equivalent planes, crushed grains of calcite and dolomite tend to be equant in shape, so that lateral dimensions of grains rarely exceed their thickness. Moreover, grains typically lie on the cleavage surface affording just one dominant crystal orientation.

The ease with which calcite cleaves may also introduce an artifact observed during imaging. Cleavages within a foil, along which only very slight separation may exist, will produce a fringe pattern if suitably oriented. The phase shift due to the void slice along the cleavage causes the fringe pattern to have contrast characteristics similar to those of a stacking fault. In most calcite samples where this is observed, the fringes are straight and parallel, as would be expected from a nearly planar cleavage. In echinoderm fragments, which are magnesian calcite single crystals, it is well known that cleavage (fracture) is "conchoidal", for reasons that are not entirely clear (however, see Berman et al., 1988). Interestingly, such artifact fringes in these crystals always show pronounced curvature.

10.2.2 Artifacts from electron-beam irradiation

Most authors have noted some radiation damage of carbonates in TEM. Barber and co-workers have indicated that it is most apparent in calcite containing substantial isomorphous substitution or impurity content (Barber and Wenk, 1984; Barber and Khan, 1987; Khan and Barber, 1990). The appearance is one of randomly distributed "spots", apparently due to very localized strain centers (Fig. 10-1). Khan and Barber (1990) have suggested that such "spotty damage" is a result of the formation of small dislocation loops or bubbles at the sites of impurities, defects, or perhaps clusters of same. Spotty damage probably occurs in most carbonates to varying degrees, and it may be associated with localized decarbonation. Barber and Khan (1987) noted especially rapid specimen damage in ferroan calcite, and the development of dislocation loops.

Not surprisingly, radiation damage is most serious for high-resolution TEM. Cater and Buseck (1985) examined degradation of stoichiometric, well-ordered dolomite in the electron beam at accelerating voltages of 100-200 kV. They demonstrated that the principal mechanism of damage *in vacuo* is decarbonation, producing initially a metastable Ca-Mg oxide and then a more stable mixture of CaO and MgO (see §10.7.3). Damage was detected within 5-15 minutes of imaging. They concluded that the interaction involves ionization within the CO_3 group and release of CO_2. Observations that beam damage is less severe at accelerating voltages of 500 kV and above (e.g., Barber et al., 1983; Freeman et al., 1983) support this view. Although Cater and Buseck reported an intermediate amorphous phase in one experiment, in general, amorphization does not appear to be common in carbonates.

Van Tendeloo et al. (1985) observed a different effect for irradiation of structurally-complex Ca-rich dolomite at 1000 kV. They found that cation ordering diminished over a period ranging from 0.5 up to 8 minutes, depending on the type of ordering. No evidence of decarbonation was reported. The strong suggestion from these studies is that different types of damage are possible depending on the nature and composition of the carbonates involved, and also depending on the conditions of electron beam irradiation.

384

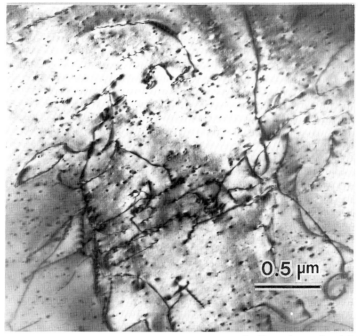

Figure 10-1. BF image illustrating "spotty constrast" that forms as a result of beam damage in calcite, especially that containing substitution of impurities. Dislocations are also in strong contrast.

10.3 STRUCTURAL ASPECTS

No effort is made here to review the structural details of various carbonate minerals in any comprehensive way. However, among the most common *rhombohedral* carbonates, certain structural relationships exist that are sufficiently important to the present chapter as to require some emphasis. The same is not true of the orthorhombic carbonate aragonite, and details of its structure are given by Speer (1983).

10.3.1 Unit cells

Although a similarity with NaCl is commonly cited, the calcite structure is best derived from an hexagonal close packing of O atoms with C and Ca atoms occupying the appropriate interstices as described by Megaw (1973) and by Reeder (1983). Layers of Ca octahedra repeat along the c axis, and share only corners with CO_3 groups and other Ca octahedra. The primitive unit cell is an acute rhombohedron (space group $R\overline{3}c$) with Z = 2 $CaCO_3$. However, virtually all modern work utilizes the hexagonal unit cell, a = 4.99 Å, c = 17.06 Å, for which the perfect cleavage indexes as $\{10\overline{1}4\}$. This cell is used exclusively in the present chapter. It is also worth noting that the use of a morphological cell, which is common in the older mineralogical literature, does not allow correct indexing of electron diffraction patterns.

It is useful to view the calcite structure as a parent structure for dolomite. Every other layer of Ca octahedra is replaced by Mg octahedra, which destroys the equivalence between successive cation layers. Because the Mg octahedra are smaller, and yet the same fundamental coordination as in calcite remains, the CO_3 groups are rotated by approximately 6° relative to their orientation in calcite (Reeder, 1983). The ordering of Ca and Mg onto nonequivalent sites and the rotation of the CO_3 groups destroys the c-glide, and the space group of dolomite is $R\bar{3}$. However, the unit cell is equivalent to that of calcite, having dimensions: a = 4.81 Å and c = 16.0 Å.

10.3.2 Cation sites

In dolomite, the nonequivalent cation sites are designated as A (Ca dominated) and B (Mg dominated) sites (cf. Reeder, 1983). The equilibrium distribution of Ca and Mg between the A and B sites is controlled primarily by temperature, but also by composition and presumably pressure. X-ray diffraction experiments on stoichiometric dolomite (using *in situ* heating) show that no detectable disorder occurs at 600°C (Reeder and Markgraf, 1986), and it is likely that significant equilibrium disorder does not occur until 900°C. Thermally-induced disorder is complete at approximately 1150°C (Reeder and Nakajima, 1982; Reeder and Wenk, 1983), at which point the space- and time-averaged cation distribution is random, and the structure and space group are the same as those of calcite. Consequences of re-ordering upon cooling below the transition are described later in this chapter.

10.3.3 Diffraction terminology for calcite-derived structures

The structures and superstructures documented in natural rhombohedral carbonates are conveniently derived from the calcite structure. It has become customary to distinguish among types of diffractions by assigning letters according to increasing departure from calcite symmetry. Diffractions characteristic of the calcite structure (space group $R\bar{3}c$) are referred to as a-type reflections (or a reflections). In dolomite (space group $R\bar{3}$), extra reflections ($h\bar{h}0l$, l = odd) occur as a result of the ordering described above; these are b reflections. Electron diffraction has revealed two additional types of diffractions—c and d reflections—which arise from different types of ordering, and each has an associated superstructure. c reflections are typically weak and exist due to ordering *within* cation layers. d reflections occur as satellites around a- and b-type spots, with diffraction vector $1/3\ 000l^*$. They result from periodic stacking defects along [0001]. Detailed descriptions of the superstructures associated with c and d reflections are given later.

10.4   HRTEM CONSIDERATIONS FOR CARBONATES

HRTEM is discussed in detail in Chapter 3 of this volume. Only one aspect related to HRTEM imaging of carbonate minerals is considered here. After examining simulations of the dolomite structure, O'Keefe and Barber (1984) concluded that a point-to-point resolution of 2.5 Å is required to resolve cations, 1.7 Å to resolve anions, and 1.3 Å to resolve carbon from oxygen. Resolution capabilities of present TEMs effectively allow imaging only of cation positions in dolomite and calcite (Meike et al., 1988). This is an important factor to bear in mind, particularly for interpretations of the different superstructures described later in this chapter.

10.5   MICROSTRUCTURES IN LOW-TEMPERATURE CARBONATES

10.5.1 Some general considerations

The low rates of solid-state diffusion that exist at low temperatures effectively preclude any significant reorganization within most crystals, even over long time scales. Many microstructures, especially if related to compositional differences, are unlikely to have been modified after their formation. Where thermal and perhaps tectonic histories preclude high temperatures and deviatoric stresses, many of the microstructures encountered in nominally sedimentary carbonates will have resulted from the crystal growth process. This could be by direct crystallization in a solution or during fine-scale, fabric-retentive replacement. The microstructure is essentially a record of the growth process, much the same as trace elements or stable isotopes record information about fluids. This is an important point to consider if one's goal is to relate microstructures to growth or diagenetic environments.

10.5.2 Dolomite

Dolomite (and isostructural carbonates) shows the greatest diversity of microstructures, and consequently has received more attention than other carbonates, including the more ubiquitous calcite. Ca-rich dolomite, in particular, may exhibit several different hetero-geneous microstructures, including modulations, coherent ribbon-like intergrowths, and ordered superstructures. Calcite also shows such microstructures, but apparently less commonly. Most workers now agree that these heterogeneous microstructures are related to compositional variations within the crystal. Furthermore, their existence can be correlated directly to deviations of *bulk crystal* composition from an ideal or stable composition. Because the occurrence of dolomite microstructures is related to compo-sition, it is important to review briefly the composition ranges of natural dolomite.

At temperatures below $500\text{-}600°C$, the stable composition range of iron-free dolomite is extremely restricted, so that any sample with deviation of greater than, perhaps, just 1 mol % from the ideal $CaCO_3:MgCO_3 = 50:50$ can be regarded as metastable (cf. Goldsmith, 1983, and references therein). The compositions of many low-temperature dolomite samples closely approximate the ideal ratio. However, many other dolomite samples contain up to 7-8 mol % $CaCO_3$ in excess of the stoichiometric amount (cf. Reeder, 1983), and by some reports even 10 mol % excess. Substantiated analyses show that Mg enrichment generally does not occur, or else it is limited to <1 mol %. Unfortunately, many dolomite compositions reported in the literature have been estimated from unit cell dimensions as determined by powder XRD. No independent calibration relating unit cell dimensions to composition has yet been established, and such correlations are made on the basis of an assumed relationship.

Significant Fe^{2+} substitution may occur in dolomite (preferentially for Mg), and the compositions of such ferroan dolomite, or ankerite, generally lie on (or are parallel to) the $CaMg(CO_3)_2\text{-}CaFe(CO_3)_2$ join. However, complete substitution of Fe^{2+} for Mg does not occur (cf. Goldsmith et al., 1962; Rosenberg, 1967; Reeder and Dollase, 1989), and maximum observed Fe^{2+} contents are approximately 70 mol % $CaFe(CO_3)_2$. Just as with iron-free dolomite, ankerite compositions range from near 50 mol % $CaCO_3$ (stoichiometric ankerite) to approximately 57-58 mol % $CaCO_3$ (calcian ankerite). This close parallel in the range of Ca enrichment between dolomite and ankerite seems to indicate that crystal chemical constraints are limiting factors.

Goldsmith and Graf (1958) were the first to recognize the range of Ca contents in sedimentary dolomite, and they further showed, by x-ray diffraction, that Ca enrichment is associated with evidence of imperfect cation order, possibly including basal stacking disorder. On the other hand, Goldsmith and Graf (1958) found that stoichiometric dolomite typically lacks such evidence, and suggest a high degree of cation order (also see Reeder, 1983). These authors also recognized that incorporation of excess Ca should not easily be accommodated in the smaller Mg sites, and, in retrospect, we see that in order to accomplish this some defect mechanism or alternate structure is generally required.

In the late 1970s, TEM revealed the first independent evidence that such defect structures exist and are indeed common (Reeder and Wenk, 1979). Barber (1977) had in fact already noted complex microstructures in some dolomite, but offered few details and no explanation since they were unrelated to the deformation and recovery mechanisms that he studied. Subsequent work (e.g., Reeder, 1981; Blake et al., 1982) showed that virtually all calcian dolomite exhibits heterogeneous microstructures, and that stoichiometric dolomite generally showed either homogeneous or dislocation microstructures. Subsequent work has since confirmed this general view, with few exceptions. Recent work has also shown that a similar situation exists for ankerite. Those containing appreciable excess Ca exhibit heterogeneous microstructures (Reksten, 1990a; Wenk et al., 1991), while those stoichiometric with respect to Ca are for the most part homogeneous (Reeder and Dollase, 1989).

Before examining these heterogeneous, composition-related microstructures, it is useful to make one further distinction among the Ca-rich dolomite samples in which they occur. The relative scarcity of dolomite demonstrated to be forming presently is commonly cited as one aspect of the so-called "dolomite problem" (cf. Zenger, 1972). Actually, the number of distinct occurrences is not that limited, but the amount of dolomite typically present is volumetrically small. Many of these Holocene dolomite samples are different in several important respects from Pleistocene and older dolomite, which will be referred to here as "ancient dolomite". The critical difference is not necessarily one of time, even though that might be implied. Rather, it presumably involves different environments or conditions of growth, which are recorded by the microstructure. In fact, most Holocene dolomite samples are very similar to one another, virtually all being extremely fine-grained and Ca rich. X-ray diffraction shows that they typically have expanded unit cells, very weak, diffuse (sometimes, almost absent) ordering reflections, and broad fundamental reflections. This metastable Holocene dolomite has been compared to Graf and Goldsmith's (1956) poorly-organized, synthetic Ca-Mg carbonates—so-called "protodolomite".

In contrast, ancient dolomite (stoichiometric and calcian) generally shows sharp X-ray diffraction maxima and clearly-defined (even if sometimes attenuated) ordering reflections (Goldsmith and Graf, 1958).

10.5.2.1 Ancient calcian dolomite (and calcian ankerite)

10.5.2.1.1 Modulated structure. The most common microstructure in ancient calcian dolomite is a fine modulation, as shown in Figure 10-2a. In diffraction contrast, it exhibits a somewhat regular, alternating dark/light contrast, very similar in appearance to spinodal decomposition microstructures (e.g., compare with Fig. 3b of McCallister and Nord, 1981). It is typically pervasive throughout a crystal, although authors have noted crystals where both modulated and homogeneous microstructure are present (e.g., Reeder, 1981). It is likely that the modulated regions are Ca rich dolomite, and elsewhere the crystal is stoichiometric, or nearly so (see §10.5.2.4).

Authors report variable degrees of development of the modulation in which the boundaries between dark and light contrast may be gradual and diffuse on the one hand (Fig. 10-2a) or fairly sharp on the other (Fig. 10-2b). The wavelength of the modulation is also variable between samples, but most commonly ranges from 75-200 Å. Wavelengths up to 400 Å have been reported.

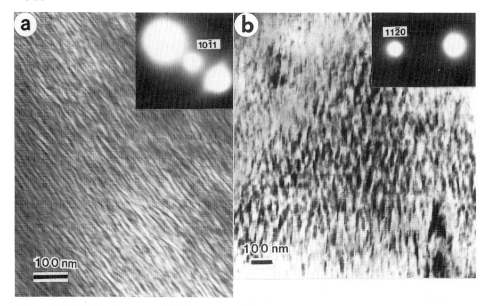

Figure 10-2. Modulated structures. (a) DF image ($\mathbf{g}$ = 10$\overline{1}$1) showing the fine modulation that is most typical of sedimentary calcian dolomite. (b) BF image of a slightly coarser modulation with sharp interfaces in calcian dolomite. From Reeder (1981).

Reeder and Wenk (1979) and Reeder (1981) interpreted the modulation as strain contrast arising from local compositional and/or ordering fluctuations. Because of its strong resemblence to spinodal microstructures, the possibility of a compositional fluctuation due to incipient exsolution was initially suggested. However, it later became apparent that the modulation formed directly during crystal growth (Reeder et al., 1984; Reeder and Prosky, 1986; Miser et al., 1987). This is demonstrated by the different orientation of the modulation within different growth sectors of a crystal and its correspondence to other growth features (see §10.5.2.1.3).

10.5.2.1.1.1 Orientation. Early studies reported the long dimension or "trend" of the modulation to be relatively close in orientation to {10$\overline{1}$4} (e.g., Reeder, 1981). The three-dimensional morphology of the modulation was unclear, but assumed to be approximately planar. Reeder and Prosky (1986) showed that the long dimension of the modulation more closely coincides with the growth normal. Therefore, for growth on a (10$\overline{1}$4) face, the trace of the modulation's long dimension is oriented approximately perpendicular to (10$\overline{1}$4). In different growth sectors of a crystal, for which growth occurred on different faces, the orientation of the modulation differs, but its long dimension is always parallel to the growth normal within any given sector (Fig. 10-3a).

Reksten (1990a) has undertaken the most extensive trace analysis of a modulated structure in calcian ankerite. She also found that the modulation trace is approximately perpendicular to {10$\overline{1}$4}. This is not surprising if one considers that {10$\overline{1}$4} is, by far, the most common form for dolomite. However, Reksten also observed that the streaking in SAED patterns showed a disk-like distribution in reciprocal space and suggested that the modulation in this calcian ankerite consisted of parallel, rod-like domains. Reksten's observations are important because previous workers had failed to consider morphology carefully. Other workers (Wenk et al., 1991) have found a platelet-like morphology for domains having an orientation parallel to {11$\overline{2}$0}. It now seems likely that different

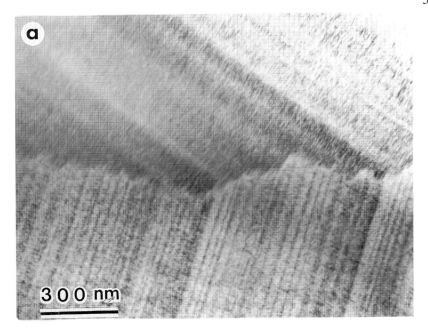

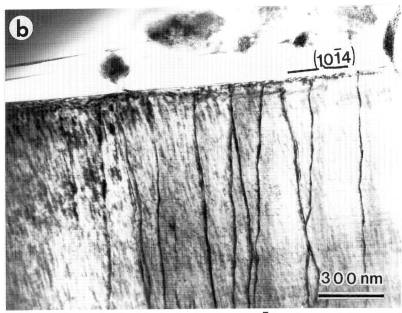

Figure 10-3. (a) BF image of calcian dolomite showing adjacent $\{10\bar{1}4\}$ sectors separated by a growth sector boundary (trending E-W). Fine-scale planar-concentric zoning indicates the orientations of the different faces on which growth occurred. A fine modulation occurs throughout both sectors, although its orientation differs between the sectors. The modulation is in strongest contrast in the sector at top, and its long dimension is parallel to the growth normal of that sector (i.e., toward the NE). In the bottom sector the modulation's long dimension is also parallel to the growth normal, which here is toward the ESE (from Reeder and Prosky, 1986). (b) BF image of calcian dolomite showing growth dislocations that are commonly associated with the modulation and the fine-scale zoning. Here, the dislocations terminate at a $(10\bar{1}4)$ crystal face. Note that the dislocations and the modulation's long dimension both are nearly parallel to the growth normal.

morphologies exist, and these are associated with growth in different directions (i.e., on nonequivalent faces).

10.5.2.1.1.2 Growth zoning. The modulation in calcian dolomite is usually associated with a fine growth banding (Figs. 10-3a and b), which is always perpendicular to it in orientation (or nearly so). Reeder and Prosky (1986) showed that this is simply planar-concentric zoning and probably reflects small compositional differences introduced during growth. Authors most commonly report orientations for the zoning parallel to $\{10\bar{1}4\}$, as might be expected since this is the dominant growth form of dolomite. Individual zones may be traced many tens of microns across an entire growth sector, and then followed into an adjacent sector (i.e., across growth sector boundaries), where its orientation corresponds to that particular face (Reeder and Prosky, 1986; Ward and Reeder, 1992). The growth zoning is typically irregular in wavelength ranging from 75 Å to greater than 1 µm. In some cases, it is remarkably repetitive, and may have an origin similar to oscillatory zoning in calcite (cf. Reeder et al., 1990).

10.5.2.1.1.3 Dislocation microstructures. Dislocations may be introduced either during crystal growth or subsequently as a result of deformation. Calcian dolomite and ankerite characteristically show a distinctive *growth* dislocation microstructure. It is well known that dislocations may be closely associated with certain growth processes. For example, the spiral growth mechanism (Burton et al., 1951) relies on the emergence of a screw dislocation (or a dislocation having a screw component) at the crystal surface where a step originates. Preferential attachment of growth units along the step causes the step to "spiral" around the dislocation allowing the dislocation and the surface to advance. However, the site of dislocation emergence need not be an active growth center, and some dislocations are simply extended as the surface advances during growth. Klapper (1980) has shown that such dislocations tend to be linear or sublinear and are typically oriented so as to lie within approximately 15° of the growth normal of a particular sector. Thus, if present, growth dislocations should exhibit systematic orientations with respect to other growth-related microstructures (Ward and Reeder, 1992).

Dislocations of the type described are quite common in calcian dolomite (Fig. 10-3b). In a given growth sector, they are oriented approximately parallel to the trace of the modulated structure and approximately perpendicular to the fine growth zoning. They may have rather high densities, and it is not clear whether any were active growth centers.

10.5.2.1.2 Electron diffraction observations. Selected-area electron diffraction (SAED) patterns from modulated calcian dolomite characteristically exhibit streaking of fundamental (a-type) and ordering (b-type) spots normal to the modulation trace (also see §10.5.2.1.1.1). Streaking is generally diffuse and no satellites have been observed that correspond to the modulation wavelength.

Many (but apparently not all) SAEDs of modulated calcian dolomite contain very weak extra reflections that are inconsistent with the dolomite space group ($R\bar{3}$). These are of the two types described earlier: c- and d-type reflections.

10.5.2.1.2.1 c-type reflections. Reeder and Wenk (1979) first described c reflections from calcian dolomite exhibiting modulated structures. Since then, nearly all authors have associated the occurrence of c reflections in calcian dolomite with the modulated structure (or the ribbon structure described later). Dark-field imaging of c reflections soon revealed that in fact they arise from narrow, elongate domains (Fig. 10-4), whose orientation generally parallels the modulation's long dimension (Wenk and Zhang,

Figure 10-4. DF image of c-type reflection in calcian ankerite showing that extra diffractions are associated with narrow domains that parallel the modulation. From Reksten (1990a).

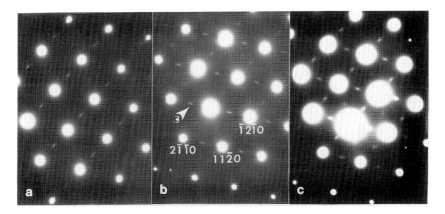

Figure 10-5. [0001] zone axis SAEDs of calcian ankerite showing c reflections in (a) one, (b) two, and (c) three orientation variants. The c reflections are elongated in orientations perpendicular to the domains. From Wenk et al. (1991).

1985; Van Tendeloo et al., 1985; Reksten, 1990b; Wenk et al., 1991). Thus, structurally distinct domains occur within a dolomite host. The association of the domains with the modulation suggests that they are Ca rich relative to the surrounding dolomite. In fact, this is essentially confirmed by evidence described below (see §10.5.4.1).

The c reflections are usually very weak, and have not been observed by single-crystal x-ray diffraction. Several authors have confirmed that they do not arise from dynamical effects. The weak intensity means that long film exposures are generally required to detect their presence. When observed, c reflections occur at positions in reciprocal space exactly midway between spots (both a- and b-type) in systematic rows having one or more of three possible equivalent orientations (Fig. 10-5). Depending on the observed number of equivalent orientation variants, the symmetry in the diffraction pattern may remain rhombohedral (all three orientations) or be lowered to monoclinic (one or two orientations).

There has been some discussion concerning allowed positions for c reflections and the question of equivalent orientations. Early findings (e.g., Reeder and Wenk 1979; Reeder, 1981; Wenk et al., 1983) indicated that c reflections occurred in only one orientation, which destroys the rhombohedral symmetry. A simple way to depict their positions in reciprocal space is by translating normal dolomite spots by one-half the reciprocal spacing along *one* of the three $\langle 11\overline{2}0\rangle^*$ directions (see Fig. 10-5a). This effectively doubles one **a** axis. Reksten (1990b) observed c reflections occurring in all three orientation variants, and Wenk et al. (1991) observed examples of all cases – one, two, and three variants (Figs. 10-5a-c). Where more than one orientation variant is observed, intensities of c spots among the different variants are not always equal. So that even when three orientation variants are present, the rhombohedral symmetry may not be strictly preserved.

HRTEM has shown that within a given domain, doubling of only one **a** dimension occurs (Fig. 10-6), so that the corresponding c reflections occur in a single orientation variant (Wenk et al., 1991). For diffraction patterns showing more than one variant, Wenk et al. argue that domains having *different* **a** axes doubled are averaged by the beam. Discussion of the ordering pattern that causes these additional reflections is deferred to §10.5.4, following a description of extra reflections in calcite and isostructural carbonates.

10.5.2.1.2.2 *d*-type reflections. At the time of this review, d-type reflections have been documented in only a few dolomite samples (Wenk and Zenger, 1983; Reksten, 1990b; Wenk, pers. comm.), and so must be considered relatively rare in comparison to c reflections. d reflections occur as satellites around a- and b-type dolomite reflections in the [0001]* direction (Fig. 10-7a). Wenk and Zenger (1983) observed the satellite spacing to be one-third the reciprocal spacing along [0001]*, or equivalent to approximately 16 Å in real space. The d reflections first described by Wenk and Zenger (cf. the more complete descriptions of Wenk and Zhang, 1985, and Van Tendeloo et al., 1985) are streaked along [0001]*, and their presence is directly associated with narrow, elongate domains approximately parallel to the long dimension of the modulation on

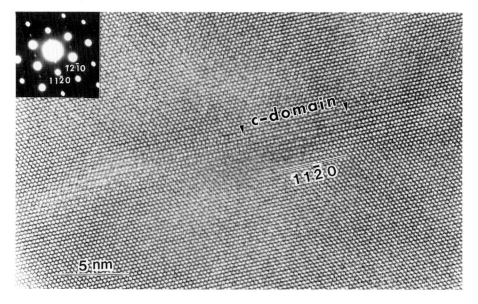

Figure 10-6. Multibeam [0001] zone axis HRTEM image showing c domain, which is coherent with surrounding ankerite. Scherzer focus. From Wenk et al. (1991).

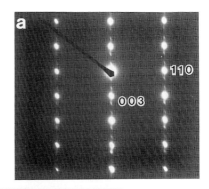

Figure 10-7. (a) SAED showing *d* satellite reflections in calcian dolomite. (b) BF image of same dolomite showing a high density of platelet-like domains parallel to (0001). These domains are arranged in bands parallel to (10$\overline{1}$4). (c) [0001] SAED of calcian ankerite showing both *c*- and *d*-type reflections. Figures 10-7a and b courtesy of H.-R. Wenk; 10-7c from Reksten (1990a).

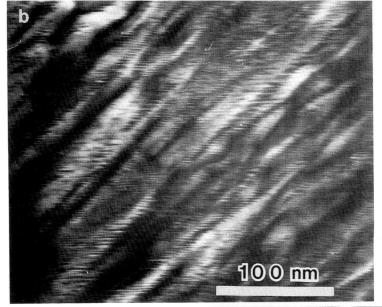

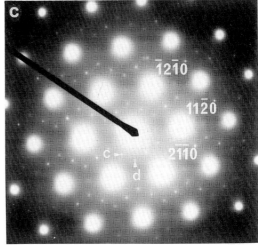

$(10\bar{1}4)$ (Fig. 10-7b). In addition to being streaked along $[0001]^*$, they are slightly elongated normal to the modulation trace, because of the orientation and shape of the domains (Fig. 10-8). In cases where both c- and d-type reflections are present, complex SAED patterns may result (Fig. 10-7c).

Based on HRTEM, Van Tendeloo et al. (1985) have interpreted the d reflections as indicating a stacking defect on $000l$ involving periodic substitution of Ca layers for Mg layers. Thus, the domains are viewed as being Ca rich. It is tempting to suggest that the host dolomite is approximately stoichiometric, but confirmation is lacking. Detailed description of the d superstructure is presented in §10.5.4.2.

10.5.2.1.3 Relationship between microstructural features. The association of the c domains with the fine modulation suggests a causal relationship. It is likely that some misfit would occur between compositionally-distinct, ordered domains and host dolomite. It is possible that the modulations represent the quasi-periodic strain field due to such misfit. HRTEM images of the c domains (see Fig. 10-6) indicate that they are strictly coherent with host dolomite, and they suggest no other disruption of the structure that might cause the modulation.

It is worth emphasizing the self consistency displayed among the various microstructural features in relation to growth, and particularly the orientational aspects. In crystals that exhibit well-defined growth sectors, the fine-scale zoning and growth dislocations faithfully record the advances of the different crystal faces. The orientation of the modulation is parallel to the growth normal within each sector, so that at growth sector boundaries, its orientation changes abruptly (Fig. 10-3a). In nonequivalent growth sectors that are compositionally different (i.e., sector zoned), the detailed appearance of the modulation differs (Reeder and Prosky, 1986; Fouke and Reeder, 1992).

The common association of these particular microstructures suggests that specific processes allow very regular heterogeneities to be introduced during growth. These processes must operate with some regularity, both because of the pervasive nature of the microstructures and because of their occurrence in so many dolomite samples.

10.5.2.1.4 Ribbon microstructure. Reeder and Barber (1982) and Barber et al. (1985) found that Ca-rich saddle dolomite—a variety of dolomite having curved crystal faces—characteristically contains curving, ribbon-like defects (Fig. 10-9). The alternating dark and light fringes exhibit complementary contrast in BF and DF and resemble those typical of stacking faults. However, unlike stacking faults, ribbon defects bifurcate into two of the same defects or terminate without any sign of a dislocation. Their curving, ribbon-like morphology indicates that they are not strictly confined to a unique crystallographic plane. Furthermore, Barber et al. (1985) found that the symmetry of the fringe pattern is not consistent with α-fringes (stacking faults).

In any given region, or possibly in a given growth sector within a crystal, ribbons tend to have a uniform overall orientation. Branching occurs preferentially in one direction, suggesting a relationship with growth direction. However, Barber and Khan (1987) suggested that branching ribbons may form a three-dimensional network within the host dolomite.

Ribbons apparently have a more variable occurrence than the modulated structure. They are quite often not pervasive in a specimen, so that ribbon-containing areas may be situated next to ribbon-free regions. Nor are ribbon microstructures restricted to dolomite having

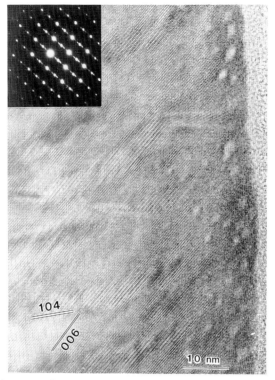

Figure 10-8. HRTEM image of *d* domains in calcian dolomite (same sample as in Fig. 10-7a,b). From Van Tendeloo et al. (1985).

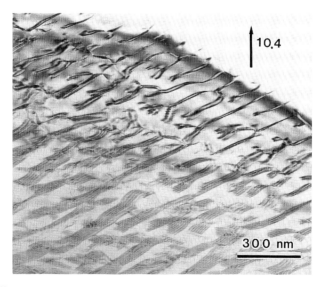

Figure 10-9. BF image showing fringed, ribbon microstructure in calcium-rich saddle dolomite. Note the branching of the defects toward the lower left. Number of fringes increases as thickness of foil increases toward lower left.

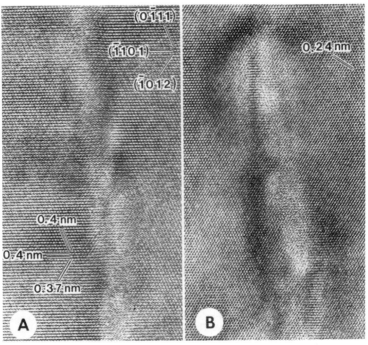

Figure 10-10. HRTEM images of ribbons viewed "edge-on" in saddle dolomite. Although the lattice fringes sometimes bend in the neighborhood of the ribbon, the ribbons are generally coherent with the host. From Barber et al. (1985).

saddle morphology. Barber and Khan (1987) characterized ribbon microstructures in ankerite and kutnahorite. Similar ribbon microstructures were also observed by Barber and Wenk (1984) in carbonatite dolomite.

The ribbons have been interpreted as very thin, coherent lamellae of Ca-rich second-phase material. Barber et al. (1985) give several lines of evidence in support of this. HRTEM images (Fig. 10-10) confirm that the material comprising the ribbons is coherent with and structurally similar to the host dolomite. Electron diffraction patterns of areas containing high densities of ribbons show slight splitting of spots, which becomes larger for higher-order reflections. DF images of higher-order split spots show that ribbon material is associated with the spot type having the smaller diffraction vector of the two (i.e., having the larger unit cell), as would be expected for a more Ca-rich carbonate. Barber et al. (1985) and Barber and Khan (1987) also carried out x-ray microanalysis which supports a Ca-rich composition for the ribbon. Owing to the very narrow width of the ribbon (just a few lattice spacings), they were unable to analyze the ribbon material uniquely (spot size is approximately 200 Å on the specimen). However, using a defocused beam they found that regions of crystal containing high densities of ribbons are characteristically Ca rich, while those regions with no ribbons are essentially stoichiometric dolomite. These observations strongly suggest that the ribbon material is calcite-like in composition. King et al. (1984) modeled the contrast resulting from a thin slab of calcite in dolomite, and found that it agreed with observations.

The c-type reflections have also been observed in specimens containing ribbon microstructure, and DF images indicate they arise from narrow domains (Fig. 10-11) that either parallel the ribbons or, more likely, coincide with them (Wenk et al., 1983).

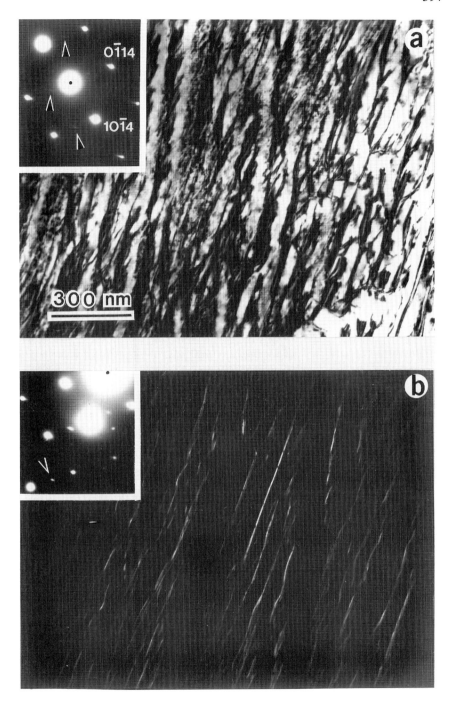

Figure 10-11. (a) BF image of saddle dolomite showing ribbon microstructure, and (b) the corresponding DF image of a *c* reflection illustrating the narrow, elongate domains. From Wenk et al. (1983).

It is not clear whether (and in what way) ribbon microstructures are related to modulated structures. It is clear from the foregoing discussions that they share many similarities. However, they apparently do not exist together in the same samples, which may indicate that environmental factors determine their occurrence. Barber et al. (1985) noted that all saddle dolomite they examined contained ribbon microstructures. Their void-filling saddle dolomite probably formed at temperatures in the range 80-150°C. Although there are no conclusive data to support the idea, one might *speculate* that ribbon microstructures tend to dominate at these slightly higher temperatures, whereas modulated structures may be more characteristic of lower temperature conditions. No doubt fluid chemistry is also a factor, but one which presently does not lend itself so easily to speculation!

10.5.2.2 Holocene dolomite

In §10.5.2, Holocene dolomite, although virtually all Ca rich, was distinguished from "ancient" calcian dolomite primarily on the basis of XRD evidence. Much Holocene dolomite exhibits a distinctive microstructure not found in ancient counterparts. A few also show variable development of the fine modulated structure that is common in ancient samples.

Reeder (1981) examined well-known occurrences of Holocene dolomite from Deep Springs Lake, Ca, the Persian Gulf, and the Coorong, Australia, and noted a lack of the fine modulations. Instead, these dolomite samples contained defects in densities sufficiently high that the exact nature of the defects could not be determined in diffraction contrast images (Fig. 10-12). In many cases, they resemble dislocation tangles in which overlapping strain fields produce strong mottled contrast. Blake and Peacor (1985) emphasized the mosaic character of Holocene dolomite and highly magnesian calcite forming as fresh-water cement. Sub-micrometer domains or crystallites exhibit very slight misorientation. Such nearly topotactic mosaics may represent multiple heterogeneous nucleation.

Figure 10-12. DF image of individual Holocene dolomite crystal within an aggregate. Note the strong, mottled contrast due to a high density of indistinguishable defects (probably dislocations).

Subsequent studies of other samples have generally confirmed these observations, but have also shown that some Holocene dolomite exhibits modulated structure (Carballo et al., 1987; Rosen et al., 1988, 1989; Wenk et al., 1992). In one sample clearly shown to be forming in marine waters, Mitchell et al. (1987) documented Ca-rich dolomite containing pervasive modulations as well as weak c-type reflections. This important observation serves to demonstrate that their young age is not the characteristic that distinguishes these Ca-rich dolomite samples from their ancient counterparts; rather it is the conditions or mechanism of their formation.

Several authors have pointed out that the high density of defects which seems to be typical of much Holocene dolomite (as in Fig. 10-12) must contribute significantly to their instability. This may partly account for the apparent lack of such microstructures in dolomite appreciably older than Holocene.

Wenk et al. (1992) described small domains having dolomite structure within a highly magnesian calcite (bulk composition $Ca_{0.7}Mg_{0.3}CO_3$) from sabkha sediments from Abu Dhabi. HRTEM images (Fig. 10-13) show spherical domains, 20-100 Å in diameter, which are coherent with the host magnesian calcite. Domains show dolomite ordering even though the bulk sample fails to show any ordering reflections.

10.5.2.3 Stoichiometric dolomite (and ankerite)

After considering the variety and complexity of the *real* structures of Ca-rich dolomite, the reader may be disappointed by stoichiometric dolomite. The vast majority of dolomite samples with $CaCO_3$ contents close to 50 mol % show predominantly homogeneous microstructures (e.g., Fig. 10-14). The most common defects are dislocations, which may be present with widely variable densities. Occasional stoichiometric dolomite samples, or those with only slight calcium excess, show regions of ribbon or modulated microstructure; but these regions are typically localized and volumetrically insignificant. Miser et al. (1987) report widely-spaced, weak modulations in a stoichiometric dolomite. However, such examples seem to be the exception (cf. Barber and Khan, 1987; Khan and Barber, 1990; Ward and Reeder, 1992).

The situation is similar for ankerite (and other carbonates isotructural with dolomite). Crystals that are essentially stoichiometric with respect to Ca show homogeneous microstructures, whereas Ca-rich samples exhibit either modulated or ribbon micro-structures (cf. Barber and Khan, 1987; Reeder and Dollase, 1989; Khan and Barber, 1990).

10.5.2.4 Relation between excess Ca and microstructures

Current opinion seems to support a correlation between the occurrence of modulated or ribbon microstructures and excess Ca in dolomite. After considering a large body of work, Khan and Barber (1990, p. 239) regarded "...the presence of calcium in excess of stoichiometric requirements as being the key factor in regard to modulated [and ribbon] microstructures in perhaps *all* dolomites.". These authors have in fact provided some of the most convincing support for their conclusion. In crystals where modulated or ribbon microstructure are intermixed with homogeneous regions, Khan and Barber have shown, using analytical electron microscopy, the former areas to be Ca-rich and the latter to be essentially stoichiometric.

Moreover, Ca-rich domains coherent with host dolomite have been identified, as

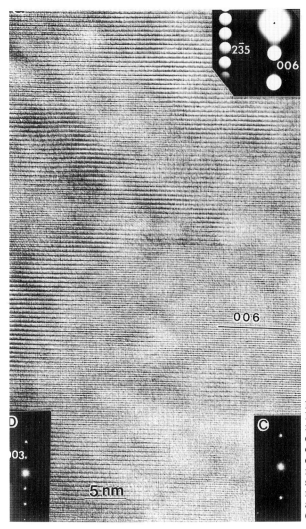

Figure 10-13. HRTEM image of Recent Ca-Mg carbonate sediment from Abu Dhabi sabkha. Spherical domains (within host magnesian calcite) show fringes characteristic of dolomite ordering. Inset optical diffraction patterns of image (at bottom) show ordering reflections for dolomite-like domains (D), but only fundamental calcite reflections for host (C). From Wenk et al. (in press).

described in previous sections. The picture that emerges is one in which at least some excess calcium is preferentially segregated into domains rather than substituting randomly within the normal dolomite structure. These domains are ordered so that rhombohedral symmetry is violated, and with different superstructures and apparently degrees of ordering possible.

This has very important implications for geochemists concerned with relationships between Ca and Mg activities in aqueous solutions and corresponding values in natural dolomite. Later in this chapter, these relationships are considered further in the context of crystal growth.

10.5.3 Calcite

Low-temperature calcite may also show heterogenous, apparently composition-related microstructures, and some quite similar to those in dolomite. This may seem rather more

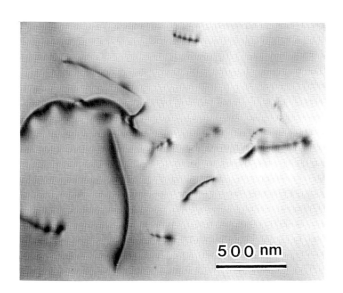

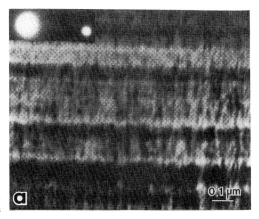

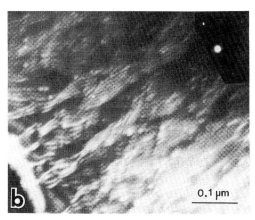

Figure 10-14 (above). BF image of stoichiometric sedimentary dolomite showing dislocations in an otherwise homogeneous microstructure.

Figure 10-15 (right). (a) BF image showing modulation and growth zoning in calcite. (b) DF image ($g = 10\bar{1}4$) of ribbon microstructure in calcite. From Gunderson and Wenk (1981).

surprising than for dolomite. At least for the case of dolómite, incorporation of excess calcium requires accommodating the relatively large Ca^{2+} ion either in the smaller B (Mg) site or, as it prefers, in a domain having a different structure. Substitution in calcite predominantly involves smaller ions (Sr^{2+} is an important-exception), which we would expect to be more easily accommodated in the Ca site. Nevertheless, some impure calcite exhibits domains having structures different than that of calcite and ordering different than that of dolomite. Before giving the impression of their being widespread and common, it is important to point out that calcite samples containing such heterogeneous microstructures are apparently not nearly as common as those lacking them. But since fewer observations have been made on calcite than dolomite, this picture might conceivably change in the future.

10.5.3.1 Modulated structure in calcite

Gunderson and Wenk (1981) were the first to identify a modulated structure in calcite. This is a fine-scale modulation (Fig. 10-15a) having an appearance somewhat similar to the modulation in calcian dolomite. However, the authors noted that the modulations commonly showed greater heterogeneity and weaker contrast than in dolomite. Irregular growth zoning may accompany the modulations, and, as in calcian dolomite, the zoning is approximately perpendicular to the modulation's long dimension. Gunderson and Wenk reported an orientation for the modulation approximately parallel to $\{10\bar{1}4\}$. It is likely that the modulation's long dimension closely parallels the growth direction, since it is normal to the growth zoning. Consequently different orientations can be expected.

Inasmuch as the calcite was presumed to have only limited cation substitution, initial explanations for the occurrence of modulations focused on aspects of the CO_3 lattice. Defects in the CO_3 lattice are difficult to assess by TEM, so that little support for this idea has been realized. More recent work has considered the minor and trace element distributions in calcite exhibiting modulations. Frisia-Bruni and Wenk (1985) documented well-developed modulations in calcite existing both as replacement of aragonite and as void-filling cement. They drew attention to the presence of minor Fe and Mg substitution in calcite exhibiting modulations, and suggested that these elements might be concentrated in ordered domains, not unlike the situation in calcian dolomite. Wenk et al. (1991) also observed modulated structure in void-filling calcite containing 5 mol % $MgCO_3$. Further support for the correlation of heterogeneous microstructures in calcite with impurity content is provided by ribbon microstructures.

10.5.3.2 Ribbon microstructure

Ribbon microstructures (Fig. 10-15b) have also been documented in several calcite samples (Gunderson and Wenk, 1981; Wenk and Zhang, 1985; Barber and Khan, 1987; Khan and Barber, 1990), and they exhibit essentially identical diffraction contrast to those in calcian dolomite. Ribbons have also been reported by Barber and Khan (1987) in siderite ($FeCO_3$) and smithsonite ($ZnCO_3$), which are isostructural with calcite. Barber and Khan's work, which included analytical electron microscopy, has been the most enlightening. Whether in calcite or an isostructural carbonate, the presence of the ribbon microstructure is clearly related to minor element substitution. In many crystals, regions of both ribbon and homogeneous microstructure are present. Energy-dispersive x-ray microanalysis shows that regions comprised of ribbon-plus-host uniformly yield higher impurity concentrations than regions lacking ribbons. This has been shown quite clearly for ferroan calcite (Fig. 10-16a) and for magnesian calcite (Fig. 10-16b), and elsewhere for calcian siderite and calcian smithsonite (Barber and Khan, 1987; Khan and Barber, 1990).

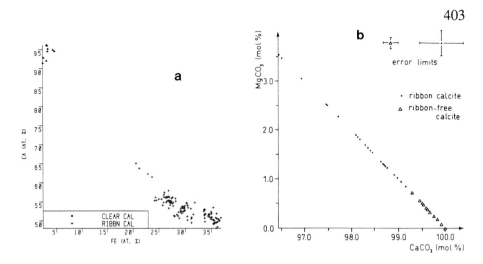

Figure 10-16. (a) AEM data from ferroan calcite showing that ribbon microstructure is associated with regions of higher Fe concentrations (from Barber and Khan, 1987). (b) AEM data from Mg- and Zn-bearing calcite showing that the occurrence of ribbon microstructure is associated with higher Mg conentrations (from Khan and Barber, 1990).

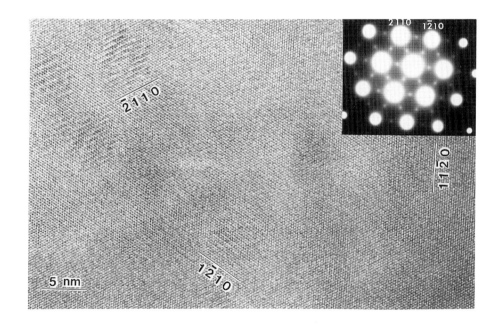

Figure 10-17. [0001] zone axis HRTEM image showing small c domains within host calcite. Three different orientation variants are observed for the domains as labeled. Inset SAED shows c reflections in three equivalent orientations. From Wenk et al. (1991).

404

10.5.3.3 c-type reflections

c-type reflections are also observed in some calcite samples (and some isostructural carbonates) that exhibit heterogeneous microstructures. Dark-field imaging of *c* reflections indicates that they arise from small domains (several nm to hundreds of nm), which are of variable morphology and distribution (Barber and Khan, 1987; Reksten, 1990c). Just as with calcian dolomite, different orientation variants have been reported for *c* reflections in calcite. Barber and Khan (1987) and Reksten (1990c) report *c* reflections at positions in all three equivalent orientations. Wenk et al. (1991) also observed *c* reflections in similar positions for calcite containing 5 mol % $MgCO_3$. However, HRTEM shows 100 Å domains in which the *orientation* of the superstructure differs among the different domains (Fig. 10-17). Wenk et al. (1991) argue that SAED patterns in which *c* reflections appear in all three orientation variants are probably an averaging of *c* domains in which the superstructures have these different orientations. This conclusion is similar to the case for dolomite. Details of the superstructure are considered in §10.5.4.

10.5.3.4 Impurity substitution and heterogeneous microstructures in calcite and isostructural carbonates

These observations from carbonate minerals having the calcite structure support a conclusion not unlike that for excess Ca in dolomite. Substitution of impurity atoms in calcite (or an isostructural carbonate) is obviously not always homogeneous and random. In the cases discussed, at least some impurity atoms are preferentially segregated into domains that are compositionally distinct from the host and exhibit cation ordering not characteristic of either calcite or dolomite. These findings bear on many aspects of the trace and minor element geochemistry of carbonates and the fluids from which they crystallize.

10.5.4 Superstructures in dolomite and calcite

It is appropriate here to return to the details of the superstructures indicated by the presence of *c*- and *d*-type reflections in electron diffraction patterns. The evidence discussed thus far has pointed to a localized distribution as domains that differ in composition from the host. The domains are coherent with the host, suggesting some fundamental structural similarities. Current models for both *c*- and *d*-type superstructures consider only cation distributions. This is certainly reasonable, although slight distortions within the CO_3 lattice that are associated with the different cation distributions can also be expected. The primary focus on cation ordering may also reflect the limitations of current HRTEMs, which are able to resolve details of the cation distribution but not of the CO_3 lattice. Consequently, little can be concluded about the latter at present.

10.5.4.1 c superstructure

10.5.4.1.1 Calcian dolomite. The first models for the *c* superstructure were based simply on observed positions of *c* reflections and a presumed doubling of an **a** axis (cf. Reeder and Wenk, 1979; Wenk et al., 1983). For *c* domains in calcian dolomite, Van Tendeloo et al. (1985) and Wenk and Zhang (1985) suggested a superstructure in which basal Mg layers contain equal amounts of Ca and Mg, which are ordered into alternating rows parallel to *one* dolomite **a** axis (Fig. 10-18a). Ca basal layers are essentially undisturbed, so that the ordered Ca-Mg layers alternate with Ca layers. Such a superstructure destroys the rhombohedral symmetry and has an ideal composition $Ca_{0.75}Mg_{0.25}CO_3$. Wenk and co-workers refer to this superstructure as γ-dolomite.

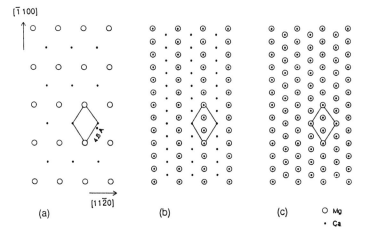

[$\bar{1}$ 100]

[11$\bar{2}$0]

(a)   (b)   (c)

○ Mg
• Ca

Figure 10-18. Schematic [0001] views of Ca and Mg distributions: (a) as proposed by Van Tendeloo et al. (1985) and Wenk and Zhang (1985) for the *c* superstructure in a single B layer of calcian dolomite; (b) projection of the cation distribution in the B layer in (a) as it alternates with A (Ca) layers. Circles with dots in centers represent columns of alternating Ca and Mg atoms. Note that columns containing Ca and Mg atoms alternate with columns of only Ca atoms in rows perpendicular to one a axis; (c) projection through A and B layers in ideal dolomite. All columns of cations are identical. The hexagonal unit cell of ideal dolomite is shown for reference. From Wenk et al. (1991).

Reksten (1990b) argued for a variation of this superstructure for *c* domains in calcian ankerite. In her model, Ca and Mg ordering occurs within a layer with the same pattern described above, but within *all* basal layers. Consequently, there is no difference between successive cation layers, and the ideal composition (neglecting Fe) is $Ca_{0.5}Mg_{0.5}CO_3$. This model contrasts with that of Van Tendeloo et al., as well as with the basic dolomite structure in which A and B layers remain distinct and alternate along [0001]. Reksten's proposed sequence of stacking identical ordered cation layers causes like cations to lie on one of the equivalent $\{10\bar{1}4\}$ planes (Fig. 10-19a,b). Neglecting any distortion in the CO_3 lattice, Reksten suggested the space group $P2/c$. Because of the 1:1 stoichiometry of this model, excess Ca would not presumably be accounted for within the domains, and it remains unclear how excess calcium would be accommodated.

HRTEM imaging of *c* domains in calcian ankerite by Wenk et al. (1991) strongly support the *c* superstructure model for dolomite of Van Tendeloo et al. Figure 10-20 shows a [0001] zone axis image of a coherent *c* domain in host ankerite. Within the *c* domain, the image shows alternating rows of dull and bright spots, which correspond to columns of Ca atoms and columns of alternating Ca and Mg atoms. Agreement with calculated multibeam simulation based on the superstructure model in Figure 10-18a and b is very good (cf. the simulation in Fig. 10-20 for ideal dolomite with Fig. 10-18c). This adds further support to prior suggestions that *c* domains are Ca-rich, possibly having a composition near $Ca_{0.75}Mg_{0.25}CO_3$.

10.5.4.1.2 Calcite. For calcite and isostructural carbonates exhibiting *c* reflections, Khan and Barber (1990) independently proposed the same superstructure that Reksten did for calcian ankerite, in which impurity atoms (Mg, Fe, Mn, etc.) are ordered along rows alternating with Ca. This model has an ideal 1:1 stoichiometry. In a separate paper, Reksten (1990c), having also observed *c* reflections in calcite, proposed two possible *c* superstructures for calcite. Reksten's Model I (Fig. 10-19a,b) is the same as the one that she described for calcian ankerite, and also the same as Khan and Barber's model. This model is consistent with *c* reflections occurring in one orientation variant. In her

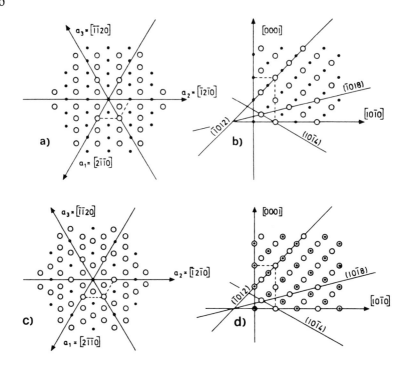

Figure 10-19. Schematic illustrations showing cation distributions for c superstructure models in calcite. (a) and (b) are projections parallel and perpendicular to c that correspond to Khan and Barber's model and Reksten's Model I (see text). Open circles are Ca atoms and small dots are substituting cations. Note that the rhombohedral symmetry is destroyed in this model. (c) and (d) are similar projections for Reksten's Model II in which rhombohedral symmetry is retained. Circles with dots in centers represent columns of Ca alternating with a substituting cation. From Reksten (1990c).

Model II (Fig. 10-19c,d), ordering of Ca and impurity atoms occurs within a basal layer so as to retain the three-fold rhombohedral symmetry. The cation stoichiometry within a basal layer is $Ca_{0.75}:M_{0.25}$ (M represents the impurity atom), and because each layer is identical, the stoichiometry of the supersturcture is 3:1. Reksten's Model II is consistent with c reflections occurring in all three orientation variants.

The HRTEM image of Mg-bearing calcite of Wenk et al. (Fig. 10-17) shows c domains having different orientation variants, but in which only one **a** axis is doubled in each. This is consistent with the 1:1 superstructure suggested by Khan and Barber (1990) and by Reksten (1990c) as her Model I. If one assumes for other calcite samples that c reflections appearing in all orientation variants actually result from an averaging of differently oriented domains, then this superstructure model would seem to be appropriate.

Wenk et al. (1991) have summarized the different superstructure models in relation to the basic calcite and dolomite structures (Fig. 10-21), and have assigned greek letters to distinguish them. These models correctly represent the idealized superstructures described by the various authors, although a few minor inconsistencies relating to Reksten's models occur in Wenk et al.'s description. In Figure 10-21, the α and β structures correspond to calcite and dolomite. The ν superstructure corresponds to the model proposed independently by Khan and Barber (1990) and by Reksten, as her Model I (Reksten, 1990c).

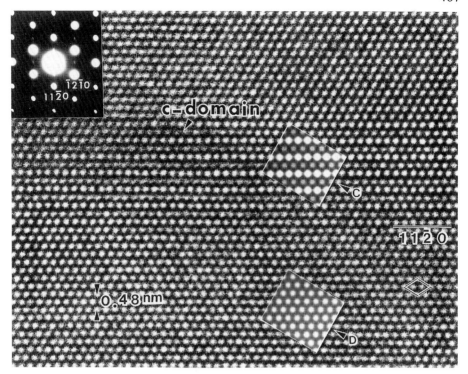

Figure 10-20. Comparison of calculated multibeam simulations with experimental images of *c* superstructure and host ankerite for the calcian ankerite shown in Figure 10-6. [0001] zone axis orientation. Calculated simulation at "c" corresponds to the superstructure model proposed by Van Tendeloo et al. (1985) and Wenk and Zhang (1985). The simulation at "d" is for ideal dolomite. An hexagonal unit cell is shown for reference. From Wenk et al. (1991).

[Wenk et al. identified Reksten's Model I with their μ superstructure.] γ is Van Tendeloo et al.'s superstructure. Both ν and γ superstructures have one **a** axis doubled so that they have been described as monoclinic. μ is Reksten's Model II superstructure in which all **a** axes are doubled and the rhombohedral symmetry is retained. (Wenk et al. inadvertently compared Reksten's Model II with the γ superstructure.)

10.5.4.2 d *superstructure*

Wenk and Zhang (1985) and Van Tendeloo et al. (1985) proposed a model for the *d* superstructure based on a periodic mistake in the Ca-Mg-Ca-Mg basal stacking sequence of dolomite. In their model, every third Mg layer is replaced by a Ca layer producing the sequence Ca-Mg-Ca-Ca-Ca-Mg-Ca (within a 16 Å repeat). This superstructure has an ideal stoichiometry of $Ca_{0.67}Mg_{0.33}CO_3$. It corresponds to the δ superstructure in Figure 10-21. Agreement of HRTEM images with calculated multibeam simulations (Meike et al., 1988) confirm the general correctness of this model. However, SAED patterns show streaking of *d* satellites, which probably indicates that some non-periodic stacking disorder also occurs.

It is very important to realize that the various models described for the *c* and *d* superstructures are intended to represent ideal superstructures. In reality, the ordering and

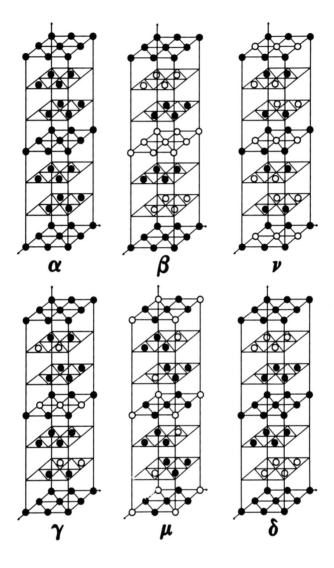

Figure 10-21. Schematic models summarizing the idealized cation distributions for various structures and superstructures of Ca-Mg carbonates. Equivalent hexagonal cells are shown with the **c** axis (vertical) corresponding to that for calcite (and dolomite). **a** axes have been doubled relative to the correct cell for calcite (α) and dolomite (β). Solid circles represent Ca and open circles Mg. See text for explanation. From Wenk et al. (1991).

stoichiometries are likely to be imperfect, which may account, in part, for variable intensities of c reflections and their apparent absence is other cases.

The existence of ordered superstructures poses interesting questions regarding their relative stabilities. Theoretical modeling of the type described by Burton (1987) for *intralayer* ordering may provide much needed insight. The superstructures present in Ca-Mg carbonates are almost certainly metastable with respect to the assemblage calcite + dolomite. Nevertheless, the formation of ordered superstructures in compositionally-distinct domains during growth has important implications, which need to be considered more carefully by carbonate petrologists and geochemists.

10.5.5 Orthorhombic carbonates

Aragonite has been examined primarily in the contexts of marine precipitates, replacement (by calcite), and as the skeletal material secreted by invertebrates. The latter topic is addressed in the following section. Gunderson and Wenk (1981) examined aragonite in several Pleistocene ooids. Multiple twinning on {110} is characteristic of the needle-like aragonite crystals, but no other microstructures were identified. There have been similar findings in the few large single crystals of aragonite that have been examined.

10.5.6 Biogenic carbonates

Calcite and aragonite are the most common carbonate minerals secreted as the hard parts of organisms, the greatest volume of which are marine invertebrates. Depending on the organism, secreted calcite can have a wide range of Mg contents: from essentially Mg-free $CaCO_3$ to greater than 25 mol % $MgCO_3$ in solid solution. Aspects such as crystal morphology, chemistry, and defects are controlled by the organism's mechanism and environment of mineralization. Microstructures introduced during biogenic crystallization have been examined in several different carbonates, although truly systematic studies have only recently been undertaken. The mechanisms by which biomineralization occurs have received considerable attention. Notable are the studies of Mann and co-workers concerning biomineralization and molecular recognition. These workers have used TEM to great advantage for characterizing carbonate crystallization on organic templates (e.g., Heywood et al., 1991). It appears that much could be gained from further TEM studies of biomineralization, including insight into mechanisms of crystallization as well as structure and morphology of skeletal material.

10.5.6.1 Aragonite

The most characteristic microstructure of biogenic aragonite is multiple twinning on {110}. Frisia-Bruni and Wenk (1985) examined aragonite secreted by calcareous sponge, coral, and gastropod, all of which showed multiple {110} twinning. Rare stacking faults and dislocations were observed in sponge aragonite. Unpublished work by the author on pelecypod (and gastropod) aragonite also reveals that twinning on {110} is ubiquitous (Fig. 10-22). Dislocations are rare. Some biogenic aragonite also contains small pores. In pelecypod aragonite, these are commonly located along the central portion of the rod-like crystals (Fig. 10-22a), but many also occur between crystals (Fig. 10-22b).

Frisia-Bruni and Wenk (1985) cautioned that {110} twinning can not be considered diagnostic of *biogenic* aragonite, because inorganic aragonite is also commonly twinned on {110} (see §10.5.5).

10.5.6.2 Calcite

Most TEM work on biogenic calcite has focused on modern echinoderm fragments. Most of these occur as large single crystals in the form of spines, plates, or columnals. Magnesium content of samples examined range from 6 to 13 mol % $MgCO_3$. Blake and Peacor (1981) and Blake et al. (1984) studied a modern deep-water crinoid containing approximately 11 mol % $MgCO_3$. The microstructure was largely homogenous, although they observed a mosaic structure as evidenced by slight misorientations between different

410

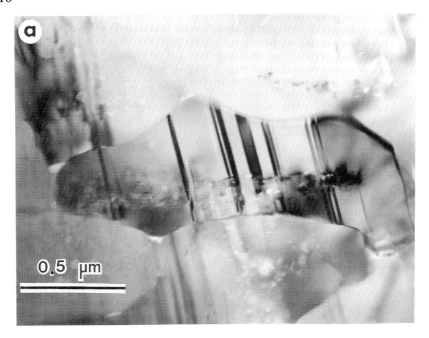

Figure 10-22. BF images of aragonite crystals comprising the skeletal architecture of unaltered pelecypod shell. (a) Multiple {110} twinning is evident in the crystal in strong contrast. Small pores occur throughout the interiors of the crystals. (b) Crystals commonly show imperfect preferred orientation, indicated here by the inset SAED pattern as well as the similar orientation of diffuse twin lamellae. Note the many small pores between and within crystals.

regions of crystal. Lattice imaging indicated coherence between regions and no defects accompanied the misorientation.

Shallow-water echinoderms show considerably more variety (Reeder, unpublished). The most distinctive features include locally high densities of dislocations and microvoids. A specimen of echinoid spine (*Lytechinus variegatus*) containing 6 mol % $MgCO_3$ from Bermuda is shown in Figure 10-23. Microstructures are highly variable within a single crystal. Homogeneous regions randomly grade into areas exhibiting strongly mottled contrast. Some of the mottled appearance is probably due to beam damage, but in other regions high dislocation densities are evident. Small voids ranging in size from 10 to 100 nm are locally common, and presumably contained fluids or organic material. The high densities of defects may contribute to the enhanced solubility or reactivity of these biogenic materials.

10.5.7 Replacement mechanisms

Calcite containing in excess of several mol % $MgCO_3$ is not stable at low temperatures, and its alteration to calcite having lower Mg content and/or to dolomite is well known to carbonate petrologists. Aragonite, also metastable at sedimentary conditions, undergoes tranformation to calcite, or in some cases dolomite. Consequently, high-magnesian calcite and aragonite are rarely preserved in ancient rocks (important exceptions do occur). It is well established that these transformations occur by dissolution and subsequent precipitation in the presence of a fluid. If the process operates on a sufficiently small scale that the product phase can be shown to have formed in space previously occupied by a precursor, it is commonly referred to as *replacement*. The detailed nature of replacement has been the subject of considerable interest. Thin solution "films" have been proposed to accomplish fabric-retentive replacement.

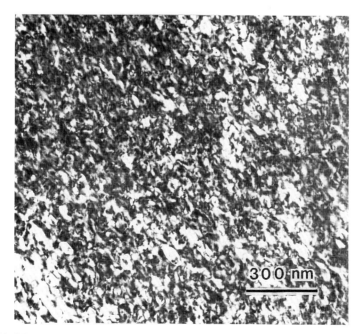

Figure 10-23. BF image of magnesian calcite echinoderm spine. Strong mottled contrast is pervasive and results from a high density of dislocations, small voids, and possibly other defects.

TEM observations are consistent with diagenetic alteration occurring by dissolution/precipitation processes and allow characterization on a finer scale than SEM. Where precursor and product carbonate exist together, interfaces between them are sharp and may or may not have intervening void space. Fluid and solid inclusions in the product may also provide evidence of fluid involvement in the alteration process.

Frisia-Bruni and Wenk (1985) examined interfaces associated with calcite replacing aragonite. They observed both topotactic and non-topotactic replacement. Replacement sometimes occurred preferentially along twin boundaries in the aragonite. Interfaces were variably straight or irregular, and no evidence was seen for microcracks which could represent a prior thin solution film.

Unpublished work by the present author (also see Reeder, 1989) confirms that fabric-retentive replacement interfaces between aragonite and calcite are typically "tight", although some unconnected microvoids do occur. It remains unclear whether these interfaces actually represent replacement "fronts" that are frozen in the rock, as many have presumed. Perhaps crack sealing (or healing) processes have closed a prior thin film zone. Alternatively, it may be possible that these interfaces represent surfaces where calcite nucleated on aragonite or else passive boundaries at which calcite simply engulfed aragonite (cf. Ward and Reeder, 1992). This is an area where additional work, particularly HRTEM, needs to be undertaken. Ness et al. (1990) described some preliminary HRTEM efforts dealing with replacement.

10.5.8 Second-phase inclusions

Electron microscopists commonly encounter second-phase inclusions within crystals. In many situations, these are viewed as a nuisance or are simply ignored. However, there are also situations where the ability to recognize that second-phase material has been incorporated during growth can be extremely useful. TEM is especially well suited for characterizing second-phase solid inclusions, not only because of its superior spatial resolution, but also because phase identification is made possible by electron diffraction and often by energy-dispersive microanalysis. The carbonate literature abounds with descriptions such as "cloudy dolomite" or "cloudy cores and clear rims", all referring to inclusions of some kind. Carbonate petrologists have long appreciated the significance of inclusions, but have rarely been able to characterize them fully, or, in the case of the smallest ones, even recognize their presence!

10.5.8.1 Solid inclusions

The advantages of TEM for identification and characterization of solid inclusions are best illustrated by a few examples.

10.5.8.1.1 Carbonate inclusions. Inclusions of calcite or dolomite within host crystals of either are quite common. Inclusions may range in size from just a few tens of nm to many μm. Although many possibilities obviously exist, one of the most interesting is the occurrence of so-called *microdolomites* in crinoidal calcite fragments. These small dolomite crystals have figured importantly in interpreting original mineralogy. Lohmann and Meyers (1977) have argued that their presence is an indication that the calcite originally was more highly magnesian and subsequently underwent stabilization to a low-magnesium state, with released Mg forming the small dolomite crystals. Figure 10-24 shows one such microdolomite in a host calcite. The SAED pattern corresponds to the microdolomite and a small portion of the host calcite. It is seen that both phases exhibit a homoaxial relation,

Figure 10-24. BF image of a microdolomite inclusion in diagenetically altered echinoderm calcite. Dolomite is Ca-rich, and displays poorly developed modulated structure. Note the spotty damage in the host calcite (see §10.2.2 in the text). Inset SAED shows homoaxial relation between dolomite and host calcite.

although the interface is not coherent. Blake et al. (1982) examined microdolomites in fossil echinoderm fragments and determined that typically they are Ca-rich and exhibit modulated structure.

Because aragonite is metastable at the temperature and pressure conditions experienced by sedimentary rocks, it is rarely preserved in ancient carbonate rocks. However, tiny relics of original aragonite are not uncommonly preserved in limestones. These relics escaped alteration to calcite. Many are likely to be of biogenic origin, and in some cases the aragonite relics show distributions and orientations that correspond to a shell's ultrastructure (Fig. 10-25a). In fossiliferous Pleistocene limestones in which the original aragonite has been incompletely altered to calcite, high densities of aragonite relics can be found near the replacement "fronts" (Fig. 10-25b).

10.5.8.1.2 Non-carbonate inclusions. Authigenic clay is not uncommon as a second-phase inclusion in carbonates. An example of clay that formed in a microcavity is shown in Figure 10-26a. Authigenic quartz is apparently not common as inclusions within calcite and dolomite, even though it may occur commonly as separate grains in limestones.

Inclusions of gypsum and halite can be especially important indicators of fluid conditions. Their presence implies certain saturation states of the fluid, which in the case of fluids derived from seawater, may have important implications for extent of evaporation. Figure 10-26b shows small inclusions of gypsum in dolomite. Phase identification is confirmed by SAED and energy-dispersive x-ray microanalysis.

10.5.8.2 Fluid inclusions

In the case of fluid inclusions, phase identification is not an issue, because the fluid has usually escaped prior to examination in the TEM. The void spaces remain, and, providing they are sufficiently small, can be easily identified in a foil. Size of the void is actually

414

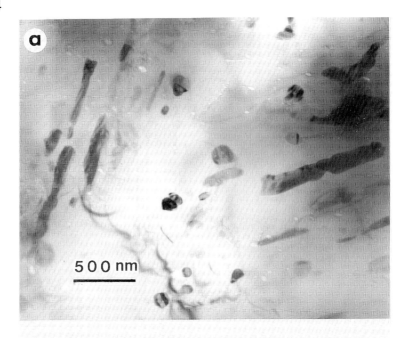

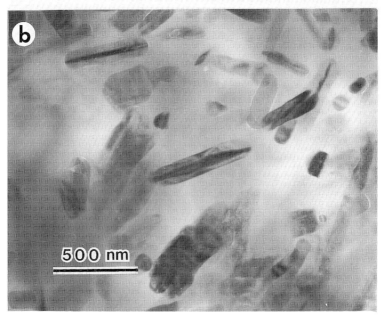

Figure 10-25. (a) BF image of neomorphic calcite resulting from fabric-retentive replacement of molluscan aragonite. Numerous aragonite relics appear in strong contrast in the single crystal calcite host. Their orientations correspond to the distribution of skeletal aragonite in the crossed-lamellar layer of a nearby, partially replaced mollusc shell. Also notice the numerous small voids (presumably fluid inclusions) throughout the calcite. (b) BF image showing a high density of aragonite relics in single crystal neomorphic calcite from the Pleistocene Miami Oolite. Much of the aragonite shows twinning on {110}.

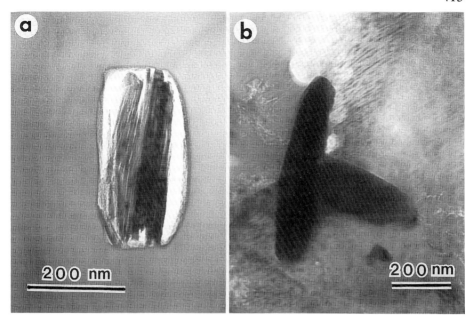

Figure 10-26. (a) BF image showing small inclusion of a clay mineral within host calcite. (b) BF image showing small gypsum inclusions in dolomite.

quite an important issue and is critically related to foil thickness. Carbonate foils suitable for conventional TEM (100-200 kV) are generally less than 300 nm thick, and for HRTEM, specimen thickness less than 100 nm is desirable (see Chapter 3). Most readers will be aware that fluid inclusion microthermometry generally relies on inclusions whose dimensions are on the order of a few micrometers and larger. Such large inclusions are usually breached and/or obliterated during ion milling that produces thin foils. Consequently, it is often impossible to identify evidence of such large inclusions. High-voltage TEM with its superior specimen penetration may provide some advantages in this regard.

On the other hand, TEM of low-temperature carbonates reveals that much smaller fluid inclusions exist, and indeed are quite common. Figure 10-27a shows fluid inclusions having a range of sizes in calcite. It is apparent that inclusions as small as 10 nm can be identified by TEM. Morphologies of inclusions vary considerably, and include rhombo-hedral to spherical forms. Densities of inclusions can be remarkably high, and bulk chemical properties may be affected by their presence.

Earlier in this chapter, it was remarked that stretching of fluid inclusions could be induced during sample preparation. In such cases, dislocations are observed to radiate from the inclusion walls and sometimes occur as loops around inclusions (Fig. 10-27b). For spherical inclusions, the distribution of dislocations due to stretching may be uniform. Stretched rhombohedral inclusions often have dislocations preferentially radiating from corners, where stresses may be higher. Although no systematic TEM studies have been undertaken of this subject, indications are that production of dislocations and/or dislocation glide accompany the volume increase associated with stretching. A potential application of TEM might be the assessment of stretching mechanisms.

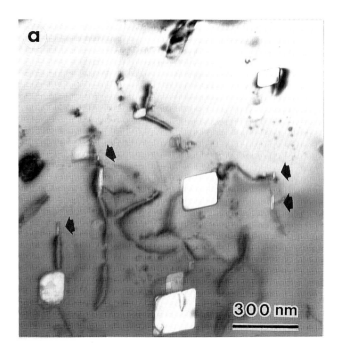

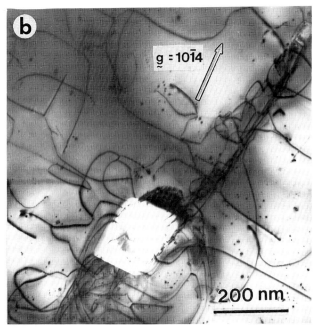

Figure 10-27. (a) BF image showing fluid inclusions having rhombohedral forms within calcite host. Arrows point to several of the smaller inclusions, just a few tens of nm in size. (b) BF image of dislocation microstructure in calcite surrounding fluid inclusion that has undergone stretching by heating *ex situ*. Many of the dislocations radiate from the inclusion walls and some appear to form loops partially surrounding the larger inclusion.

10.5.9 Growth microstructures and relationship to processes

Essentially all of the microstructural features that have been described thus far, as well as the superstructures, formed during crystal growth, whether by direct crystallization from solution or by replacement of a pre-existing carbonate. If one assumes that no significant changes have subsequently occurred to modify the microstructures, they can be regarded as a very detailed record of the growth process. In a broad sense, different microstructures should be representative, or perhaps even diagnostic, of different growth conditions. The difficulty, however, lies in determining which microstructures and which growth conditions are uniquely related to one another. A second challenge will be determining the growth conditions that prevail in different geological environments. Some insight is already being gained, and a few examples illustrate how microstructural information might be applied.

Ward and Reeder (1992) considered how growth microstructures bear on the question of replacement *vs* primary growth. Although some early observations suggested that modulations in dolomite and calcite formed as a result of replacement of precursor carbonate, continued work has shown that they exist in crystals that form by direct crystallization from solution (i.e. pore-filling cement) as well as replacement crystals (cf. Frisia-Bruni and Wenk, 1985; Ward and Reeder, 1992). In fact in calcian dolomite containing the fine modulation, microstructures are generally indistinguishable among cement and replacement dolomite. From this, it might be concluded that essentially the same mechanism operated to produce the microstructure, irrespective of whether replacement or free precipitation occurred. Ward and Reeder noted that one of the most useful results that growth microstructures provide is direct evidence of the growth direction.

It may be possible to speculate about the nature of a growth mechanism if we assume that the modulation results from strain associated with Ca-rich domains. The fine modulation that occurs in low-temperature calcian dolomite forms during growth with an orientation parallel to the growth direction (see §10.5.2.1.1.1 and §10.5.2.1.3). In adjacent growth sectors, the orientation of the modulation differs, but remains consistent with growth direction. The formation of Ca-rich domains would seem to require some kind of chemical segregation on the crystal surface. Barber et al. (1985) and Reksten (1990a) suggested that cellular growth might allow localization of Ca or impurities at the growth surface. The details of the process presumably vary with fluid conditions. In the case of ribbon microstructures, segregation seems to occur on a larger scale and produces more continuous, ribbon-like features. Segregation and cellular growth are well known among other types of growth processes, but are relatively unexplored in growth from aqueous solutions. Future studies should examine these aspects in relation to formation of microstructures.

10.6 CARBONATES OF HIGH(ER)-TEMPERATURE ORIGIN

10.6.1 Metamorphic calcite and dolomite

Most of the TEM work on metamorphic carbonates has been done in the context of natural deformation. From these studies and other observations it seems clear that the heterogeneous composition-related microstructures observed in some low-temperature carbonates are absent. Furthermore, *c*- and *d*-type superstructures are not evident. The absence of such features in metamorphic samples is consistent with equilibrium considerations. In calcite and dolomite samples from virtually all metamorphic grades, microstructures are restricted to those expected for plastic deformation, recovery, or recrystallization (cf. Barber and Wenk, 1976, 1979; Barber, 1977; White and White, 1980). These include a wide range of dislocation microstructures and twinning, and have been reviewed by Wenk et al. (1983).

10.6.2 Carbonatite dolomite and calcite

Barber and Wenk (1984) report that dolomite from carbonatite shows a variety of microstructures. Among the most interesting are branching, ribbon-like defects similar to those in sedimentary calcian dolomite, particularly saddle dolomite. Diffraction contrast characteristics of these ribbons reveal bifurcation and termination. Barber and Wenk also showed that ribbons commonly connect with second-phase inclusions of calcite. Microanalysis revealed that thicker portions of ribbons were Ca rich relative to the host, and by inference, the authors concluded that the ribbon material is calcite-like. The similarity of composition-related microstructures in dolomite samples having origins as distinct as low-temperature solution growth and high-temperature melt-related growth is quite remarkable, and suggests that the development of these microstructures is constrained by crystallographic factors.

Calcite that coexists with the dolomite exhibits platelets and rods of second-phase material crystallographically oriented with respect to host calcite. Barber and Wenk (1984) were unable to establish compositions for these precipitates, but observed that their distribution correlated with Fe content in the calcite.

10.7   TRANSFORMATION MICROSTRUCTURES

Several high-temperature transformations occur in various carbonates. These contrast sharply with the predominately low-temperature transformations, which occur by dissolution and precipitation in the presence of an aqueous phase. The transformations discussed here have been studied in the dry state.

10.7.1 Aragonite-calcite transition

Aragonite is metastable at ambient conditions, and in the absence of water, it shows no evidence of inverting to the more stable polymorph calcite, even on geologic time scales. However, the dry transformation to calcite can be readily induced by heating. This polymorphic transformation had received considerable attention prior to examination *in situ* by TEM (cf. Carlson, 1983). Burrage and Pitkethly (1969) relied on beam heating in a TEM to affect the transformation with some success. McTigue and Wenk (1985) examined the transition *in situ* using a heating stage in a 1.5 MeV TEM. Heating above 400°C, McTigue and Wenk observed buckling of the foil, perhaps reflecting the large volume change associated with the tranformation — 8%. At 500°C, calcite nucleation was observed preferentially at defects and at the thin edge of the foil. Calcite nuclei grew as their sharp, crystallographically-oriented boundaries swept across the foil consuming the aragonite (Fig. 10-28). During heating, McTigue and Wenk noted that some regions of aragonite developed heterogeneous contrast (resembling modulated structure) that become unobservable with further heating. Essentially defect-free calcite formed topotactically with $(100)_{arag} = (11\bar{2}0)_{calc}$, $(010)_{arag} \sim (\bar{1}104)_{calc}$, and $(001)_{arag} \sim [221]_{calc}$ [the latter direction in calcite is the zone containing both $(\bar{1}104)$ and $(\bar{1}01\bar{2})$]. McTigue and Wenk (1985) show that these orientation relationships can be explained on structural grounds.

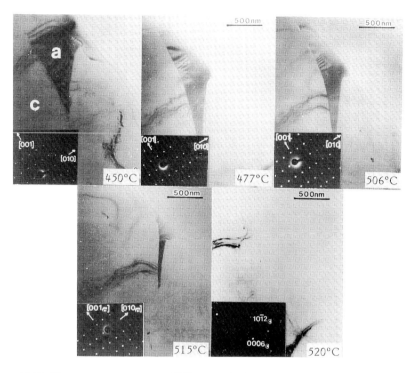

Figure 10-28. Time-temperature sequence of BF images showing transformation of aragonite (dark triangular region) to calcite as grain boundary sweeps across thin foil. From McTigue and Wenk (1985).

10.7.2 Cation ordering transition in dolomite

It was noted previously that dolomite undergoes a cation order-disorder transition near 1150°C. The transition temperature is well above even high-grade metamorphic conditions, so that applications are probably limited among natural dolomite samples. Upon cooling through the transition temperature, Reeder and Nakajima (1982) showed that a pervasive domain microstructure forms in dolomite (Fig. 10-29a). Smoothly-curving domain boundaries strongly resemble antiphase boundaries (cf. Fig. 12-34a in Chapter 12 by Nord). However, contrast analysis reveals that domains are not related by translation as required for antiphase domains. Rather they are related by a 180° rotation operation, and hence should be regarded as twin domains (Reeder and Nakajima, 1982). The authors showed that twin domains are predicted for the $R\overline{3}c \rightarrow R\overline{3}$ transition on the basis of group theoretical considerations. HRTEM (Fig. 10-29b) shows that Ca and Mg layers are juxtaposed across the domain boundary owing to the rotation operation. The CO_3 lattice remained largely undisturbed. Additional details are given by Nord in Chapter 12.

10.7.3 Decomposition transformations

Carbonate decomposition has been studied by many different techniques. It is an important issue for petrologists, because decomposition to oxides is the major limitation for the stability of carbonates at high temperatures. Both calcite and dolomite decomposition

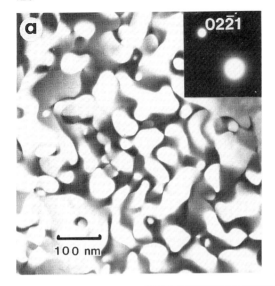

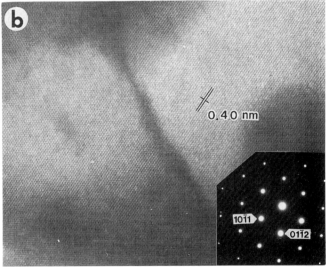

Figure 10-29. (a) DF image showing domain microstructure in dolomite heated above the cation ordering transition temperature and quenched rapidly. Domains are related by a twin operation. (b) Lattice image showing offset in fringe pattern across domain boundary. From Reeder and Nakajima (1982).

have been studied *in situ* by TEM. This work has also been done *in vacuo*, so that details of the transformation may not be entirely applicable to other situations.

Towe (1978) used beam heating at 100 kV to affect decomposition of calcite. He observed the development of a porous network of fine CaO needles having several different preferred orientations. McTigue and Wenk (1985) studied decomposition of calcite following its transformation from aragonite, and found similar results. Their experiment was conducted in a heating stage, and their observations were made at 1.5 MeV. Heating above 500°C showed development of CaO particles having preferred orientation. Topotactic relations were given as: $(11\bar{2}0)_{calc} = (1\bar{1}0)_{lime}$, $(\bar{1}104)_{calc} \sim (001)_{lime}$, and $(0001)_{calc} = (\bar{1}\bar{1}1)_{lime}$. McTigue and Wenk showed that this topotaxy preserves the basic structure within the anion lattice.

Kim et al. (1987) examined decomposition of magnesite *in situ* using both beam heating and a hot stage. Decomposition occurs topotactically yielding a porous network of MgO particles pseudomorphing the $MgCO_3$. The dominant orientation relationship observed was $(000\bar{1})_{mc} = (111)_{MgO}$ and $[11\bar{2}0]_{mc} = [2\bar{1}\bar{1}]_{MgO}$.

Not surprisingly, decomposition of dolomite is more complex. Cater and Buseck (1985) examined decomposition induced by the electron beam using HRTEM (200 kV). Decomposition resulted initially in the formation of a face-centered cubic (fcc) phase, interpreted to be an equimolar solid solution, $Ca_{0.5}Mg_{0.5}O$. Cater and Buseck noted a preferred orientation with $[0001]_{dolo} = [111]_{fcc}$. Interestingly, the Ca-Mg ordering in dolomite is apparently not retained in the fcc solid solution. The process is accompanied by a volume contraction, but a porous network like that formed in calcite does not appear. After prolonged, intense exposure to the electron beam the fcc solid solution altered to CaO and MgO particles, which were randomly oriented. The authors noted that the CaO and MgO crystallites were smaller (10-100 Å) than the CaO needles (100-1000 Å) reported by McTigue and Wenk for calcite decomposition.

10.8 AREAS FOR FUTURE WORK

Although the studies reviewed in this chapter demonstrate how important the application of TEM has been to understanding natural carbonates, many questions remain unanswered. This is particularly true regarding the compositional and structural aspects of Ca-Mg carbonates. Further application of TEM, including AEM and HRTEM, should help to resolve many of these. It seems most appropriate to end this chapter with a partial list of topics that deserve further attention.

1. Study of possible relation between the fine modulation and ribbon microstructures, and their conditions of formation.
2. Determination of actual compositions of *c*- and *d*-domains and their host carbonate.
3. Further characterization of recent Ca-Mg carbonate sediments, particularly biogenic magnesian calcite.
4. Examination of reaction mechanisms and correlation with microstructures, particularly crystal growth, biomineralization, and fabric-retentive replacement.
5. Characterization of carbonates produced in synthesis and replacement experiments.
6. Study of calcite/dolomite exsolution in metamorphic carbonates.

ACKNOWLEDGMENTS

I thank David Barber and Rudy Wenk for many years of collaboration on TEM studies of carbonates. David Barber, Dave Blake, Don Peacor, and Rudy Wenk provided valuable comments on an earlier version of this chapter. The author's work on carbonates has been supported by NSF grants, including EAR-8803423, EAR-9003915 and EAR-9204809.

REFERENCES

Bahnck, D. and Hull, R. (1990) Experimental measurement of transmission electron microscope specimen temperature during ion milling. In Anderson, R., ed., Specimen Preparation for Transmission Electron Microscopy of Materials II, Mat. Res. Soc. Symp. Proc., v. 199, Mat. Res. Soc., Pittsburgh, PA, pp. 253-261.

Barber, D.J. (1977) Defect microstructures in deformed and recovered dolomites. Tectonophysics 39, 193-213.

Barber, D.J. and Wenk, H.-R. (1984) Microstructures in carbonates from the Alno and Fen carbonatites. Contrib. Mineral. Petrol.,88, 233-245.

Barber, D.J. and Wenk, H.-R. (1976) Defects in deformed calcite and carbonate rocks. In Wenk, H.-R., ed., Electron Microscopy in Mineralogy, Springer-Verlag, Berlin, pp. 428-442.

Barber, D.J. and Wenk, H.-R. (1979) On geological aspects of calcite microstructure. Tectonophysics 54, 45-60.

Barber, D.J. and Khan, M.R. (1987) Composition-induced microstructures in rhombohedral carbonates. Mineral. Mag. 51, 71-86.

Barber, D.J., Freeman, L.A. and Smith, D.J. (1983) Analysis of high-voltage, high-resolution images of lattice defects in experimentally-deformed dolomite. Phys. Chem. Minerals 9, 102-108.

Barber, D.J., Reeder, R.J. and Smith, D.J. (1985) A TEM microstructural study of dolomite with curved faces (saddle dolomite). Contrib. Mineral. Petrol. 91, 82-92.

Berman, A., Addadi, L. and Weiner, S. (1988) Interactions of sea-urchin skeleton macromolecules with growing calcite crystals—a study of intracrystalline proteins. Nature 331, 546-548.

Blake, D.F. and Peacor, D.R. (1981) Biomineralization in crinoid echinoderms: characterization of crinoid skeletal elements using TEM and STEM microanalysis. Scanning Electron Microscopy 3, 321-328.

Blake, D.F. and Peacor, D.R. (1985) TEM/STEM microanalysis of Holocene fresh-water magnesian carbonate cements from the Coast Range of California. Am. Mineral. 70, 388-394.

Blake, D.F., Peacor, D.R. and Wilkinson, B.H. (1982) The sequence and mechanism of low-temperature dolomite formation: Calcian dolomites in a Pennsylvanian echinoderm. J. Sedim. Petrol. 52, 59-70.

Blake, D.F., Peacor, D.R. and Allard, L.A. (1984) Ultrastructural and microanalytical results from echinoderm calcite: Implications for biomineralization and diagenesis of skeletal material. Micron and Microscopica Acta 15, 85-90.

Burrage, B.J. and Pitkethly, D.R. (1969) Aragonite transformations observed in the electron microscope. Phys. Status Solidi 32, 399-405.

Burton, B.P. (1987) Theoretical analysis of cation ordering in binary rhombohedral carbonate systems. Am. Mineral. 72, 329-336.

Burton, W.K., Cabrera, N. and Frank, F.C. (1951) The growth of crystals and the equilibrium structure of their surfaces. Phil. Trans. Roy. Soc. London A243, 299-358.

Carballo, J.D., Land, L.S. and Miser, D.E. (1987) Holocene dolomitization of supratidal sediments by active tidal pumping, Sugarloaf Key, Florida. J. Sedim. Petrol. 57, 153-165.

Carlson, W.D. (1983) The polymorphs of $CaCO_3$ and the aragonite-calcite transformation. In Reeder, R.J., ed., Carbonates: Mineralogy and Chemistry, Rev. Mineral. 11, 191-225.

Cater, E.D. and Buseck, P.R. (1985) Mechanism of decomposition of dolomite, $Ca_{0.5}Mg_{0.5}CO_3$, in the electron microscope. Ultramicroscopy 18, 241-252.

Fouke B.W. and Reeder, R.J. (1992) Surface controls on dolomite composition: evidence from sectoral zoning. Geochim. Cosmochim. Acta (in press)

Freeman, L.A., Barber, D.J., O'Keefe, M.A. and Smith, D.J. (1983) High resolution studies of defects in natural dolomites. Proc. 7th Int'l Conf. on HVEM, Lawrence Berkeley Lab, Univ. of California, Berkeley, CA, 377-380.

Frisia-Bruni, S. and Wenk, H.-R. (1985) Replacement of aragonite by calcite in sediments from the San Cassiano Formation (Italy). J. Sedim. Petrol. 55, 159-170.

Goldsmith, J.R. (1983) Phase relations of rhombohedral carbonates. In Reeder, R.J., ed., Carbonates: Mineralogy and Chemistry, Rev. Mineral. 11, 49-76.

Goldsmith, J.R. and Graf, D.L. (1958) Structural and compositional variations in some natural dolomites. J. Geol. 66, 678-693.

Goldsmith, J.R. and Heard, H.C. (1961) Subsolidus phase relations in the system $CaCO_3$-$MgCO_3$. J. Geol. 69, 45-74.

Goldsmith, J.R., Graf, D.L., Witters, J. and Northrop, D.A. (1962) Studies in the system $CaCO_3$-$MgCO_3$-$FeCO_3$: 1. Phase relations, 2. A method for major element spectral analysis, 3. Compositions of some ferroan dolomites. J. Geol. 70, 659-688.

Graf, D.L. and Goldsmith, J.R. (1955) Dolomite-magnesian calcite relations at elevated temperatures and CO_2 pressures. Geochim. Cosmochim. Acta 7, 109-128.

Graf, D.L. and Goldsmith, J.R. (1956) Some hydrothermal syntheses of dolomite and protodolomite. J.

Geol. 64, 173-187.

Gunderson, S.H. and Wenk, H.-R. (1981) Heterogeneous microstructures in oolitic carbonates. Am. Mineral. 66, 789-800.

Heywood, B.R., Rajam, S. and Mann, S. (1991) Oriented crystallization of $CaCO_3$ under compressed monolayers. Part 2. - Morphology, structure and growth of immature crystals. J. Chem. Soc. Faraday Trans. 87, 735-743.

Khan, M.R. and Barber, D.J. (1990) Composition-related microstructures in zinc-bearing carbonate assemblages from Broken Hill, New South Wales. Mineralogy and Petrology 41, 229-245.

Kim, M.G., Dahmen, U. and Searcy, A.W. (1987) Structural transformations in the decomposition of $Mg(OH_2)$ and $MgCO_3$. J. Am. Ceram. Soc. 70, 146-154.

King, A.H., Chen, F.-R., Reeder, R.J. and Barber, D.J. (1984) Calculated images of dolomite-calcite interfaces: Feature identification in saddle dolomite. In Bailey, G.W., ed., Proc. 42nd Ann. Mtg. Elec. Micr. Soc. Am., 586-587.

Klapper, H. (1980) Defects in non-metal crystals. In Tanner, B.K. and Bowen, D.K., eds., Characterization of Crystal Growth Defects by X-ray Methods, Plenum, New York, pp. 130-166.

Lohmann, K.C. and Meyers, W.J. (1977) Microdolomite inclusions in cloudy prismatic calcite: a proposed criterion for former high magnesian calcites. J. Sedim. Petrol. 47, 1078-1088.

McCallister, R.H. and Nord, G.L. (1981) Subcalcic diopsides from kimberlites: Chemistry, microstructures, and thermal history. Contrib. Mineral. Petrol. 78, 118-125.

McTigue, J.W. and Wenk, H.-R. (1985) Microstructures and orientation relationships in the dry-state aragonite-calcite and calcite-lime phase transformations. Am. Mineral. 70, 1253-1261.

Megaw, H.D. (1973) Crystal Structures: A Working Approach, Saunders, Philadelphia.

Meike, A., Wenk, H.-R., O'Keefe, M.A. and Gronsky, R. (1988) Atomic resolution microscopy of carbonates. Interpretation of contrast. Phys. Chem. Minerals 15, 427-437.

Miser, D.E., Swinnea, J.S. and Steinfink, H. (1987) TEM observations and X-ray crystal-structure refinement of a twinned dolomite with a modulated microstructure. Am. Mineral. 72, 188-193.

Mitchell, J.T., Land, L.S. and Miser, D.E. (1987) Modern marine dolomite cement in a north Jamaican fringing reef. Geology 15, 557-560.

Ness, S.E., Haywick, D.W. and Cuff, C. (1990) Overcoming the problems of high-resolution transmission electron microscopy of biogenic aragonite. Mineral. Mag. 54, 589-592

O'Keefe, M.A. and Barber, D.J. (1984) Interpretation of HREM images of dolomite. Int'l Physics Conf. Ser. No. 68, Chap. 5, Inst. Phys., pp. 177-180.

Reeder, R.J. (1981) Electron optical investigation of sedimentary dolomites. Contrib. Mineral. Petrol. 76, 148-157.

Reeder, R.J. (1983) Crystal chemistry of the rhombohedral carbonates. In Reeder, R.J., ed., Carbonates: Mineralogy and Chemistry, Rev. Mineral. 11, 1-47.

Reeder, R.J. (1989) TEM observations of aragonite-calcite replacement fronts. Abstracts with Programs, 1989, Geol. Soc. Am. Ann. Mtg., A258.

Reeder, R.J. and Wenk, H.-R. (1979) Microstructures in low temperature dolomites. Geophysical Rev. Ltr. 6, 77-80.

Reeder, R.J. and Wenk, H.-R. (1983) Structure refinements of some thermally disordered dolomites. Am. Mineral. 68, 769-776.

Reeder, R.J. and Barber, D.J. (1982) Lattice defects in saddle dolomites: an explanation for crystal distortion. Abstracts with Programs, Ann. Mtg. Geol. Soc. Am. 14, 597.

Reeder, R.J. and Nakajima, Y. (1982) The nature of ordering and ordering defects in dolomite. Phys. Chem. Minerals 8, 29-35.

Reeder, R.J. and Markgraf, S.A. (1986) High-temperature crystal chemistry of dolomite. Am. Mineral. 71, 795-804.

Reeder, R.J. and Prosky, J.L. (1986) Compositional sector zoning in dolomite. J. Sedim. Petrol. 56, 237-247.

Reeder, R.J. and Dollase, W.A. (1989) Structural variation in the dolomite-ankerite solid-solution series: An x-ray, Mössbauer, and TEM study. Am. Mineral. 74, 1159-1167.

Reeder, R.J., Prosky, J.L. and Meyers, W.J. (1984) Correlation of crystal growth defects with cathodoluminescent zoning in calcian dolomite crystals. Abstracts with Programs, Ann. Mtg. Geol. Soc. Am. 16, 631.

Reeder, R.J., Fagioli, R.O. and Meyers, W.J. (1990) Oscillatory zoning of Mn in solution-grown calcite crystals. Earth-Science Reviews 29, 39-46.

Reksten, K. (1990a) Modulated microstructures in calcian ankerite. Am. Mineral. 75, 495-500.

Reksten, K. (1990b) Superstructures in calcian ankerite. Phys. Chem. Minerals 17, 266-270.

Reksten, K. (1990c) Superstructures in calcite. Am. Mineral. 75, 807-812.

Rosen, M.R., Miser, D.E. and Warren, J.K. (1988) Sedimentology, mineralogy, and isotopic analysis of Pellet Lake, Coorong region, South Australia. Sedimentology 35, 105-122.

424

Rosen, M.R., Miser, D.E., Starcher, M.A. and Warren, J.K. (1989) Formation of dolomite in the Coorong region, South Australia. Geochim. Cosmochim. Acta 53, 661-669.

Rosenberg, P.E. (1967) Subsolidus relations in the system $CaCO_3$-$MgCO_3$-$FeCO_3$ between 350° and 550°C. Am. Mineral. 52, 787-796.

Speer, J.A. (1983) Crystal chemistry and phase relations of orthorhombic carbonates. In Reeder, R.J., ed., Carbonates: Mineralogy and Chemistry, Rev. Mineral. 11, 145-189.

Towe, K.M. (1978) Ultrastructure of calcite decomposition in vacuo. Nature 274, 239-240.

Van Tendeloo, G., Wenk, H.-R. and Gronsky, R. (1985) Modulated structures in calcian dolomite: A study by electron microscopy. Phys. Chem. Minerals 12, 333-341.

Ward, W.B. and Reeder, R.J. (1992) The use of growth microfabrics and transmission electron microscopy in understanding replacement processes in carbonates. In Rezak, R. and Lavoie, D., eds., Carbonate Microfabrics, Springer-Verlag, New York.

White, J.C. and White, S.H. (1980) High-voltage electron microscopy of naturally deformed polycrystalline dolomite. Tectonophysics 66, 35-54.

Wenk, H.-R. (1985) Carbonates. In Wenk, H.-R., ed., Preferred Orientation in Deformed Metals and Rocks: An Introduction to Modern Texture Analysis, Academic Press, Orlando, Florida, pp. 361-384.

Wenk, H.-R. and Zenger, D.H. (1983) Sequential basal faults in Devonian dolomite, Nopah Range, Death Valley area, California. Science 222, 502-504.

Wenk, H.-R. and Zhang, F. (1985) Coherent transformations in calcian dolomites. Geology 13, 457-460.

Wenk, H.-R., Barber, D.J. and Reeder, R.J (1983) Microstructures in carbonates. In Reeder, R.J., ed., Carbonates: Mineralogy and Chemistry, Rev. Mineral. 11, 301-367.

Wenk, H.-R., Meisheng, H., Lindsey, T. and Morris, J.W. (1991) Superstructures in ankerite and calcite. Phys. Chem. Mineral. 17, 527-539.

Wenk, H.-R., Meisheng, H. and Frisia, S. (in press) Partially disordered dolomites: Microstructural characterization of Abu Dhabi sabkha carbonates. Am. Mineral.

Zenger D.H. (1972) Significance of supratidal dolomitization in the geologic record. Geol. Soc. Am. Bull. 83, 1-12.

Chapter 11. ANALYSIS OF DEFORMATION IN GEOLOGICAL MATERIALS

Harry W. Green, II Department of Geology,
University of California, Davis, California 95616, U.S.A.

11.1 INTRODUCTION

Experimental work on time-dependent geological processes always requires extrapolation from the time scale available in the laboratory to the very much larger time scales of natural processes. The great pitfall of such extrapolations is the possibility of a mechanism change between the laboratory time scale where quantitative measurements are made and the natural time scale where results are applied. If such a mechanism change occurs, the extrapolation is worthless, regardless of how precisely and carefully the experiments are conducted. For example, if flow has occurred in the crust by diffusive mass transfer mechanisms, extrapolation of laboratory data on a similar rock type deformed in the dislocation creep field provides nothing but an upper bound on the stresses involved, even if the extrapolation were perfect. The most reliable way to avoid such an error is to "ask the rocks," which amounts to examining field relations and microstructures in search of evidence reflecting the mechanisms that have operated under natural conditions. The microstructures of rocks that have been naturally deformed at elevated temperatures can serve as particularly fruitful evidence to constrain experimental and theoretical pronouncements about their history because they often preserve aspects of the processes by which they have flowed. For example, the dislocation types and configurations and the preferred crystallographic orientations found in mantle xenoliths and the basal portions of ophiolites are virtually indistinguishable from those produced in laboratory experiments on peridotites at high temperature and pressure (e.g., compare Zeuch and Green, 1984, with Jin et al., 1990; see also Nicolas and Poirier, 1976). Such correspondence between natural and experimental microstructures establishes that subsolidus mantle flow occurs by dislocation creep to at least as deep as the origin of these mantle samples (10-200 km). On the other hand, natural deformation microstructures in crustal rocks in many cases indicate that two or more different mechanisms have been active at different stages in the tectonic history (e.g., Knipe, 1990).

A wide variety of instruments is now available for investigation of the microstructures of rocks. First among these remains the petrographic (optical) microscope that allows quick, easy survey of structures on the grain-scale and larger where indicators of the geometry (strain and rotation) of deformation are most commonly preserved. In olivine, dislocations can be made visible in the optical microscope by heating in air and consequent preferential oxidation along dislocation lines (Kohlstedt et al., 1976). Delineation of structures can be very detailed when dislocation densities are low (e.g., Gueguen, 1977; Jin et al., 1990). Unfortunately, this powerful technique is not available for other materials. It is from optical examination that regions are selected for high-resolu-tion imaging. Scanning electron microscopy (SEM) using back-scattered electron (BSE) imaging can also be used to advantage. For example: (1) it has been used to extend the resolution of oxidized sections to higher dislocation densities in olivine (Karato, 1986); (2) it is very useful for detailing the complicated distribution of melt in materials deformed in the partially molten state (R.S. Borch, Z.M. Jin and H.W. Green, unpublished results); (3) it allows determination of the distribution of the spinel phase of Mg_2GeO_4 in an olivine matrix of the same composition, providing critical evidence for the anticrack mechanism of high pressure, transformation-induced, faulting (Green and Burnley, 1989; 1990; Burnley et al., 1991); (4) when combined with electron channeling it allows imaging of subgrain structures. Lastly, and most importantly, transmission electron microscopy (TEM) provides direct imaging of many kinds of defects and, when combined with selected-area electron diffraction, allows determination of the crystallographic character of the defects. Each of these techniques provides information different or on a different scale from the others, hence their complimentary use is especially powerful (e.g., Burnley et al., 1991).

This chapter summarizes briefly the nature of the defects responsible for flow in crystalline materials and illustrates what can be learned from them and how they can be used to interpret geological phenomena. For more detailed discussion of defects and flow mechanisms, the reader is referred to the excellent little book *Creep of Crystals*, by Poirier (1985) and more technical presentations referenced therein. A more formal intro-duction to dislocation theory, with emphasis on metals, can be found in *Introduction to Dislocations* by Hull and Bacon (1984). Similarly, the reader interested in additional infor-mation on the physics of imaging crystal defects may wish to consult *Electron Microscopy of Thin Crystals* by Hirsch et al. (1965). An excellent all-around introduction to TEM for geologists is *Transmission Electron Microscopy of Minerals and Rocks* by McLaren (1991). The latter also includes brief summaries and illustrations of the deformation microstructures of minerals different from those used as examples here.

One thing that cannot be covered by the present discussion is the very important problem of integration of microstructural observations, field measurements, etc. into a coherent discussion of the geological *history* of an area. More than summarizing the record of a specific series of events, such histories can serve to provide models of particular styles of tectonic evolution. An outstanding example of such a synthesis for the Assynt region of the Moine Thrust Zone, Scotland is given by Knipe (1990).

11.2   DEFECTS IN CRYSTALS AND THEIR IMAGES

Defects in crystals fall into three major classes (point, line and surface), having dimensions of 0, 1 and 2, respectively. Each can be created by and/or participate in deformation in different ways. The first can be directly imaged by high resolution electron microscopy (HREM), but such analysis rarely is performed in deformation studies. However, the images of line and surface defects in materials that have flowed at high temperatures clearly reflect the activity of point defect motion (diffusion). Moreover, experimentation on oxides and silicates has demonstrated that variation of the point defect populations can have important consequences for flow rates. In addition, at very high temperatures or in the presence of an intergranular fluid, point defect migration can become a significant flow mechanism by itself (diffusion creep or solution-transfer creep). For all of these reasons we begin our discussion with point defects.

11.2.1   Point defects

Point defects are individual atoms occupying positions in the crystal structure that are not normally occupied (interstitials) or atoms missing from sites that are normally occupied (vacancies). Creation of either a vacancy or an interstitial involves changing the free energy of the crystal. To understand this, imagine that we remove an atom from the interior of a crystal and place it on the surface, thereby creating a vacancy. On the surface, the atom will occupy a volume approximately the same as it did within the crystal and the vacant site will also be approximately the same size as it was when the site was filled (it will not be exactly the same size because the imbalance of bonding will cause the crystal to relax around the vacancy and either collapse slightly into it or withdraw slightly from it). Therefore, the creation of a vacancy is always a *dilatant* process; the volume of a crystal containing vacancies must be larger than the volume of a perfect crystal. This change in volume is referred to as the *volume of formation* of the vacancy, ΔV_f^v. Similarly, a vacancy has an *energy of formation*, ΔE_f^v (corresponding to the work involved in breaking bonds during its creation plus the energy associated with relaxation around the vacancy) and an *entropy of formation*, ΔS_v^f, due to changes in the atomic vibrations in the distorted region around the vacancy. Therefore,

$$\Delta G_f^v = \Delta E_f^v + p\Delta V_f^v - T\Delta S_f^v \tag{11.1}$$

where ΔG_f^v is the Gibbs Free Energy of formation of the vacancy, p is pressure and T is the (absolute) temperature. Strictly speaking, thermodynamics is a macroscopic phenomenon, hence the quantities in (11.1) must be applied to a statistically significant population of vacancies. The change of free energy of a crystal containing n moles of vacancies, n_v, is given, therefore, by the sum of the free energy of the individual vacancies minus a term encompassing the entropy of mixing $\Delta \overline{S}_m^v$.

$$\Delta \overline{G}^v = n_v \Delta \overline{G}_f^v - T\Delta \overline{S}_m^v \tag{11.2}$$

where the bar over all thermodynamic quantities indicates molar quantities. $\Delta \overline{G}_f^v$ is always a positive quantity, but for small quantities of vacancies, the first term in (11.2) is smaller than the second, hence crystals containing low concentrations of vacancies are stable relative to perfect crystals. Indeed, it can be shown (e.g., Poirier, 1985) that, at equilibrium, the concentration of vacancies is given by:

$$X_v = \exp\left(-\frac{\Delta \overline{G}_f^v}{RT}\right) \tag{11.3}$$

where X_v is the mole fraction of vacancies and R is the gas constant. This is a fundamental result because it means that vacancies will be spontaneously created at high temperatures; their *equilibrium* concentration can reach 10^{-3} near the melting point. Therefore, because all diffusion-related processes are a direct function of the vacancy concentration and because the vacancies are also very mobile at high temperatures, diffusion can be an important contributor to flow processes at high temperatures. Creation of interstitials can be described in analogous terms, but the much greater distortion of the lattice around them causes $\Delta E_f^i \gg \Delta E_f^v$. Therefore, the equilibrium concentration of interstitials is always very much smaller than that of vacancies. Since ΔV_f^i is negative whereas ΔV_f^v is positive, the relative importance of interstitials may increase at very high pressure.

In rocks and ceramics, most phases consist of more than one atomic species and have more than one type of lattice site. Moreover, most phases exhibit a large component of ionic bonding, making maintenance of electrical neutrality an important constraint on the relative concentrations of defects of the various types. Thus, for a pure binary compound of the form $A_\alpha B_\beta$ (e.g., Al_2O_3 or Fe_2O_3), at equilibrium we must have:

$$\alpha X_A^v = \beta X_B^v, \tag{11.4}$$

where α and β are the stoichiometric coefficients of the compound. We also must consider in a case such as (11.4) the effect of substitution into the lattice of foreign atoms with a different valence (e.g., Ti^{4+} for Al^{3+} in Al_2O_3). The requirement of electrical neutrality means that for the additional charge associated with each Ti atom, some compensating change must occur. This is generally accommodated by creation of vacancies in the cation sublattice such that, in this case, one Al^V is created for every 3 Ti atoms substituted. These *extrinsic* vacancies are in addition to the *intrinsic* (thermal) vacancies described in (11.4).

Similarly, in transition metal oxides and silicates, change of the oxygen fugacity, fO_2, will change the oxidation state of some ions. For example, in the case of Fe_2O_3, reduction of fO_2 would cause some ferric ions to be reduced to ferrous, again introducing an electrical imbalance. In this case, for each three Fe^{3+} ions reduced to Fe^{2+} the number of Fe^V must be reduced by one to maintain electrical neutrality. The sensitivity of the point defect populations of such materials to the fO_2 implies that any material properties dependent on such populations (diffusion, electrical conductivity, high temperature deformation, etc.) will also be a function of fO_2.

11.2.2 Line Defects

A one-dimensional defect in a crystal is called a *dislocation*—a (generally curved) line of distortion within the crystal with perfect crystal around it. Dislocations cannot end in perfect crystal, hence they must either extend between other dislocations, between 2-dimensional defects (such as grain boundaries) or they must close on themselves, forming a loop.

11.2.2.1 Dislocation geometry

A *glide* dislocation loop surrounds a region that has been displaced relative to the host crystal in a specific crystallographic direction by an amount equal to the dimension of the unit cell in that direction or some fraction thereof. That vector, called the *Burgers vector*, **b**, characterizes the dislocation. Expansion of the loop fully to the margins of the host crystal results in disappearance of the dislocation and production of a step on the surface such that the portion above the plane of the loop has been displaced by **b** relative to the portion below. Figure 11-1 shows an idealized loop drawn in the form of a rectangle to illustrate the two fundamental aspects of such loops. Portions of the loop that lie perpendicular to **b** are *edge dislocations*; each constitutes the edge of a half-plane of atoms that lies between two continuous atomic planes. Note that in any loop there are two such edge dislocations, of opposite sign, such that if they were slid together their extra half planes would connect, restoring the crystal to its perfect structure. Portions of the loop that lie parallel to **b** are *screw dislocations*; the screw dislocation transforms the crystal structure surrounding it into a helix in which adjacent planes of the crystal structure are connected together in a ramp-like geometry (much as levels of a parking garage are connected by ramps). The two screw dislocations of a loop are also of opposite sign; if they were slid together, the right-hand and left-hand helices would cancel, again restoring the perfect crystal structure. Glide dislocation loops expand in the glide plane by individual line elements moving normal to themselves, but the net displacement that describes the difference between the interior of the loop and the host crystal is always **b**. Of course, dislocation loops in most materials will not be rectangles composed of pure edge and screw components. In general, loops will be smoothly curved and consist of regions of approximately pure edge and screw and large areas of *mixed* character.

Prismatic dislocation loops surround regions where there is an extra atomic layer within a crystal (Fig. 11-2a) or define a region in which part of a plane is missing (Fig. 11-2b). In simple metals these loops are referred to as interstitial or vacancy loops; they

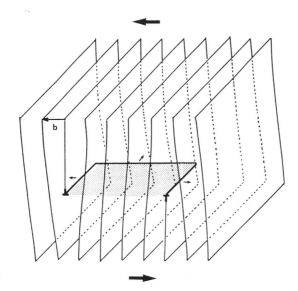

Figure 11-1. Idealization of a glide dislocation loop consisting of pure edge and screw components (heavy line). A set of crystal lattice planes perpendicular to the glide plane of the dislocation and normal to its Burgers vector, **b**, are shown outlined in light solid and dashed lines. Only half of the loop is shown to enable better visualizaton of the three-dimensional structure of the loop. Note that edge dislocatons (inverted T's) mark the end of "half planes" of atoms and that the single screw dislocation shown (parallel to **b**) converts the lattice planes it traverses into a ramp with a right-handed sense. The other screw dislocaton (not shown) converts the lattice planes into a left-handed ramp. Expansion of the loop (small arrows) in response to a shear stress (bold arrows) results in translation of the portion of the crystal above the glide plane by **b** with respect to the portion below the glide plane. [Used by permission of Cambridge University Press, from Poirier (1985), Fig. 2.4, p. 52.]

represent a planar grouping of extra atoms or vacant sites, respectively. These dislocations have **b** perpendicular to the plane of the loop; they are sessile (immobile) under shear stress but under appropriate conditions they can operate as sources for glide dislocations. Prismatic dislocations are created and move in response to pressure gradients, such as arise at the interface between phases of different thermal expansivity or compressibility when temperature or pressure is changed (Fig. 11-2c).

Dislocations are most profitably studied by conventional TEM, rather than HRTEM, because in deformation studies it is the interaction of dislocations with the applied shear stress and with each other that is of principal interest. Both of those interactions occur on the scale of the elastic continuum, hence higher resolution is rarely required. Only when the focus is impurity segregation, misfit between phases, etc. does HREM become important.

The "image" of a dislocation produced by diffraction contrast is actually an image of the distortion of the crystal structure in the vicinity of the dislocation. Consider the edge dislocation shown in Figure 11-1. Because the Bragg angle, 2θ, for electrons is very small (due to the very short wavelength associated with electrons), strong diffraction occurs from planes that are essentially parallel to the electron beam of the microscope. In the vicinity of the dislocation, the atomic planes of the crystal are distorted, hence diffraction will occur from the bent region when the crystal as a whole is not in an orientation for Bragg diffraction from the crystal. Thus, when the crystal is in almost any orientation with

430

respect to the beam, diffraction will occur adjacent to the core of the dislocation, producing an "image" of the dislocation. Depending on the particular orientation, the image may be a sharp, clear line, a fuzzy, broad line, or even a pair of lines, one on either side of the core of the dislocation. However, for crystals that approximate isotropic elasticity, planes which are parallel to the Burgers vector experience little or no distortion and thus if these are the only planes diffracting strongly, the dislocation "vanishes". If we designate the diffraction vector in the reciprocal lattice as $\mathbf{g}$, we have for the invisibility criterion $\mathbf{g} \cdot \mathbf{b} = 0$ from the well-known relationship that the scalar (dot) product of two vectors is zero if and only if they are perpendicular to each other. Strictly speaking, this invisibility criterion applies only to screw dislocations; for edges there will be residual contrast unless the additional criterion $\mathbf{g} \cdot \mathbf{b} \times \mathbf{u} = 0$ (where $\mathbf{u}$ is a unit vector parallel to the dislocation line) is also satisfied. The problem is compounded when the elasticity of the crystal is significantly anisotropic. In such cases, the best procedure to follow for Burgers vector determination is computer-simulation of the image for a variety of $\mathbf{g}$ vectors (see McLaren, 1991, p. 161 for more detailed discussion of determination of $\mathbf{b}$).

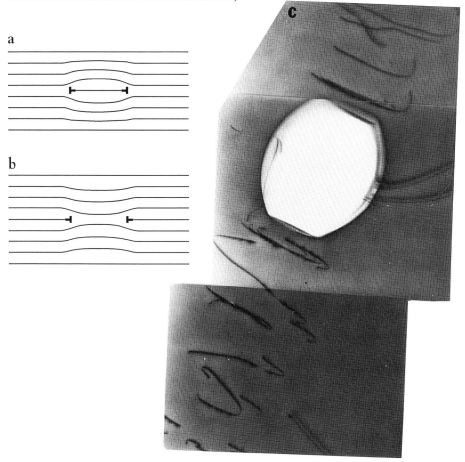

Figure 11-2. Prismatic dislocation loops. (a). Sketch of interstitial loop consisting of an extra disk of atoms one unit cell thick. (b). Sketch of vacancy loop consisting of a region within which a unit-cell-thick disk of atoms is missing. (c) Interstitial loops punched out parallel to [100] in olivine from an included CO_2 bubble in a mantle xenolith. [(c) used by permission of *American Journal of Science*, from Kirby and Green (1980) Plate 6D, p. 565.]

Fortunately, in most materials the residual contrast when **g·b** = 0 but **g·b** × **u** ≠ 0 is sufficiently weak that ambiguity does not arise. Thus, we have a relatively simple method for determination of **b**: One tilts the specimen in the microscope to different orientations, some that yield invisibility and others that do not (Fig. 11-3). Each of these is obtained by

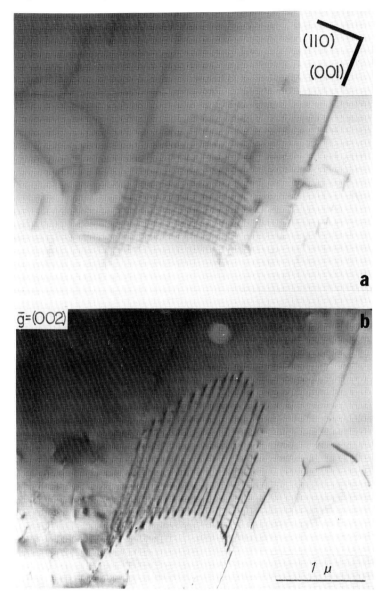

Figure 11-3. Low-angle (twist) boundary in olivine approximately parallel to (010). (a) With several reflections operating, **b** = [100] (E-W) and **b** = [001] (N-S) screw dislocations are visible as a rectangular array. (b) When tilted in the microscope so that only the single reciprocal lattice vector, **g** = 002, is operating, only **b** = [001] screws remain in contrast because **g·b** = 0 for dislocations with **b** = [100]. Note that the [001] dislocations are distinctly straighter than those with **b** = [100]. [Used by permission of *Tectonophysics*, from Zeuch and Green (1984) Fig. 3, p. 273.]

finding a two-beam condition (orientation in which almost all of the diffracted electrons are diffracted from a single set of planes). Determination of **g** from the selected-area electron diffraction pattern (SAEDP) for each orientation allows calculation of **b** through $g_1 \times g_2 = kb$, where k is a constant. In practice, one often knows the range of Burgers vectors present in a material from previous studies or knows the likely possibilities from the shortest lattice vectors, allowing the full process outlined here to be shortened (as in Fig. 11-3).

Like a vacancy or an interstitial, a dislocation distorts (strains) the crystal in its vicinity and imparts an additional energy to the crystal. However, the misfit is much larger around a dislocation, giving rise to a large energy per unit length. As a consequence, the energy per unit volume associated with a dislocation is much larger than that associated with a point defect and entropy effects are too small to stabilize a finite equilibrium concentration of dislocations.

11.2.2.2 Dislocation stress/strain fields

The elastic distortion of the crystal around a dislocation generates a stress field through Hooke's law and provides the mechanism whereby dislocations interact with an applied macroscopic stress. It is that interaction which causes dislocations to move and thereby for the crystal to deform. In addition, the stress fields around dislocations also cause them to interact with each other at a distance, giving rise to the progressive hardening observed during straining of materials at low temperature and the softening that can be induced by heating the material during or after deformation. These processes give rise to characteristic microstructures (discussed below) that, if preserved, allow inference of the conditions under which a rock was deformed. Moreover, the interactions between the dislocations and the applied stress field lead to a steady-state microstructure that, to a first order, is dependent only on the magnitude of the applied stress, creating the possibility, under favorable conditions, of inferring that stress (cf. Gueguen, 1977).

11.2.2.3 Dislocation multiplication

Each dislocation loop, when expanded to the boundaries of the crystal produces an offset of only one unit cell. Thus, it is obvious that production of a significant strain requires passage of very large numbers of dislocations through each of the crystals in a material. Examination of any ductile material in the undeformed state shows that the number of dislocations present ($\sim 10^4$ to 10^7 mm of dislocation line per cubic mm of crystal) is vastly too few to accomplish the large strains that can be produced easily. Therefore, an efficient process is necessary for continual generation of new dislocations to allow deformation to proceed. As might be expected from the very large numbers of dislocations necessary to produce even moderate strains, the dominant multiplication process is very simple and can operate in all crystalline materials. Figure 11-4 shows this process, based on a *Frank-Read source*, named after two of the pioneers of dislocation theory who predicted the existence of this source mechanism before it was observed. If a segment of dislocation line (edge, screw or mixed, it doesn't matter) is pinned at its ends (Fig. 11-4a), that segment will be immobile below some critical shear stress, τ_c. As the stress is raised on the segment, it bows out (Fig. 11-4b) until at $\tau = \tau_c$ it describes a semicircle (Fig. 11-4c). At that stage all components of the loop have been created and the loop can continue to expand at that or any higher stress. As the loop expands, while still anchored at its ends, it goes through the configurations shown in Figure 11-4d and e. When the two portions of the loop touch (Fig. 11-4e), two segments, of precisely opposite signs, are brought into coincidence and they cancel each other. The result is separation into two parts—a loop that is now freed from the pinning points and can expand indefinitely

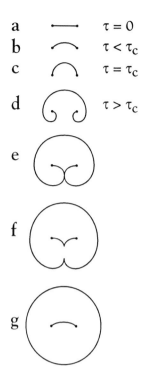

a $\qquad$ $\tau = 0$

b $\qquad$ $\tau < \tau_c$

c $\qquad$ $\tau = \tau_c$

d $\qquad$ $\tau > \tau_c$

e

f

g

Figure 11-4. Frank-Reed dislocation source. A segment of dislocation line pinned at its ends (a) can operate as a continual source of dislocation loops if subjected to sufficient shear stress. As the stress is raised, the dislocation segment bows out progressively (b) until it describes a semicircle (c). At that stress, $\tau = \tau_c$, the loop can continue to expand (d) if the stress is maintained and will evolve until it closes upon itself (e). The elements of the loop that touch are of opposite sign, hence they annihilate each other and separate the structure into two parts (f), one of which then expands freely and the other "snaps" back to the original position where it bows out again (g) and the process repeats.

under the applied stress, plus a pinned segment identical to the original one that can begin the process anew (Fig. 11-4f,g). τ_c varies inversely with the length of the pinned dislocation segment so that, as the stress is raised, longer segments are activated first and shorter ones later. If a constant strain-rate is imposed upon the material, the stress will rise until a sufficient population of dislocation sources has been activated to allow extensive flow to proceed (steady state). If the loops generated by the more easily-activated sources encounter obstacles and are stopped before the stress rises sufficiently to activate more difficult sources, it may be that steady-state flow will not be achieved and eventually the fracture stress will be reached. For discussion of other, related, dislocation multiplication mechanisms, the reader is referred to Hull and Bacon (1984).

11.2.2.4 Glide of dislocations

When at rest, a dislocation sits in an energy valley such that the surrounding crystal is distorted the minimum amount possible. In order for it to move (glide), it must be "pushed" by the applied stress up the energy hill representing the increased distortion of bonds as it is displaced from the energy valley. At the top of the energy hill, the atomic bonds across the dislocation line are broken and the dislocation slides down the hill into the next valley as bonds are established across the new position adjacent to the next line of atoms. This process must repeat as the dislocation glides. The resisting force represented by these periodic energy barriers in the crystal lattice is called the *Peierls force* and provides the fundamental strength of the crystal. At low homologous temperatures (temperatures far from the melting point), the Peierls force is large compared to thermal oscillations, hence it figures prominently in control of flow.

434

As is clear from its geometry (Fig. 11-1), an edge dislocation is confined to its glide plane unless it can shorten or lengthen its half plane. Thus, at low homologous temperatures where self-diffusion is inhibited, it is common for edge dislocations to encounter obstacles that stop their advance. Succeeding edges will "pile up" behind the leader, thereby multiplying the force on the obstacle, but if the obstacle cannot be broken, the pile-up will grow all the way back to the source and shut it down. The geometric restrictions are not so stringent for the movement of screw dislocations; a screw dislocation represents a continuous distortion that does not tie it irrevocably to a single glide plane. True, screws tend to move most easily on low index planes, but if a barrier is encountered they have the

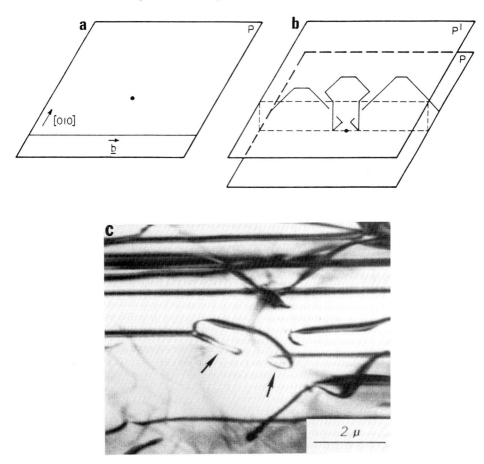

Figure 11-5. Double cross-slip. (a) A screw dislocation moving on its primary glide plane P encounters an obstacle symbolized by a dot, cross-slips onto a secondary glide plane until it bypasses the obstacle and then cross-slips again back onto P' parallel to P. (b) The cross-slipping part of the dislocation "drags out" edge dislocations on the secondary glide plane and then begins to expand again on P'. In the meantime, the portion of the dislocation not affected by the obstacle continues on P. (c) [Next page] TEM micrograph of double cross-slip of **b** = [001] screw dislocation in experimentally deformed olivine. (d) Optical micrograph of oxidized specimen of naturally deformed olivine showing multiple cross-slip of **b** = [100] dislocations. Arrows point to two dislocations that have cross-slipped several times. Arrowhead labels small prismatic loops produced by climb and partial annihilation of edge dipoles produced in the process (see Jin et al, 1990 for discussion). [(a) and (b) used by permission of *Contrib. Mineral Petrol.,* from Zeuch and Green (1977) Fig. 6a,b, p. 149; (c) used by permission of *Tectonophysics,* from Zeuch and Green (1984) Fig. 6C, p. 277; (d) used by permission of *Tectonophysics,* modified after Jin et al. (1990), Fig. 9, p. 35.]

potential to cross slip onto a different plane inclined to the primary slip plane. Diffusion is not required for this to occur so it imparts an additional mobility to screws not possessed by edges. Therefore, screws are able to circumvent obstacles by cross slip and then continue on their way by cross-slipping again back onto a plane parallel to the original one. This process of *double cross slip* not only provides for avoidance of barriers to glide but also can operate as a Frank-Read source (Fig. 11-5b). Nevertheless, a cross-slipping screw dislocation must "drag out" immobile edges as it goes, so the inability of edges to leave their glide plane remains a major impediment to extensive plasticity at low homologous temperatures.

11.2.2.5 Climb of dislocations

The restriction of edge dislocations to their glide planes is relaxed at high homologous temperatures (temperatures reasonably close to the melting point), where diffusion is active. Under those conditions, vacancies can migrate to or away from the dislocations, causing their extra half planes to shrink (positive climb) or extend (negative climb). With this extra degree of freedom, the ability of edge dislocations to circumvent obstacles to their movement is greatly increased. The progres-sive reduction of flow stress with increasing temperature in virtually all materials is a reflection of such greater ease of movement. The greater mobility conferred by climb also allows dislocations to respond more easily to the stresses they exert on each other, leading at high homologous temperature to much more mutual annihilation of dislocations of opposite sign and to ordered, lower-energy, arrangements of those that remain.

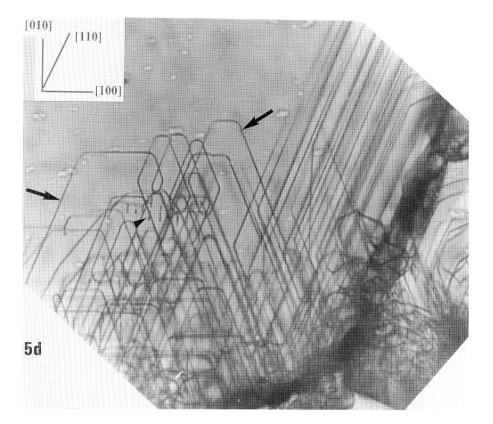

The concept of homologous temperature introduced in this section requires some clarification for application to the natural deformations of interest to geologists. Strictly speaking, the homologous temperature is the fraction of the melting temperature in Kelvins ($T_h = T/T_m$). It is a very useful concept for semiquantitative comparison of different materials because it scales various properties such as diffusion. Thus, ice at 245K is "hot" ($T_h = 0.90$), self-diffusion is rapid, and climb of edge dislocations is active. On the other hand, olivine at 1000K is "cold" ($T_h = 0.46$) and self-diffusion is slow; dislocation climb is not possible on laboratory time scales. However, if one compares ice at 245K and olivine at 1950K ($T_h = 0.90$ for both), they are much more similar in many of their properties. A commonly-cited rule-of-thumb is that diffusive processes become significant at $T_h > 0.5$. The complication in extrapolation of this concept to natural processes is that many of the effects of time and temperature are similar. For example, to enhance the effectiveness of a diffusive process, either one can wait longer at a given temperature or raise the temperature. In the former case, one allows a longer time for a given jump frequency to be effective and in the other case, one increases the jump frequency and hence the rate of the process. Thus, when comparing two materials at a given homologous temperature or a single material at different homologous temperatures, one must require the time scales of the two situations to be the same. The concept of homologous temperature loses any usefulness without that constraint. It is easy to see from this discussion that any thermally-activated process will occur at lower homologous temperatures in the earth, where the time scale is long, than it will in the laboratory, where the time scale may be more than ten orders of magnitude shorter.

11.2.3 Surface defects

Some surfaces in polycrystalline materials (such as grain and phase boundaries) exist from the creation of the material . Several other types can arise during deformation and they provide information on the amount of deformation, the mechanisms by which it was accomplished and the temperature, strain rate and stress involved.

11.2.3.1 Stacking faults

In discussion of the nature of dislocations in §11.2.2, it was assumed that the Burgers vector was a vector of the Bravais lattice, *ie* having a length equal to a dimension of the unit cell. If that is the case, then passage of a dislocation leaves perfect crystal behind it. However, even the unit cells of simple metals consist of more than one plane of atoms; oxides and silicates have large unit cells with multiple atomic planes. The larger the unit cell, the greater is the strain around a dislocation and the greater the energy of misfit. If the dislocation were to split into two or more *partial dislocations* (Fig. 11-6a), the total misfit energy would be reduced, but between the partials the crystal above the slip plane would be out of register with respect to the region below the slip plane. The surface separating these two regions is called a *stacking fault*. Thus, dislocations will spontaneously separate into partials if and only if the sum of the misfit energies of the two (or more) partials plus the energy of the stacking fault(s) between them is less than the misfit energy of the perfect (unit) dislocation. Since the interaction energy between partials decreases with their separation and the energy of a stacking fault increases with its area, there will exist an equilibrium spacing of partials that minimizes the total energy. For systems with low stacking-fault energy (e.g., orthopyroxene), the equilibrium spacing is large and stacking faults dominate the microstructure (Fig. 11-6b), whereas for systems for which the stacking fault energy is higher the separation can be less than the size of the image in the TEM, making it difficult to verify if separation into partials occurs at all. For example, the straightness of **b** = [001] screws in olivine (Fig. 11-6c) strongly suggests they are dissociated, but the results of attempts to image the separation with weak beam techniques

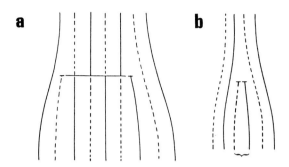

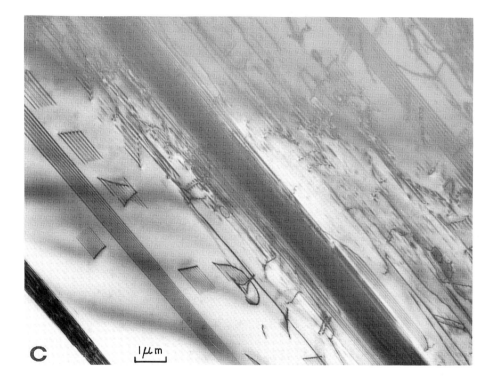

Figure 11-6. Splitting of dislocations. The crystal lattice is idealized as containing two different planes of atoms, the sum of which constitute the unit repeat in that direction. That is, the unit cell of the crystal consists of one plane of each type. If the dislocation is a unit dislocation (that is, if the Burgers vector is a unit-cell dimension), an edge dislocation has the structure shown in (a). If, however, the misfit energy is sufficiently high, the dislocation can split into two or more *partial* dislocations separated by a *stacking fault* as shown in (b). Both the dislocations and the stacking fault can be imaged in the TEM; an example from a naturally deformed orthopyroxene (bronzite) is shown in (c). The width of the stacking fault depends on its energy; if the combined energy of the fault and the partials is only slightly lower than that of a unit dislocation, the separation of the partials can be less than the image width of the dislocations. [(a) and (b) used by permission of Cambridge University Press, from Poirier (1985), Fig. 2.12c and d, p. 60; (c) used by permission of American Geophysical Union, from Green and Radcliffe (1972), Fig. 12, p. 151.

438

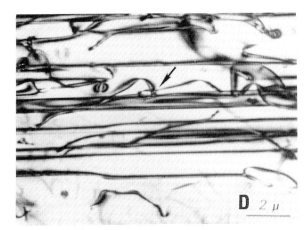

Figure 11-6 (d) shows extremely straight **b** = [001] screw dislocations in olivine, suggesting that they fall into the latter category. [Used by permission of *Tectonophysics*, from Zeuch and Green (1984) Fig. 6D, p. 277.]

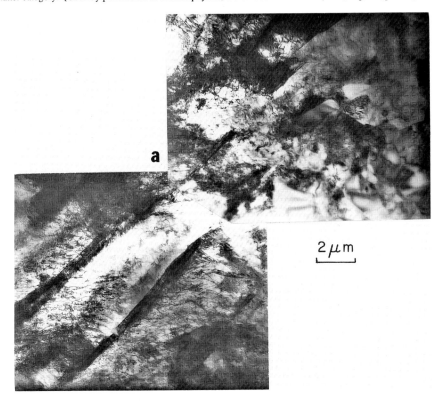

Figure 11-7. (a) At relatively low temperatures, dislocation mobility is low, overcoming of obstacles is difficult and materials are strong. As a consequence, they exhibit high densities of poorly-organized dislocations. Especially strong obstacles can lead to pile-ups of dislocations in certain glide planes, as shown here (NE-SW) in olivine deformed in the laboratory at 900°C, 150 MPa, e = 10^{-4} sec^{-1}. Despite the moderate temperature, the very high dislocation densities can drive recrystallization (upper right). (b) [See next page.] At higher temperatures, diffusion-controlled processes become much easier, leading to climb of edge dislocations into low angle boundaries, as shown here (NE-SW) in olivine deformed as in (a), except at 1300°C. [Used by permission of American Geophysical Union, from Green and Radcliffe (1972), Figs. 13 and 14, p. 152.]

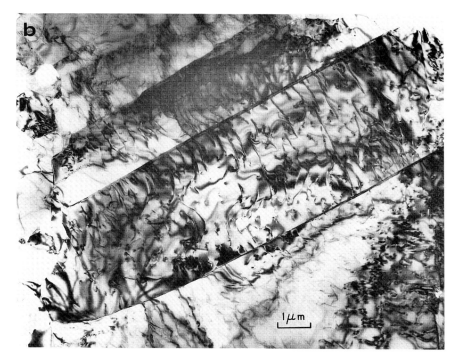

have been controversial. When separation of the partials is significant, tilting experiments in the microscope analogous to those for Burgers vector determination can determine the fault vector (e.g., see Kohlstedt and Vander Sande, 1973).

The presence of stacking faults places constraints on the mobility of partial dislocations that do not exist for unit dislocations. In particular, dissociated screw dislocations cannot cross-slip unless the stacking fault is pinched closed (eliminated). The screw can then move onto the cross-slip plane as a unit dislocation or it can dissociate again on the new plane. Pinching the stacking fault closed is a difficult process and therefore separation of dislocations into partials severely restricts cross-slip as a mechanism for circumventing obstacles to dislocation motion.

11.2.3.2 Subgrain boundaries

The availability of climb and cross-slip at elevated temperatures allows dislocations to move relatively freely in response to the stresses imposed upon them by each other and by an applied stress. As a consequence, dislocations of the same type but of opposite sign can attract and annihilate each other and those of like sign can organize themselves into low energy configurations. Figure 11-7a shows the pileups and dense, tangled configuration of distributed dislocations typical of moderate temperature ("cold-work") deformation of olivine. In contrast, Figure 11-7b shows the high-temperature development of simple, ordered arrays that divide the crystal up into small, more perfect, blocks called subgrains that differ in orientation by up to 10°. These two specimens were deformed at the same strain rate, hence the number of dislocations moving at any given time had to be the same. The markedly lower total density observed in Figure 11-7b (and the much lower stress required to deform this specimen) is due to the ability of the dislocations to surmount most barriers by climb and/or cross-slip and keep going. As a consequence, the only disloca-tions not moving in that specimen are those that have been "knitted" into the subgrain boundaries, whereas in the colder specimen, large numbers have become immobilized and trapped, leading to increase in the flow stress.

In deformation at high homologous temperatures, as more dislocations are generated and move within individual subgrains, some of them enter the subgrain walls, thereby decreasing the spacing between dislocations in the boundary and increasing the misorientation across it. Analysis of the dislocations in subgrain boundaries generally shows them to be restricted to one or two Burgers vectors and for those of each Burgers vector to be mostly of the same sign. Moreover, adjacent boundaries are in most cases also made of dislocations of the same sign. These observations show that subgrain walls are not just graveyards for dislocations; the specific geometry of these walls shows that they constitute a means whereby a crystal can be bent extensively while still remaining relatively perfect in most regions. This is easily understood if we step back for a minute and consider that flow of an aggregate of crystals (a rock) inevitably produces incompatibilities at grain- and phase boundaries, causing the deformation to become heterogeneous. Crystals are forced to bend and rotate by the constraints placed on them by their neighbors. Under conditions where diffusion is active, dislocations that are necessary for production of the strain in the crystals mostly either meet up with others of opposite sign and annihilate each other or they are pushed to grain boundaries and vanish. The remaining dislocations are the ones that accommodate the strain inhomogenities imposed on a crystal by the other crystals adjacent to it. It is these "geometrically necessary" dislocations (Ashby, 1970) that collect into the subgrain boundaries as the bending and twisting progress. The number of dislocations within subgrain walls changes and the misorientation across them either increases or decreases with progressive deformation depending on whether the local, time-varying, boundary constraints require increase or decrease of the heterogeneity of deformation in that crystal. Thus, the dislocations found in the subgrain walls tell us which Burgers vectors were responsible for the intracrystalline deformation, and the distribution and types of walls provides information about how the crystal responded to the stresses applied to its boundaries by its neighbors. Development of these inhomogeneities is the principal process by which (under conditions where climb and cross slip are active) crystals of symmetry less than isometric accomodate to the fact that they generally do not have sufficient numbers and types of slip systems to produce the shape changes necessary to relieve the locally-applied stresses. The *size* of subgrains varies with the applied stress, reflecting the relative kinetics of dislocation motion in response to the applied stress and to interaction stresses.

In contrast to this process of subgrain formation, crystals deformed at lower homologous temperatures do not have the additional elemental deformation mechanisms introduced by climb and cross slip. As a consequence, when the local stresses imposed by incompatibilities at grain boundaries require strains which cannot be achieved by the currently active glide systems, the crystals can only respond by activation of additional glide systems or by development of displacement discontinuities at grain boundaries (grain-boundary sliding). The latter process is also difficult at low homologous temperatures, hence, if additional glide systems cannot be activated at reasonable stresses, microfracture will occur, or even faulting.

11.2.3.3 *Grain boundaries*

At moderate to high homologous temperatures, dislocation rearrangement processes can lead to small regions of essentially perfect crystal surrounded by less perfect crystal. If the misorientation across the boundary of such a nucleus becomes greater than about 10°, the boundary becomes very mobile, enabling the more perfect region to grow, replacing the high-energy (deformed) material with essentially perfect crystal of a different orientation. This process of *recrystallization*, in which a boundary sweeps through a region and removes most or all of the dislocations, contrasts with the *recovery* processes described above wherein individual dislocations migrate to clear out local regions and establish boundaries around them. In fact, the two processes can operate either in competition (in which sufficiently rapid recovery can inhibit recrystallization) or in cooperation, in which progressive rotation of subgrains (by continual incorporation of more and more dislocations of the same sign into subgrain walls) leads to development of the high-angle boundaries necessary for rapid boundary migration.

11.2.3.4 Phase boundaries

Most phase boundaries are simply the result of element partitioning into the various phases of the stable assemblage and have little to do with deformation. However, if changing physical conditions induce exsolution of one phase from another or isochemical phase transformations, deformation can be affected in a variety of ways. Examples are the interference with dislocation motion of precipitated fluid or solid phases, production of a martensitic phase transformation that contributes to straining, or development of incoherent nucleation and growth of a denser phase that also can contribute to straining or lead to failure (see below).

11.3 MECHANISMS OF FLOW AND
CHARACTERISTIC MICROSTRUCTURES

Most processes by which rocks and minerals flow are the same as in metals: (1) dislocation glide; (2) mechanical twinning; (3) grain-boundary sliding; (4) diffusion creep. In fine-grained aggregates the latter two processes are in many cases combined in the phenomenon called *superplasticity*. In addition, at low temperatures rocks can flow by *cataclasis* (distributed microcracking).

11.3.1 Cataclasis

Cataclasis is macroscopically ductile flow by pervasive microfracturing. It produces optical microstructures with wide-spread small shear offsets and finer-scale deformation that appears continuous and therefore can be mistaken for evidence of plastic (dislocation) flow (Fig. 11-8a). TEM of such materials shows in some cases a mixture of dislocation mechanisms and microcracking but under appropriate conditions, at least in plagioclase feldspar aggregates, evidence for dislocation activity is absent and the flow is entirely by microcracking (Fig. 11-8b). With increasing temperature and pressure, cataclastic flow gives way to dislocation flow. In some minerals this transition may be induced by pressure alone (e.g., calcite; Fredrich et al., 1989), but in others, an increase in both temperature and pressure is required (e.g., quartz; Hirth and Tullis, 1991). This transition in the micromechanics of flow is not what is usually referred to as the "brittle-to-ductile" transition because the latter refers to the change from macroscopically brittle behavior (faulting) to macroscopically ductile flow (distributed strain). This usually has been interpreted as the activation of dislocation flow as a consequence of changing environmental conditions that reduced the stress a rock could support to below the fracture stress. However, as pointed out by Tullis and Yund (1992), in regions of the crust where feldspar is the stress-supporting matrix, this transition may reflect the conditions under which distributed microcracking (cataclasis) replaces localized cracking (faulting). The geological implications for the two kinds of brittle-ductile transition are significantly different .

442

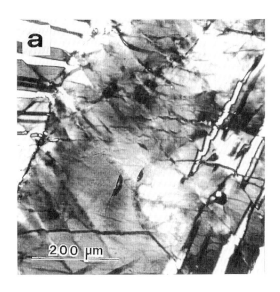

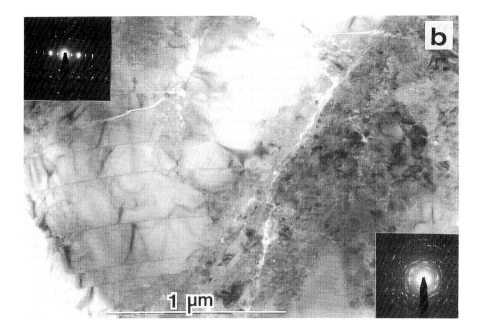

Figure 11-8. Cataclastic microstructures. Under appropriate conditions, at least in plagioclase feldspar, purely cataclastic deformation can appear in the optical microscope to have been accomplished by plastic (dislocation) flow. (a) Optical micrograph (crossed polarizers) of sample of Bushveld anorthosite shortened 10% at 300°C, 1000 MPa. In addition to the microcracks separated by 20-50 mm, undulatory extinction is prominent. (b) TEM micrograph of the same specimen shows submicron-spaced microcracks and local crushing. SAEDPs show that shear displacements have occurred on the microcracks (upper left) and that extensive rotations have taken place between fragments of the crushed zone (lower right). No dislocations were reported in either material. [Used by permission of Academic Press, from Tullis and Yund (1992) Figs. 4B and 6A.]

11.3.2 Dislocation glide and creep

At lower homologous temperatures in all crystalline materials and extending to quite high temperatures in some (e.g., those with the diamond-cubic structure), the Peierls force dominates the resistance to dislocation motion. In such cases it is glide of the dislocations that controls the rate of deformation.

At higher homologous temperatures, thermal agitation is increased and the tendency of the Peierls barriers to restrict glide to the easiest systems is reduced, thereby promoting cross slip of screw dislocations. In addition, processes involving diffusion become much more rapid and leave their mark on the microstructure, principally through enhancement of climb of edge dislocations. In simple materials like metals, the evidence is very strong that in this high temperature, *creep*, regime, the rate-controlling step in deformation is the climb of edge dislocations. In more complicated materials such as minerals, although the high temperature microstructure is similarly recovered, there is still considerable controversy over whether or not diffusion is the rate-controlling step.

The nucleation and growth of new crystals during high temperature flow provides an alternative mechanism for the maintenance of a relatively defect-free, soft, crystalline aggregate. In cases where such dynamic recrystallization is abundant, it could conceivably be the rate-controlling step. The migration of grain boundaries also involves diffusion, so if flow is controlled by climb of edge dislocations or by recrystallization, the ultimate control on flow rates will be provided by the rate of self-diffusion.

11.3.3 Mechanical twinning

Mechanical twinning cannot produce large strains because it is a naturally self-limiting process. That is, once a crystal has been completely twinned (a process involving only small strains), it cannot flow further by this mechanism unless the stress system is changed so that it can untwin, thereby reducing the finite strain. Nevertheless, twinning can dominate over all other deformation mechanisms for small strains if the critical resolved shear stress for twinning is sufficiently low. Calcite and some of the hexagonal metals provide examples of this phenomenon. In other systems, e.g., Si, mechanical twinning occurs only at very high stresses and so it is observed only under conditions where dislocation mechanisms are restricted by a very high Peierls force.

Although mechanical twinning is known to take place by glide of partial dislocations, details have been lacking as to how the particular partials that produce the twinning are generated without the accompanying partials that would return the structure to its original orientation, and how the twinning partials are coordinated to produce a perfect twin many atomic layers thick. A model developed to explain the process in Si (Pirouz, 1987) now also appears to satisfy the observations on twinning in corundum (A.H. Heuer, personal comm., 1992) and may prove to be applicable to many minerals. Figure 11-9 shows an example of mechanical twinning in diopside (Kirby and Christie, 1977).

11.3.4 Superplasticity and diffusion creep

Superplastic flow of fine-grained materials occurs by a process involving extensive grain-boundary sliding; it has now been demonstrated to occur in a wide variety of metals, ceramics and rocks. It has been proposed that dislocation processes (e.g., Suery and Mukherjee 1985) or diffusion (e.g., Ashby and Verrall, 1973) accommodate the misfits that must invariably occur if grain boundaries are sliding. It may well be that both accom-

444

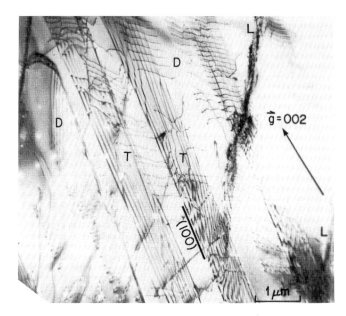

Figure 11-9. Mechanical twinning in diopside; **g** = 002. Deformation of polycrystalline diopside at 400°C, 1600 MPa, produced dislocation arrays (D) and microtwins (T) on (100), and lamellar features of unknown origin (L) on (101). The twins can be recognized by the fringe contrast on their boundaries. [Used by permission of *Physics and Chemistry of Minerals*, from Kirby and Christie (1977), Fig. 4A, p. 151.]

modation processes can operate in different materials or in the same material under different conditions.

It is a common misconception in the geological literature that superplastic flow is a low strain-rate phenomenon. That perception has arisen because in most studies that have shown this flow mechanism to operate, the grain size of the experimental material has been constant and a transition in flow mechanism is observed with decreasing strain rate (e.g., Schmid et al., 1977). However, in studies where grain size is a variable, it has been shown that superplasticity can be produced at very high strain rates ($>10^2$ sec^{-1}) in sufficiently fine-grained material (Mukherjee et al., 1989). That is, for a fixed grain size, grain-boundary sliding is favored by decrease in strain rate, but the maximum strain rate at which grain-boundary sliding is important increases with decreasing grain size. An extreme case of this phenomenon is the superplastic flow of extraordinarily fine-grained γ-Mg_2GeO_4 in fault zones developed in the α-phase (olivine structure) deformed in the γ-phase (spinel structure) stability field (Tingle et al., 1992a,b).

One of the difficulties of microstructural study of superplastic materials is that, rather than exhibiting specific microstructures, they are characterized by the absence of structures typical of other flow mechanisms. For example, Vaughan and Coe (1981) showed that Mg_2GeO_4 polycrystals, when transformed from the olivine to the high pressure spinel phase, exhibited a rheology indicative of superplastic flow, but the microstructures of their specimens just showed fine-grained material with low defect densities (Fig. 11-10). Negative evidence does not make for a powerful argument and interpretation of naturally deformed materials as superplastic simply because they are fine-grained often is met with skepticism, and rightly so. For example, relatively defect-free crystals can be found in natural mylonites of various minerals. Many of these mylonites, however, show strong preferred crystallographic orientations, which argues very strongly that the strain was

Figure 11-10. Fine-grained (1-3 mm) polycrystalline aggregate of γ-Mg$_2$GeO$_4$ (spinel structure) with low defect densities produced by transformation from the α-phase (olivine structure) at ~1300K, 1500 MPa. Vaughan and Coe (1981) showed that such material has superplastic rheology. [Used by permission of American Geophysical Union, from Green et al. (1992) Fig. 3g.]

produced by dislocation processes; superplastic deformation produces little preferred orientation. Other fine-grained mylonites that lack preferred orientation remain candidates for natural superplastic flow (e.g., Schmid, 1982), but they still lack *positive* evidence for the process.

When superplastic flow is accompanied by extensive self-diffusion, it becomes a chicken-and-egg argument as to whether the flow process is grain-boundary sliding accommodated by self-diffusion or diffusion creep accompanied by grain-boundary sliding. Neither of these elemental processes can operate in isolation. If diffusion creep were to operate completely by itself, every crystal would have to deform identically to its neighbors and the crystals should be greatly flattened and elongated after large strains. Such microstructures are never observed; strong grain-shape fabrics in minerals with little inherent structural anisotropy (e.g., quartz, calcite, olivine) develop only from dislocation flow processes and are accompanied by characteristic preferred orientations.

11.4 PHASE TRANSFORMATION UNDER STRESS

The presence of nonhydrostatic stress can affect polymorphic phase transformations in several ways (Green et al., 1992). The stress can have a direct effect, activating a shear-induced (martensitic) mechanism or inducing anisotropic nucleation and growth of the new phase. Alternatively, it can have an indirect effect through production of very high defect densities that, in turn, enhance the kinetics of transformation. These various stress effects can lead to (1) growth of a phase outside of its stability field; (2) enhanced kinetics of

446

growth of a phase inside its stability field; (3) change in transformation mechanism from one type to another; (4) triggering of an instability leading to faulting.

Direct production of a polymorph outside its stability field with high stresses has been demonstrated by Coe and Kirby (1975; cf. Kirby, 1976) and by Pirouz et al. (1990), shown in Figure 11-11. Indirect growth of metastable phases after large strains at high strain rates was reported by Newton et al. (1969) for calcite → aragonite and by Green (1972) for quartz → coesite. Fortunately, the high stresses and strain rates characteristic of these experiments that have produced high pressure polymorphs outside of their stability fields are not likely to occur under natural metamorphic conditions other than meteorite impact.

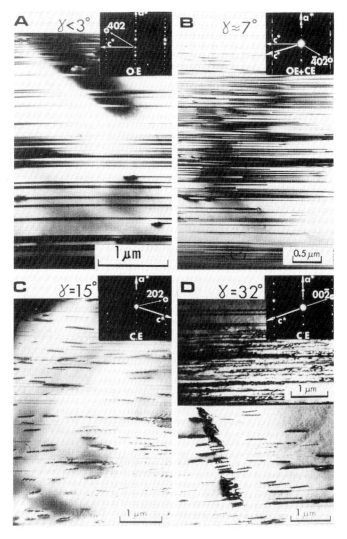

Figure 11-11. Martensitic transformations. (a-d) Progressive transformation from ortho- to clino-enstatite with increasing shear strain.

447

Figure 11-11 (e-f) Mechanical twinning (narrow lamellae oriented E-W) and martensitic lamella of hexagonal silicon in cubic host crystal. High resolution image in (f) shows two hexagonal lamellae (H) and a microtwin between them (T), with the thinner hexagonal lamella terminating near the left of the micrograph.

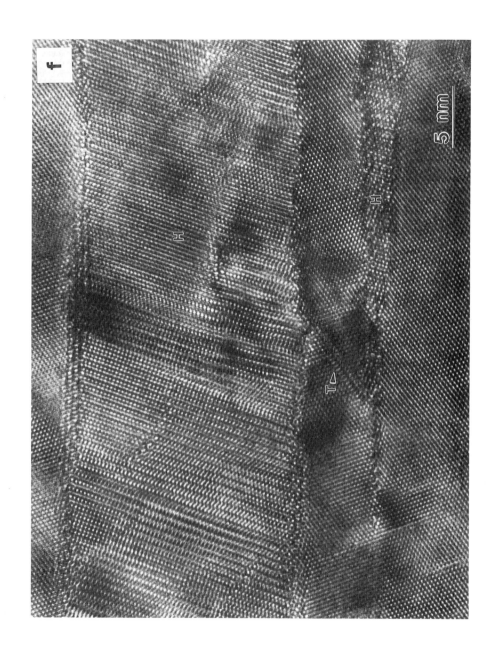

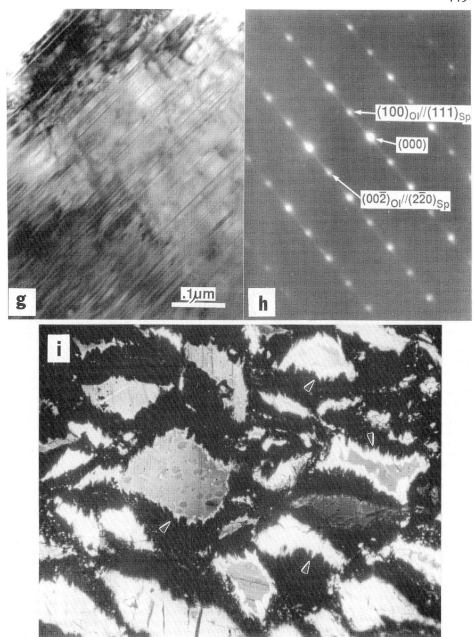

Figure 11-11 (g-i) Lamellae of γ-Mg₂GeO₄ (spinel structure; cubic oxygen sublattice) in a host of α-Mg₂GeO₄ (olivine structure; pseudohexagonal oxygen sublattice) and SAEDP showing streaking due to the high density of lamellae. Burnley et al. (1991) showed that this transformation mechanism operates only under differential stresses >900 MPa; (i) Optical micrograph (crossed polarizers) showing anisotropic growth of isotropic γ-Mg₂GeO₄ spinel (black), into anisotropic olivine (various shades of gray). Arrowheads point out examples of cusps of residual olivine (bright) between growing crystals of spinel. Note that the cusps of olivine are strongly aligned, indicating that the spinel crystals are growing preferentially parallel to the maximum compressive stress (N-S).

450

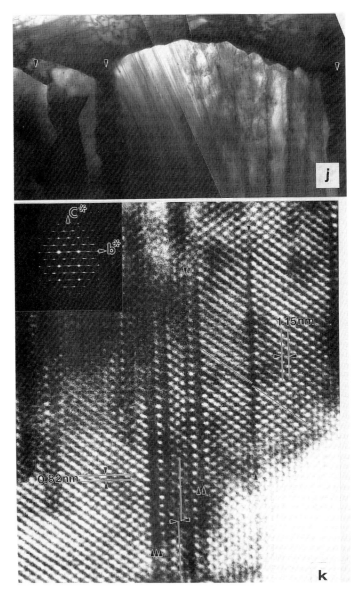

Figure 11-11 (j) TEM mosaic showing the detailed relationships between olivine (top) and spinel under conditions of anisotropic growth of γ-Mg$_2$GeO$_4$. The spinel crystals are riddled with defects, and have nucleated by the incoherent mechanism, as indicated by the lack of topotaxy between the single-crystal olivine host and the randomly-oriented spinel crystals. Arrowheads point out cusps of olivine (black) between growing crystals of spinel. This transformation mechanism operates at differential stresses <900 MPa and sufficiently high temperatures (reproduced with permission from Green et al., 1992). (k) High density of lamellae of β-Mg$_2$SiO$_4$ (pseudocubic oxygen sublattice) in host of γ-phase (cubic oxygen sublattice). Brearley et al. (1992) interpret this structure as partial transformation of metastable γ to stable β by a martensitic mechanism under essentially hydrostatic pressure. [(a-d) used by permission of *Contributions to Mineralogy and Petrology,* from Coe and Kirby (1975) Fig. 2, p. 35. (e-f) used by permission of *Acta Metall. Mater.,* from Pirouz et al. (1990) Figs. 5 and 7, pp. 317-318. (g-j) used by permission of American Geophysical Union, from Green et al. (1992), Figs. 1a,b; 2b,c. (k) used by permission of *Physics and Chemistry of Minerals,* from Brearley et al. (1992) Fig.7a.]

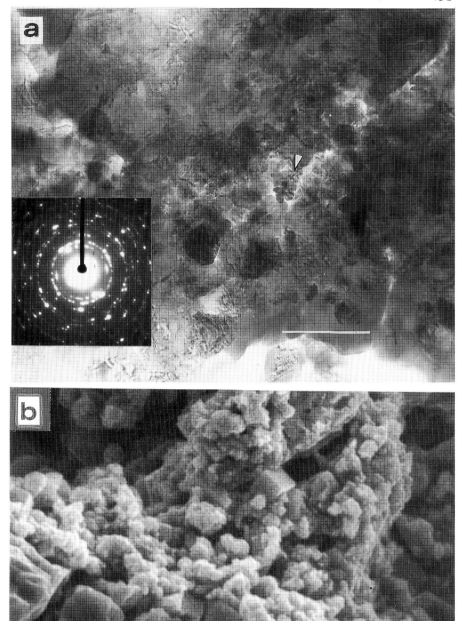

Figure 11-12. Anticrack faulting microstructures in Mg_2GeO_4. (a) TEM micrograph showing extremely fine-grained γ-phase grown within anticracks that precede and trigger high-pressure faulting. SAEDP shows the random orientation of the γ crystals. Scale bar = 0.5 μm. [Used by permission of *Journal of Geophysical Research,* from Burnley et al. (1991) Fig. 10, p. 434.] (b) SEM micrograph of etched specimen showing fine-grained γ-phase lining fault zone and fragment of included olivine (larger, etched, feature in right center). The low value of the sliding resistance on such spinel-lined faults coupled with its small pressure dependence and high sensitivity to sliding rate (Tingle et al., 1992a,b) shows this material to be superplastic. Scale bar = 10μm. [Micrograph used by permission of the Geological Society, from Green and Burnley (1990) Fig. 3a.]

For conditions in which a phase is transported out of its own stability field and into that of a different polymorph under stress, the stress may control the mechanism by which the transformation is accomplished. For example, under differential stresses greater than 900 MPa, the olivine $\rightarrow$ spinel transformation in Mg_2GeO_4 occurs by a martensitic mechanism (Burnley and Green, 1989; Burnley, 1990), shown in Figure 11-11g,h. Under lower stresses, the same transformation occurs by incoherent nucleation and growth (Vaughan et al., 1984). Furthermore, in this system, and in the olivine $\rightarrow$ β-phase transformation in natural olivine, reaction under moderate stresses induces a bulk anisotropy of the transformation (Fig. 11-11i,j); the nucleation rate of the high density phase is greater on grain-boundaries normal to the greatest compressive stress, σ_1, and growth occurs fastest parallel to σ_1 (Vaughan et al., 1984; Green, 1986; Green et al., 1992). The anisotropy effect should occur only during transformations for which a significant volume change occurs and the martensitic mechanism should be favored for systems in which the volume change is small. Indeed, when the volume change is small, it is possible for martensitic mechanisms to operate in the absence of stress (e.g., in cobalt; Votava, 1960; Poirier, 1981. An example in which a martensitic mechanism may be important in the earth is the spinel $\rightarrow$ β-phase transformation in Mg_2SiO_4 (Brearley et al., 1992), shown in Figure 11-11k.

A faulting instability triggered by phase transformation has been documented for the olivine $\rightarrow$ spinel transformation in Mg_2GeO_4 (Green and Burnley, 1989; 1990; Burnley et al., 1991) and for the olivine $\rightarrow$ β-phase transformation in natural olivine (Green et al., 1990), and is implicated for the ice lh $\rightarrow$ II transformation (Kirby et al., 1991). In this phenomenon, the effect of stress is to restrict the initial distribution of the high density phase to Mode I microanticracks (normal to maximum compression) before failure begins. The material within the anticracks has a grain size of a few tens of nanometers (Fig. 11-12a) and hence it is superplastic (Tingle et al., 1992a,b) and much weaker than the metastable olivine. An instability is triggered as the anticracks increase in size and number, leading to faulting and incorporation of the superplastic "lubricant" into the fault zone (Fig. 11-12b). This process is the probable mechanism of deep-focus earthquakes and could perhaps operate similarly at shallower depths, for example during the calcite $\rightarrow$ aragonite transformation.

ACKNOWLEDGMENTS

I thank all of my colleagues and students who have contributed so greatly to the aspects of my own work that are discussed and illustrated here, and to the NSF for its generous support of those projects. I also thank the many authors and publishers of the other figures reproduced here for permission to use their work. The manuscript has benefited significantly from reviews by Gordon Nord and an anonymous reviewer.

REFERENCES

Ashby, M.F. (1970) The deformation of plastically non-homogeneous crystals. Phil. Mag. 21, 399-424.

Ashby, M.F. and Verrall, R.A. (1973) Micromechanisms of flow and fracture and their relevance to the rheology of the upper mantle. Phil. Trans. Roy. Soc. London 288A, 59-95.

Brearley, A.J., Rubie, D.C. and Ito, E. (1992) Mechanisms of the transformations between the α, β and γ polymorphs of Mg_2SiO_4 at 900°C and 15 GPa. Phys. Chem. Minerals (in press).

Burnley, P.C. (1990) The effect of nonhydrostatic stress on the olivine-spinel transformation in Mg_2GeO_4, Ph.D. dissertation, Univ. of California, Davis, CA.

Burnley, P.C. and Green, H.W. (1989) Stress dependence of the mechanism of the olivine-spinel transformation. Nature 338, 753-756.

Burnley, P.C., Green, H.W. and Prior, D.J. (1991) Faulting associated with the olivine to spinel transformation in Mg_2GeO_4 and its implications for deep-focus earthquakes. J. Geophys. Res. 96, 425-443.

Coe, R.S. and Kirby, S.H. (1975) The orthoenstatite to clinoenstatite transformation by shearing and reversion by annealing: Mechanism and potential applications. Contrib. Mineral. Petrol. 52, 29-55.

Fredrich, J.T., Evans, B. and Wong, T.-f. (1989) Micromechanics of the brittle to plastic transition in Carrara marble. J. Geophys. Res. 94, 17607-17617.

Green, H.W. (1972) Metastable growth of coesite in highly strained quartz. J. Geophys. Res. 77, 2478-2482.

Green, H.W. (1986) Phase transformation under stress and volume transfer creep. In Mineral and Rock Deformation: Laboratory Studies–The Paterson Volume, B.E. Hobbs and H.C. Heard, eds., Geophys. Monograph 36, 201-211.

Green, H.W. and Burnley, P.C. (1989) A new, self-organizing mechanism for deep-focus earthquakes. Nature 341, 733-737.

Green, H.W. and Burnley, P.C. (1990) The failure mechanism for deep-focus earthquakes. In Deformation Mechanisms, Rheology and Tectonics. R.J. Knipe and E.H. Rutter eds., Geol. Soc. Special Pub. 54, 133-141.

Green, H.W. and Radcliffe, S.V. (1972) Deformation processes in the upper mantle. In Flow and Fracture of Rocks – The Griggs Volume. H.C. Heard et al., eds., Geophys. Monograph 16, 139-156.

Green, H. W., Young, T.E., Walker, D. and Scholz, C.H. (1990) Anticrack-associated faulting at very high pressure in natural olivine. Nature 348, 720-722.

Green, H.W., Young, T.E., Walker, D. and Scholz, C.H. (1992) The effect of nonhydrostatic stress on the $\alpha \rightarrow \beta$ and $\alpha \rightarrow \gamma$ olivine phase transformations. In High-Pressure Research: Applications to Earth and Planetary Sciences. Y. Syono and M.H. Manghnani, eds., Geophys. Monograph (in press).

Gueguen, Y. (1977) Dislocations in mantle peridotite nodules. Tectonophysics 39, 231-254.

Hirsch, P.B., Howie, A., Nicholson, R.B., Pashley, D.W. and Whelan, M.J. (1965) Electron Microscopy of Thin Crystals. Butterworths, London.

Hirth, G. and Tullis, J. (1991) Mechanisms responsible for the brittle-ductile transition in experimentally deformed quartz aggregates. EOS, Trans. Am. Geophys. Union 72, 286.

Hull, D. and Bacon, D.J. (1984) Introduction to Dislocations, 3rd ed. Pergamon Press, Oxford and New York.

Jin, Z.-M., Green, H.W. and Borch, R.S. (1990) Microstructures of olivine and stresses in the upper mantle beneath Eastern China. Tectonophysics 169, 23-50.

Karato, S.-I. (1986) Scanning electron microscope observation of dislocations in olivine. Phys. Chem. Minerals 14, 245-248.

Kirby, S.H. (1976) The role of crystal defects in the shear-induced transformation of orthoenstatite to clinoenstatite. In Applications of Electron Microscopy in Mineralogy. H.-R. Wenk, ed., 465-472, Springer, Berlin-Heidelberg-New York.

Kirby, S.H. and Christie, J.M. (1977) Mechanical twinning in diopside $Ca(Mg,Fe)Si_2O_6$: Structural mechanism and associated crystal defects. Phys. Chem. Minerals 1, 137-163.

Kirby, S.H., Durham, W.B. and Stern, L. (1991) Mantle phase changes and deep-earthquake faulting in subducting lithosphere. Science 252, 216-225.

Kirby, S.H. and Green, H.W. (1980) Dunite xenoliths from Hualalai volcano: Evidence for mantle diapiric flow beneath the island of Hawaii. Am. J. Science 280-A, 550-575.

Knipe, R.J. (1990) Microstructural analysis and tectonic evolution in thrust systems: examples from the Assynt region of the Moine Thrust Zone, Scotland. In Deformation Processes in Minerals, Ceramics and Rocks. D.J. Barber and P.G. Meredith, eds., Unwin Hyman, London.

Kohlstedt, D.L., Goetze, C., Durham, W.G. and Vander Sande, J.B. (1976) A new technique for decorating dislocations in olivine. Science 191, 1045-1046.

Kohlstedt, D.L. and Vander Sande, J.B. (1973) Transmission electron microscopy investigation of the defect microstructure of four natural orthopyroxenes, Contrib. Mineral. Petrol. 42, 169-180.

McLaren, A.C. (1991) Transmission Electron Microscopy of Minerals and Rocks, Cambridge Univ. Press, Cambridge and New York.

Mukherjee, A.K., Bieler, T.R. and Chokshi, A.M. (1989) Superplasticity in metals and ceramics. In Materials Architecture. J.B. Bilde-Sorensen, ed., 207-233, Riso National Laboratory, Roskilde, Denmark.

Newton, R.C., Goldsmith, J. R. and Smith, J.V. (1969) Aragonite crystallization at reduced pressures and its bearing on aragonite in low-grade metamorphism. Contrib. Mineral. Petrol. 22, 335-348.

Nicolas, A. and Poirier, J-P (1976) Crystalline Plasticity and Solid State Flow in Metamorphic Rocks. Wiley, London and New York.

Pirouz, P. (1987) Deformation mode in silicon, slip or twinning? Scripta Metall. 21, 1463-1468.

Pirouz, P., Chaim, R., Dahmen, U. and Westmacott, K.H. (1990) The martensitic transformation in silicon - I. Experimental observations. Acta Metall. Mater. 38, 313-322.

Poirier, J.-P. (1981) Martensitic olivine-spinel transformation and plasticity of the mantle transition zone. In Anelastic Properties and Related Properties in the Earth's Mantle. Geodynamics Ser. 4, 113-117, Am. Geophys. Union, Washington, D.C.

Poirier, J.-P. (1985) Creep of Crystals. Cambridge Univ. Press, Cambridge, UK.

Schmid, S. (1983) Microfabric studies as indicators of deformation mechanisms and flow laws operative in mountain building. In Mountain Building Processes. K.J. Hsü, ed., 95-110, Academic Press, London and New York.

Schmid, S., Boland, J.N. and Paterson, M.S. (1977) Superplastic flow in fine grained limestone. Tectonophysics 43, 257-291.

Suery, M. and Mukherjee, A.K. (1985) Superplasticity—Correlation between structure and properties. In Creep Behavior of Crystalline Solids. B. Wilshire ed., 137-200, Pineridge Press, Swansea, UK.

Tingle, T.N., Green, H.W., Scholz, C.H., Koczynski, T.A. and Burnley, P.C. (1992a) Pressure independence of the sliding stress on faults in Mg_2GeO_4 generated by the anticrack mechanism. EOS, Trans. Am. Geophys. Union 73, Spring Meeting Suppl., 297.

Tingle, T.N., Green, H.W. and Scholz, C.H. (1992b) The rheology of faults triggered by the olivine $\rightarrow$ spinel transformation in Mg_2GeO_4 and its implications for the mechanism of deep-focus earthquakes. In J.M. Christie issue of the J. Structural Geol. (submitted).

Tullis, J. and Yund, R. (1992) The brittle-ductile transition in feldspar aggregates: An experimental study. [in the Brace volume] B. Evans, ed., Academic Press, N.Y. (in press).

Vaughan, P.J., Green, H.W. and Coe, R.S. (1984) Anisotropic growth in the olivine-spinel transformation of Mg_2GeO_4 under nonhydrostatic stress. Tectonophysics 108, 299-322.

Vaughan, P.J. and Coe, R.S. (1981) Creep mechanism in Mg_2GeO_4: Effects of a phase transition. J. Geophys. Res. 86, 389-404.

Votava, E. (1960) Electron microscopic investigation of the phase transformation of thin cobalt samples. Acta Metall. 8, 901-904.

Zeuch, D.H. and Green, H.W. (1977) Naturally decorated dislocations in olivine from peridotite xenoliths. Contrib. Mineral. Petrol. 62, 141-151.

Zeuch, D.H. and Green, H.W. (1984) Experimental deformation of a synthetic dunite at high temperature and pressure. II. Transmission electron microscopy. Tectonophysics 110, 263-296.

Chapter 12. IMAGING TRANSFORMATION-INDUCED MICROSTRUCTURES

Gordon L. Nord, Jr. United States Geological Survey,
Mail Stop 969, Reston, Virginia 22092, U.S.A.

12.1 INTRODUCTION

Transformation-induced microstructures include *twin domains, antiphase domains* and *tweed textures*. These microstructures are generated during phase transformations that involve a change in space group symmetry. *Domains, in this context,describe chemically and structurally identical regions in the transformed crystal.* The reader is cautioned that the term "domains" is also commonly used to describe a variety of features such as differences in composition or even diffusion rates. The geometrical relationships of one domain to another are a function of the point symmetry (twin domains) or translation symmetry (antiphase domains) that are lost during the transformation. Tweed textures are metastable microstructures that arise from orthogonal lattice distortions associated with the softening of acoustic modes at the phase transition. The subject of this chapter is the imaging of these microstructures by amplitude or phase contrast transmission electron microscopy.

Microscopically, phase transformations are driven by the creation of order. This order can arise in several ways, such as (1) changes in site occupancies (order-disorder transitions); (2) displacements in atomic positions (displacive transitions); and (3) alignment of magnetic or electrical dipoles. In the case of twinning, if the twin boundaries can be moved by the application of a field, then the crystal is considered *ferroic* (Newnham, 1974). If it is difficult or impossible to move the twin boundaries by applying a field the material is known as *hard*. In many materials, twins and antiphase domains can be induced by an external field. As the external field is removed the domains disappear in some substances but persist in others. The order that occurs in the absence of an inducing field is known as *spontaneous* order. The terms *ferroelectric, ferroelastic* and *ferromagnetic* are used for primary ferroic crystals in which the twin or antiphase domains differ respectively in the sign of spontaneous polarization, spontaneous strain, and spontaneous magnetization. All of these materials exhibit a hysteresis effect. Other materials that possess a macroscopic spontaneous strain but in which there is no possible physical mechanism to reorient that strain are called *co-elastic* (Salje, 1990) or improper ferroelastics. Our discussion will be limited to microstructures resulting from ferroelastic and co-elastic structural phase transitions in minerals although keep in mind that domain microstructures resulting from magnetic (Chapman, 1989) or electric transitions can also be imaged in the TEM .

The relationship of microstructures to phase transitions is interesting historically. Some transformation twins are visible by eye, and the fact that they are generated by heating and cooling through a critical temperature was well known by the early part of this century. Interestingly, in Buerger's (1951) review of the crystallographic aspects of phase transitions, he stated that displacive transitions almost invariably result in twins, but he made no mention of antiphase microstructures in order-disorder transitions. At the same conference, Siegel (1951) discussed "out-of-phase" regions in AB alloys, although the only evidence for antiphase domains (APDs) in alloys at this time was broadening of x-ray lines. Images of antiphase domain boundaries (APBs) had to wait until electron diffraction contrast theory was developed. The first images of APBs were seen in AuCu by Glossop and Pashley (1959), in Cu₃Au by Fisher and Marcinkowski (1961), and in Fe₃Al by Marcinkowski and Brown (1962).

TEM studies of transformation-induced microstructures in minerals followed shortly with the investigations of tweed microstructures in adularia and orthoclase imaged by McConnell (1965) and Dauphiné twins in quartz, imaged by McLaren and Phakey (1969). The development of the ion-thinner and funding by the NASA Lunar and Planetary Science Program saw a spurt of activity in the 1970s which led to the imaging of antiphase domains in the lunar minerals bytownite, anorthite and pigeonite. Transmission electron microscopy in mineralogy had undeniably "come of age" by the second NATO meeting on feldspars at Manchester in 1972 (MacKenzie and Zussman, 1974). Minerals that once could be deciphered only through complex x-ray diffraction patterns suddenly could be seen in direct images, and it became immediately apparent that feldspars and other minerals are alive with microstructures. The first textbook relating TEM to mineralogy appeared, *Electron Microscopy in Mineralogy* (Wenk, 1976), which to this day is required reading for geologists intent on using transmission electron microscopy.

Why are transformation microstructures useful to the Earth sciences? On one hand the Earth scientist is looking for information of historical interest, such as markers along pressure-temperature-time paths of rocks or even large slabs of the crust that are now at the Earth's surface. The mere presence of domain microstructures may indicate that the material passed through the transition temperature or pressure, and the size and morphology of the domains may reveal the rate of descent through and below the transition point. On the other hand, the Earth scientist who is interested in active processes such as earthquakes, volcanic eruptions and magnetic or electrical activity needs information about the effect of domain microstructures on the physical properties of rocks. The mobility of domain boundaries or walls is particularly important for magnetic, electrical and elastic hysteresis properties. In addition, domain boundaries or walls are structurally different than the domains themselves and if the boundaries are volumetrically large they can modify or even reverse the physical properties of the low symmetry phase (i.e., the domains). Moreover, because the boundaries represent metastable structure, they provide a surface for enhanced nucleation of exsolution products. Finally, the presence of stranded metastable microstructures, such as tweed textures, may alter physical properties, such as diffusion rates and closure temperatures.

Information with regard to domain microstructure that might prove useful for addressing problems in the Earth sciences would include:

(1) Crystallographic relationship between adjacent domains.

(2) The size and shape of domains and surface area of domain boundaries.

(3) The chemistry, structure and width of domain boundaries or walls

(4) The mobility and geometric changes of domain boundaries or walls with respect to a field (temperature, pressure, stress, magnetic or electrical)

(5) Interactions with other domain boundaries and with defects

(6) The energy or stability of domain boundaries.

12.2 STRUCTURAL PHASE TRANSITIONS, CRYSTAL SYMMETRY AND MICROSTRUCTURE

Structural phase transitions are the result of changes in crystal structure which can be either *symmetry breaking* or *non-symmetry breaking*. Examples of non-symmetry breaking transitions are Al/Si ordering in high to low albite and Mg/Fe ordering in orthopyroxene. Although these transitions are thermodynamically important, they involve no change in space group and thus produce no domain microstructures. Symmetry-breaking transitions, however, do produce domain microstructures, and their character can be predicted from an examination of the space group relationships between the high and low symmetry phases. When the low symmetry phase is a subgroup of the high symmetry phase, two fundamentally different types of microstructures result: twin domains and antiphase domains. Twin domains arise from the loss of a *point symmetry* element whereas antiphase domains arise from the loss of a *translational symmetry* element (or lattice point) (Fig. 12-1). For the purposes of this review, phase transitions in minerals are grouped according to the type of symmetry reduction without regard to whether they are continuous or discontinuous; heterogeneous or homogeneous; diffusional or diffusionless; or first order, second order or tricritical. The only criteria is a group-subgroup relationship. Table 12-1 tabulates the space group, point group and Bravais Lattice changes for some transitions in minerals. The table and this review follow the conventions set out in the 1983 edition of the *International Tables for Crystallography*. The prediction of transformation-induced microstructures can in general be deduced by consideration of changes in the 14 Bravais lattices, the 32 point groups and the order of the 18 abstract point groups. These groups are included in the Appendix, §12.7.

Salje (1990) warns us about using *just* symmetry considerations for predicting phase transitions and transition microstructures. He states that "the concept that all ferroelastic phase transitions always follow the simple track of direct group-theoretical predictions of the simplest manner ... died some years ago." The main emphasis now is on the physics of elastic instabilities and the coupling of several different physical processes that combine to lower the Gibbs free energy of a crystal. However, the group-subgroup relationship is still useful for classification purposes and appears to work well with respect to transformation-induced microstructures in minerals.

There are many useful reviews of phase transitions in the literature that generally range from straightforward treatments of the observed crystallographic changes (Buerger, 1951) and sometimes the resulting microstructures (Heuer and Nord, 1976; Delaey, 1991) to full blown group theoretical treatments (Van Tendeloo and Amelinckx, 1974; Janovec, 1976; Toledano and Toledano, 1987). In the early part of this century Gibbs's thermodynamic treatment of phase transitions was extended by Ehrenfest to include continuous transitions and in the mid-twentieth century Landau combined thermodynamics with symmetry considerations in his treatment of phase transitions. Thermodynamics and crystallography must now be considered together in the study of phase transitions in earth sciences (Newnham, 1974). Recent applications of Landau's treatment to phase transitions in silicates as discussed in Carpenter (1988) and Salje (1990).

Phase transitions have been classified by group theory into group/subgroup relationships. For any space group listed in the *International Tables for Crystallography* there is a table of maximal non-isomorphic subgroups. Space groups with different space group numbers are non-isomorphic except for the members of the eleven enantiomorphic pairs of space groups which are isomorphic. Thus, the relationship between the high and low symmetry phases can be identified by looking at the table of subgroups for the high symmetry phase.

The reduction of symmetry operations from the high symmetry phase to the low symmetry phase can occur in three different ways:

(i) by reducing the order of the point group through eliminating all symmetry operations of some kind. The order of a point group is the number of symmetry operations in the group and is tabulated in the Appendix, §12.7. For example the β to α quartz transition entails an inversion from 622 (β-quartz) to 32 (α-quartz). This inversion eliminates the twofold axis parallel to **c** and reduces the order of the point group from 12 to 6, resulting in two domains. These are the Dauphiné twins which are related by the two-fold axis parallel to **c** that is lost in the transition.

(ii) by loss of translational symmetry. This can occur either by the loss of a centering translation (the vector between the origin and a lattice point) and/or by a change in the size of the unit cell (a unit lattice translation). For example, a variety of transitions involving pyroxenes and amphiboles are in this category where a C-centered monoclinic cell is reduced to a primitive monoclinic cell. The lost centering translation is $1/2(\mathbf{a} + \mathbf{b})$, which is the vector from the origin to the lattice point centered on the C face. The size of the unit cell remains the same. The result is two domains that are antiphase with respect to one another.

(iii) by combination of (i) and (ii). The cubic to tetragonal transition in cristobalite may be an example of this type. This transition involves two reductions in point group symmetry; the three-fold axes as well as the center of symmetry are lost. In addition, a loss of lattice translations occurs by the transition from a face-centered to primitive cell.

Subgroups of the first kind (i) are called *translationengleiche* or *t subgroups* because the set of all translations is retained. Subgroups of the second kind (ii) are called *klassengleiche* or *k subgroups*. The Theorem of Hermann (1929) states that subgroups of the kind (iii) can never occur among the maximal subgroups but can be derived by a stepwise process of linking maximal subgroups of types (i) and (ii) in a chain.

The transitions listed in Table 12-1 are grouped according to this classification used in the *International Tables* (1983). The reader can then go to the space group of the high symmetry phase in the *International Tables* and examine the possible maximal subgroups. There are some clear concepts present in Table 12-1 and Figure 12-1 that should be useful to those studying phase transitions in the TEM.

(1) Twins arise from transitions in which the order of the point group is reduced but the number of lattice points per unit volume remains the same.

(1a) Twins with coincident lattices (crystallographic axes are parallel but not necessarily of the same sign) arise from transitions in which the crystal family remains the same (merohedral twinning).

(1b) Twins with noncoincident lattices (some crystallographic axes are not parallel) arise from transitions in which the crystal family is changed (pseudo-merohedral twinning).

(2) Antiphase domains arise from transitions in which the point group remains the same but the number of lattice points per unit volume is reduced. The displacement vector relating the domains is the vector from the origin to the lost lattice point.

Figure 12-1. Transformation-induced microstructures

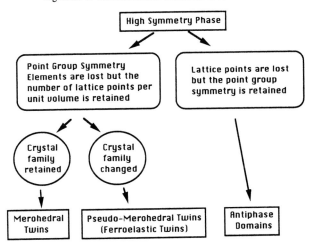

Table 12-1. Symmetry changes during phase transitions in minerals

Phase Transition	Space Group Change	Point Group Change	Bravais Lattice Change	Symmetry Operation(s) Relating Domains
Type (i) - *Translationengleich* - Same Primitive Cell but Lower Point Group Symmetry				
Twinning by Merohedry				
Quartz	$P6_422 \to P3_121$	$622 \to 321$	$hP \to hP$	twofold axis
Dolomite/Ilmenite	$R\bar{3}c \to R\bar{3}$	$\bar{3}m \to \bar{3}$	$hR \to hR$	twofold axis
Leucite	$I4_1/acd \to I4_1/a$	$4/mmm \to 4/m$	$tI \to tI$	mirror
Vesuvianite	$P4/nnc \to P4/n$	$4/mmm \to 4/m$	$tP \to tP$	mirror
Twinning by Pseudo-Merohedry				
Alkali Feldspar	$C2/m \to C\bar{1}$	$2/m \to \bar{1}$	$mC \to a(C)$	twofold axis, mirror
Leucite	$Ia3d \to I4_1/acd$	$m\bar{3}m \to 4/mmm$	$cI \to tI$	threefold axis
Cordierite	$P6/mcc \to Cccm$ cell doubled	$6/mmm \to mmm$	$hP \to oC$	threefold axis
Type (ii) - *Klassengleich* - Same Point Group but Lost Translation Symmetry				
Pigeonite	$C2/c \to P2_1/c$	$2/m \to 2/m$	$mC \to mP$	1/2 (**a+b**)
Bytownite	$C\bar{1} \to I\bar{1}$, cell doubled	$\bar{1} \to \bar{1}$	$a(C) \to a(I)$	1/2 (**a+b**), 1/2**c**
Anorthite	$I\bar{1} \to P\bar{1}$	$\bar{1} \to \bar{1}$	$a(I) \to aP$	1/2 (**a+b+c**)
Omphacite	$C2/c \to P2/n$	$2/m \to 2/m$	$mC \to mP$	1/2 (**a+b**)
Cummingtonite	$C2/m \to P2_1/m$	$2/m \to 2/m$	$mC \to mP$	1/2 (**a+b**)
Joesmithite	$C2/m \to P2/a$	$2/m \to 2/m$	$mC \to mP$	1/2 (**a+b**)
Scapolite	$I4/m \to P4_2/n$	$4/m \to 4/m$	$tI \to tP$	1/2 (**a+b+c**)
Sphene	$A2/a \to P2_1/a$	$2/m \to 2/m$	$mA \to mP$	1/2 (**b+c**)
Type (iii) - Combination of Types (i) and (ii)				
Cristobalite	$Fd\bar{3}m \to P4_32_12 \ (P4_12_12)$ cell halved	$m\bar{3}m \to 422$	$cF \to tP$	threefold axis, center of symmetry, 1/2 (**a+b+c**)

The number of domains (or variants) produced during a phase transition from a high symmetry (HS) phase to a low symmetry (LS) phase can be predicted by the relationship:

$$\text{Domains} = \frac{\text{point group order} \times \text{number of lattice points HS Phase}}{\text{point group order} \times \text{number of lattice points LS Phase}} \times \text{change in cell size}$$

The point group order is tabulated in the appendix. The number of lattice points per unit cell is 4 for face-centered; two for C-centered; two for body-centered; one for primitive and three for rhombohedral centering. Furthermore, the change in the size of the unit cell, must be considered. For example, the face-centered cubic to primitive tetragonal transition in cristobalite has 48 point symmetry elements (order 48) and 4 lattice points in the high temperature phase and 8 point symmetry elements (order 8) and 1 lattice point in the low temperature phase. In addition the size of the unit cell is reduced by 1/2. This gives $[(48\times4)/(8\times1)] \times 0.5 = 12$ domains, the same result obtained by Hatch and Ghose (1991) using group theoretical arguments.

12.3 TYPES OF BOUNDARIES

12.3.1 Twin domain boundaries

A twin can be defined as a crystal made up of two or more homogeneous individuals of the same phase in juxtaposition and oriented with respect to each other according to well-defined laws (Cahn, 1954). Twins can originate during growth, deformation, or transformation. We will be only concerned with transformation twins here even though geologic samples are likely to contain twins that have formed by a variety of mechanisms. It is important therefore to understand the crystallographic basis of twinning as well as their morphology, mobility and interactions with other microstructures.

The crystallographic basis of twinning can be characterized by two empirical concepts (after Cahn, 1954, and attributed to Friedel). (1) A pair of twin individuals together possess an element of macroscopic symmetry over and above that which is present in the individuals alone. For example, albite twins are related by a mirror plane which is not an allowed symmetry element in triclinic albite. (2) This extra symmetry must be of the kind encountered in crystal morphology. The possibilities are a center of symmetry, a plane of symmetry, or a rotational axis of symmetry corresponding to *inversion twins*, *reflection twins* or *rotation twins*, respectively. If the macroscopic symmetry element is a plane then it must be parallel to a lattice plane in both twin individuals. If it is an axis it must be parallel to a lattice row in both twin individuals. For example, the macroscopic symmetry element in albite twinning is a (010) pseudo mirror plane and (010) is shared by each twin individual. Likewise, the macroscopic symmetry element in pericline twinning is a pseudo twofold axis parallel to [010], and the axis is shared by each twin individual.

Friedel classified twins as four types of which two are pertinent to solid-solid phase transformations, and we will review these briefly (following Cahn, 1954). This classification is based on lattices alone and assumes nothing about the structure. It only represents necessary, but not sufficient, conditions for twinning but it does divide transformation twinning into that which produces spot-splitting and that which produces no spot-splitting in diffraction patterns. This is the relevant consideration for imaging the twin domains and twin boundaries in the transmission electron microscope.

12.3.1.1 Twinning by merohedry

If the high and low symmetry phases are of the same Bravais lattice but have different point group symmetries then the twinning is by *merohedry*. Examples of transformations that result in rotation or reflection twinning by merohedry are: (1) the $P6_422 \rightarrow P3_121$ transition in quartz in which both polymorphs share the Bravais lattice hP (hexagonal family, primitive lattice); (2) the $R\bar{3}c \rightarrow R\bar{3}$ transition in dolomite and ferrian ilmenite, which shares the Bravais lattice hR (hexagonal family, rhombohedral lattice), and; (3) the $I4_1/acd \rightarrow I4_1/a$ transition in leucite, which shares the Bravais lattice tI (tetragonal family, body-centered lattice). In the case of merohedral twinning, the reciprocal lattices of both twin individuals are coincident; no spot-splitting is seen in electron diffraction patterns (Fig. 12-2). They are constrained by symmetry to be coincident because they share the same Bravais lattice with the high symmetry phase.

Inversion twins (or inversion domains) are merohedral twins for which the domains are related by a center of symmetry. They are also known as enantiomorphic twins or domains

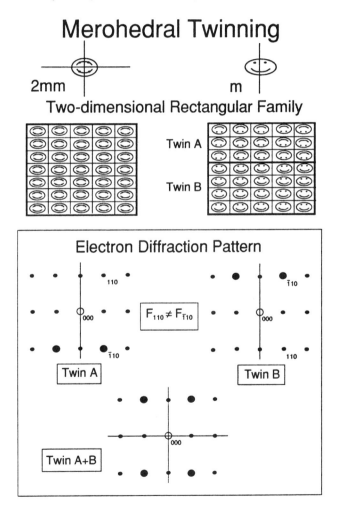

Figure 12-2. Merohedral twinning. The transition $2mm \rightarrow m$ in the two-dimensional rectangular family is shown to form merohedral twins. A "smiling face" structure is superimposed on the m phase which results in unequal structure factors, F_{hk}.

if the twins belong to one of the eleven pairs of enantiomorphic space groups. Only non-centrosymmetric crystals can form inversion twins. All lattice vectors of each individual are antiparallel to the adjacent twin and therefore the lattices of both individuals coincide; no spot-splitting is seen in electron diffraction patterns. Brazil twins in quartz are a good example of inversion twins, since the Brazil law defines right- and left-handed individuals (enantiomorphic twins). However, this type of twinning does not arise by a phase transformation because β-quartz does not have a center of symmetry and therefore can not lose it via a phase transition. Inversion twinning appears to be rare during phase transitions in minerals; cristobalite is one example. Inversion twinning has been studied by TEM in $LiFe_5O_8$ (Van der Biest and Thomas, 1975) and in $LiAl_5O_8$ (Portier et al., 1975), which contain the enantiomorphic pairs, $P4_332$ ($P4_132$).

12.3.1.2 Twinning by pseudo-merohedry.

Phase transitions that involve a point group reduction from one crystal family to another result in a change of Bravais lattice. As a result of the differences in unit cell dimensions, the real space (and reciprocal space) lattices of adjacent twins are rotated slightly with respect to each other in order to achieve an exact lattice fit along the composition plane. However, the lattice mismatch typically is quite small. Consequently, the relative rotation of the lattices is minor, and diffraction spots produced by adjacent twins appear to be slightly split (Fig. 12-3). Because the lattices of such twins are almost, but not exactly, coincident, these twins are known as *pseudo-merohedral twins*. Ferroelastic twins are pseudo-merohedral twins. As in the case of merohedral twins, pseudo-merohedral twins are also related by a symmetry element that existed in the high symmetry phase but was lost during the phase transition. Because this symmetry element is not a possible symmetry element in the crystal family of the low symmetry phase it is often referred to as a pseudo-symmetry element.

Examples of transformations that result in twinning by pseudo-merohedry are: (1) the $C2/m \rightarrow C\bar{1}$ transition in alkali feldspar where albite twins are related by the (010) pseudo-symmetry plane and pericline twins are related by the [010] pseudo-symmetry twofold axis, both of which are symmetry elements in the monoclinic phase but not in the triclinic phase; (2) the $Ia\bar{3}d \rightarrow I4_1/acd$ transition in leucite where twins are related by the $\{101\}_t$ pseudo-symmetry planes which are $\{110\}_c$ mirror planes in the cubic phase but not in the tetragonal phase; (3) the $P6/mcc \rightarrow Cccm$ transition in cordierite where twins are related by the $\{110\}_o$ and $\{310\}_o$ pseudo-symmetry planes which are $\{10\bar{1}0\}_h$ and $\{11\bar{2}0\}_h$ mirror planes in the hexagonal phase but not in the orthorhombic phase.

12.3.1.3 Tweed textures

Tweed textures are experimentally found to be common features in co-elastic materials that form macroscopic twins in the co-elastic or low symmetry phase. They have been observed in alkali feldspars and cordierite. Salje (1990) describes tweed textures as the result of softening of acoustic modes near the critical temperature of the transition. The modes are perpendicular to the twin boundaries and the resulting vibration has the same characteristics as a shear motion or a rotation of the lattice about a junction of the twin boundaries. The periodic superposition of these movements leads to a dynamical tweed pattern. Tweed textures exist because they lead to a decrease in the total energy of the crystal. The orientation of the tweed modulations is constrained by the symmetry of the high symmetry phase. Thus, the modulations in alkali feldspar lie parallel and perpendicular to (010) and the modulations in cordierite lie parallel and perpendicular to $\{10\bar{1}0\}_h$.

Pseudo-Merohedral Twinning

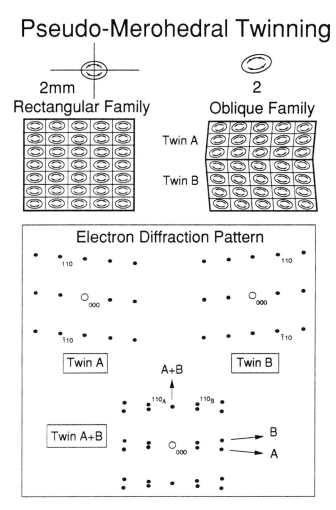

Figure 12-3. Pseudo-merohedral twinning. A slight lattice distortion of the two-dimensional 2*mm* rectangular lattice reduces the symmetry to the two-dimensional oblique lattice, 2, resulting in pseudo-merohedral twins. The electron diffraction pattern shows spot splitting.

12.3.1.4 *Twin boundary morphology*

Usually transformation twins have a well-defined morphology with planar boundaries, such as the albite and pericline twins of the transition in alkali feldspar. Sometimes, however, they have curved boundaries with no obvious crystallographic orientation, such as the twins formed in the $R\overline{3}c \rightarrow R\overline{3}$ transitions of dolomite and ferrian ilmenite. The morphology of the twin boundaries follows directly from the temperature evolution of the spontaneous strain. For transitions involving pseudo-merohedral twinning, in which the lattices are not coincident, the strain can be large (~2%). The orientation of the twin boundary is then determined by the condition that the low-symmetry phase tends to maintain the total symmetry of the high-symmetry phase as a statistical average (Salje, 1990). Therefore the twin boundary tends to lie parallel to a symmetry element in the high symmetry phase. However, for transitions involving merohedral twinning the lattices are coincident and elastic theory predicts only one elastic domain. Thus in these cases one must consider the crystal structure. For example, in merohedrally twinned quartz, Dauphiné twins are constrained by the atomic structure to be planar, although the orientations are not rational. In merohedrally twinned leucite, ferrian ilmenite, dolomite, and vesuvianite the twins are weakly constrained by the atomic structure and the twin boundaries are curved although there is some tendency toward a shape effect.

12.3.1.5 *Twin boundary mobility*

Twin boundary mobility is dependent on the crystallographic differences between the adjacent twin individuals. If the twin-inducing transition involves cation order-disorder, any movement of the twin boundary will require diffusion of cations at the boundary. Boundary mobility therefore will be dependent on the mobility of the diffusing species. Consequently, twin boundaries are more mobile for phases with high transition temperatures, and mobility usually is limited to a range of only a few hundred degrees or less below the transition temperature. If, however, the transition only involves atomic displacements of a few tenths of Ångströms, then twin boundaries can move very rapidly, as in the ferroelastics.

12.3.1.6 *Twin boundary interactions*

Twin boundary interactions are particularly important in mineral systems with more than one twin law. Microcline for instance exhibits considerable interaction between albite and pericline twins. Salje (1990) describes rounded intersections, S-shape domain boundaries, triple junctions and needle-shaped domains along with the appropriate theory.

12.3.2 Antiphase domain boundaries

As mentioned above, antiphase domains arise from the loss of translational symmetry. In all cases the point group remains the same but the number of lattice points per unit volume is reduced. The number of different variants therefore is a function of the number of different translation vectors that are lattice vectors in the disordered structure but are not lattice vectors in the ordered structure. In alloys, most transitions that give rise to antiphase domains and domain boundaries involve cation order-disorder, such as Cu_3Au. However, in silicates many displacive transitions also give rise to antiphase domains and domain boundaries. For example, in the C-centered monoclinic cell of high pigeonite (Fig. 12-4) there are two lattice points, one at $0,0,0$ and one at $1/2,1/2,0$. Low pigeonite has a primitive monoclinic cell with only one lattice point at $0,0,0$ but nearly the same unit cell size as high pigeonite. Both the high and low forms have $2/m$ point group symmetry. Upon cooling through the displacive high-low transition temperature, the low symmetry phase can originate at either of the two crystallographically equivalent lattice points of the high temperature phase. One region can grow with an origin at $0,0,0$ of the C-centered phase and an adjacent region can grow with an origin at $1/2,1/2,0$ of the C-centered phase. Upon impingement a boundary is formed separating the two regions, and the two regions will be out-of-phase by the vector $\mathbf{R} = 1/2(\mathbf{a}+\mathbf{b})$. This vector is known as the *displacement vector* and it represents the translation vector lost during the transition from the C-centered lattice to the primitive lattice.

When translation symmetry is lost extra spots are observed in the electron diffraction pattern that either occupy positions that were systematically absent in the disordered phase or as extra spots that change the volume of the unit cell. In the case of pigeonite (Fig. 12-4) the reflections with $h+k$ = odd are absent in the high temperature phase. This is

Antiphase Domains

Transition from C-Centered to Primitive Latttice

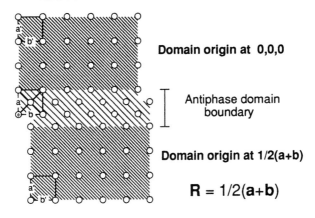

Domain origin at 0,0,0

Antiphase domain
boundary

Domain origin at 1/2(a+b)

$$R = 1/2(a+b)$$

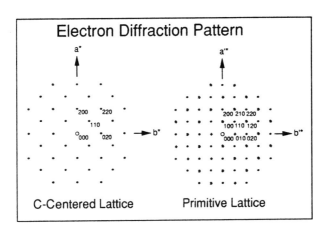

Electron Diffraction Pattern

C-Centered Lattice   Primitive Lattice

Figure 12-4. Origin of antiphase domains. Antiphase domains and domain boundaries are generated during the C-centered to primitive transition. The C-centered unit cell of the high temperature phase is shown in the boundary as **a,b**. The primitive unit cells of the antiphase domains in the low temperature phase are shown as **a',b'**. The antiphase domain boundary has been left with some thickness to point out the high temperature structure of the boundary. The displacement vector is $R = 1/2(a+b)$, which is the same as the vector between the origin and the C-centered lattice point of the high temperature unit cell. The electron diffraction pattern, **a*b***, of the C-centered cell shows the absence of reflections, $h+k$, odd. The **a'*b'*** pattern of the primitive cell contains all reflections. The antiphase domain boundaries are imaged in two-beam dark-field conditions using an $h+k$ odd reflection.

consistent with the C-centered cell. In the low temperature primitive cell all reflections must be present (except for those absent because of the **c**-glide in pyroxene), so that reflections with $h+k$ = odd appear during the transition. The formation of antiphase domains, therefore, is usually accompanied by the appearance of extra reflections in the electron diffraction pattern.

12.3.3 Combination

In a few phase transitions both point symmetry and translation symmetry elements are lost. In some instances the transformation may entail multiple transitions such that point symmetry and translation symmetry are lost at separate temperatures. Alternately, it also is possible that both types of symmetry are lost simultaneously. Cristobalite is one example in which the high temperature phase is face-centered cubic ($m\bar{3}m$) and the low temperature phase is primitive tetragonal (422). The point symmetry elements lost are the three-fold axes and the center of symmetry, and the translation symmetry lost is the translation vector $1/2(110)_c$. The loss of these symmetry elements results in pseudo-merohedral twins, inversion twins and antiphase domains, respectively. The observed transformation-induced microstructure is delightfully complex.

12.4 METHOD OF DOMAIN BOUNDARY
IDENTIFICATION IN THE TEM

Antiphase domain boundaries are made visible in the transmission electron microscope because of a change in the phase of the electron beam as it crosses the boundary (Fig. 12-5). As a result, the boundary appears as a set of fringes that are parallel to the top and bottom surfaces of the thin sample. Twin domains and boundaries are made visible in the transmission electron microscope because of a difference in amplitude and/or phase of the diffracted beams between each twin individual. This phenomenon results in differential contrast among adjacent twins and/or fringes at the twin domain boundaries. Contrast from inversion twin domains is more complex than contrast from rotation or reflection twins. It is the task of the microscopist to sort out these contrast mechanisms by a series of bright-field (BF) and dark-field (DF) image experiments. If the space groups or point groups of the high and low symmetry phases are known, then the job is considerably easier. Thankfully, there are a number of clear explanations to guide the beginner. Some of the most useful are: Hirsch et al. (1965), Amelinckx (1972), Amelinckx and Van Landuyt (1976), Van der Biest and Thomas (1976), Edington (1976) and recently McLaren (1991).

The identification of domains and domain boundaries in transmission electron microscopy requires an understanding of diffraction contrast as discussed in earlier chapters of this book. In ideal kinematic conditions the visibility or invisibility of domains or domain boundaries is a function of (1) the operating diffraction vector $\mathbf{g}_{hkl}$, (2) the deviation, s, from the exact Bragg angle, (3) the extinction distance, ξ_{hkl}, and (4) the phase angle, ϕ, of the structure factor. The structure factor for the reflection hkl is

$$F_{hkl} = \sum_n f_n \exp[2\pi i(hx_n + ky_n + lz_n)]$$

where f_n is the atomic scattering factor for electrons of the nth atom, and x_n, y_n and z_n are the fractional coordinates of the nth atom in the unit cell. The phase angle $\phi_{hkl} = 2\pi(hx_n + ky_n + lz_n)$, and therefore

$$F_{hkl} = \sum_n f_n \exp(i\phi_{hkl})$$

or in the trigonometric form

$$F_{hkl} = \sum_n f_n \cos\phi_{hkl} + i\sum_n f_n \sin\phi_{hkl}$$

For centrosymmetric crystals the sine terms vanish.

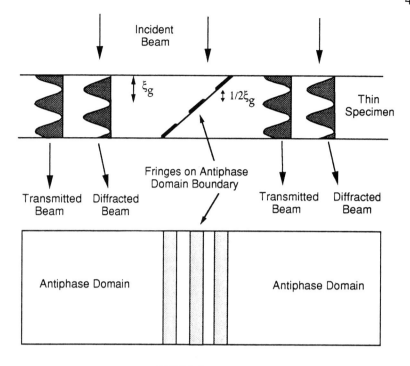

Figure 12-5. Schematic diagram of the phase shift at an antiphase domain boundary. A thin sample is shown with the intensity distribution of the transmitted and diffracted beams as a function of depth. A phase shift of π at the APB produces fringes at the boundary with a spacing of $1/2\xi_g$. The fringes are symmetrical in both bright- and dark-field images.

The extinction distance, ξ_{hkl}, (ξ_g in Fig. 12-5) is the critical distance in a perfect crystal over which the intensity of the diffracted beam varies from zero to maximum intensity and back to zero. The extinction distance depends on the operative reflection and may be calculated from the relationship (Hirsch et al., 1965)

$$\xi_{hkl} = \frac{\pi V_c \cos\theta}{\lambda F_{hkl}}$$

where V_c is the unit-cell volume, θ is the Bragg angle, and λ is the electron wavelength calculated at the operating voltage of the TEM.

12.4.1 Visibility criteria for twin domains and boundaries

For the case of merohedral twinning, the reciprocal lattices of each twin individual are coincident. In general, merohedral twins are made visible because certain coincident reflections that had identical structure factors in the high temperature phase have non-identical structure factors in the low temperature phase, as illustrated in Figure 12-2. This changes the extinction distance in adjacent twins for the same operating reflection and results in dark-light contrast.

Under some circumstances the difference in the phase angle, ϕ, for adjacent twin domains will result in fringe contrast at the twin boundary or in weak dark-light contrast. This is generally because the low symmetry phase has lost a center of symmetry and the sine terms are non-zero. Dauphiné twins show this contrast under certain conditions (McLaren and Phakey, 1969).

For the case of pseudo-merohedral twinning, the diffraction pattern exhibits split reflections. By careful positioning of a small objective aperture over only one of the split reflections, one twin will appear bright and the other will remain dark. Similarly, the twin boundary will be invisible if the row of coincident reflections (see Fig. 12-3, row A+B) is used for the BF or DF images.

12.4.2 Visibility criteria for antiphase domain boundaries

The structure and orientation of antiphase domains (APDs) are identical; the only difference is a translation across the antiphase domain boundary (APB). Electron beams passing through adjacent domains will undergo an identical change in phase and amplitude. The diffraction contrast will be identical for each domain whether imaged in bright-field or dark-field conditions. However, at an inclined boundary between the domains, the beam is transmitted or diffracted first through one domain and then through the adjacent domain (Fig. 12-5). Upon crossing the domain boundary the phase of the diffracted beam is shifted. Under two-beam conditions it is the presence of this phase shift that makes antiphase domain boundaries visible as a set of fringes parallel to the center line of the thin specimen. Not all diffracted beams will undergo a phase shift and therefore under some conditions the antiphase microstructures will be invisible.

In general, for antiphase boundaries produced by phase transitions the displacement across the boundary introduces a phase shift, $\alpha = 2\pi\mathbf{g} \cdot \mathbf{R}$, in the diffracted beam where $\mathbf{g}$ is the operating diffraction vector and $\mathbf{R}$ is the displacement vector (Fig. 12-4). For all cases considered in this chapter the phase shift is either 0 or an integral multiple of π. Therefore, these boundaries are called π-boundaries (There are however, other types of boundaries such as stacking faults in fcc metals that give rise to $\alpha = \pm 2\pi/3$ but in this paper we only need to consider $\alpha = \pi$). In general, we assume that there are no deviations, $\pm \mathbf{s}$, from the exact Bragg orientation for the two-beam condition (i.e., $\Delta s = 0$).

In order to define the visibility criteria for an antiphase domain boundary, one constructs a simple table of operating vectors and displacement vectors. For example if $\mathbf{g} = 130$ and $\mathbf{R} = 1/2(110)$ then $\mathbf{g} \cdot \mathbf{R} = 2$ and $\alpha = 4\pi$ or 0. If $\alpha = 0$ then no fringes are visible and the boundary is not visible. However, if $\mathbf{g} = 120$ and $\mathbf{R} = 1/2(110)$, then $\mathbf{g} \cdot \mathbf{R} = 3/2$ and $\alpha = 3\pi$ or π and the antiphase boundary is visible as a set of dark and light fringes depending or the thickness of the sample and the extinction distance. More generally, the phase shift is π for $h+k$ odd and 0 for $h+k$ even when the displacement vector is $1/2(\mathbf{a} + \mathbf{b})$. This is the visibility criterion for APBs in all phases that undergo transitions from a C-centered to a primitive cell , providing that there is no change in cell size. The fundamental reflections are $h+k$ even. These are present in both the high symmetry and low symmetry phases. The $h+k$ odd reflections are only present in the low symmetry phase. The APBs are only visible in dark-field using an $h+k$ odd reflection as the operating vector.

π-fringes have the following set of characteristics (after McLaren, 1991) when $\mathbf{s} = 0$:

(1) A bright fringe appears at the center of the thin specimen in the bright-field image and a dark fringe appears at the center of the thin specimen in the dark-field image.

(2) Over the whole depth of the specimen and for any specimen thickness, the bright-field and dark-field images are complementary.

(3) The fringes are parallel to the center line of the specimen, and the bright/dark nature of the outer fringes is determined by the thickness z_0 of the specimen. In bright-field they are dark for $z = k\xi_g$ (where k is an integer and ξ_g is the extinction distance) and bright for $z = (k + 1/2)\xi_g$. The opposite contrast is found in the dark-field images. Therefore with increasing thickness a new fringe is added to each side of the boundary fringe pattern at a thickness contour.

(4) The fringe separation corresponds to a depth periodicity of $\xi_g/2$ in both bright-field and dark-field images.

12.5 TRANSFORMATION-INDUCED MICROSTRUCTURES IN MINERALS

In this section I have briefly outlined the details of phase transitions in a number of well studied and important minerals. In particular, I have chosen examples where the relationship between the symmetry changes of the phase transition and the crystallography of the induced microstructure is well known and illustrated by TEM images. At the end of each description, I have discussed the possible application of the TEM studies to problems in earth science.

12.5.1 Twin domains

12.5.1.1 Quartz transition

The high temperature, β form of quartz (SiO_2) crystallizes in either of two enantiomorphic space groups where $P6_422$ is left-handed and $P6_222$ is right-handed (see Megaw, 1973, for the subtleties of nomenclature and handedness). Quartz undergoes two transitions, $P6_422 \rightarrow IC \rightarrow P3_121$ (or $P6_222 \rightarrow IC \rightarrow P3_221$), upon cooling through 573°C (846 K). where IC is an incommensurate phase with a narrow stability field of only 1.3-1.5 K (Dolino et al., 1984). The α-IC-β quartz transition is particularly fascinating, and TEM has played a leading role in its elucidation. The incommensurate phase was discovered by hot-stage TEM (Van Tendeloo et al., 1976). The review by Heaney and Veblen (1991) of the α–β quartz transition gives a good historical and modern synopsis of the structural and thermodynamic details.

The point group changes from 622 in β-quartz to 321 in α-quartz, and the Bravais lattice remains hexagonal primitive. Dauphiné twins form during the transition and are related by the twofold axes parallel to [0001] that are present in the high temperature phase but absent in the low temperature phase (Fig. 12-6). Another common twin in quartz is the Brazil twin, which relates right and left handed lamellae by reflection across the mirror plane parallel to (11$\bar{2}$0). Brazil twins are formed during growth or during deformation and are not related to the α–β transition. McLaren and Phakey discussed the contrast conditions necessary to image Dauphiné twins (1969) and Brazil twins (1966) in α-quartz. These two papers are very informative and students are well advised to read them carefully.

470

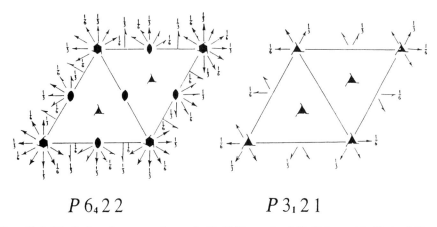

$P\,6_4\,2\,2$ $P\,3_1\,2\,1$

Figure 12-6. Distribution of symmetry elements in $P6_422$ (β-quartz) and $P3_121$ (α-quartz). The twofold axes parallel to **c** (normal to plane of drawing) in the β-phase are lost during the transition and relate the Dauphiné twins of the α-phase [Note that the origin is displaced by 1/3**c** between $P6_422$ and $P3_121$ as given in the *International Tables*, 1983.]

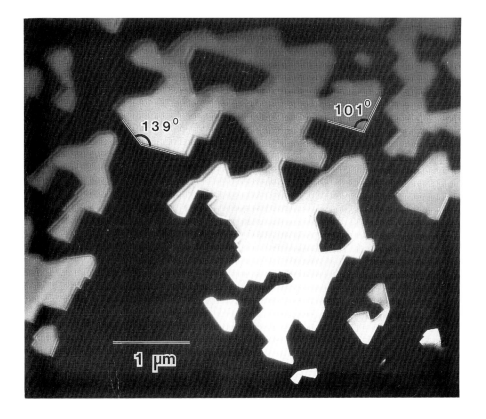

Figure 12-7. A dark-field transmission electron micrograph of α-quartz with **g** = 30$\bar{3}$1. The two Dauphiné twin orientations appear as light and dark areas. The domain boundaries intersect at 101° or 139°. [Fig 3. from Heaney and Veblen, 1991.]

Dauphiné twins are examples of merohedral twinning where the lattices of the two twin individuals coincide exactly. The 180° rotation about the **c**-axis results in the coincidence of $hkil$ and $\bar{h}\bar{k}il$ reflections. Figure 12-7 shows a dark-field image, $\mathbf{g} = 30\bar{3}1$ of one twin and $\mathbf{g} = \bar{3}031$ of the adjacent twin, where the dark-light contrast is the result of different structure factors. McLaren and Phakey (1969) calculated model structure factors, phases and extinction distances for a pair of Dauphiné twins (Table 12-2).

Table 12-2. Structure factors, phases and extinction distances
in Dauphiné-twinned quartz

	$P3_221$ α-Quartz, Twin 1			Twin 2				
g	$\|F_{hkil}\|$	ϕ_{hkil}	ξ_{hkil}	$\|F_{hkil}\|$	ϕ_{hkil}	ξ_{hkil}	$\Delta\phi$	ξ_{hkil}/ξ_{hkil}
$\{10\bar{1}0\}$	5.2	180°	1554Å	5.2	180°	1554Å	0°	1
$\{10\bar{1}1\}$	11.6	120°	690Å	7.5	120°	1068Å	0°	2/3
	7.5	0°	1068Å	11.6	0°	690Å	0°	3/2
$\{10\bar{1}2\}$	4.6	0°	3954Å	2.2	180°	1730Å	180°	2.3
	2.2	180°	1730Å	4.6	0°	3954Å	180°	0.43
$\{11\bar{2}0\}$	4.7	180°±18°	1720Å	4.7	180°±18°	1720Å	±36°	1
$\{11\bar{2}2\}$	5.3	±109°	1500Å	5.3	±129°	1500Å	±20°	1
		±10°			m10°			
		±129°			±109°			

The model values are compared to experimental dark-field images in the TEM. For the reflection $\{10\bar{1}0\}$ both the extinction distances and phases are equal; therefore neither the twin domains nor the boundaries will be visible. For the reflection $\{11\bar{2}0\}$ there is a phase difference of 36° but the extinction distances are equal. Thus, there will be weak fringes visible on the boundary but both twins will have equal contrast. For the reflection $\{10\bar{1}1\}$ there is an extinction distance difference but no phase difference. In dark-field images using this type of reflection, one twin will be bright and one dark. This is the type of reflection, $h0\bar{h}1$, used by Heaney and Veblen (1991) to create the dark-field image in Figure 12-7. McLaren and Phakey (1969) point out that in images where $\mathbf{g} = \{10\bar{1}1\}$ there are also fringes on the boundary even though $\Delta\phi=0$. They attribute these fringes to the large differences in the extinction distances. Boundary fringes are also visible in Figure 12-7. Finally, for the reflection $\{10\bar{1}2\}$ there are both differences in extinction distance and phase that give rise to both contrast between the twins and fringes at the boundary.

The trigonal symmetry of α-quartz forces the twin boundaries to lie parallel to the **c**-axis. The angles of the boundary wall intersections, however, are: 2ε, 60°, 120-2ε°, and 120+2ε°, where ε is the angle between the boundary wall and **a**. Therefore, there are four distinct sets of domain walls but only two have been observed (Heaney and Veblen, 1991). Interestingly enough, these two boundary wall angles are the direct result of the intervening incommensurate phase (Fig. 12-8). Walker (1983) reviewed the model for a second order transition from the β-phase to the incommensurate phase in which the incommensurate wave vector, which is along the **a*** direction at the β-IC transition temperature, gradually rotates away from the **a*** direction as the temperature is lowered. It rotates 9-10° in a positive or negative fashion resulting in two sets of domain boundaries with the same energy.

Theoretically, because Dauphiné twins form during the β to α transition, the presence of such twins in quartz could be a possible geothermometer indicating that in the past the crystal had been at temperatures above 573°C. In reality, however, Dauphiné twins can also form as growth twins and during deformation. In fact, mechanical Dauphiné

472

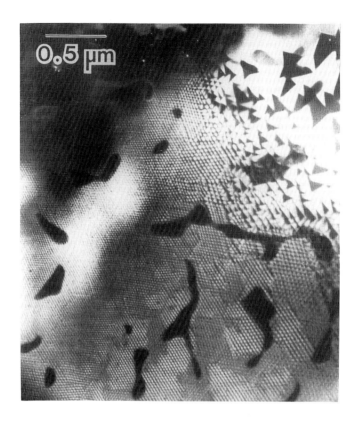

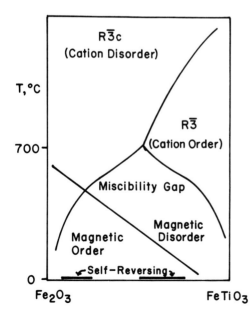

Figure 12-8 (above). The incommensurate phase of quartz is seen as a distorted hexagonal array of lines in the lower left. A temperature gradient exists across the specimen with temperature increasing toward the lower left. Large Dauphiné twins can be seen to grade into triangular microtwins. Macrodomains appear as regions of differing contrast within the mosaic. The large dark areas are grains of copper. The c-axis is normal to the image. [Fig. 4a from Heaney and Veblen, 1991.]

Figure 12-9 (left). Schematic ilmenite-hematite binary phase diagram showing the cation order-disorder transition, the miscibility gap between disordered hematite and ordered ilmenite and a magnetic order-disorder transition. [Fig. 1 from Nord and Lawson, 1989.]

twinning is thought to be responsible for preferred orientation of deformed and recrystallized quartzites (Tullis, 1970). Not much TEM work has been done on the distribution or style of Dauphiné twins in rocks arising from different geologic processes.

Finally, Dauphiné twins are known as "electric" twins because their piezoelectric coefficient, d_{111}, reverses sign from one orientation to another. They are, therefore, considered a nuisance which detracts from the performance of a piezoelectric oscillator. However, electric fields that presumably arise from piezoelectric effects commonly occur as precursors to earthquakes. The interaction between electric twins in quartz-rich rocks and earthquake precursors may be of interest in the future.

12.5.1.2 Ferrian ilmenite, dolomite transition

Both ferrian ilmenite and dolomite undergo an ordering transition from the high temperature space group, $R\bar{3}c$, to the low temperature space group, $R\bar{3}$. Ferrian ilmenite is a solid solution phase in the hematite-ilmenite binary (Fe_2O_3-$FeTiO_3$) in the composition range Ilm_{50} to Ilm_{75} (Fig. 12-9). In ferrian ilmenite the high temperature phase is disordered with respect to Fe^{2+}, Fe^{3+} and Ti^{4+} while below T_c, Fe^{2+} and Ti^{4+} order into alternate A and B cation layers parallel to (0001); Fe^{3+} is distributed equally between the two layers. Similarly, in dolomite ($CaMg(CO_3)_2$) the high temperature phase is disordered with respect to Mg^{2+} and Ca^{2+}, whereas below T_c these cations order into alternate layers as in ferrian ilmenite.

The transition involves a reduction of point symmetry from the point group $\bar{3}m$ (order 12) to the point group $\bar{3}$ (order 6). The 6 point group symmetry elements lost are the 3 mirror planes normal to $\mathbf{a}$ and the 3 twofold axes parallel to $\mathbf{a}$. The Bravais lattice remains rhombohedral (3 lattice points), and the crystal family remains hexagonal. Therefore, the unit cell size is unchanged, and no translational symmetry is lost. The microstructures that arise during both the ferrian ilmenite and dolomite transitions are twin domains that are related by the lost twofold rotation. According to the formula the number of domains is $(12 \times 3)/(6 \times 3) = 2$. The domains could be equally related by the $\mathbf{c}$-glide plane lost during the transition. Because there is no change in crystal family the domains are twinned by merohedry and the diffraction pattern shows a single reciprocal lattice with no spot splitting. The twin domains were first imaged in dolomite by Reeder and Nakajima (1982) (see Fig. 10-29a, Chapter 10) and in ferrian ilmenite by Lawson et al. (1981).

Because the high temperature disordered phase has a $\mathbf{c}$-glide plane, reflections of the type $h\bar{h}0l$, where $l=2n+1$, are absent. Upon cation ordering, the $\mathbf{c}$-glide is lost and reflections appear of the type $h\bar{h}0l$, where $l=2n+1$ (Fig. 12-10a). These are *ordering reflections*, not superlattice reflections, because there is no change in lattice type. Almost all of the intensity in the ordering reflections comes from the difference in the cation population on the A and B layers. Therefore, the change of intensity with temperature can be directly related to the change in the degree of cation order with temperature.

Similarly, the intensity from reflections of the type $hkil$, where $l=2n+1$, arises almost totally from the oxygen positions, whereas the intensity from reflections of the type $hkil$, where $l=2n$, arises from both the oxygen and cation positions. We can, therefore, use dark-field images of these three different types of reflections to probe both the cation and oxygen positions of the ordered phase.

The model structure factors, phases and extinction distances have been calculated by Nord and Lawson (1989) for ordered ilmenite, 200 keV electrons, and at zero deviation

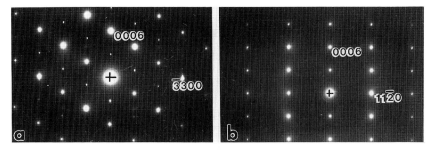

Figure 12-10. Selected area electron diffraction patterns in Ilm70 for (a) the zone [11$\bar{2}$0] and (b) 90° away for the zone [1$\bar{1}$00] . In (a) the ordering reflections are seen as rows of weak spots trending upper left to lower right. In (b) the only ordering reflections are 0003 and 0009 which are intense because of double diffraction. [Fig. 4e and f from Nord and Lawson, 1989.]

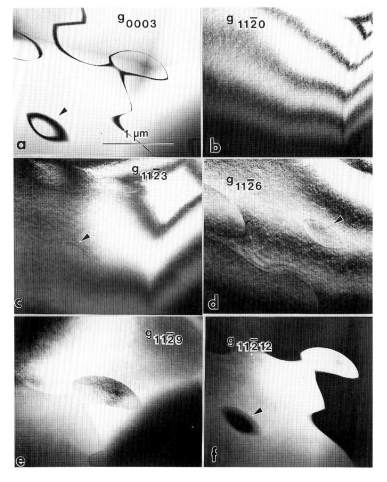

Figure 12-11. Dark-field transmission electron micrographs of domain boundaries from the same area of Ilm70 annealed at 900°C for 1 hour. The operating reflection, **g**, indicated on each micrograph corresponds to the reflections in the [1$\bar{1}$00] diffraction pattern in Figure 12-10b. An arrow on each micrograph indicates the position of the boundary of a single closed twin domain. [Fig. 5 from Nord and Lawson, 1989.]

from the exact Bragg condition. Similarly, these values were also calculated for a rotation twin (t) with the atomic coordinates of the unrotated twin $(X,Y,Z)_{ilm}$ transformed by the relationship $X_t = Y_{ilm}$, $Y_t = X_{ilm}$, and $Z_t = 0.5 - Z_{ilm}$.

The selected area diffraction pattern containing these reflections is shown in Figure 12-10b, and the dark-field images using the reflections tabulated above are seen in Figure 12-11. The only ordering reflections are 0003 and 0009 where for $h\bar{h}0l$, $l = 2n+1$. For both of these reflections the structure factors of the twin domains are identical but the phases are 180° out-of-phase. The contrast should be that of a π-fault. Indeed in the dark-field micrograph the boundary shows up as a dark fringe. Only one fringe appears because the extinction distance is large. In order to see the accompanying bright fringes a thicker area of the sample is required. For the reflections 0006, 00012, and $11\bar{2}0$ there is no contrast because the structure factor and phases are identical. For the reflections $11\bar{2}3$, $11\bar{2}6$, $11\bar{2}9$ and $11\bar{2}12$, the phases are identical but the extinction distances are different. See Table 12-3. As expected, dark-field images with $\mathbf{g} = 11\bar{2}12$ display exceptionally strong black-white contrast.

Annealing the quenched samples coarsens the twin microstructure (Fig. 12-12). Initially the domains are somewhat flattened parallel to the basal plane (best illustrated in Fig. 12-12b). Coarsening straightens out curved domain boundaries and pinches off highly curved areas as a precursor to complete elimination. In ferrian ilmenite the twins become somewhat football shaped (American Football, Fig. 12-11a) with the least eccentric portion of the curve nearly parallel to the basal plane. As the boundaries pass through the orientation parallel to $\mathbf{c}$, a small series of steps appears, presumably to minimize the area of the twin boundary parallel to $\mathbf{c}$. This geometry is controlled by the strain anisotropy of the boundary.

Ferrian ilmenite, Ilm_{50} to Ilm_{75}, is a common constituent in dacitic volcanics such as the ash flows of Mt. Shasta, Nevado del Ruiz and Mt. Pinatubo. Unlike quartz, the presence of the merohedral twins in ferrian ilmenite is positive evidence that the sample was above the transition temperature, T_c. Because T_c is close to the crystallization temperatures of dacitic magmas, the twin domains are a sensitive geothermometer for that system Lawson et al., 1987).

Finally, the twin domain boundary in ferrian ilmenite can act as a second magnetic phase and as a special type of magnetic domain wall. Ferrian ilmenite is ferrimagnetic when in the ordered ($R\bar{3}$) phase, but the twin boundary is disordered ($R\bar{3}c$), and exhibits only a weak ferromagnetic moment like hematite. Statistically, part of the boundary will be enriched in Fe and part enriched in Ti. Therefore, the Fe-enriched, disordered boundary will have a higher Curie temperature than the twin domain proper, and the twin boundary acts as a second magnetic component. This has been experimentally verified (Hoffman, 1992; Nord and Lawson, 1992). The twin boundary can also act as a magnetic domain wall because the ordered Fe-rich and Ti-rich layers are antiphase across the twin boundary and therefore the ferrimagnetic moments are antiphase across the twin boundary (Nord and Lawson, 1992). The presence of the magnetic twin boundary has been shown to be the cause of self-reversing thermoremanent magnetization and increased magnetic hardness.

12.5.1.3.   *Leucite transitions*

High temperature cubic leucite, $KAlSi_2O_6$, undergoes two transitions upon cooling (Fig. 12-13). The higher temperature transition involves tetrahedral rotations associated with the collapse of the framework around the K-cavities, and the low temperature transition may be caused by localization of the K cation. There are some suggestions that Al-Si ordering may also play a primary role (Hatch et al., 1990).

Table 12-3. Structure factors, phases and extinction distances in ilmenite

hkil	Twin 1			Twin 2			Δφ	ξ_{hkil}/ξ_{hkil}
	$|F_{hkil}|$	ϕ_{hkil}	ξ_{hkil}	$|F_{hkil}|$	ϕ_{hkil}	ξ_{hkil}		
0003	5.53	180°	7147Å	5.53	0°	7147Å	180°	1.0
0006	29.45	180	1342	29.45	180	1342	0	1.0
0009	10.38	0	3809	10.38	180	3809	180	1.0
00012	24.70	0	1600	24.70	0	1600	0	1.0
11$\bar{2}$0	24.40	0	1620	24.40	0	1620	0	1.0
11$\bar{2}$3	59.96	0	659	49.04	0	806	0	0.82
11$\bar{2}$6	44.30	0	892	58.61	0	674	0	0.75
11$\bar{2}$9	31.92	180	1238	21.93	180	1802	0	0.69
11$\bar{2}$12	4.62	180	8558	17.95	180	2202	0	0.26

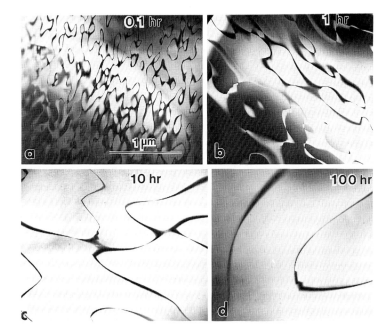

Figure 12-12. Dark-field transmission electron micrographs of Ilm$_{70}$ annealed at 800°C for the times indicated. **g** = 0003 in all micrographs. [Fig. 8 from Nord and Lawson, 1989.]

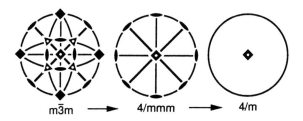

Figure 12-13. Point group changes in leucite, $m\bar{3}m \rightarrow 4/mmm \rightarrow 4/m$. The $m\bar{3}m \rightarrow 4/mmm$ transition results in loss of threefold axes parallel to [111]$_c$, and the $4/mmm \rightarrow 4/m$ transition entails loss of the {100}$_t$ and {110}$_t$ mirror planes.

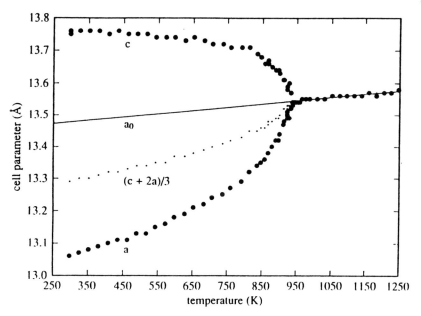

Figure 12-14. Change of lattice constants with temperature in leucite. The cubic cell, a_o, is extrapolated to low temperatures. If the transition were purely ferroelastic then $(c+2a)/3 = a_o$, which is clearly not the case. [Used by permission of the editor of *Physics and Chemistry of Minerals*, from Palmer et al.(1989), Fig. 1, p. 715.]

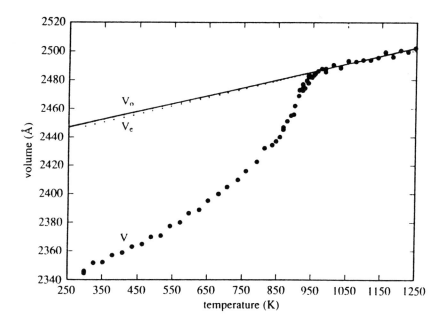

Figure 12-15. Change of unit cell volume with temperature in leucite. The volume change for a purely ferro-elastic transition, V_e, and the extrapolated volume of the cubic cell, V_o, are indicated. [Used by permission of the editor of *Physics and Chemistry of Minerals*, from Palmer et al. (1989), Fig. 2, p. 716.]

The higher temperature transition is an "impure ferroelastic" cubic to tetragonal transition, $Ia\overline{3}d \rightarrow I4_1/acd$, that occurs at approximately 938 K (665°C) (Palmer, 1990). This transition is characterized by the expansion of the **c**-axis, contraction of the **a**-axis, and reduction in volume upon cooling (Figs. 12-14 and 12-15) and involves a point group change from $m\overline{3}m$ (order 48) to $4/mmm$ (order 16). There is no change in crystal family or unit cell; both cells are body-centered. The reduction of the point group order by a factor of 3 gives rise to 3 twin orientations, each related to the other by the lost threefold symmetry axis of the cubic phase. The three fourfold axes of the cubic phase become the fourfold axes for the three tetragonal twins.

Typically, leucites are seen by optical microscopy to consist of repeated lamellar twins in several orientations. The composition planes of each pair of twins are the $\{110\}_c$ mirror planes of the high temperature, $m\overline{3}m$, phase that are lost during the transition (Fig. 12-13). These become $\{101\}_t$ pseudo mirror planes in the tetragonal phase (pseudo because they are not allowed in the tetragonal crystal family). The twins, therefore, are pseudo-merohedral with $\{101\}_t$ composition planes. Because there are two possible $\{101\}_t$ composition planes (90° from each other) for each pair of twin individuals the resultant microstructure could consist of three twin orientations with six different interfaces (Amelinckx and Van Landuyt, 1976; Palmer et al., 1988). Local stresses however will enhance the growth of one variant so that small areas will show one set of twin lamellae while a centimeter size crystal may contain all three.

There is a 5% difference between the **a** and **c** axes of the tetragonal leucite phase and in selected area electron diffraction patterns the reflections from both twin individuals will be easily separated. Figure 12-16a shows a [001] model diffraction pattern for the high temperature cubic phase. The $h+k+l$ odd reflections are absent because of the body-centered lattice. Upon cooling, the **a**-axis of the cubic phase can become either an **a** or **c** axis of the tetragonal phase. This relationship gives rise to two types of low index diffraction patterns for the twinned low symmetry phase. In one type, the $[100]_t$ axes from adjacent twins are coincident (Fig. 12-16b), and in the other type, the $[100]_t$ axis from one twin is coincident with the $[001]_t$ axis from the adjacent twin (Fig. 12-16c). The extent of spot splitting is considerable in Figure 12-16b and less so in Figure 12-16c. It is easy to obtain a strong contrast difference between the twin lamellae (Fig. 12-17) by either tilting the specimen until only one of the split reflections is strongly diffracting (bright-field) or by placing the objective aperture over only one of the split reflections (dark-field).

The second leucite transition, $I4_1/acd \rightarrow I4_1/a$, at 918 K (645°C) is within 20 K of the cubic to tetragonal transition, requires no change of crystal system, Bravais lattice, or unit cell (Palmer et al., 1989). It involves a point group change from $4/mmm$ to $4/m$ and the loss of the $\{100\}_t$ and $\{110\}_t$ symmetry planes (**c** and **d** glide planes in the space group) and $<100>_t$ and $<110>_t$ twofold axes (Fig. 12-13). In the $\mathbf{a^*c^*}$ diffraction net, $(h,l$ odd) **c**-glide reflections that were absent in the high temperature phase are now present in the low temperature phase because of the loss of the **c**-glide in $I4_1/a$ (compare Fig. 12-16a to Figs. 12-16b and 12-16c). In the $\mathbf{a^*a^*}$ diffraction nets, h,k odd reflections remain absent because of the continuing presence of the **a**-glide (compare Fig. 12-16a to Fig. 12-16d). These model diffraction patterns do not include any spots that might appear because of double diffraction, so be careful.

The microstructure that results from the $I4_1/acd \rightarrow I4_1/a$ transition consists of twins related by the lost symmetry elements of the $I4_1/acd$ phase (Fig. 12-18). Because the crystal family is unchanged, the reciprocal lattices will be coincident, and the twins are merohedral. TEM images of the merohedral twins generally show curved boundaries but some samples may show relatively straight segments (see Fig. 12-14 of Amelinckx and Van Landuyt, 1976). The irregular morphology of the merohedral twin boundaries is a

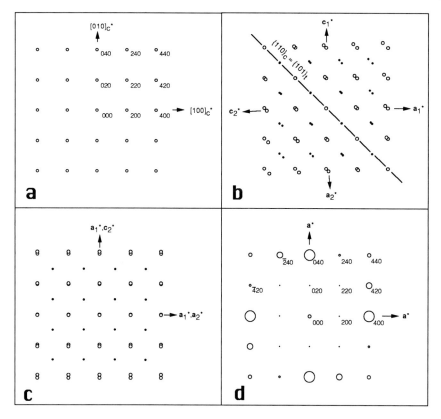

Figure 12-16 Model electron diffraction patterns for leucite. (a) is a [001] pattern for the high temperature cubic phase. (b) is a combined pattern of the [100] zones from two twin-related individuals, 1 and 2. Each diffraction net shares the $\{101\}_t$ twin plane of the tetragonal twins which was the $\{110\}_c$ symmetry plane of the cubic phase. (c) is a combination of the [001] and [100] zones from two twin individuals. (d) is a [001] diffraction pattern from only one of the merohedral twins with space group $I4_1/a$. The calculated amplitudes of the diffraction spots are indicated by the relative size of the circles.

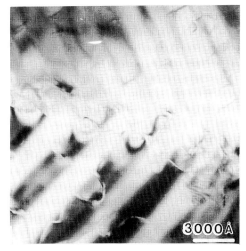

Figure 12-17. Two orientations of pseudo-merohedral twins in leucite. [Used by permission of the editor of *Physics and Chemistry of Minerals*, from Palmer et al.(1988), Fig. 3a, p. 301].

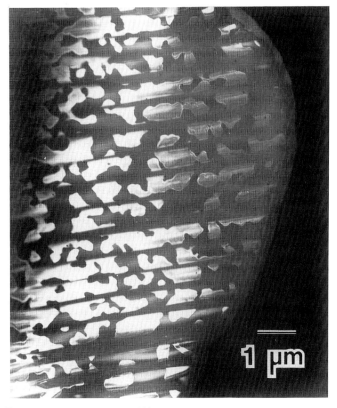

Figure 12-18. Dark-field micrograph using the 420 reflection from one set of pseudo-merohedral twin lamellae. Merohedral twins occur as irregularly shaped domains within the set of pseudo-merohedral twin lamellae that is in contrast. [Fig. 4b from Heaney and Veblen, 1990.] The selected area diffraction pattern for this image will be similar to Figure 12-16c.

consequence of the small lattice strain at the boundaries. However, the planar $\{011\}_t$ orientation of the pseudo-merohedral twin boundaries is related to the large spontaneous strain associated with the $Ia\overline{3}d \rightarrow I4_1/acd$ transition.

The merohedral twin domains can be made to exhibit black-white contrast (Fig. 12-18) because the intensity distribution of the reflections about $[001]_t$ for one twin individual (Fig. 12-16d) is asymmetric. In this regard, the merohedral twins of leucite are similar to Dauphiné twins and the merohedral twins in ferrian ilmenite and dolomite. The asymmetry can be described as the direct result of the loss of the $\{110\}_t$ mirror planes in the $4/mmm$ phase. For example, the intensity of the 420 reflection is about one-third that of the 240 reflection (Fig. 12-16d). Because the twins can be related by the lost mirror planes, in selected area diffraction patterns (SAED) taken across the twin boundaries, the 420 reflection of one twin individual will be superimposed on the 240 reflection of the adjacent twin individual. The resulting SAED of the two twins will appear to have a symmetrical intensity distribution, but if one forms an image using only the superimposed 420-240 reflections, the merohedral twins will appear in dark-field images with a black-white contrast. Similarly, the twins can be imaged in bright-field if the 420 or 240 reflections are tilted into strong illumination. One cannot do this with $hh0$ and $h00$ reflections, for instance, because they lie on the lost $\{110\}$ and $\{100\}$ mirrors.

Black-white contrast for the merohedral twins cannot be obtained using reflections in the diffraction pattern containing the a^*c^* nets for both twin individuals (Fig. 12-16b). This is because in the a^*c^* net (b^*c^* net) the structure factors for reflections made coincident by merohedral twinning are identical. For instance, $F_{103} = F_{013}$, by virtue of the fourfold c-axis in the point group $4/m$.

No applications have been made on observations of leucite twinning although the twinning can only occur upon cooling through the transition. Nothing is known about the coarsening behaviour of either the pseudo-merohedral or the merohedral twins.

12.5.1.4 *Vesuvianite transition*

Vesuvianite, $Ca_{10}(Mg,Fe)_2Al_4[Si_2O_7]_2[SiO_4]_5(OH,F)_4$ crystallizes in the space group $P4/nnc$. Recent studies have found that when vesuvianite is cooled below 800°C, the cations order and the space group symmetry is reduced to $P4/n$. The point group change, $4/mmm$ to $4/m$, reduces the point group order from 16 to 8. The transition gives rise to merohedral twin domains, since no translation vectors are lost and the subgroup remains within the same crystal family. Veblen and Wiechmann (1991) were able to image the domain microstructure in dark-field by using reflections hkl and $h\bar{k}\bar{l}$, which exhibit large differences in structure factor. Veblen and Wiechmann (1991) also suggest that the space group actually may be $P2/n$. This proposed transition from tetragonal to monoclinic symmetry would generate a second set of twins that would be pseudo-merohedral because of the change in crystal family. In part this suggestion is based on the contrast reversal between the twins when the specimen was tilted <0.1°. Presumably, this contrast reversal upon low-angle tilting arises from spot splitting, which in turn implies that the twins are pseudo-merohedral and not merohedral. However, the microstructure is complex and they were not able to resolve the issue of the space group definitively. Allen and Burnham (1992) propose a more complex domain microstructure from x-ray structure refinements.

12.5.1.5 *Cordierite transition*

Cordierite, $Mg_2Al_4Si_5O_{18}$, undergoes an Al-Si ordering transition at 1450°C from the hexagonal form, $P6/mcc$, to the orthorhombic form, $Cccm$. The point group change from $6/mmm$ to mmm entails a reduction from sixfold to twofold rotational symmetry parallel to c (This also reduces the six mirror planes parallel to c to only two orthogonal mirror planes parallel to hexagonal c). The order of the point group is reduced from 24 to 8. The Bravais lattice changes from primitive with one lattice point per unit cell to C-centered with two lattice points per unit cell, and the size of the unit cell doubles from 777 Å^3 to 1570 Å^3. Therefore, the number of domains is $[(24\times1)/(8\times2)] \times 2 = 3$. The relationship between the hexagonal unit cell and the orthorhombic unit cell is illustrated in Figure 12-19. The hexagonal a cell edge is parallel to the orthorhombic b cell edge. Because there are three equivalent a cell edges it is fairly easy to see that three orientations of the orthorhombic cell will result from cooling through the transition. These are illustrated in Figure 12-20.

The orthorhombic twins will be related by the threefold symmetry lost during the transition. The composition planes (or contact planes) of the twins are the $\{110\}_O$ or $\{310\}_O$ pseudo-mirror planes of the orthorhombic phase (Fig. 12-21) which are $\{10\bar{1}0\}$ or $\{1\bar{2}10\}$ mirror planes, respectively, in the hexagonal phase. The hexagonal mirror planes become pseudo-symmetry elements in the orthorhombic phase because neither plane is an allowed symmetry plane in the orthorhombic crystal family. As temperature is decreased below the transition, the orthorhombic lattice departs from the hexagonal symmetry of the high temperature structure, thereby inducing a spontaneous

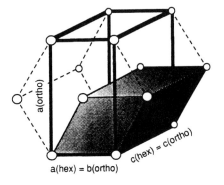

Figure 12-19. Transition from the hexagonal primitive Bravais lattice of the high temperature phase to the orthorhombic C-centered Bravais lattice of the low temperature phase of cordierite. The hexagonal lattice is shaded and a hexagonal prism is shown dashed to help visualize the angles of the unit cell. Thick lines outline the orthorhombic lattice; only one of the three possible orientations is shown.

Figure 12-20. Three orientations of twin domains in cordierite sharing {110}$_o$ composition planes. The original hexagonal unit cell is in the center. A phase transition producing a geometric arrangement such as this is thought to give rise to cordierite trillings in metamorphic rocks (Zeck, 1972; Putnis and Holland, 1986).

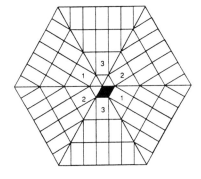

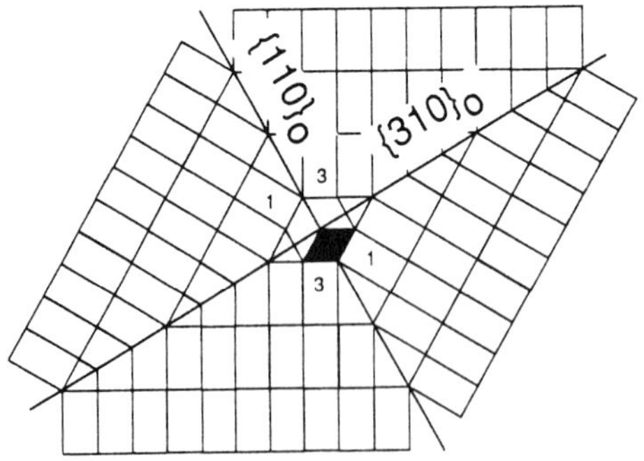

Figure 12-21. Two orientations of twin domains in cordierite showing both the {110}$_o$ and {310}$_o$ composition planes.

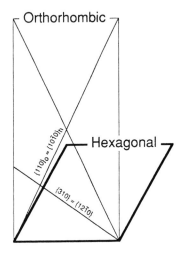

Figure 12-22. The relationship between hexagonal and orthorhombic lattices of cordierite, with exaggerated distortion of the orthorhombic lattice.

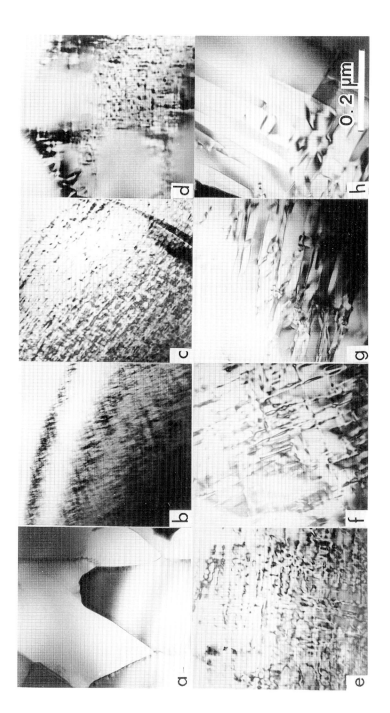

Figure 12-23. Two sequences of microstructures associated with hexagonal to orthorhombic transition in cordierite. At 1400°C: (a). homogeneous hexagonal cordierite; (b,c) development and coarsening of tweed texture parallel to {10$\bar{1}$0} and {1$\bar{1}$20}; (e) orthorhombic cordierite nucleates within the tweed structure. At lower temperatures (1180°C and 1290°C) the sequence a,b,c is the same however the orthorhombic phase nucleates at a finer scale giving the appearance that it formed by coarsening of the tweed microstructure. Figures 12-23f, g and h display the {110} and {310} twinning. The {310} orientation is lost during further coarsening. [Used by permission of the editor of *Physics and Chemistry of Minerals*, from Putnis et al.(1987), Figs. 2a-h, p. 449.]

484

strain (Fig. 12-22). As a result of this lattice distortion, the angle between the {110} planes increases from 120° (hexagonal) to 120°38' (orthorhombic). Therefore, the twinning is by pseudo-merohedry, and diffraction patterns of the twinned orthorhombic phase will be characterized by spot splitting (see Figs. 3a,b of Kitamura and Hiroi, 1982). Indeed, all of the published images of cordierite pseudo-merohedral twinning utilize spot splitting for image contrast. Transmission electron microscopy observations (Figs. 12-23 and 12-24) of the actual microstructure reveal that the transition involves far more than simple twinning.

The first microstructure to form upon annealing hexagonal cordierite in the stability field of orthorhombic cordierite is a modulated tweed texture (Putnis, 1980 and Putnis et al., 1987). This tweed texture consists of two orthogonal distortion waves normal to {10$\bar{1}$0} and {1$\bar{2}$10}, which are the mirror planes of the hexagonal phase (Fig. 12-23). Streaks without satellites observed in the electron diffraction pattern attests to the poor periodicity of the tweed texture. No spot splitting is observed in diffraction patterns of the tweed texture, suggesting no distortion of the orthorhombic lattice. Therefore, in its initial stages of annealing, the structure remains metrically hexagonal. Upon annealing at 1400°C, the orthorhombic phase nucleates in the tweed texture. At lower annealing temperatures,

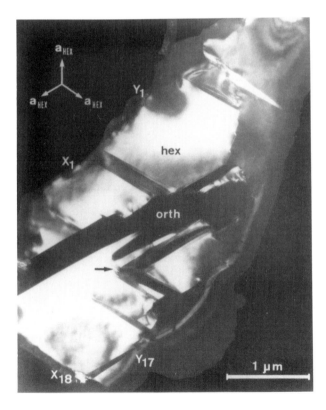

Figure 12-24. Two orientations of orthorhombic cordierite lamellae in a hexagonal host. The hexagonal-orthorhombic interface is along {10$\bar{1}$0}$_h$. The two orthorhombic lamellae are twin-related and where they meet the twin plane is {310}$_o$. The dark-field image was produced by allowing only a reflection from the hexagonal phase through the objective aperture. [Used by permission of the editor of *Contributions to Mineralogy and Petrology*, from Kitamura and Hiroi (1982), Fig. 4, p. 113.]

the orthorhombic phase nucleates at a finer scale in the tweed microstructure giving the appearance of a coarsening process, The orthorhombic phase is twinned with the twin composition planes {110} and {310} parallel to the pre-existing hexagonal mirror planes {10$\bar{1}$0} and {1$\bar{2}$10} of the tweed microstructure. Commonly, the {1$\bar{2}$10} modulation of the tweed can weaken during annealing, and consequently the twin plane {310} of the final product can be uncommon. The final product is a coarsely twinned microstructure with {110} twin planes predominating.

Upon growth of the orthorhombic phase, spot splitting will occur because of the distortion of the orthorhombic lattice. Kitamura and Hiroi (1982) show an example from an indialite, which is hexagonal cordierite partly inverted to the orthorhombic form. The sample contains two orientations of orthorhombic lamellae within the hexagonal host. The interface between the hexagonal and orthorhombic phase is the same as the orthorhombic composition plane, {10$\bar{1}$0}‖{110}. However, the orthorhombic twin domains share {310}.

TEM studies on cordierites have shown (1) the first order nature of the phase transition; (2) the relationship of the "distortion index" to the development of microstructure; (3) the degree of metastability of the high symmetry phase at low temperature; and (4) the relative stability of {110}$_0$ twin boundaries over {310}$_0$ twin boundaries. One geologic application has been to relate the development of the hexagonal to orthorhombic transition to the formation of sector-twinned cordierites which previously were thought to arise from twinning in the orthorhombic phase only (Zeck, 1972; Putnis and Holland, 1986).

12.5.1.6 Feldspar transition

Alkali feldspars ($KAlSi_3O_8$-$NaAlSi_3O_8$) undergo a transition from a high temperature monoclinic phase, sanidine-monalbite ($C2/m$), to a low temperature triclinic phase, microcline-albite ($C\bar{1}$) (Figs. 12-25a and 12-26, C-centered cell). The transition temperature is approximately 480°C for $KAlSi_3O_8$ and 978°C for $NaAlSi_3O_8$. Microstructures produced during the transition are albite and pericline twins. In addition, metastable intermediate structural states are common in the K-rich feldspars (orthoclase and adularia) and they are characterized by a tweed texture as seen with the transmission electron microscope. The nomenclature of alkali feldspars is complicated by the common occurrence of stranded metastable states, the broad miscibility gap, and the long history of study by mineralogists. Smith and Brown (1988) give an extensive review of both nomenclature and TEM observations.

The transition is characterized by a point group change from $2/m$ (order 4) in sanidine-monalbite to $\bar{1}$ (order 2) in microcline-albite. Two symmetry elements of the point group $2/m$ are lost during the transition: the twofold axis parallel to [010] and the mirror plane parallel to (010). Because of the change in crystal family these elements become pseudo-symmetry elements in the $C\bar{1}$ phase. There is little change in the unit cell size and the C-centering is retained by convention; the number of domains is $[(4\times2)/(2\times2)] \times 1 = 2$. These two domains can be related by either one of two twin laws, the albite law with (010) as the twin plane and the pericline law with [010] as the twin axis. The twin plane of the pericline law is an irrational plane called the *rhombic section* which is parallel to [010] but varies in its orientation with composition. Both the albite and pericline twin planes in alkali feldspar (microcline and anorthoclase) are illustrated in Figure 12-27a.

Model a^*b^* diffraction patterns ([001] zone axis) for both twin laws are shown in Figure 12-27. It is important to remember that in the high temperature monoclinic phase the [010] axis is perpendicular to the (010) plane and as twin elements they will remain

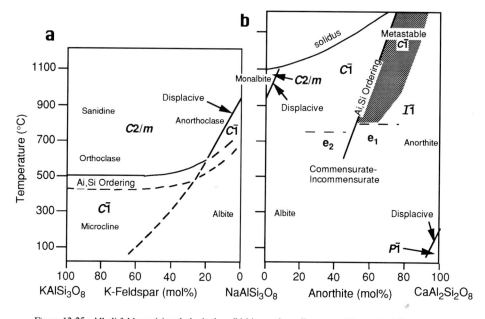

Figure 12-25. Alkali feldspar (a) and plagioclase (b) binary phase diagrams without miscibility gaps.

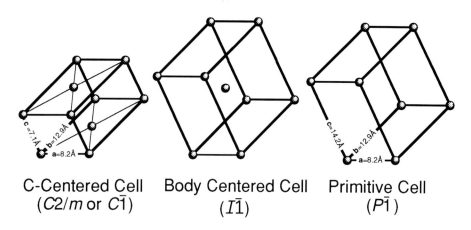

Figure 12-26. Alkali feldspar and plagioclase Bravais lattices.

exactly orthogonal in the triclinic phase. When both albite and pericline twinned areas contribute to the $\mathbf{a^*b^*}$ electron diffraction pattern a symmetrical distribution of four diffraction spots (Fig. 12-27d) results from the orthogonality of the twin elements. This type of transformation twinning in alkali feldspar has been called "M" twinning (arising from the monoclinic phase). "M" twinning is quite common in natural microcline because the transition temperature is below the crystallization temperature of most K-feldspars, but it is never seen in natural albite because the transition temperature is higher than the crystallization temperature of albite (Fig. 12-25). As an added complication, albite and

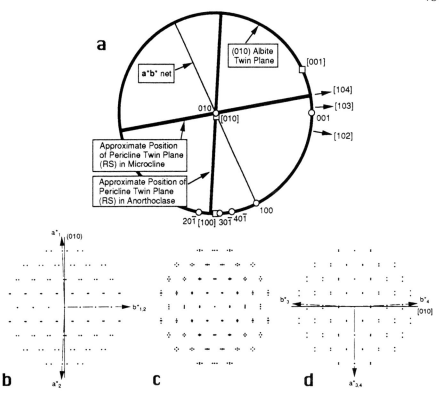

Albite Twinned Microcline   M - Twinned Microcline   Pericline twinned Microcline

Figure 12-27. (a) is a stereographic projection of one twin individual in microcline showing crystallographic axes (open squares) and diffracted beams (open circles), where $\alpha = 90.5°$, $\beta = 116°$, $\gamma = 87.5°$. The albite twin plane (010) is the primitive circle. The pericline twin plane is the irrational rhombic section (RS), which varies with composition and structural state (Smith and Brown, 1988, p. 531). For low microcline the rhombic section lies +100° away from [100]. In anorthoclase, however, the rhombic section lies -3° away from [100]. Model $\mathbf{a^*b^*}$ electron diffraction patterns are shown of albite (b), pericline (c), and "M"-twinning (d) in K-feldspar. Note the absence of $h+k$, odd reflections as required by the C-centered lattice.

pericline twins can also form in triclinic alkali feldspar during crystallization or deformation. These can be differentiated from "M" twinning because in the triclinic phase the twin elements (010) and [010] will deviate about 3.5° from orthogonality. This gives rise to only three diffraction spots in the $\mathbf{a^*b^*}$ pattern and has been called "T" twinning (arising from the triclinic phase) (not illustrated in Fig. 12-27).

The loss of the twofold axis and mirror plane during the $C2/m \rightarrow C\bar{1}$ transition in alkali feldspar can be accomplished by either small atomic displacements (displacive transition) or ordering of aluminium and silicon on tetrahedral sites (order-disorder transition). Figure 12-25a, after Brown and Parsons (1989), shows the supposed equilibrium temperature variation of these two transitions as a function of composition. The temperature of the displacive transition falls off rapidly with increasing orthoclase content, while the critical temperature associated with Al-Si ordering decreases less steeply, creating a crossover at approximately Ab$_{75-80}$. It appears, therefore, that the monoclinic to

488

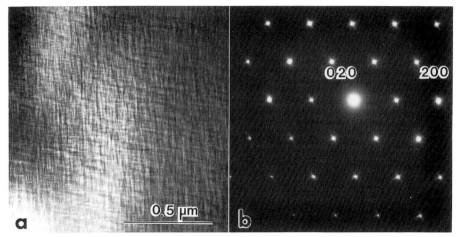

Figure 12-28. Benson orthoclase. A weak tweed texture (a) with $g = hk0$, produces streaks parallel and normal to b^* in the a^*b^* diffraction pattern in (b).

triclinic transition in Na-feldspar goes via a displacive transition, whereas the same transition in K-feldspar goes via an order-disorder transition. In reality, these two transitions are strongly coupled, and this coupling affects the thermodynamic character of the transition as well as the induced microstructures (Salje, 1985 and Harris et al., 1989).

The first TEM observations of K-feldspar were made on adularia (McConnell, 1965) and orthoclase (Nissen, 1967). The adularia was shown by single crystal x-ray diffraction to be monoclinic with intense streaks parallel to b^* on single crystal photographs. In the TEM, McConnell observed streaks both parallel and perpendicular to b^*. He also observed that $0k0$ diffraction maxima showed streaks normal to b^* with little intensity parallel to b^* and inferred from $h0l$ patterns that $h00$ diffraction maxima showed streaks parallel to b^* but are absent normal to b^*. Dark-field images taken with $hk0$ diffraction maxima showed a tweed texture with bright nodes at the intersections of the wavy microstructure. Because of the observation that streaks parallel to b^* are absent for $0k0$ maxima and streaks normal to b^* are absent for selected reflections in the ($h0l$) pattern (therefore absent for $h00$ diffraction maxima) McConnell interpreted the tweed microstructure as two orthogonal transverse wave systems. He concludes that the structure of orthoclase and adularia is a "homogeneously perturbed monoclinic lattice" and that the streaks are normal to (010) and [010].

The tweed orientations in adularia and orthoclase correspond to the albite twin plane (010) and the pericline twin plane (rhombic section) when the lattice parameters are essentially monoclinic. Figure 12-28a shows such a tweed microstructure in orthoclase, Or_{91}, collected from the Benson Iron Mine, St. Lawrence County, New York (Stewart and Wright, 1974). The a^*b^* electron diffraction pattern (Fig. 12-28b) shows streaks parallel and normal to b^*. There are now many examples of this tweed texture in adularia and orthoclase (McConnell, 1965; Nissen, 1967; Eggleton and Buseck, 1980; McLaren, 1978, 1984; and FitzGerald and McLaren, 1982). It even persists after other parts of the K-feldspar have transformed to microcline.

McConnell (1965) and Eggleton and Buseck (1980) suggest that there exists an energy balance between the reduction in free energy due to Al/Si ordering and that gained by the strain energy of the lattice distortion (tweed texture). The degree of Al/Si order is charac-

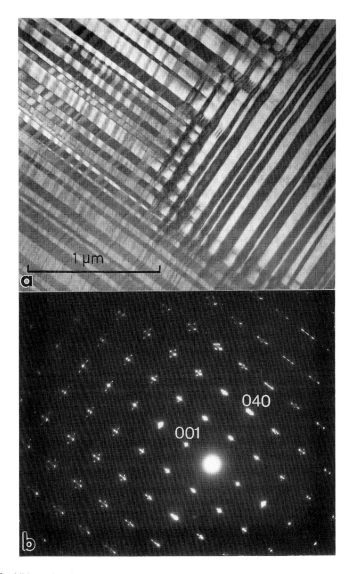

Figure 12-29. Albite and pericline twinning in anorthoclase using the [100] zone axis orientation. [Used by permission of the editor of *Chemical Physics of Solids and their Surfaces*, from McLaren (1978), Fig. 18a and b, p. 25.]

terized by the Al content in the t_1 site of the monoclinic phase. The degree of order corresponding to the presence of tweed textures in orthoclase is $2t_1 = 0.7$ to 0.8 where complete disorder equals 0.5 and complete order equals 1.0 (Stewart and Wright, 1974). Thus orthoclase ($2t_1 = 0.7$ to 0.8) is stabilized by the tweed microstructure. Any further ordering requires a nucleation step initiated by external stress or by fluids.

Anorthoclase crystallizes above the $C2/m \rightarrow C\bar{1}$ transition (Fig. 12-25a) and also exhibits "M-twinning". Unlike orthoclase and adularia, anorthoclase displays only coarse domains of albite and pericline twins (note in Fig. 12-27a, that the orientation of the rhombic section in anorthoclase is very different than that in microcline). McLaren (1978) in a TEM study showed that large domains of only albite and pericline twins exist. Near the edges of these domains, the pericline twins form simple lamellar twins whereas the albite twins also show a second set of low contrast zig-zag lamellae (Fig. 12-29a). SAED patterns from the latter area show only albite twin reflections. McLaren interpreted this second set of lamellae as degraded pericline twin lamellae. The four-spot diffraction pattern (Fig. 12-29b) indicative of M-twinning was only obtained when the aperture was placed over a boundary area containing both albite and pericline twins.

Microcline is the stable low temperature form of $KAlSi_3O_8$ and commonly exhibits a cross-hatched pattern in the optical microscope that usually is interpreted as transformation-twinning (M-twinning). McLaren (1978) has conducted extensive TEM studies of microcline with cross-hatched patterns by closely correlating optical microscopy and transmission electron microscopy. He was particularly interested in the observation that in the optical microscope the cross-hatched patterns did not extinguish as four orientations but only as two orientations (Smith, 1974, p.335). Smith (1974) suggests that this may be caused by "recrystallization" of pericline twin lamellae into the albite twin orientation while retaining the cross-hatched pattern.

FitzGerald and McLaren (1982) and McLaren (1984) have discussed the best imaging conditions for observing both pericline and albite twin planes in microcline. They suggest that the zone [104] is superior because the [104] directions from both twin individuals are within 0.14° of one another and the twin boundaries are parallel to the electron beam. However, zone axes between [001] and [102] (Fig. 12-27a) can be used with success (see Figs. 12-30 and 12-31).

McLaren (1978), reported observations on four different microclines and indicated that albite twinning is more common than pericline twinning. He was able to differentiate between albite and pericline twins by using dark-field images. If an operating reflection is normal to a twin plane the twin plane will be out-of-contrast because there is no distortion of the lattice plane parallel to the twin plane. Therefore, if one uses the operating reflection, 040, albite twin boundaries will be out of contrast and only pericline twin boundaries will be in contrast. The operating reflection, $20\bar{1}$, is nearly normal to pericline twin boundaries (the rhombic section) and the pericline twin boundaries will be out-of-contrast when dark-field images are formed with it (Fig. 12-30).

The surprising result of McLaren's observations is that coexisting albite and pericline twins are rare in microcline. Moreover, selected area diffraction patterns of areas of cross-hatching on the TEM scale show only the presence of albite twins (Fig. 12-31). He concludes that the original pericline twins have changed their orientation to the albite law and that the pericline twin boundaries have become serrate producing an array of lens-shaped albite-twin lamellae. The degradation of the pericline twins into albite twins also gives rise to a range of twin orientations between pericline and albite. This results in complex diffraction patterns for both electron and x-ray diffraction. The driving force for the degradation is thought to be related to the reduction of the higher stresses associated with pericline twin walls in comparison with albite twin walls.

TEM studies of transformation-induced microstructures in alkali feldspars have contributed primarily to their potential for unraveling P-T-t paths in complex terrains (see Brown and Parsons, 1989, for a review). Recently, the TEM has been required to characterize the microstructure of alkali feldspars used for $^{40}Ar/^{39}Ar$ dating and thermo-chronometry (FitzGerald and Harrison, 1991; and Parsons et al., 1991). In this technique,

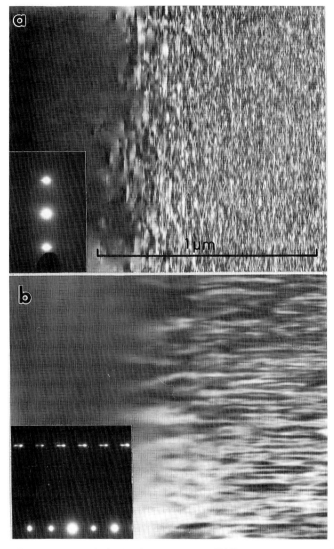

Figure 12-30. Albite and pericline twinning in microcline using the [102] zone axis orientation. In (a), $g = 20\overline{1}$ and the albite twins are in contrast. In (b), $g = 040$ and the albite twins are out of contrast; however, fine scale lamellae approximately parallel to **b** are now visible. The orientation of these lamellae suggests that they are pericline twins, but there is no evidence for pericline twins in the associated diffraction pattern. [Used by permission of the editor of *Chemical Physics of Solids and their Surfaces*, from McLaren (1978), Fig. 13a and b, p. 20.]

Ar is released by stepwise heating an irradiated sample to high temperatures in vacuum. The measured isotopic ratios provide radiometric ages. K-feldspars, however, behave as if comprised of a discrete distribution of "diffusion domains". Because these domains contain a range of closure temperatures, a single K-feldspar sample may reveal a broad segment of a cooling history (Harrison et al. 1991). Tweed textures, twinning, exsolution lamellae, grain and subgrain boundaries, inclusions and dislocations are important features that may describe the size and character of "diffusion domains" and their boundaries in alkali feldspar. These "diffusion domains" are sites of differing Ar retentivity (diffusion

492

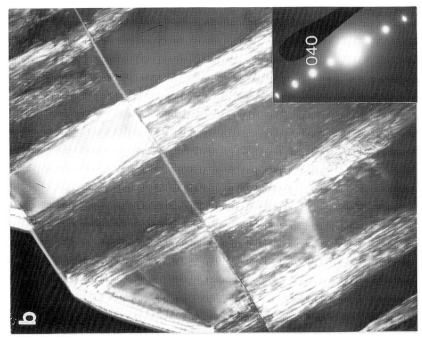

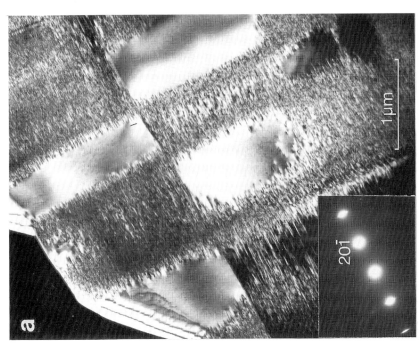

Figure 12-31. Chess-board pattern in microcline. (a) Dark-field micrograph with **g** = 040 of the same area. (b) Dark-field micrograph with **g** = 20$\bar{1}$ showing well-developed chess board pattern. (b) Dark-field micrograph with **g** = 20$\bar{1}$ showing well-developed chess board pattern, from McLaren (1978), Figs. 16a and b, p. 23.]

rate and activation energy) during degassing experiments. In light of the variety of transformation-induced microstructures and the degradation of pericline twin walls during cooling, significant insight into the origin of $^{40}Ar/^{39}Ar$ spectra will be gained by carefully correlated isotopic and TEM studies.

12.5.2 Antiphase domains

12.5.2.1 Plagioclase transitions

Sodic plagioclase (albite), $NaAlSi_3O_8$, crystallizes in the space group $C\bar{1}$. This is a nonstandard space group in the triclinic crystal family but has been chosen historically for ease of comparison to the $C2/m$ space group of disordered alkali feldspar (Fig. 12-25 and 26). The number of lattice points per unit cell is identical to that of $C2/m$. As discussed above, the $C2/m \rightarrow C\bar{1}$ transition occurs via Al/Si ordering in potassium feldspar or via a displacive transition in sodium feldspar. Both transitions result in pseudo-merohedral twinning, commonly known as albite or pericline twinning. Sodium feldspar also undergoes Al/Si ordering at temperatures below the monoclinic to triclinic transition, from disordered high albite to ordered low albite. In low albite, Al is fully ordered on one of four available tetrahedral sites (the T_1O site). There is no change in space group, Bravais lattice, or unit cell size during ordering, and thus no microstructure forms during the transition. The high to low albite transition is an example of a non-symmetry breaking transition; such transitions are popularly known in mineralogy as *non-convergent*.

Calcium feldspar (anorthite), $CaAl_2Si_2O_8$, contains equal amounts of Al and Si and experiences Al/Si ordering as well as polyhedral distortion with decreasing temperature. Al/Si ordering results in a $C\bar{1} \rightarrow I\bar{1}$ transition, and polyhedral tilting produces a displacive $I\bar{1} \rightarrow P\bar{1}$ transition (Fig. 12-25 and 26). The $C\bar{1} \rightarrow I\bar{1}$ transition is characterized by a doubling of the unit cell and a change in Bravais lattice from C-centered to body-centered; the triclinic crystal family and the point group, $\bar{1}$ (order = 2), remain the same. According to our rule the number of domains = $[(2\times2)/(2\times2)] \times 2 = 2$. Because a translation vector is lost we expect the formation of antiphase domains. The lost translation vector is the C-centering vector of the high temperature phase, $1/2[110]$. This vector has the same length in the low temperature $I\bar{1}$ unit cell since it is unaffected by the doubling along **c**. Similarly, the lattice vector $1/2[001]$ is lost during the transition because of the doubling of **c**. In the body-centered lattice these two translation vectors, $1/2[110]$ and $1/2[001]$ are identical because they are related by the body-centering vector $1/2[111]$. However, they are not equivalent in the primitive phase as shown below.

Anorthite undergoes further symmetry reduction by the displacive $I\bar{1} \rightarrow P\bar{1}$ transition, which occurs at 510 K (237°C). The transition only involves the change from a body-centered to a primitive unit cell; the point group, crystal family and unit cell size remain the same. Therefore the number of domains is $[(2x2)/(2x1)]x1 = 2$. Again, the domains are antiphase because of the loss of the body-centering translation vector, $1/2[111]$.

The order/disorder and displacive transitions in calcic plagioclase produce extra reflections in electron diffraction patterns, and these spots are used for dark-field imaging of the antiphase domain boundaries. The (a) reflections are present in the C-centered lattice; the (b) reflections appear during the formation of the body-centered lattice; and the (c) and (d) reflections appear during the formation of the primitive lattice. Table 12-4 indicates the visibility criteria $\alpha = 2\pi g \bullet R$ for the antiphase boundaries. If $\alpha = \pi$, then the boundary will be visible since the diffracted beam undergoes a 180° phase change upon passing through the boundary.

From Table 12-4 it is easily seen that no boundaries can be imaged using (a) reflections, whereas only the domain boundaries that arise during Al-Si ordering can be imaged using (b) reflections (Figs. 12-32a,c,d). These domains are commonly called "b-domains". The displacive domain boundaries are visible with either (c) or (d) reflections, and these are commonly called "c-domains" (Figs. 12-32b,e,f). The b-domains are also visible using (d) reflections, and if b-domains were related by 1/2[001] then they would be visible using (c) reflections. However, no b-domains have ever been imaged using (c) reflections (Müller et al., 1973; McLaren and Marshall, 1974), supporting the notion that the 1/2[001] displacement vector is not common. Structural refinements by Wenk and Kroll (1984) support this observation; after the $I\bar{1} \to P\bar{1}$ transition, the atomic distances between atoms separated by 1/2[001] will be greater than the distance between atoms separated by 1/2[110], thereby creating an energy difference between the two domain boundaries.

Table 12-4. Visibility criteria for antiphase domain boundaries in plagioclase

	R = 1/2[110]	R = 1/2[001]	R = 1/2[111]
(a) $h+k$ even, l even	2π	2π	4π
$\mathbf{g} = 112$			
(b) $h+k$ odd, l odd	3π	π	4π
$\mathbf{g} = 121$	visible	visible	
(c) $h+k$ even, l odd	2π	π	3π
$\mathbf{g} = 111$		visible	visible
(d) $h+k$ odd, l even	3π	2π	5π
$\mathbf{g} = 122$	visible		visible

The micrographs in Figure 12-32 illustrate only the tip of the iceberg when it comes to interesting features of antiphase domain boundaries in plagioclase. The following are some general observations: (1) b-domain boundaries are smooth curves in both quenched samples of synthetic bytownite and in lunar anorthite. By contrast, c-domain boundaries are rough and finely serrated. This difference presumably is a reflection of the transition mechanism. Order-disorder transitions commonly result in smooth boundaries, and displacive transitions commonly form rough boundaries. Even c-domains imaged in highly ordered Val Pasmeda anorthite from a contact metamorphic marble show finely irregular boundaries (Müller and Wenk, 1973). (2) For bytownite and labradorite compositions, b-domain boundaries reorient parallel to the directions of exsolution and of the intermediate plagioclase incommensurate structure (Nord et al., 1974; Grove, 1977; Grove et al., 1983). (3) b-domains also directly interact with the incommensurate phase of intermediate plagioclase. Wenk and Nakajima (1980) as well as Grove (1977) have shown well documented evidence for the continuity between the incommensurate phase and b-domain boundaries. (4) b-domains can form in synthetic An_{100} and in natural plagioclase up to compositions as rich in anorthite component as An_{95} (Nord et al., 1973). However, the proposed $C\bar{1} \to I\bar{1}$ equilibrium transition is close to the center of the plagioclase binary (Fig. 12-25). The fact that calcic plagioclase crystallizes metastably as $C\bar{1}$ and then inverts to $I\bar{1}$ behind the growing interface has been shown by Kroll and Müller (1980) experimentally and discussed for natural samples by Nord (1983) and Müller et al. (1984).

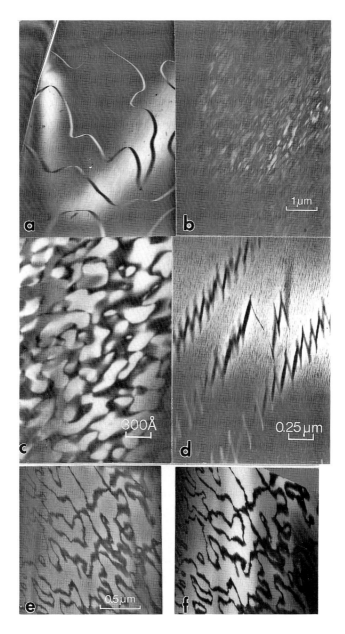

Figure 12-32. Antiphase domain boundaries in plagioclase. (a) shows b-domains ($\mathbf{g} = 03\overline{1}$) and (b) shows c-domains ($\mathbf{g} = 041$) in the same area in anorthite from Apollo 68415, dark-field, 1,000 keV. (c) shows b-domains in An$_{78-80}$, crystallized at 1100°C and cooled at 2°C/hour ($\mathbf{g} = 031$, dark-field, 1,000 keV). (d) shows "zig-zag" b-domains and a background of exsolution parallel to ($03\overline{1}$) in an An$_{82}$ Stillwater bytownite (from Nord et al, 1974) ($\mathbf{g} = 01\overline{1}$, dark-field, 800 keV). (e) and (f) are a stereo pair of c-domains from Apollo 15415 anorthite ($\mathbf{g} = 021$, dark-field, 800 keV). [Figs.12-32a,b,c,e and f are used by permission of the editor of *Electron Microscopy in Mineralogy*, from Heuer and Nord (1976), Figs. 7b and c, p. 291, Fig. 5b, p. 287 and Fig. 6b, p. 289. Fig. 12-32d used by permission of the editor of *The Feldspars*, from Nord et al. (1974), Fig. 1, p. 524.]

12.5.2.2 Pigeonite transition

Pigeonite, $(Mg,Fe,Ca)SiO_3$, undergoes a displacive phase transition from $C2/c$ to $P2_1/c$ at temperatures controlled by composition but generally between 500 and 1100°C. The transition principally involves a change in the coordination of cations on the M2 sites, and only translational symmetry is lost as the unit cell changes from C-centered to primitive. The point group remains $2/m$, and the unit cell size is constant. Consequently, two distinct APDs form during the transition. Because the domains are related by the lost C-centering vector, $1/2[110]$, they will be antiphase in character (Fig. 12-4). The domain boundaries are visible with $h+k$ odd reflections. A typical example for the visibility expression, $\alpha = 2\pi\mathbf{g} \cdot \mathbf{R}$, where $\mathbf{g} = 120$ ($h+k$ odd) and $\mathbf{R} = 1/2[110]$ yields $\alpha = 3\pi$ for a phase change of 180°

Studies of antiphase domains in pigeonite have demonstrated the following:

(1) Antiphase domain boundaries of quenched pigeonites are rough in comparison with the boundaries of more slowly cooled pigeonites (Fig. 12-33a). This is consistent with the displacive transition mechanism. Shimobayashi (1992) observed the formation of pigeonite APBs using heating stage experiments, and he suggests that the transition occurs via an athermal martensitic mechanism.

(2) Boundaries of more slowly cooled pigeonites tend to orient themselves along planes of minimum strain. Carpenter (1978) found sigmoidal segments that ranged in orientation from (120) to (110) and straight segments parallel to (211) or $(\bar{2}11)$.

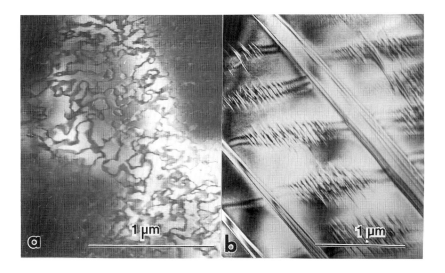

Figure 12-33. Antiphase domains in pigeonite. (a) shows rough antiphase domain boundaries in synthetic pigeonite (~$Wo_{15}En_{50}Fs_{35}$) which was quenched from 1225°C and 20 kbars (McCallister and Nord, unpublished). (b) shows (001) augite lamellae with intervening straight antiphase domain boundaries. Small (001) augite lamellae have precipitated on the APBs. [Fig. 12-33b is used by permission of the editor of *Physics and Chemistry of Minerals*, from Carpenter (1978), Fig. 1a, p. 239.]

(3) The antiphase domain boundaries can act as nucleation sites for calcium-rich clinopyroxene because of the excess free energy associated with the APBs, calcium enrichment at the boundaries, and the high temperature structure of the boundaries. Carpenter (1978) found that augite preferentially precipitates on (211) and ($\bar{2}$11) relative to (120) and (110) because of the favorable size of the M2 sites along the (211) and ($\bar{2}$11) planes (Fig. 12-33b).

12.5.2.3   Omphacite transition

Omphacite, $(Na,Ca)(Mg,Fe^{2+},Al^{3+},Fe^{3+})Si_2O_6$, undergoes an ordering transition from $C2/c$ to $P2/n$ at relatively high temperatures, up to 850°C. Because only the Bravais lattice type changes (point group $2/m$ and unit cell size are constant), two domains related by $1/2[110]$ form at the transition. Domain boundaries were first imaged by Champness (1973) using $h+k$ odd reflections (Fig. 12-34a) and by Phakey and Ghose (1973). Extensive work by Carpenter (reviewed in 1983) showed a wide variety of domain textures from smooth equidimensional domains to irregular streaky domains with large variations in size over small distances.

The ordering temperature is higher than the crystallization temperature of most metamorphic omphacites; however, almost all omphacites studied have antiphase domain microstructures. Carpenter (1983) concludes that the crystals grow with metastable short range order, and long-range order occurs after crystallization with the development of antiphase domain microstructures.

Electron diffraction evidence for the space groups $P2/c$ and $P2$ also has been reported (Carpenter, 1983). If $P2$ exists then the point group is reduced as well, and merohedral twins would be expected in addition to the antiphase domains.

The size of regular equidimensional antiphase domains in omphacites appears to show a distinct trend such that size increases with peak metamorphic temperature (Fig. 12-34b), from ≤ 50 Å at ~300°C to 3500 Å at ~800°C (Carpenter, 1981). Carpenter notes that this relationship is a "strange" geothermometer because it records only the peak temperature and is not reset during cooling. In other words, the sluggishness of the ordering makes domain coarsening insensitive to time and sensitive only to temperature. Recent work, however, indicates that APD size can be modified (Fig. 12-34b) by dislocation glide and/or creep processes followed by coarsening which results in a wide variation in domain size (van Roermund and Lardeaux, 1991). Thus, some caution is necessary to utilize APD size in omphacite as a geothermometer.

12.5.2.4   Cummingtonite transition

Prewitt et al. (1970) first reported the $C2/m \rightarrow P2_1/m$ transition at about 45°C in a manganese-rich cummingtonite from the Gouverneur, New York, talc district. The transition is displacive and similar to that in pigeonite; therefore, two antiphase domains are expected with a displacement vector of $1/2[110]$. Independent attempts by P. Champness and this author failed to find any domain microstructure in the Gouverneur sample and even failed to detect the primitive reflections in the TEM, even though the reflections are present in x-ray precession photographs. Carpenter (1982) was more successful with manganese-free cummingtonite exsolution lamellae in anthophyllite from the Wright Talc Mine, Adirondacks, New York. He was able to image antiphase domain boundaries with $h+k$ odd reflections confirming the displacement vector and the existence of the transition in cummingtonite.

498

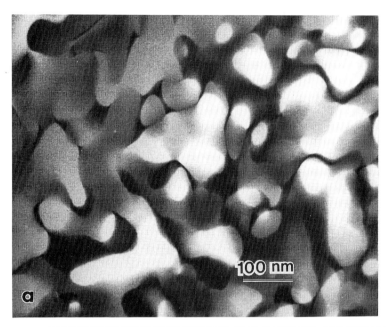

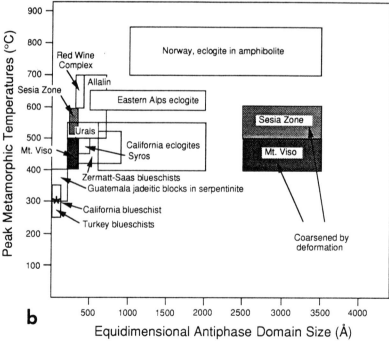

Figure 12-34. Antiphase domains in omphacite. (a) shows domains imaged in omphacite ($Jd_{31}Di_{52}Ag_{17}$) from Jenner, California, $g = 101$ (Champness, 1973). (b) is a graph relating the measured size of equidimensional APDs to the estimated peak metamorphic temperature. The Sesia Zone and Mt Viso samples have two APD sizes because of enhanced coarsening by deformation. [Modified after Carpenter, 1981 and van Roermund and Lardeaux, 1991.]

12.5.2.5 Scapolite transition

The established space group for endmember scapolites, marialite ($Na_4Al_3Si_9O_{24}Cl$) and meionite ($Ca_4Al_6Si_6O_{24}CO_3$), is $I4/m$, whereas intermediate scapolites are found with the space group $P4_2/n$. A transition from body-centered to primitive with no change in point group, crystal family or unit cell size results in two antiphase domains with a displacement vector of 1/2[111]. Smooth domain boundaries have been imaged by Phakey and Ghose (1972) and irregular ones by Oterdoom and Wenk (1983), who used "b" reflections ($h+k+l$, odd) as is consistent with the displacement vector. Oterdoom and Wenk (1983) interpret the formation of the domains as the result of Al-Si ordering, although intermediate scapolites remain ordered above 1000°C. It appears that the domain microstructure is the result of disordered scapolite crystallizing metastably at metamorphic temperatures and ordering in a similar fashion to that observed in the $C\bar{1} \rightarrow I\bar{1}$ transition in plagioclase.

12.5.3 Combinations of translation and twin interfaces

12.5.3.1 Cristobalite transition

Cristobalite undergoes a transition from face-centered cubic (β) to primitive tetragonal (α), $Fd\bar{3}m \rightarrow P4_12_12(P4_32_12)$, at approximately 268°C (Dollase, 1965). The high temperature point group $m\bar{3}m$ has 48 point symmetry elements and the face-centered cubic lattice has 4 lattice points per unit cell. The low temperature point group, 422, has 8 point group symmetry elements and the primitive tetragonal lattice has one lattice point per unit cell. The size of the unit cell is reduced by 1/2 during the transition. The number of domains therefore is $[(48×4)/(8×1)] × 0.5 = 12$ (see also Hatch and Ghose, 1991). These domains will consist of three pseudo-merohedral twin domains related by the lost threefold axis (cubic to tetragonal), two inversion domains related by the lost center of symmetry and two antiphase domains related by the lost translation vector $1/2[110]_c$ (= $1/2[111]_t$). Transmission electron microscopy observations of cristobalite from lunar basalts reveal the presence of antiphase domain boundaries within the ferroelastic twins (Figs. 12-35a and b), but no inversion twins have been imaged to date (Lally et al., 1978). This suggests that the sequence of transformation is face-centered cubic to body-centered tetragonal to form the three pseudo-merohedral twins followed by a transition to primitive tetragonal forming inversion twins and APBs. Figure 12-36 shows the sequence of changes in the Bravais lattices and the unit cell. Note that because of the reduction in cell size no lattice points are lost in the $F \rightarrow I$ transition, but the body-centered lattice point is lost in the $I \rightarrow P$ transition. The loss of the center of symmetry could come before or after the formation of the twins, depending upon the path of the transition.

The antiphase domain boundaries can be imaged in dark-field with reflections of the type $h+k+l$, odd. The domains are cylindrical in shape (Fig. 12-35b) with finely serrated boundaries similar to those from other displacive transitions in pigeonite and anorthite. Cristobalite usually contains a variety of other microstructures such as stacking faults and lamellar features interpreted as retained β-phase (J. Scott Lally, personal comm.)

12.5.3.2 Coarsening experiments

Antiphase domain boundaries and merohedral twin domain boundaries have an excess free energy known as the interfacial energy, and at suitably high temperatures the area of the domain boundaries will be reduced by increasing the size of the domain. Domains should coarsen according to the relation

$$d^n - d_o^{\ n} = nK_o e^{-(Q/RT)}t$$

where d is the average domain diameter, d_o is the initial domain diameter, K_o is a constant, Q is the activation energy for domain coarsening, R is the universal gas constant, T is the absolute temperature, t is time, and n is a constant, ideally equal to 2 (Poquette and Mikkola, 1969). For small initial domain sizes, isothermal coarsening fits $d^2 \propto t$. Deviations from $n = 2$ have been ascribed to the concentration of excess or impurity atoms at the antiphase domain boundary (Carpenter, 1979).

The slow coarsening rates in pigeonite and omphacite are attributed by Carpenter to segregation of Ca^{2+} at the boundary in the case of pigeonite and Ti^{4+} at the boundary in some omphacites. See Table 12-5 for a list of microstructure coarsening experiments.

Table 12-5. Microstructure coarsening experiments

Mineral	n	K_o	Q	Reference
Pigeonite	10		100-120 kcal/mole	Carpenter (1979)
Omphacite	8±2	5×10^{36}	75 kcal/mole	Carpenter (1981)
Ferrian Ilmenite	2.44		77 kcal/mole	Nord and Lawson (1988)
Anorthite b-domains	2.75			Müller et al. (1984)

12.6   SOME CONCLUSIONS

Most of the work by transmission electron microscopy on transformation-induced microstructures was initiated to verify hypotheses developed from x-ray diffraction observations. For example, antiphase domain microstructures were proposed for pigeonite and plagioclase long before they were observed in the TEM. Perhaps TEM has contributed most importantly in demonstrating the widespread occurrence of metastable, disordered phases outside their stability fields (Carpenter and Putnis, 1985). Cordierite, omphacite, body-centered plagioclase, adularia and others, all involving Al-Si ordering, have been shown by TEM to crystallize as the disordered phase, hundreds of degrees below their stability fields.

TEM also has shown how microstructures change with time and temperature. A good example is the relative stability of boundaries. In K-feldspar, albite twin boundaries are preferred over pericline twin boundaries; in cordierite, {110} twin boundaries are preferred over {310} twin boundaries; and in $P\bar{1}$ anorthite, the displacement vector 1/2[110] is preferred over 1/2[001]. These boundaries and vectors are identical in the high symmetry phase but acquire different energies in the low symmetry phase; these energy differences favor the formation of one orientation over the other.

Finally, one should consider how transformation-induced microstructures affect physical properties. There are of course many examples of the switching properties of domain boundaries in ferroic materials, but these are rare in natural materials. Nevertheless, domain microstructures in minerals have been shown to affect diffusion rates, magnetic properties and elastic properties.

Future work will certainly discover twins and APDs that form in multi-anvil and diamond cell high pressure experiments, especially in perovskites (Hu et al. 1992) and other silicates. Hot stage TEM will challenge our concepts of phase transitions based on

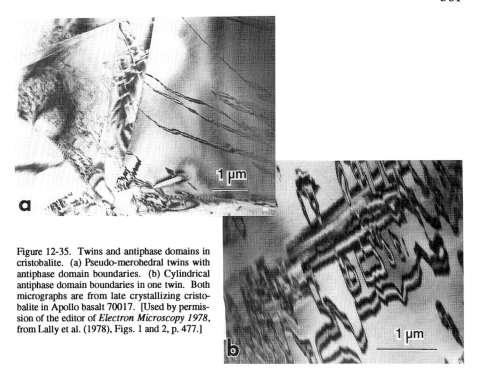

Figure 12-35. Twins and antiphase domains in cristobalite. (a) Pseudo-merohedral twins with antiphase domain boundaries. (b) Cylindrical antiphase domain boundaries in one twin. Both micrographs are from late crystallizing cristobalite in Apollo basalt 70017. [Used by permission of the editor of *Electron Microscopy 1978*, from Lally et al. (1978), Figs. 1 and 2, p. 477.]

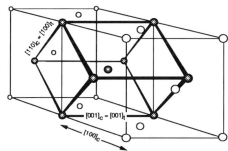

Face Centered Cubic → Body Centered Tetragonal

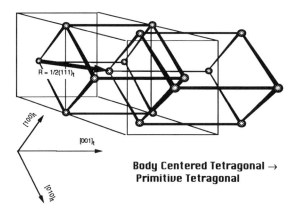

**Body Centered Tetragonal →
Primitiue Tetragonal**

Figure 12-36. Lattice transitions in cristobalite. One possible sequence of transitions is cF to tI to tP. The transition cF to tI results in three twins, one is shown. All lattice points are retained but the number per unit cell is reduced from four to two and the size of the cell is reduced by half because it is redefined. The tI to tP transition results in antiphase domains because of the loss of the body-centering vector 1/2[111].

502

quenched results. An exciting and important trend will be correlating magnetic properties of Fe-Ti oxides and TEM microstructures, correlating conductivity measurements with microstructures, and conducting isothermal experiments on coarsening and growth of microstructures. One only needs an appreciation of the possible types of domains that can arise during phase transitions and how to image those domains and domain boundaries using the transmission electron microscope.

12.7 APPENDIX

12.7.1 Conventional cells of the three-dimensional Bravais lattices

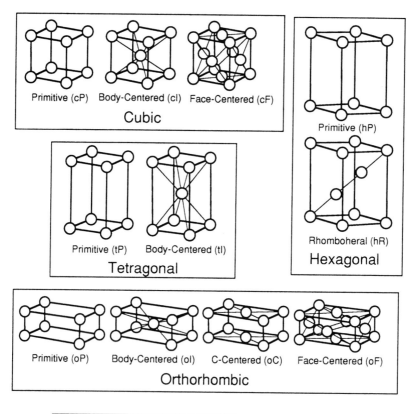

Primitive (cP)   Body-Centered (cI)   Face-Centered (cF)
Cubic

Primitive (hP)

Rhomboheral (hR)
Hexagonal

Primitive (tP)   Body-Centered (tI)
Tetragonal

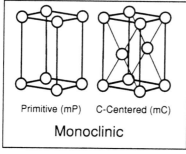

Primitive (oP)   Body-Centered (oI)   C-Centered (oC)   Face-Centered (oF)
Orthorhombic

Primitive (mP)   C-Centered (mC)
Monoclinic

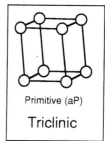

Primitive (aP)
Triclinic

12.7.2 The 32 point groups

Explanation of symbols:

◖	Twofold rotation axis: 2
▲	Threefold rotation axis: 3
◆	Fourfold rotation axis: 4
⬤	Sixfold rotation axis: 6
○	Center of symmetry: $\bar{1}$

▲	Threefold inversion axis: $\bar{3}$
◖◗	Fourfold inversion axis: $\bar{4}$
⬣	Sixfold inversion axis: $\bar{6}$

◖ Twofold rotation axis with center of symmetry: 2/m

◆ Fourfold rotation axis with center of symmetry: 4/m

⬣ Sixfold rotation axis with center of symmetry: 6/m

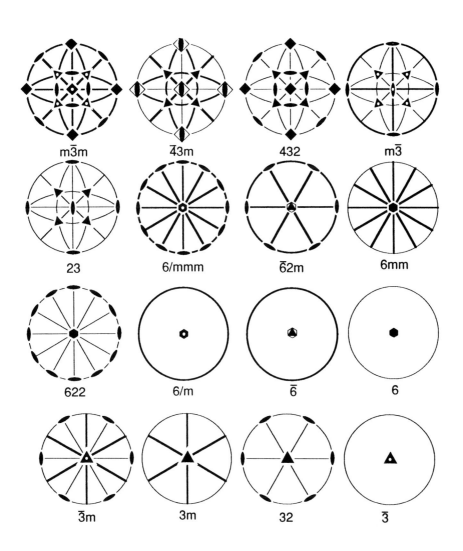

504

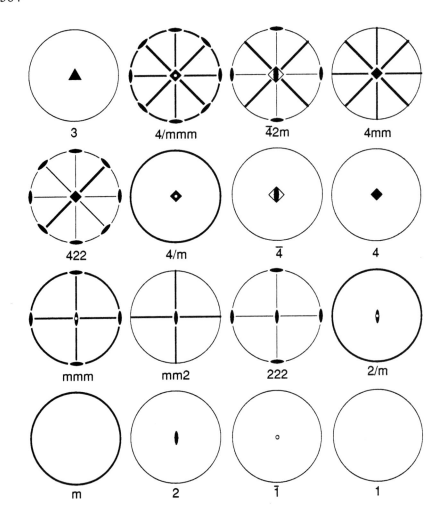

12.7.3 The abstract point groups

Order	Point Group	Order	Point Group
1	1	8	$422, 4mm, \overline{4}2m$
2	$\overline{1}, 2, m$	12	$6/m$
3	3	12	$\overline{3}m, 622, 6mm, \overline{6}2m$
4	$2/m, 222, mm2$	12	23
4	$4, \overline{4}$	16	$4/mmm$
6	$\overline{3}, 6, \overline{6}$	24	$6/mmm$
6	$32, 3m$	24	$m\overline{3}$
8	mmm	24	$432, \overline{4}3m$
8	$4/m$	48	$m\overline{3}m$

ACKNOWLEDGMENTS

Reviews by Howard Evans, Peter Heaney, Richard Reeder and Malcolm Ross greatly improved the readability of the manuscript in addition to the elmination of some errors. Particular thanks, goes to Howard Evans for answering endless questions, and to Michael Carpenter, Andrew Putnis and Ekhard Salje for introducing this author to some of the broader applications of phase transitions and microstructures to Earth science.

REFERENCES

Allen, F. and Burnham, C.W. (1992) A comprehensive structure-model for vesuvianite: symmetry variations and crystal growth. Can. Mineral. 30, 1-18.

Amelinckx, S. (1972) The geometry and interfaces due to ordering and their observation in transmission electron microscopy and electron diffraction. Surface Sci. 31, 296-354.

Amelinckx, S. and Van Landuyt, J. (1976) Contrast effects at planar interfaces. In H-R. Wenk, ed., Electron Microscopy in Mineralogy, 68-112. Springer-Verlag, New York.

Brown, W.L. and Parsons, I. (1989) Alkali feldspars: ordering rates, phase transformations and behaviour diagrams for igneous rocks. Mineral. Mag. 53, 25-42.

Buerger, M.J. (1951) Crystallographic aspects of phase transformations: In R. Smoluchowski, ed., Phase Transformations in Solids, 183-211. J. Wiley & Sons, New York.

Cahn, R.W. (1954) Twinned crystals. Adv. Phys. 3, 363-445.

Carpenter, M.A. (1978) Nucleation of augite at antiphase boundaries in pigeonite. Phys. Chem. Minerals 2, 237-251.

Carpenter, M.A. (1979) Experimental coarsening of antiphase domains in a silicate mineral. Science 206,681-683.

Carpenter, M.A. (1981) Omphacite microstructures as time-temperature indicators of blueschist and eclogite-facies metamorphism. Contrib. Mineral. Petrol. 78, 441-451.

Carpenter, M.A. (1982) Amphibole microstructures: some analogies with phase transformations in pyroxenes. Mineral. Mag. 46, 395-397.

Carpenter, M.A. (1983) Microstructures in sodic pyroxenes: implications and applications. Periodico di Mineralogia - Roma, 52, 271-301.

Carpenter, M.A. (1988) Thermochemistry of aluminium/silicon ordering in feldspar minerals. In E.K.H. Salje, ed., Physical Properties and Thermodynamic Behaviour of Minerals, 265-323. NATO ASI Series C255, Reidel, Boston.

Carpenter, M.A. and Putnis, A. (1985) Cation order and disorder during crystal growth: some implications for natural mineral assemblages. In Thompson, A.B. and Rubie, D.C., eds., Metamorphic Reactions, Advances in Physical Geochemistry, 4, 1-26.

Champness, P.E. (1973) Speculation on an order-disorder transformation in omphacite. Am. Mineral. 58, 540-542.

Chapman, J.N. (1989) High resolution imaging of magnetic structures in the transmission electron microscope. Materials Sci. Engineering B3, 355-358.

Delaey, L. (1991) Diffusionless transformations. In P. Haasen, ed., Phase Transformations in Materials, Material Science and Technology, Vol. 5. VCH, New York.

Dolino, G., Bachheimer, J.P., Berge, B., Zeyen, C.M.E. (1984) Incommensurate phase of quartz: I. Elastic neutron scattering. J. Physique 45, 361-371.

Dollase, W.A. (1965) Reinvestigation of the structure of low cristobalite. Z. Krist. 121, 369-377.

Edington, J.W. (1976) Practical Electron Microscopy in Material Science, Van Nostrand Reinhold Co., New York.

Eggleton, R.A. and Buseck, P.R. (1980) The orthoclase-microcline inversion: A high-resolution transmission electron microscope study and strain analysis. Contrib. Mineral. Petrol. 74, 123-133.

Fisher, R.M. and Marcinkowski, M.J. (1961) Direct observation of antiphase boundaries in the AuCu3 superlattice. Phil. Mag. 6, 1385-1405.

FitzGerald, J.D. and McLaren, A.C. (1982) The microstructures of microcline from some granitic rocks and pegmatites. Contrib. Mineral. Petrol. 80, 219-229.

FitzGerald, J. D. and Harrison, T. M. (1991) Microstructures in MH-10 and their implications for domain models in K-feldspars. Abs., Trans. Am. Geophys. Union 72, 290

Glossop, A.B. and Pashley, D.W. (1959) The direct observation of anti-phase domain boundaries in ordered copper-gold (CuAu) alloy. Proc. R. Soc. London A250, 132-146.

506

Grove, T.L. (1977) Structural characterization of labradorite-bytownite plagioclase from volcanic, plutonic and metamorphic environments. Contrib. Mineral. Petrol. 64, 273-302.

Grove, T.L., Ferry, J.M. and Spear, F.S. (1983) Phase transitions and decomposition relations in calcic plagioclase. Am. Mineral. 68, 41-59.

Harris, M.J., Salje, E.K.H., Güttler, B.K. and Carpenter, M.A. (1989) Structural states of natural potassium feldspar: An infrared spectroscopic study. Phys. Chem. Minerals 16, 649-658.

Harrison, T. M., Lovera, O. M. and Heizler, M.T. (1991) $^{40}Ar/^{39}Ar$ results for alkali feldspars containing diffusion domains with differing activation energy. Geochim. Cosmochim. Acta 55, 1435-1448.

Hatch, D.M., Ghose, S. and Stokes, H.T. (1990) Phase transitions in leucite, $KAlSi_2O_6$. I. Symmetry analysis with order parameter treatment and the resulting microscopic distortions. Phys. Chem. Minerals 17, 220-227.

Hatch, D.M. and Ghose, S. (1991) The $\alpha-\beta$ phase transition in cristobalite, SiO_2. Phys. Chem. Minerals 17, 554-562.

Heaney, P.J. and Veblen, D.R. (1990) A high-temperature study of the low-high leucite phase transition using the transmission electron microscope. Am. Mineral. 75, 464-476.

Heaney, P.J. and Veblen, D.R. (1991) Observations of the $\alpha-\beta$ phase transition in quartz: A review of imaging and diffraction studies and some new results. Am. Mineral. 76, 1018-1032.

Hermann, C. (1929) Zur systematischen strukturtheorie. IV Untergruppen. Z. Krist. 69, 533-555.

Heuer, A.H. and Nord, Jr., G.L . (1976) Polymorphic phase transitions in minerals. In H-R. Wenk, ed., Electron Microscopy in Mineralogy, 274-303. Springer-Verlag, New York.

Hirsch, P.B., Howie, A., Nicholson, R.B., Pashley, D.W. and Whelan, M.J. (1965) Electron microscopy of thin crystals. Butterworths, London.

Hoffman, K.A. (1992) Self-reversal of thermoremanent magnetization in the ilmenite-hematite system: order-disorder, symmetry, and spin alignment. J. Geophys. Res. 97B, 10883-10896.

Hu, M., Wenk, H.-R., and Sinitsyna, D. (1992) Microstructures in natural perovskites. Am. Mineral. 77, 359-373.

International Tables for Crystallography (1983) T. Hahn, ed., D. Reidel, Boston, 854 p.

Janovec, V. (1976) A symmetry approach to domain structures. Ferroelectrics 12, 43-53.

Kitamura, M. and Hiroi, Y. (1982) Indialite from Unazuki pelitic schist, Japan, and its transition texture to cordierite. Contrib. Mineral. Petrol. 80, 110-116.

Kroll, H. and Müller, W. F. (1980) X-ray and electron optical investigation of synthetic high temperature plagioclases. Phys. Chem. Mineral. 5, 255-277.

Lally, J.S., Nord, Jr., G.L., Heuer, A.H. , and Christie, J.M. (1978) Transformation-induced defects in α-cristobalite. Proc. 9th Int'l Congress on Electron Microscopy, Electron Microscopy, 1, 476-477

Lawson, C.A., Nord, Jr., G.L., Dowty, E. and Hargraves, R.B. (1981) Antiphase domains and reverse thermoremanent magnetism in ilmenite-hematite minerals. Science, 213, 1372-1374.

Lawson, C.A., Nord, Jr., G.L. and Champion, D.E. (1987) Fe-Ti oxide mineralogy and the origin of normal and reverse remanent magnetization in dacitic pumice blocks from Mt. Shasta, California. Phys. Earth Planet. Interiors 46, 270-288.

MacKenzie, W.S. and Zussman, J., eds. (1974) The Feldspars, Manchester Univ. Press, 717 p.

Marcinkowski, M.J. and Brown, N. (1962) Direct observation of antiphase boundaries in the Fe_3Al superlattice. J. Appl. Phys. 33, 537-552.

McConnell, J.D.C. (1965) Electron optical study of effects associated with partial inversion of a silicate phase. Phil. Mag. 11, 1289-1301.

McLaren, A.C. and Phakey, P.P. (1966) Electron microscope study of Brazil twin boundaries in amethyst quartz. Physica Status Solidi, 13, 413-422.

McLaren, A.C. and Phakey, P.P. (1969) Diffraction contrast from Dauphine twin boundaries in quartz. Physica Status Solidi, 31, 723-737.

McLaren, A.C. and Marshall, D.B. (1974) Transmission electron microscope study of the domain structure associated with the b-, c-, d-, e- and f-reflections in plagioclase feldspars. Contrib. Mineral. Petrol. 44, 237-249.

McLaren, A.C. (1978) Defects and microstructures in feldspars. In Chemical Physics of Solids and Their Surfaces, Vol 7, edited by M.W. Roberts and J.M. Thomas, pp. 1-30. London: The Chemical Society.

McLaren, A.C. (1984) Transmission Electron Microscope investigations of the microstructures of microclines. In Feldspars and Feldspathoids. Brown, W.L., ed., Reidel, Dordrecht, 373-409.

McLaren, A.C. (1991) Transmission electron microscopy of minerals and rocks. Cambridge University Press, New York, 387 p.

Megaw, H.D. (1973) Crystal Structures: A Working Approach. W.B. Saunders Co., Philadelphia, 563 p.

Müller, W.F. and Wenk, H.-R. 1973 Changes in the domain structure of anorthites induced by heating. N. Jahrb. Mineral. Monatshefte H-1, 17-26.

Müller, W.F., Wenk, H.-R., Bell, W.L. and Thomas, G. (1973) Analysis of the displacement vectors of antiphase domain boundaries in anorthites ($CaAl_2Si_2O_8$). Contrib. Mineral. Petrol. 40, 63-74.

Müller, W.F., John, R. J. and Kroll, H. 1984 On the origin and growth of antiphase domains in anorthite. Bull. Minéral. 107, 489-494.

Newnham, R.E. (1974) Domains in minerals. Am. Mineral. 59, 906-918.

Nissen, H.-U. (1967) Direct electron-microscopic proof of domain texture in orthoclase ($KAlSi_3O_8$). Contrib. Mineral. Petrol. 16, 354-360.

Nord, Jr., G.L. (1983) The $C\bar{1} \rightarrow I\bar{1}$ transition in lunar and synthetic anorthite. (abstract) 3rd NATO Advanced Study Institute on Feldspars, Feldspathoids and their Parageneses, Rennes, France, 28.

Nord, Jr., G.L., Lally, J.S. Heuer, A.H., Christie, JM., Radcliffe, S.V., Griggs, D.T. and Fisher, R.M. (1973) Petrologic study of igneous and metaigneous rocks from Apollo 15 and 16 using high-voltage transmission electron microscopy. Proc. 4th Lunar Sci. Conf., Geochim. Cosmochim. Acta, Suppl. 4, 1, 953-970.

Nord, Jr., G.L., Heuer, A.H. and Lally, J.S. (1974) Transmission electron microscopy of substructures in Stillwater bytownites. In W.S. MacKenzie and J. Zussman, eds., The Feldspars, Manchester University Press, 522-535.

Nord, Jr., G.L. and Lawson, C.A. (1988) Order-disorder transition in Fe_2O_3-$FeTiO_3$: structure and migration kinetics of transformation-induced twin domain boundaries. in G.W. Lorimer, ed., Phase Transformations '87. The Institute of Metals, London, 578-580.

Nord, Jr., G.L. and Lawson, C.A. (1989) Order-disorder transition-induced twin domains and magnetic properties in ilmenite-hematite. Am. Mineral. 74, 160-176.

Nord, Jr., G.L. and Lawson, C.A. (1992) Magnetic properties of $ilmenite_{70}$-$hematite_{30}$: Effect of transformation-induced twin boundaries. J. Geophys. Res. 97B, 10897-10910.

Oterdoom, W.H. and Wenk, H-R. (1983) Ordering and composition of scapolite: field observations and structural interpretations. Contrib. Mineral. Petrol. 83, 330-341.

Palmer, D.C., Salje, E. and Putnis, A. (1988) Twinning in tetragonal leucite. Phys. Chem. Minerals 16, 298-303.

Palmer, D.C., Salje, E., and Schmahl, W.W. (1989) Phase transitions in leucite: X-ray diffraction studies. Phys. Chem. Minerals 16, 714-719.

Palmer, D. C. (1990) Volume anomaly and the impure ferroelastic phase transition in leucite. In Salje, E.K.H., ed., Phase transitions in ferroelastic and co-elastic crystals. Cambridge University Press, New York, 350-366.

Parsons, I., Waldron, K.A., Walker, F.D.L., Burgess, R. and Kelley, S.P. (1991) Microtextural controls of [40]Ar loss and [18]O exchange in alkali feldspars. Abs., Trans. Am. Geophys. Union 72, 290

Phakey, P.P. and Ghose, S. (1972) Scapolite: Observation of anti-phase domain structure. Nature Physical Science, 238, 78-80.

Phakey, P.P. and Ghose, S. (1973) Direct observation of antiphase domain structure in omphacite. Contrib. Mineral. Petrol. 39, 239-245.

Poquette, G.E. and Mikkola, D.E. (1969) Antiphase domain growth in Cu_3Au. Trans. Metal. Soc. AIME, 245, 743-751

Prewitt, C.T., Papike, J.J. and Ross, M. (1970) Cummingtonite: A reversible, nonquenchable transition from $P2_1/m$ to $C2/m$ symmetry. Earth Planet. Sci. Lett. 8, 448-450.

Putnis, A. (1980) Order-modulated structures and the thermodynamics of cordierite reactions. Nature, 287, 128-131.

Putnis, A. and Holland, T.J.B. (1986) Sector trilling in cordierite and equilibrium overstepping in metamorphism. Contrib. Mineral. Petrol. 93, 265-272.

Putnis, A., Salje, E., Redfern, S.A.T., Fyfe, C.A. and Strobl, H. (1987) Structural states of Mg-cordierite I: Order parameters from synchrotron x-ray and NMR data. Phys. Chem. Minerals, 14, 446-454.

Portier, R., Lefebvre, S. and Fayard, M. (1975) Enantiomorphous ordered domains in stoichiometric lithium aluminate $LiAl_5O_8$. Mat. Res. Bull. 10, 883-888.

Reeder, R.J. and Nakajima, Y. (1982) The nature of ordering and ordering defects in dolomite. Phys. Chem. Minerals 8, 29-35.

Salje, E.K.H. (1985) Thermodynamics of sodium feldspar I: order parameter treatment and strain induced coupling effects. Phys. Chem. Minerals 12, 93-98.

Salje, E.K.H. (1990) Phase transitions in ferroelastic and co-elastic crystals. Cambridge University Press, New York, 366 p.

Shimobayashi, N. (1992) Direct observation on the formation of antiphase domain boundaries in pigeonite. Am. Mineral. 77, 107-114.

Siegel, S. (1951) Order-disorder transitions in metal alloys: In R. Smoluchowski, ed., Phase Transformations in Solids, J. Wiley & Sons, New York, 183-211.

Smith, J.V. (1974) Feldspar Minerals, Vol. 2, Springer-Verlag, New York.

Smith, J.V. and Brown, W.L. (1988) Feldspar Minerals, Vol. 1, (2nd ed.). Springer-Verlag, Berlin, 828 p.

508

Stewart, D.B. and Wright, T.L (1974) Al/Si order and symmetry of natural alkali feldspars, and the relationship of strained cell parameters to bulk composition. Bull. Soc. fr. Mineral. Cristallogr. 97, 356-377.

Tolédano, J.C. and Tolédano, P. (1987) The Landau Theory of Phase Transitions. World Scientific Lecture Notes in Physics, Vol. 3, Singapore, 451 p.

van Roermund, H.L.M. and Lardeaux, J.M. (1991) Modification of antiphase domain sizes in omphacite by dislocation glide and creep mechanisms and its petrological consequences. Mineral. Mag. 55, 397-407.

Van Tendeloo, G. and Amelinckx, S. (1974) Group theoretical considerations concerning domain formation in ordered alloys. Acta Cryst. A30, 431-440.

Van Tendeloo, G., Van Landuyt, J. and Amelinckx, S. (1976) The $\alpha-\beta$ phase transition in quartz and AlPO$_4$ as studied by electron microscopy and diffraction. Physica Status Solidi, 33, 723-735.

Van der Biest, O. and Thomas, G. (1975) Identification of enantiomorphism in crystals by electron microscopy. Acta Cryst. A31, 70-76.

Van der Biest, O. and Thomas, G. (1976) Fundamentals of electron microscopy. In H-R. Wenk, ed., Electron Microscopy in Mineralogy, 18-51. Springer-Verlag, New York.

Veblen, D.R. and Wiechmann, M.J. (1991) Domain structure of low-symmetry vesuvianite from Crestmore, California. Am. Mineral. 76, 397-404.

Walker, M.B. (1983) Theory of domain structures and associated defects in the incommensurate phase of quartz. Phys. Rev. B, 28, 6407-6410.

Wenk, H-R., ed. (1976) Electron Microscopy in Mineralogy, Springer-Verlag, New York, 564 p.

Wenk, H.-R. and Kroll, H. (1984) Analysis of $P\bar{1}$, $I\bar{1}$ and $C\bar{1}$ plagioclase structures. Bull. Minéral. 107, 467-487.

Wenk, H-R. and Nakajima, Y. (1980) Structure, formation and decomposition of APBs in calcic plagioclase. Phys. Chem. Minerals 6, 169-186.

Zeck, H.P. (1972) Transformation trillings in cordierite. J. Petrol. 13, 367-380.